# HANDBOOK OF OPTIMIZATION IN MEDICINE

# Springer Optimization and Its Applications

## VOLUME 26

*Aims and Scope*
Optimization has been expanding in all directions at an astonishing rate during the last few decades. New algorithmic and theoretical techniques have been developed, the diffusion into other disciplines has proceeded at a rapid pace, and our knowledge of all aspects of the field has grown even more profound. At the same time, one of the most striking trends in optimization is the constantly increasing emphasis on the interdisciplinary nature of the field. Optimization has been a basic tool in all areas of applied mathematics, engineering, medicine, economics and other sciences.

The *Springer Series in Optimization and Its Applications* publishes undergraduate and graduate textbooks, monographs and state-of-the-art expository works that focus on algorithms for solving optimization problems and also study applications involving such problems. Some of the topics covered include nonlinear optimization (convex and nonconvex), network flow problems, stochastic optimization, optimal control, discrete optimization, multi-objective programming, description of software packages, approximation techniques and heuristic approaches.

# HANDBOOK OF OPTIMIZATION IN MEDICINE

Edited By

PANOS M. PARDALOS
University of Florida, Gainesville, Florida, USA

H. EDWIN ROMEIJN
The University of Michigan, Ann Arbor, Michigan, USA

 Springer

*Editors*

Panos M. Pardalos
University of Florida
Department of Industrial and Systems
   Engineering
USA
pardalos@ise.ufl.edu

H. Edwin Romeijn
The University of Michigan
Department of Industrial and Operations
   Engineering
USA
romeijn@umich.edu

ISBN: 978-0-387-09769-5      e-ISBN: 978-0-387-09770-1
DOI: 10.1007/978-0-387-09770-1

Library of Congress Control Number: 2008938075

Mathematics Subject Classifications (2000): 90-XX, 92C50, 90C90

There are in fact two things, science and opinion; the former begets knowledge, the later ignorance.

— Hippocrates (460–377 BC), Greek physician

# Preface

In recent years, there has been a dramatic increase in the application of optimization techniques to the delivery of health care. This is in large part due to contributions in three fields: the development of more and more efficient and effective methods for solving large-scale optimization problems (operations research), the increase in computing power (computer science), and the development of more and more sophisticated treatment methods (medicine). The contributions of the three fields come together because the full potential of the new treatment methods often cannot be realized without the help of quantitative models and ways to solve them. As a result, every year new opportunities unfold for obtaining better solutions to medical problems and improving health care systems.

This handbook of optimization in medicine is composed of carefully refereed chapters written by experts in the fields of modeling and optimization in medicine and focuses on models and algorithms that allow for improved treatment of patients. Examples of topics that are covered in the handbook include:

- Optimal timing of organ transplants;
- treatment selection for breast cancer based on new classification schemes;
- treatment of head-and-neck, prostate, and other cancers; using conventional conformal and intensity modulated radiation therapy as well as proton therapy;
- optimization in medical imaging;
- classification and data mining with medical applications;
- treatment of epilepsy and other brain disorders;
- optimization for the genome project.

We believe that this handbook will be a valuable scientific source of information to graduate students and academic researchers in engineering, computer science, operations research, and medicine, as well as to practitioners who can tailor the approaches described in the handbook to their specific needs and applications.

We would like to take the opportunity to express our thanks to the authors of the chapters, the anonymous referees, and Springer for making the publication of this volume possible.

Gainesville, Florida                                    *Panos M. Pardalos*
March 2008                                              *H. Edwin Romeijn*

# Contents

# List of Contributors

**Oguzhan Alagoz**
Department of Industrial and
Systems Engineering
University of Wisconsin-Madison
Madison, Wisconsin 53706
alagoz@engr.wisc.edu

**Markus Alber**
Radioonkologische Klinik
Universitätsklinikum Tübingen
Hoppe-Seyler-Strasse 3, D-72076
Tübingen
Germany
msalber@med.uni-tuebingen.de

**Fernando Alonso**
Department of Optimization
Fraunhofer Institut for Industrial
Mathematics (ITWM)
D-67663 Kaiserslautern
Germany

**Ioannis P. Androulakis**
Department of Biomedical
Engineering and Department
of Chemical and Biochemical
Engineering, Rutgers, The State
University of New Jersey, Piscataway,
New Jersey 08854
yannis@rci.rutgers.edu

**Thomas Bortfeld**
Department of Radiation Oncology
Massachusetts General Hospital and
Harvard Medical
School Boston, Massachusetts 02114
tbortfeld@partners.org

**W. Art Chaovalitwongse**
Department of Industrial and
Systems Engineering
Rutgers, The State University of
New Jersey
Piscataway, New Jersey 08854
wchaoval@rci.rutgers.edu

**Pando G. Georgiev**
University of Cincinnati
Cincinnati, Ohio 45221
pgeorgie@ececs.uc.edu

**Kapil Gupta**
Center for Operations Research in
Medicine and HealthCare
School of Industrial and Systems
Engineering
Georgia Institute of Technology
Atlanta, Georgia 30332-0205
kgupta@isye.gatech.edu

**Michael Hintermüller**
Department of Mathematics
University of Sussex
Brighton BN1 9RF
United Kingdom
m.hintermueller@sussex.ac.uk

**Julie Simmons Ivy**
Edward P. Fitts Department of
Industrial and Systems Engineering,
North Carolina State University,
Raleigh, North Carolina 27695-7906
jsivy@ncsu.edu

**Srijit Kamath**
Department of Computer and
Information Science and Engineering
University of Florida
Gainesville, Florida 32611-6120
srijitk@ufl.edu

**Stephen L. Keeling**
Department of Mathematics and
Scientific Computing
University of Graz
A-8010 Graz
Austria
stephen.keeling@uni-graz.at

**Karl-Heinz Küfer**
Department of Optimization
Fraunhofer Institut for Industrial
Mathematics (ITWM)
D-67663 Kaiserslautern
Germany
kuefer@itwm.fhg.de

**Eva K. Lee**
Center for Operations Research in
Medicine and HealthCare
School of Industrial and Systems
Engineering
Georgia Institute of Technology
Atlanta, Georgia 30332-0205
evakylee@isye.gatech.edu

**Jonathan Li**
Department of Radiation Oncology
University of Florida
Gainesville, Florida 32610-0385
lijg@ufl.edu

**Gino J. Lim**
Department of Industrial
Engineering University of Houston
Houston, Texas 77204
ginolim@uh.edu

**Michael Monz**
Department of Optimization
Fraunhofer Institut for Industrial
Mathematics (ITWM)
D-67663 Kaiserslautern
Germany
monz@itwm.fhg.de

**Jatinder Palta**
Department of Radiation Oncology
University of Florida
Gainesville, Florida 32610-0385
paltajr@ufl.edu

**Sanjay Ranka**
Department of Computer and
Information Science and Engineering
University of Florida
Gainesville, Florida 32611-6120
ranka@cise.ufl.edu

**Rembert Reemtsen**
Institut für Mathematik
Brandenburgische Technische
Universität Cottbus
Universitätsplatz 3–4, D-03044
Cottbus
Germany
reemtsen@math.tu-cottbus.de

**Mark S. Roberts**
Section of Decision Sciences and
Clinical Systems Modeling, Division
of General Internal Medicine
School of Medicine
University of Pittsburgh
Pittsburgh, Pennsylvania 15213
robertsm@upmc.edu

**Sartaj Sahni**
Department of Computer and
Information Science and Engineering
University of Florida
Gainesville, Florida 32611-6120
sahni@cise.ufl.edu

**Andrew J. Schaefer**
Departments of Industrial Engineer-
ing and Medicine
University of Pittsburgh
Pittsburgh, Pennsylvania 15261
schaefer@ie.pitt.edu

**Alexander Scherrer**
Department of Optimization
Fraunhofer Institut for Industrial
Mathematics (ITWM)
D-67663 Kaiserslautern
Germany
scherrer@itwm.fhg.de

**Ahmad Saher Azizi Sultan**
Department of Optimization
Fraunhofer Institut for Industrial
Mathematics (ITWM)
D-67663 Kaiserslautern
Germany

**Philipp Süss**
Department of Optimization
Fraunhofer Institut for Industrial
Mathematics (ITWM)
D-67663 Kaiserslautern
Germany
suess@itwm.fhg.de

**Fabian J. Theis**
University of Regensburg
D-93040 Regensburg
Germany
fabian@theis.name

**Christian Thieke**
Clinical Cooperation Unit Radiation
Oncology
German Cancer Research Center
(DKFZ)
D-69120 Heidelberg
Germany
c.thieke@dkfz-heidelberg.de

**Tsung-Lin Wu**
Center for Operations Research in
Medicine and HealthCare
School of Industrial and Systems
Engineering
Georgia Institute of Technology
Atlanta, Georgia 30332-0205
tlwu@isye.gatech.edu

# 1

# Optimizing Organ Allocation and Acceptance

Oguzhan Alagoz[1], Andrew J. Schaefer[2], and Mark S. Roberts[3]

[1] Department of Industrial and Systems Engineering, University of
  Wisconsin-Madison, Madison, Wisconsin 53706
  `alagoz@engr.wisc.edu`
[2] Departments of Industrial Engineering and Medicine, University of Pittsburgh,
  Pittsburgh, Pennsylvania 15261
  `schaefer@ie.pitt.edu`
[3] Section of Decision Sciences and Clinical Systems Modeling, Division of General
  Internal Medicine, School of Medicine, University of Pittsburgh, Pittsburgh,
  Pennsylvania 15213
  `robertsm@upmc.edu`

## 1.1 Introduction

Since the first successful kidney transplant in 1954, organ transplantation
has been an important therapy for many diseases. Organs that can safely be
transplanted include kidneys, livers, intestines, hearts, pancreata, lungs, and
heart-lung combinations. The vast majority of transplanted organs are kidneys
and livers, which are the focus of this chapter. Organ transplantation is the
only viable therapy for patients with end-stage liver diseases (ESLDs) and
the preferred treatment for patients with end-stage renal diseases (ESRDs).
As a result of the the urgent need for transplantations, donated organs are
very scarce. The demand for organs has greatly outstripped the supply. Thus
organ allocation is a natural application area for optimization. In fact, organ
allocation is one of the first applications of medical optimization, with the
first paper appearing 20 years ago.

The United Network for Organ Sharing (UNOS) is responsible for managing the national organ donation and allocation system. The organ allocation
system is rapidly changing. For instance, according to the General Accounting Office, the liver allocation policy, the most controversial allocation system
[14], has been changed four times in the past six years [17, 28]. The multiple
changes in policy over a short time period is evidence of the ever-changing
opinions surrounding the optimal allocation of organs. For example, although
the new liver allocation policy is anticipated to "better identify urgent patients
and reduce deaths among patients awaiting liver  transplants" [28], anecdotal

P.M. Pardalos, H.E. Romeijn (eds.), *Handbook of Optimization in Medicine*,
Springer Optimization and Its Applications 26, DOI: 10.1007/978-0-387-09770-1_1,
© Springer Science+Business Media LLC 2009

evidence suggests that there is some question among the transplant community as to whether the new allocation rules are satisfactory [10, 26].

UNOS manages organ donation and procurement via Organ Procurement Organizations (OPOs), which are non-profit agencies responsible for approaching families about donation, evaluating the medical suitability of potential donors, coordinating the recovery, preservation, and transportation of organs donated for transplantation, and educating the public about the critical need for organ donation. There are currently 59 OPOs that operate in designated service areas; these service areas may cover multiple states, a single state, or just parts of a state [28]. The national UNOS membership is also divided into 11 geographic regions, each consisting of several OPOs. This regional structure was developed to facilitate organ allocation and to provide individuals with the opportunity to identify concerns regarding organ procurement, allocation, and transplantation that are unique to their particular geographic area [28].

Organs lose viability rapidly once they are harvested, but the rate is organ-specific. The time lag between when an organ is harvested and when it is transplanted is called the *cold ischemia time* (CIT). During this time, organs are bathed in storage solutions. The limits of CIT range from a few hours for heart-lung combinations to nearly three days for kidneys. Stahl et al. [24] estimated the relationship between CIT and liver viability. The Scientific Registry of Transplant Recipients states that the acceptable cold ischemia time limit for a liver is 12 to 18 hours [22], whereas the Center for Organ Recovery and Education gives the maximum limit as 18 to 24 hours [5].

There are two major classes of decision makers in organ allocation. The first class of decision makers is the individual patient, or the patient and his or her physician. Typically, the objective for such a perspective is to maximize some measure of that patient's benefit, typically life expectancy. The second class may be described as "society," and its goal is to design an organ allocation system so as to maximize some given criteria. Some examples of these criteria include total clinical benefit and some measure of equity. Equity is a critical issue in the societal perspective on organ allocation as there is considerable evidence that certain racial, geographic, and socioeconomic groups have greater access to organs than do others [27].

We limit our discussion to the U.S. organ allocation system. The remainder of this chapter is organized as follows. In Section 1.2, we describe the kidney allocation system, and in Section 1.3, we detail the liver allocation system. These two organs comprise the vast majority of organ transplantations; the details for other organs are described on the UNOS webpage [28]. Previous research on the patient's perspective is discussed in Section 1.4, and the societal perspective is described in Section 1.5. We provide conclusions and directions for future work in Section 1.6.

## 1.2 Kidney Allocation System

More than 60,000 patients are on the nationwide kidney waiting list. In 2003, 15,000 patients received a kidney transplant, of which more than 40% were from living donors [29]. The kidney waiting list and number of transplants are larger than those of all other organs combined. However, this need is somewhat mitigated by the fact that an alternate kidney replacement therapy (dialysis) is widely available. We describe the kidney allocation system as of late 2004 below. This allocation system is subject to frequent revision; readers are referred to the UNOS webpage [28] for updates to these and other allocation policies.

Kidneys are typically offered singly; however, there are certain cases when a high risk of graft failure requires the transplant of both kidneys simultaneously. UNOS defines two classes of cadaveric kidneys: standard and expanded. Kidneys in both classes have similar allocation mechanisms, as described below. Expanded-criteria kidneys have a higher probability of graft failure and are distinguished by the following factors:

1. Age: kidneys from some donors between 50 and 59 years and kidneys from every donor older than 60 years are expanded-criteria kidneys.
2. Level of creatinine in the donor's blood, which is a measure of the adequacy of kidney function: kidneys from donors with higher creatinine levels may be considered expanded-criteria kidneys.
3. Kidneys from donors who died of cardiovascular disease may be considered expanded-criteria.
4. Kidneys from donors with high hypertension may be considered expanded-criteria.

Patients who are willing to accept expanded-criteria kidneys do not have their eligibility for regular kidneys affected.

The *panel-reactive antibody* (PRA) level is a measure of how hard a patient is to match. It is defined as the percentage of cells from a panel of donors with which a given patient's blood serum reacts. This estimates the probability that the patient will have a negative reaction to a donor; the higher the PRA level, the harder the patient is to match.

A *zero-antigen mismatch* between a patient and a cadaveric kidney occurs when the patient and donor have compatible blood types and have all six of the same HLA-A, B, and DR antigens. There is mandatory sharing of zero-antigen-mismatched kidneys. When there are multiple zero-antigen-mismatched kidneys, there is an elaborate tie-breaking procedure that considers factors including the recipient's OPO, whether the patient is younger than 18, and certain ranges of PRA level. One interesting concept is that of debts among OPOs. Except in a few cases, when a kidney is shared between two OPOs, the receiving OPO must then share the next standard kidney it harvests in that particular blood type category. This is called a *payback debt*.

An OPO may not accumulate more than nine payback debts at any time. Priority for matching zero-antigen-mismatched kidneys is given to patients from OPOs that are owed payback kidneys. The full description of the tie-breaking procedure is available from the UNOS webpage [28].

If a kidney has no zero-antigen mismatches, kidneys with blood type O or B must be transplanted into patients with the same blood type. In general, kidneys are first offered within the harvesting OPO, then the harvesting region, and finally nationally. Within each of these three categories, patients who have an ABO match with the kidney are assigned points, and each kidney is offered to patients in decreasing order of points. A patient has the opportunity to refuse a kidney for any reason without affecting his or her subsequent access to kidneys.

Once minimum criteria are met, patients begin to acquire waiting time. One point is given to the patient who has been on the waiting list the longest amount of time. All other patients are accorded a fractional point equal to their waiting time divided by that of the longest-waiting patient. A patient receives four points if she has PRA level 80% or greater. Patients younger than 11 years old are given four points, and patients between 11 and 18 years of age are given three points. A patient is given four points if he or she has donated a vital organ or segment of a vital organ for transplantation within the United States. For the purposes of determining the priority within the harvesting OPO, a patient's physician may allocate "medical priority points." However, such points are not considered at the regional or national levels.

It is interesting to note that, excluding medical priority points, points based on waiting time can only be used to break ties among patients with the same number of points from other factors. In other words, kidneys are allocated lexicographically: the first factors are PRA level, age, and so on. Only among tied patients in the first factors is waiting time considered.

## 1.3 Liver Allocation System

This section describes the current liver allocation system. Basic knowledge of this system is necessary to understand the decision problem faced by the ESLD patients and the development of the decision models. The UNOS Board of Directors approved the new liver allocation procedure for implementation as of February 28, 2002 [28].

UNOS has different procedures for adult and for pediatric patients. Because researchers consider only the adult patients, we describe only the adult liver allocation procedure. UNOS maintains a patient waiting list that is used to determine the priority among the candidates. Under the current policy, when a liver becomes available, the following factors are considered for its allocation: liver and patient OPO, liver and patient region, medical urgency of the patient, patient points, and patient waiting time.

The medical urgency of the adult liver patients is represented by UNOS Status 1 and Model for End Stage Liver Disease (MELD) scores. According to the new UNOS policy, a patient listed as Status 1 "has fulminant liver failure with a life expectancy without a liver transplant of less than seven days" [28]. Patients who do not qualify for classification as Status 1 do not receive a status level. Rather, these patients will be assigned a "probability of pre-transplant death derived from a mortality risk score" calculated by the MELD scoring system [28]. The MELD score, which is a continuous function of total bilirubin, creatinine, and prothrombin time, indicates the status of the liver disease and is a risk-prediction model first introduced by Malinchoc et al. [16] to assess the short-term prognosis of patients with liver cirrhosis [30]. Wiesner et al. [30] developed the following formula for computing MELD scores:

$$\text{MELD Score} =$$
$$10 \times [0.957 \times \ln(\text{creatinine mg/dl}) + 0.378 \times \ln(\text{bilirubin mg/dl})$$
$$+ 1.120 \times \ln(\text{INR}) + 0.643 \times I_c]$$

where INR, international normalized ratio, is computed by dividing prothrombin time (PT) of the patient by a normal PT value, mg/dl represents the milligrams per deciliter of blood, and $I_c$ is an indicator variable that shows the cause of cirrhosis, i.e., it is equal to 1 if the disease is alcohol or cholestatic related and it is equal to 0 if the disease is related to other etiologies (causes). As Wiesner et al. [30] note, the etiology of disease is removed from the formula by UNOS. In addition to this, UNOS makes several modifications to the formula: any lab value less than 1 mg/dl is set to 1 mg/dl, any creatinine level above 4 mg/dl is set to 4 mg/dl, and the resulting MELD score is rounded to the closest integer [28]. By introducing these changes, UNOS restricts the range of MELD scores to be between 6 and 40, where a value of 6 corresponds with the best possible patient health and 40 with the worst.

Kamath et al. [15] developed the MELD system to more accurately measure the liver disease severity and to better predict which patients are at risk of dying. However, there are concerns about the accuracy of the MELD system. First, there were some biases in the data used to develop the model. For instance, the data available to the researchers were mostly based on patients with advanced liver disease [16]. Furthermore, the MELD system was validated on the patients suffering from cirrhosis [30], therefore it is possible that the MELD system does not accurately measure the disease progression for other diseases, e.g., acute liver diseases. Moreover, as stated, although they presented data to indicate that the consideration of patient age, sex, and body mass is unlikely to be clinically significant, it is possible that other factors, including a more direct measurement of renal function (iothalamate clearance), may improve the accuracy of the model [15]. Additionally, the MELD system was validated on only three laboratory values: creatinine and bilirubin levels and prothrombin time. Thus, it is possible that the MELD system does not accurately consider patients with liver cancer because they would score as if they were healthy [10]. Consequently, relying mainly on laboratory results may not be the best solution for all patients [9].

Patients are stratified within Status 1 and each MELD score using patient "points" and waiting time. Patient points are assigned based on the compatibility of their blood type with the donor's blood type. For Status 1 patients, candidates with an exact blood type match receive 10 points; candidates with a compatible, though not identical, blood type receive 5 points; and a candidate whose blood type is incompatible receives 0 points. As an exception, though type O and type $A_2$ (a less common variant of blood type A) are incompatible, patients of type O receive 5 points for being willing to accept a type $A_2$ liver. For non–Status 1 patients with the same MELD score, a liver is offered to patients with an exact blood type match first, compatible patients second, and incompatible patients last. If there are several patients having the same blood type compatibility and MELD scores, the ties are broken with patient waiting time. The waiting time for a Status 1 patient is calculated only from the date when that patient was listed as Status 1. Points are assigned to each patient based on the following strategy: "Ten points will be accrued by the patient waiting for the longest period for a liver transplant and proportionately fewer points will be accrued by those patients with shorter tenure" [28]. For MELD patients, waiting time is calculated as the time accrued by the patient at or above his or her current score level from the date that he or she was listed as a candidate for liver transplantation.

Figure 1.1 shows a schematic representation of the liver allocation system. In summary, the current liver allocation system works as follows: every liver available for transplant is first offered to those Status 1 patients located within the harvesting OPO. When more than one Status 1 patient exists, the liver is offered to those patients in descending point order where the patient with the highest number of points receives the highest priority. If there are no suitable Status 1 matches within the harvesting OPO, the liver is then offered to Status 1 patients within the harvesting region. If a match still has not been found, the liver is offered to all non–Status 1 patients in the harvesting OPO in descending order of MELD score. The search is again broadened to the harvesting region if no suitable match has been found. If no suitable match exists in the harvesting region, then the liver is offered nationally to Status 1 patients followed by all other patients in descending order of MELD scores.

UNOS maintains that the final decision to accept or decline a liver "will remain the prerogative of the transplant surgeon and/or physician responsible for the care of that patient" [14]. The surgeon and/or the physician have very limited time, namely one hour, to make their decision [28] because the acceptable range for cold ischemia time is very limited. Furthermore, as the Institute of Medicine points out, there is evidence that the quality of the organ decreases as cold ischemia time increases [14]. In the event that a liver is declined, it is then offered to another patient in accordance with the above-described policy. The patient who declines the organ will not be penalized

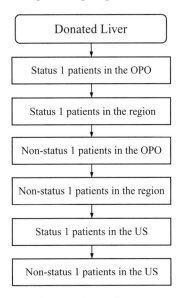

**Fig. 1.1.** Current liver allocation system.

and will have access to future livers. Organs are frequently declined due to low quality of the liver. For example, the donor may have had health problems that could have damaged the organ or may be much older than the potential recipient, making the organ undesirable [13].

## 1.4 Optimization from the Patient's Perspective

This section describes the studies on the optimal use of cadaveric organs for transplantation that maximizes the patient's welfare. Section 1.4.1 summarizes studies that consider the kidney transplantation problem. Section 1.4.2 describes studies that consider the liver transplantation problem.

### 1.4.1 Optimizing kidney transplantation

David and Yechiali [6] consider when a patient should accept or reject an organ for transplantation. They formulate this problem as an optimal stopping problem in which the decision maker accepts or reject offers $\{X_j\}_0^\infty$ that are available at random times $\{t_j\}_0^\infty$, where $\{X_j\}_0^\infty$ is a sequence of independent and identically distributed positive bounded random variables with distribution function $F(x) = P(X \leq x)$. If the patient accepts the offer at time $t_j$, the patient quits the process and receives a reward $\beta(t_j)X_j$, where $\beta(t)$ is a continuous nonincreasing discount function with $\beta(0) = 1$. If the patient does not accept the offer, then the process continues until the next offer, or patient death. The probability that the decision maker dies before the new offer arrives at time $t_{j+1}$ is given by the variable $1 - \alpha_{j+1} = P(T \leq t_{j+1} | T > t_j)$ defined by $T$, the lifetime of the underlying process. Their objective is to find a stopping rule that maximizes the total expected discounted reward from any time $t$ onward.

They first consider the case in which the offers arrive at fixed time points and there are a finite number of offers $(n)$ available. In this case, they observe that the optimal strategy is a control-limit policy with a set of controls $\{\lambda_n^j\}_{j=0}^n$, and an offer $X_j$ at time $t_j$ is accepted if and only if $\beta_j X_j > \lambda_n^j$, where $\lambda_n^j$ is the maximum expected discounted reward if an offer at time $t_j$ is rejected. Because for each $j \leq n$, $\{\lambda_n^j\}_{n=0}^\infty$ is a nondecreasing bounded sequence of $n$, it has a limit $l_j$.

They extend their model to the infinite-horizon problem in which the offers arrive randomly. They prove that if the lifetime distribution of the decision maker is increasing failure rate (IFR) [4], then the optimal policy takes the form of a continuous nonincreasing real function $\lambda(t)$ on $[0, \infty)$, such that an offer $x$ at time $t$ is accepted if and only if $\beta(t)x \geq \lambda(t)$. $\lambda(t)$ is equal to the future expected discounted reward if the offer is rejected at time $t$, and an optimal policy is applied thereafter. They show that the IFR assumption is a necessary assumption in this setting.

David and Yechiali also consider the case where the arrivals follow a nonhomogeneous Poisson process. They consider several special cases of this model such as the organ arrival is nonhomogeneous Poisson with nonincreasing intensity and the lifetime distribution is IFR. In this case, they prove that the control limit function $\lambda(t)$ is nonincreasing, so that a patient becomes more willing to accept lower quality organs as time progresses. They obtain a bound for the $\lambda(t)$ for this special case.

They provide an explicit closed form solution of the problem when the lifetime distribution is Gamma with homogenous Poisson arrivals. They present a numerical example for this special case using data related to the kidney transplant problem.

Ahn and Hornberger [1] and Hornberger and Ahn [11] develop a discrete-time infinite horizon discounted Markov decision process (MDP) model for deciding which kidneys would maximize a patient's total expected (quality-adjusted) life. In their model, the patient is involved in the process of determining a threshold kidney quality value for transplantation. They use expected one-year graft survival rate as the criterion for determining the acceptability of a kidney. The state space consists of the patient state and includes five states: alive on dialysis and waiting for transplantation $(S_1)$; not eligible for transplantation $(S_2)$; received a functioning renal transplant $(S_3)$; failed transplant $(S_4)$; and death $(S_5)$. They assume that the patient assigns a quality-of-life score to each state. They use months as their decision epochs because of the sparsity of their data. The patient makes the decision only when he or she is in state $(S_1)$. The quality-adjusted life expectancy (QALE) of the patient in state $(S_1)$ is a function of (1) QALE if a donor kidney satisfying eligibility requirements becomes available and the patient has the transplantation, (2) QALE if an ineligible donor kidney becomes available and the patient is not transplanted, and (3) the quality of life with dialysis in that month. Because of the small number of states, they provide an exact analytical solution for threshold kidney quality.

They use real data to estimate the parameters and solve the model for four representative patients. The minimum one-year graft survival rate, $d^*$, differs significantly among the four patients. They compare their results with what might be expected by using the UNOS point system for four representative donor kidneys. They also perform a one-way sensitivity analysis to measure the effects of the changes in the parameters. Their results show that the important variables that affect the minimum eligibility criterion are quality of life assessment after transplant, immunosuppressive side effect, probability of death while undergoing dialysis, probability of death after failed transplant, time preference, and the probability of being eligible for retransplantation.

### 1.4.2 Optimizing liver transplantation

Howard [12] presents a decision model in which a surgeon decides to accept or reject a cadaveric organ based on the patient's health. He frames the organ acceptance decision as an optimal stopping problem. According to his model, a surgeon decides whether or not to accept an organ of quality $q \in (0, \bar{q}]$ for a patient in health state $h \in (0, \bar{h}]$, where the state $q = 0$ describes a period in which there is no organ offer and the state $h = 0$ corresponds with death. The organ offers arrive with distribution function $f(q)$. If the surgeon rejects the organ, the patient's health evolves according to a Markov process described by $f(h'|h)$, where $f(h'|h)$ is IFR. If the surgeon accepts an organ offer, then the probability that the operation is successful in period $t + 1$ is a function of current patient health $h$ and organ offer $q$ and is denoted by $p(h, q)$. If the patient's single period utility when alive is $u$ and the immediate reward of a successful operation is $B$, the total expected reward from accepting an organ at time $t$, $EV^{TX}(h, q)$ and from rejecting an organ at time $t$, $EV^W(h)$ are as follows:

$$EV^{TX}(h, q) = p(h, q)B,$$

and

$$EV^W(h) = \int_q \int_h V^W(h', q') f(h'|h) f(q') dh' dq',$$

where $V^W(h, q)$ is defined by the following set of equations:

$$V^W(h, q) = u + \delta \max \left\{ EV^{TX}(h, q), EV^W(h) \right\},$$

where $\delta$ is the discount factor.

Howard estimates the parameters in his decision model using liver transplantation data in the United States. However, he does not provide any structural insights or numerical solutions to this decision model. Instead, he provides statistical evidence that explains why a transplant surgeon may reject a cadaveric liver offer. His statistical studies show that as the waiting list has grown over time, the surgeons have faced stronger incentives to use lower quality organs. Similarly, the number of organ transplantations has increased dramatically in years when the number of traumatic deaths decreased.

Howard also discusses the trends in organ procurement in light of his findings and describes some options to the policy makers who believe that too many organs are discarded. One option is to use the results of a decision that calculates the optimal quality cutoff and enforce it via regulations. Another option is to penalize hospitals that reject organs that are subsequently transplanted successfully by other transplant centers. It is also possible to implement a dual list system in which the region maintains two waiting lists, one for patients whose surgeons are willing to accept low-quality organs and one for patients whose surgeons will accept only high-quality organs.

Alagoz et al. [2] consider the problem of optimally timing a living-donor liver transplant in order to maximize a patient's total reward, for example, life expectancy. Living donors are a significant and increasing source of livers for transplantation, mainly due to the insufficient supply of cadaveric organs. Living-donor liver transplantation is accomplished by removing an entire lobe of the donor's liver and implanting it into the recipient. The non-diseased liver has a unique regenerative ability so that a donor's liver regains its previous size within two weeks. They assume that the patient does not receive cadaveric organ offers.

In their decision model, the decision maker can take one of two actions at state $h \in \{1, \ldots, H\}$, namely, "Transplant" or "Wait for one more decision epoch," where 1 is the perfect health state and $H$ is the sickest health state. If the patient chooses "Transplant" in health state $h$, he or she receives a reward of $r(h, T)$, quits the process, and moves to absorbing state "Transplant" with probability 1. If the patient chooses to "Wait" in health state $h$, he or she receives an intermediate reward of $r(h, W)$ and moves to health state $h' \in S = \{1, \ldots, H+1\}$ with probability $P(h'|h)$, where $H+1$ represents death. The optimal solution to this problem can be obtained by solving the following set of recursive equations:

$$V(h) = \max\left\{r(h,T), r(h,W) + \lambda \sum_{h' \in S} P(h'|h)V(h')\right\}, h = 1, \ldots, H,$$

where $\lambda$ is the discount factor, and $V(h)$ is the maximum total expected discounted reward that the patient can attain when his or her current health is $h$.

They derive some structural properties of this MDP model including a set of intuitive sufficient conditions that ensure the existence of a control-limit policy. They prove that the optimal value function is monotonic when the transition probability matrix is IFR and the functions $r(h, T)$ and $r(h, W)$ are nonincreasing in $h$. They show that if one disease causes a faster deterioration in patient health than does another and yet results in identical post-transplant life-expectancy, then the control limit for this disease is less than or equal to that for the other. They solve this problem using clinical data. In all of their computational tests, the optimal policy is of control-limit type. In some of the examples, when the liver quality is very low, it is optimal for the patient to choose never to have the transplant.

Alagoz et al. [3] consider the decision problem faced by liver patients on the waiting list: should an offered organ of a given quality be accepted or declined? They formulate a discrete-time, infinite-horizon, discounted MDP model of this problem in which the state of the process is described by patient state and organ quality. They consider the effects of the waiting list implicitly by defining the organ arrival probabilities as a function of patient state.

They assume that the probability of receiving a liver of type $\ell$ at time $t + 1$ depends only on the patient state at time $t$ and is independent of the type of liver offered at time $t$. According to their MDP model, the decision maker can take one of two actions in state $(h, \ell)$, where $h \in \{1, \ldots, H + 1\}$ represents patient health and $\ell \in S_L$ represents current liver offer. Namely, "Accept" the liver $\ell$ or "Wait for one more decision epoch." If the patient chooses "Accept" in state $(h, \ell)$, he or she receives a reward of $r(h, \ell, T)$, quits the process, and moves to absorbing state "Transplant" with probability 1. If the patient chooses to "Wait" in state $(h, \ell)$, then he or she receives an intermediate reward of $r(h, W)$ and moves to state $(h', \ell') \in S$ with probability $\mathcal{P}(h', \ell'|h, \ell)$. The optimal solution to this problem is obtained by solving the following set of recursive equations [18]:

$$
V(h, \ell) = \max \left\{ r(h, \ell, T), r(h, W) + \lambda \sum_{(h', \ell') \in S} \mathcal{P}(h', \ell'|h, \ell) V(h', \ell') \right\},
$$
$$
h \in \{1, \ldots, H\}, \ell \in S_L, \quad (1.1)
$$

where $\lambda$ is the discount factor, and $V(h, \ell)$ is the maximum total expected discounted reward that the patient can attain when his or her current state is $h$ and the current liver offered is $\ell$.

Alagoz et al. derive structural properties of the model, including conditions that guarantee the existence of a liver-based and a patient-based control-limit optimal policy. A *liver-based control-limit policy* is of the following form: for a given patient state $h$, choose the "Transplant" action and "Accept" the liver if and only if the offered liver is of type $1, 2, \ldots, i(h)$ for some liver state $i(h)$ called the *liver-based control limit*. Similarly, a *patient-based control-limit policy* is of the simple form: for a given liver state $\ell$, choose the "Transplant" action and "Accept" the liver if and only if the patient state is one of the states $j(\ell), j(\ell) + 1, \ldots, H$, for some patient state $j(\ell)$ called the *patient-based control limit*.

The conditions that ensure the existence of a patient-based control-limit policy are stronger than those that guarantee the existence of a liver-based control-limit policy. They compare the optimal control limits for the same patient listed in two different regions. They show that if the patient is listed in region A where he or she receives more frequent and higher quality liver offers than in region B, then the optimal liver-based control limits obtained when he or she is listed in region A are lower than those obtained when he or she is listed in region B.

They use clinical data to solve this problem, and in their experiments the optimal policy is always of liver-based control-limit type. However, some optimal policies are not of patient-based control-limit type. In some regions, as the patient gets sicker, the probability of receiving a better liver increases significantly. In such cases, it is optimal to decline a liver offer in some patient states even if it is optimal to accept that particular liver offer in better patient states. Their computational tests also show that the location of the patient has a significant effect on liver offer probabilities and optimal control limits.

## 1.5 Optimization from the Societal Perspective

This section describes the studies on optimal design of an allocation system that maximizes the society's welfare. Section 1.5.1 summarizes studies that consider the general organ allocation problem. Section 1.5.2 describes studies that consider the kidney allocation problem.

### 1.5.1 Optimizing general organ allocation system

Righter [19] considers a resource allocation problem in which there are $n$ activities each of which requires a resource, where resources arrive according to a Poisson process with rate $\lambda$. Her model can be applied to the kidney allocation problem, where resources represent the organs and activities represent the patients. When a resource arrives, its value $X$, a nonnegative random variable with distribution $F(\cdot)$, becomes known, and it can either be rejected or assigned to one of the activities. Once a resource is assigned to an activity, that activity is no longer available for further assignments. Activities are ordered such that $r_1 \geq r_2 \geq \cdots \geq r_n \geq 0$, where $r_i$ represents the activity value. Each activity has its own deadline that is exponentially distributed with rate $\alpha_i$ and is independent of other deadlines. When the deadline occurs, the activity terminates. The reward of assigning a resource to an activity is the product of the resource value and the activity value. The objective is to assign arriving resources to the activities such that the total expected return is maximized. If all activity deadlines are the same, i.e., $\alpha_i = \alpha$ for all $i$, then the optimal policy has the following form: assign a resource unit of value $x$ to activity $i$ if $v_i(\alpha) < x \le v_{i-1}(\alpha)$, where each threshold $v_i(\alpha)$ represents the total expected discounted resource value when it is assigned to activity $i$ under the optimal policy. She defines $v_0(\alpha) = \infty$ and $v_{n+1}(\alpha) = 0$. Furthermore, $v_0(\alpha) > v_1(\alpha) > \cdots > v_n(\alpha) > v_{n+1}(\alpha)$, where $v_i(\alpha)$ does not depend on $n$ for $n \ge i$, and $v_i(\alpha)$ does not depend on $r_j$ for any $j$.

Righter analyzes the effects of allowing the parameters to change according to a continuous time Markov chain on the structural properties of the optimal value function. She first assumes that the arrival rate of resources change according to a continuous Markov chain whereas all other model parameters are fixed and proves that the optimal policy still has the same structure, where the thresholds do not depend on the $r_j$ but depend on the current system state (environmental state). She then considers the case in which the activity values and deadline rates change according to a random environment and proves that the thresholds and the total returns are monotonic in the parameters of the model. In this case, the thresholds depend on the $r_j$'s as well as the environmental state. She also provides conditions under which model parameters change as functions of the environmental state that ensure the monotonicity of the total returns.

David and Yechiali [7] consider allocating multiple organs to multiple patients where organs and patients arrive simultaneously. That is, an infinite random sequence of pairs (patient and organ) arrive sequentially, where each organ and patient is either of Type I with probability $p$ or of Type II with probability $q = 1 - p$. When an organ is assigned to the candidate, it yields a reward $R > 0$ if they match in type or a smaller reward $0 < r \le R$ if there is a mismatch. If an organ is not assigned, it is unavailable for future assignments, however, an unassigned patient stays in the system until he or she is assigned an organ. The objective is to find assignment policies that maximize various optimality criteria.

David and Yechiali first consider the average reward criterion. A policy $\pi$ is average-reward optimal if it maximizes the following equation:

$$\phi_\pi(s) = \liminf_{t \to \infty} \frac{E\left[\sum_{n=0}^{t-1} r_\pi(n) | \text{initial state} = s\right]}{t},$$

where $r_\pi(n)$ is the average reward earned in day $n$, and states are represented by pairs $(i, j)$ denoting $i$ Type I and $j$ Type II candidates waiting in the system $(0 \le i, j < \infty)$. They prove that when there are infinitely many organs and patients, the optimal policy is to assign only perfect matches for any $0 \le p \le 1$ and $0 \le r \le R$, and the optimal gain is the perfect-match reward, $R$. If there exist at most $k$ patients, then the reasonable policy of order $k$ is the optimal policy, where a reasonable policy of order $k$ is defined as follows. A policy is a reasonable policy of order $k$ if it satisfies the following conditions: (i) assign a match whenever possible and (ii) assign a mismatch when $n_1$ candidates are present prior to the arrival, with $k$ being the smallest number $n_1$ specified in (ii).

David and Yechiali then consider the finite- and infinite-horizon discounted models. They show that for a finite-horizon model, the optimal policy has the following form: assign a perfect match when available, and assign a mismatch if and only if $r > r^*_{n,N}$, where $r^*_{n,N}$ is a control limit that changes with the optimal reward-to-go function when there are $n$ Type I candidates and $N$ periods to go. Unfortunately, they could not find a closed-form solution for $r^*_{n,N}$. They also show that the infinite-horizon discounted-reward optimal policy is of the following form: assign a perfect match when available, and assign a mismatch according to a set of controls

$$r^*_1 \geq r^*_2 \geq \cdots \geq r^*_{k-1} \geq r^*_k \geq \cdots$$

on $r$ and according to $k$, where $k$ represents the number of mismatching candidates in the system and $r_k$ are a set of control limits on $r$.

David and Yechiali [8] consider allocating multiple ($M$) organs to multiple ($N$) patients. Assignments are made one at a time, and once an organ is assigned (or rejected), it is unavailable for future assignments. Each organ and patient is characterized by a fixed-length attribute vector $X = (X_1, X_2, \ldots, X_p)$, where each patient's attributes are known in advance, and each organ's attributes are revealed only upon arrival. When an offer is assigned to a patient, the two vectors are matched, and the reward is determined by the total number of matching attributes. There are at most $p + 1$ possible match levels. The objective is to find an assignment policy that maximizes the total expected return for both discounted and undiscounted cases. They assume that $p$ equals 1, so that each assignment of an offer to a candidate yields a reward of $R$ if there is a match and a smaller reward $r \leq R$ if there is a mismatch.

They first consider the special case in which $M \geq N$, each patient must be assigned an organ, and a fixed discount rate ($\alpha$) exists. They assume that $f_1 \leq f_2 \leq \cdots \leq f_N$, where $f_1, \ldots, f_N$ are the respective frequencies $P\{X = a_1\}, \ldots, P\{X = a_N\}$, the $N$ realizations of the attribute vector. Using the notation $(\mathbf{f})$ for $(f_1, \ldots, f_{N+1})$ and $(\mathbf{f}_{-1})$ for $(f_1, \ldots, f_{i-1}, f_{i+1}, \ldots, f_{N+1})$, the optimality equations are

$$V_{N+1,M+1}(\mathbf{f})|X_1 = \max \begin{cases} R + \alpha V_{N,M}(\mathbf{f}_{-1})|\{X_1 = a_i\} & \text{(match)} \\ r + \alpha \max_k V_{N,M}(\mathbf{f}_{-k}) & \text{(a mismatch)} \\ \alpha V_{N+1,M}(\mathbf{f}) & \text{(rejection)}, \end{cases}$$

where $V_{N,M}(\mathbf{f})$ is the maximal expected discounted total reward when there are $N$ waiting patients with $N$ attribute realizations $(a_1, \ldots, a_N)$ and $M$ offers available. They prove that if $N < M$ and $a_1, \ldots, a_N$ are distinct, the optimal policy is to assign a match whenever possible and to reject a mismatch or assign it to $a_1$ depending on whether $\alpha \xi_1 \geq r$ or $\alpha \xi_1 < r$, where $\xi_1 = f_1 R + (1 - f_1)r$.

David and Yechiali then consider the case where $M = N$ and no rejections are possible. In this case, the optimal policy is as follows: if an offer matches one or more of the candidates, it is assigned to one of them. Otherwise it is assigned to a candidate with the rarest attribute.

Finally, they relax the assumption that all candidates must be assigned and $M \geq N$. In this case, they prove that the optimal policy is to assign the organs to one of the candidates if a match exists and to assign to $a_1$ when $f_1 < \varphi$, where $\varphi$ is a function of $f_i$'s and can be computed explicitly for some special cases.

Stahl et al. [23] use an integer programming model to formulate and solve the problem of the optimal sizing and configuration of transplant regions and OPOs in which the objective is to find a set of regions that optimizes transplant allocation efficiency and geographic equity. They measure efficiency by the total number of intra-regional transplants and geographic equity by the minimum OPO intra-regional transplant rate, which is defined as the number of intra-regional transplants in an OPO divided by the number of patients on the OPO waiting list.

They model the country as a simple network in which each node represents an OPO, and arcs connecting OPOs indicate that they are contiguous. They assume that a region can consist of at most nine contiguous OPOs, an OPO supplies its livers only to the region that contains it, and both transplant allocation efficiency and geographic equity could be represented as factors in a function linking CIT and liver transport distance. They also assume that the probability of declining a liver offer, which is measured by the liver's viability, is solely dependent on its CIT. Primary nonfunction occurs when a liver fails to work properly in the recipient at the time of transplant. They use two functional relationships between primary nonfunction and CIT: linear and polynomial.

Stahl et al. solve an integer program to find the optimal set of regions such that the total number of intra-regional transplants are maximized. They define the binary variable $x_j$ for every possible region $j$ such that it is equal to 1 if region $j$ is chosen and is equal to 0 if region $j$ is not chosen. Then, the integer program is as follows:

$$\max \left\{ \sum_{j \in J} c_j x_j : \sum_{j \in J} a_{ij} x_j = 1, i \in I; x_j \in \{0,1\}, j \in J \right\}, \quad (1.2)$$

where $I$ is the set of all OPOs; $J$ is the set of all regions; $a_{ij} = 1$ if region $j$ contains OPO $i$, and 0 otherwise; and $c_j$ represents the total number of intra-regional transplants for region $j$. They provide a closed-form estimate of $c_j$. If the number of regions is constrained to be equal to 11, then the constraint $\sum_{j \in J} x_j = 1$ is added. The integer program defined in (1.2) does not consider the geographic equity. Let $f_{ij}$ and $\lambda_{min}$ represent the intra-regional transplant rate in OPO $i$ contained in region $j$ and the minimal local transplant rate, respectively. Then, the integer program considering the geographic equity can be reformulated as follows:

$$\max \left\{ \sum_{j \in J} c_j x_j + \rho \lambda_{\min} : \sum_{j \in J} a_{ij} x_j = 1, i \in I; \right.$$

$$\left. \sum_{j \in J} f_{ij} x_j - \lambda_{\min} \geq 0, i \in I; \; x_j \in \{0, 1\}, j \in J \right\}, \qquad (1.3)$$

where $\rho$ is a constant that indicates the importance the decision makers place on the minimum transplant rate across OPOs versus intra-regional transplants. Hence, changing $\rho$ will provide a means for balancing the two conflicting factors, transplant allocation efficiency and geographic equity.

Stahl et al. conduct computational experiments using real data to compare the regional configuration obtained from their model to the current configuration. The optimal sets of regions tend to group densely populated areas. Their results show that the proposed configuration resulted in more intra-regional transplants. Furthermore, for all values of $\rho$, the minimum intra-regional transplant rate across OPOs is significantly higher than that in the current regional configuration. However, as $\rho$ increases, the increase over the current configuration diminishes. They also perform sensitivity analyses, which show that the outcome is not sensitive to the relationship between CIT and primary nonfunction.

### 1.5.2 Optimizing the kidney allocation system

Zenios et al. [31] consider the problem of finding the best kidney allocation policy with the three-criteria objective of maximizing total quality-adjusted life years (QALYs) and minimizing two measures of inequity. The first measures equity across various groups in terms of access to kidneys, and the second measures equity in waiting times. They formulate this problem using a continuous-time, continuous-space deterministic fluid model but do not provide a closed-form solution.

In their model, there are $K$ patient and $J$ donor classes. They assume that patients of class $k = 1, \ldots, K_W$ are registered on the waiting list and patients of class $k = K_W + 1, \ldots, K$ have a functioning graft. The state of the system at time $t$ is described by the $K$-dimensional column vector $x(t) = (x_1(t), \ldots, x_K(t))^T$, which represents the number of patients in each class. Transplant candidates of class $k \in \{1, \ldots, K_W\}$ join the waiting list at rate $\lambda_k^+$ and leave the waiting list with rate $\mu_k$ due to death or due to organ transplantation. Organs of class $j \in \{1, \ldots, J\}$ arrive at rate $\lambda_j^-$, from which a fraction $v_{jk}(t)$ is allocated to transplant candidates $k$. Note that $v_{jk}(t)$ is a control variable and $u_{jk}(t) = \lambda_j^- v_{jk}(t)$ is the transplantation rate of class $j$ kidneys into class $k$ candidates. When a class $j$ kidney is transplanted into a class $k \in \{1, \ldots, K_W\}$ patient, the class $k$ patient leaves the waiting list and becomes a patient of class $c(k, j) \in \{K_W + 1, \ldots, K\}$. Furthermore, $c(k, j)$ patients depart this class at rate $\mu_{c(k,j)}$ per unit time; a fraction $q_{c(k,j)} \in [0, 1]$

of these patients are relisted as patients of class $k$ as a result of graft failure, whereas a fraction $1 - q_{c(k,j)}$ of them exit the system due to death.

The system state equations are given by the following linear differential equations:

$$\frac{d}{dt}x_k(t) = \lambda_k^+ - \mu_k x_k(t) - \sum_{j=1}^{J} u_{jk}(t) + \sum_{j=1}^{J} q_{c(k,j)}\mu_{c(k,j)}x_{c(k,j)}(t);$$

$$k = 1, \ldots, K_W, \qquad (1.4)$$

$$\frac{d}{dt}x_k(t) = \sum_{j=1}^{J}\sum_{i=1}^{K_W} u_{ji}(t)1_{\{c(i,j)=k\}} - \mu_k x_k(t); \quad k = K_W + 1, \ldots, K, \quad (1.5)$$

and are subject to the state constraints

$$x_k(t) \geq 0; \quad k = 1, \ldots, K. \qquad (1.6)$$

The organ allocation rates $u(t)$ must satisfy the following constraints:

$$\sum_{k=1}^{K_W} u_{jk}(t) \leq \lambda_j^-; \quad j = 1, \ldots, J, \qquad (1.7)$$

$$u_{jk}(t) \geq 0; \quad k = 1, \ldots, K_W \quad \text{and} \quad j = 1, \ldots, J. \qquad (1.8)$$

Zenios et al. note that this model ignores the three important aspects of the kidney allocation problem: crossmatching between donor and recipient, unavailability of recipients, and organ sharing between OPOs. The model assumes that the system evolution is deterministic. They use the QALY to measure the efficiency of the model. Namely, they assume that UNOS assigns a quality of life (QOL) score $h_k$ to each patient class $k = 1, \ldots, K$, and the total QALY over a finite time horizon $T$ is found using

$$\int_0^T \sum_{k=1}^{K} h_k x_k(t)dt.$$

For a given allocation policy $u(t) = (u_{1\cdot}(t)^T, \ldots, u_{J\cdot}(t)^T$, where $u_{j\cdot}(t) = (u_{j1}(t), \ldots, u_{jK_W}(t))^T$, their first measure of equity, *waiting time inequity*, is calculated by

$$\frac{1}{2}\int_0^T \sum_{k=1}^{K_W}\sum_{i=1}^{K_W} \lambda_k(t, u(t))\lambda_i(t, u(t)) \cdot \left(\frac{x_k(t)}{\lambda_k(t, u(t))} - \frac{x_i(t)}{\lambda_i(t, u(t))}\right)^2 dt,$$

where $\lambda(t, u(t)) = (\lambda_1(t, u(t)), \ldots, \lambda_{K_W}(t, u(t)))$ represents the instantaneous arrival rate into class $k$ under allocation policy $u(t)$.

The second measure of equity considers the likelihood of transplantation. They observe that

$$\lim_{T \to \infty} \frac{\int_0^T \sum_{j=1}^J u_{jk}(t)dt}{\lambda_k^+ T}$$

gives the percentage of class $k$ patients who receive transplantation. Then the vector of likelihoods of transplantation is given by

$$\frac{\int_0^T \tilde{D}u(t)dt}{\lambda^+ T},$$

where $\tilde{D} \in \mathcal{R}^{K_W \times K_W J}$ is a matrix with components

$$\tilde{D}_{ki} = \begin{cases} 1 \text{ if } i \bmod K_W = k; \\ 0 \text{ otherwise.} \end{cases}$$

Because this form is not analytically tractable, they insert the Lagrange multipliers $\gamma = (\gamma_1, \ldots, \gamma_{K_W})^T$ into the objective function using the following expression in the objective function:

$$\int_0^T \gamma^T \tilde{D}u(t)dt.$$

They combine the three objectives and the fluid model to obtain the following control problem: choose the allocation rates $u(t)$ to maximize the tricriteria objective of

$$\int_0^T \left( \beta \sum_{k=1}^K h_k x_k(t) \right.$$

$$- (1-\beta) \sum_{k=1}^{K_W} \sum_{i=1}^{K_W} \lambda_k(t, u(t)) \lambda_i(t, u(t)) \cdot \left( \frac{x_k(t)}{\lambda_k(t, u(t))} - \frac{x_i(t)}{\lambda_i(t, u(t))} \right)^2$$

$$\left. + \gamma^T \tilde{D}u(t) \right) dt,$$

subject to (1.4)–(1.8), where $\beta \in [0, 1]$.

Because there does not appear to be a closed-form solution to this problem, they employ three approximations to this model and provide a heuristic *dynamic index policy*. At time $t$, the dynamic index policy allocates all organs of class $j$ to the transplant candidate class $k$ with the highest index $G_{jk}(t)$, which is defined by

$$G_{jk} = \pi_{c(k,j)}(x(t)) - \pi_k(x(t)) + \gamma_k,$$

where $\pi_{c(k,j)}(x(t))$ represents the increase in

$$\beta \sum_{k=1}^K h_k x_k(t) - (1-\beta) \sum_{k=1}^{K_W} \sum_{i=1}^{K_W} \lambda_k(t, u(t)) \lambda_i(t, u(t)) \cdot \left( \frac{x_k(t)}{\lambda_k(t, u(t))} - \frac{x_i(t)}{\lambda_i(t, u(t))} \right)^2$$

if an organ of class $j$ is transplanted into a candidate of class $k$ at time $t$.

Zenios et al. construct a simulation model to compare the dynamic index policy to the UNOS policy and an FCFT (first-come first-transplanted) policy. They evaluate the effects of the dynamic index policy on the organ allocation system for several values of $\beta$ and $\gamma$. They consider two types of OPOs: a typical OPO and a congested OPO, where the demand-to-supply ratio is much higher than that of a typical OPO. Their results show that the the the dynamic index policy outperforms both the FCFT and UNOS policy.

Su and Zenios [25] consider the problem of allocating kidneys to the transplant candidates who have the right to refuse the organs. They use a sequential stochastic assignment model to solve variants of this problem. They assume that the patients do not leave the system due to pre-transplant death.

Their first model considers the case when the patient does not have the right to reject an organ. This model also assumes that there are $n$ transplant candidates with various types to be assigned to $n$ kidneys, which arrive sequentially—one kidney in each period. The type of kidney arriving at time $t$ is a random variable $\{X_t\}_{t=1}^n$, where $\{X_t\}_{t=1}^n$ are independent and identically distributed with probability measure $P$ over the space of possible types $\mathcal{X}$. There are $m$ patient types where the proportion of type $i$ candidates is denoted by $p_i$. When a type $x$ kidney is transplanted into a type $i$ patient, a reward of $R_i(x)$ is obtained. The objective is to find an assignment policy $I = (i(t))_{t=1,\ldots,n}$ that maximizes total expected reward, $E\left[\sum_{t=1}^n R_{i(t)}(X_t)\right]$, where $i(t)$ denotes the candidate type that is assigned to the kidney arriving at time $t$. The optimization problem is to find a partition $\{A_i^*\}_{i=1}^m$ to

$$\max_{\{A_1,\ldots,A_m\}} \sum_{i=1}^m E[R_i(X)1_{\{X\in A_i\}}]$$

$$\text{such that } P(A_i) = p_i \; i = 1,\ldots,m,$$

where $1_{\{X\in A_i\}}$ is the indicator function, which takes the value of 1 if $X \in A_i$ and 0 if $X \notin A_i$, and $\{A_i\}_{i=1}^m$ is a partition of the kidney space $\mathcal{X}$.

They analyze the asymptotic behavior of this optimization problem and prove that the optimal partitioning policy is asymptotically optimal as $n \to \infty$. This result reduces the sequential assignment problem into a set partitioning problem. If the space $\mathcal{X}$ consists of $k$ discrete kidney types with probability distribution $(q_1,\ldots,q_k)$, then the partition policy can be represented by the set of numbers $\{a_{ij}\}_{1\leq i\leq m, 1\leq j\leq k}$ such that when a kidney of type $j$ arrives, it is assigned to a candidate of type $i$ with probability $a_{ij}/\sum_{i=1}^m a_{ij}$, where $a_{ij}$ is the joint probability of a type $i$ candidate being assigned a type $j$ kidney. Then the optimal partition policy is given by the solution $\{a_{ij}^*\}$ to the following assignment problem:

$$\max_{\{a_{ij}\}} \sum_{i=1}^m \sum_{j=1}^k a_{ij} r_{ij}$$

$$\text{such that } \sum_{i=1}^m a_{ij} = q_j \qquad j = 1,\ldots,k$$

$$\sum_{j=1}^{k} a_{ij} = p_i \qquad i = 1, \ldots, m.$$

They derive the structural properties of the optimal policy under different reward functions including multiplicative reward structure and a match-reward structure, in which if the patient and kidney types match the transplantation results in a reward of $R$, and if there is a mismatch then the transplantation results in a reward of $r < R$. They show that if the reward functions satisfy the increasing differences assumption, i.e., $R_i(x) - R_j(x)$ is increasing in $x$, then the optimal partition is given by $A_i^* = [a_{i-1}, a_i)$, where $a_o = -\infty, a_m = \infty$, and

$$Pr(X \leq a_i) = p_1 + \cdots + p_i.$$

Su and Zenios then consider the problem of allocating kidneys to the patients when the patients have the right to refuse an organ offer and measure the effects of patient autonomy on the overall organ acceptance and rejection rates. In this model, they assume that an organ rejected by the first patient will be discarded. They define a partition policy $A = \{A_i\}$ as incentive-compatible if the following condition holds for $i = 1, \ldots, m$:

$$\inf_{x \in A_i} R_i(x) \geq \frac{\delta}{p_i} \cdot E[R_i(X)] 1_{\{X \in A_i\}},$$

where $\delta$ is the discount rate for future rewards. Intuitively, a partition policy will be incentive-compatible if each candidate's reward from accepting a kidney offer is no less than their expected reward from declining such an offer. They add the incentive-compatibility (IC) constraint to the original optimization problem to model candidate autonomy. They find that the inclusion of candidate autonomy increases the opportunity cost each candidate incurs from refusing an assignment and make such refusals unattractive.

They perform a numerical study to evaluate the implications of their analytical results. Their experiments show that as the heterogeneity in either the proportion of candidates or the reward functions increases, the optimal partitioning policy performs better. They compared the optimal partitioning policy to a random allocation policy with and without the consideration of candidate autonomy. In general, the optimal partition policy performed much better than a random allocation policy. Additionally, candidate autonomy can have a significant impact on the performance of the kidney allocation system. However, the optimal partitioning policy with the inclusion of IC constraints performs almost as well as the optimal policy when candidates are not autonomous. This is because the inclusion of IC constraints eliminates the variability in the stream of kidneys offered to the same type of candidates.

Roth et al. [20] consider the problem of designing a mechanism for direct and indirect *kidney exchanges*. A *direct kidney exchange* involves two donor-patient pairs such that each donor cannot give his or her kidney to his or her own patient due to immunological incompatibility, but each patient can receive a kidney from the other donor. An *indirect kidney exchange* occurs when a donor-patient pair makes a donation to someone waiting for a kidney, and the patient receives high priority for a compatible kidney when one becomes available. The objective is to maximize the number of kidney transplants and mean quality of match.

Let $(k_i, t_i)$ be the donor-recipient pair, where $k_i$ denotes kidney $i$ from live donor and $t_i$ denotes patient $t_i$, and $K$ denotes the set of living donors at a particular time. Each patient $t_i$ has a set of compatible kidneys, $K_i \subset K$, over which the patient has heterogenous preferences. Let $w$ denote the option of entering the waiting list with priority reflecting the donation of his or her donor's kidney $k_i$. Let $P_i$ denote the patient's strict preferences over $K_i \cup \{k_i, w\}$, where $P_i$ is the ranking up to $k_i$ or $w$, whichever ranks higher. A kidney exchanging problem consists of a set of donor-recipient pairs $\{(k_1, t_1), \ldots, (k_n, t_n)\}$, a set of compatible kidneys $K_i \subset K = \{k_1, \ldots, k_n\}$ for each patient $t_i$, and a strict preference relations $P_i$ over $K_i \cup \{k_i, w\}$ for each patient $t_i$. The objective is to find a *matching* of kidneys/wait-list option to patients such that each patient $t_i$ is either assigned a kidney in $K_i \cup \{k_i\}$ or the wait-list option $w$, while no kidney can be assigned to more than one patient but the wait-list option $w$ can be assigned to more than one patient. A kidney exchange mechanism selects a matching for each kidney exchange problem.

Roth et al. [20] introduce the *Top Trading Cycles and Chains* (TTCC) mechanism to solve this problem and show that the TTCC mechanism always selects a matching among the participants at any given time such that there is no other matching weakly preferred by all patients and donors and strictly preferred by at least one patient-donor pair. They use a Monte Carlo simulation model to measure the efficiency of the TTCC mechanism. Their results show that substantial gains in the number and match quality of transplanted kidneys might result from the adoption of the TTCC mechanism. Furthermore, a transition to the TTCC mechanism would improve the utilization rate of potential unrelated living-donor kidneys and Type O patients without living donors.

In another work, Roth et al. [21] consider the problem of designing a mechanism for pairwise kidney exchange, which makes the following two simplifying assumptions to the model described in [20]: (1) they consider exchanges involving two patients and their donors and (2) they assume that each patient is indifferent between all compatible kidneys. These two assumptions change the mathematical structure of the kidney exchange problem, and the problem becomes a cardinality matching problem. Under these assumptions, the kidney exchange problem can be modeled with an undirected graph whose vertices represent a particular patient and his or her incompatible donor(s) and whose edges connect those pairs of patients between whom an exchange is possible, i.e., pairs of patients such that each patient in the pair is compatible with a donor of the other patient. Finding an efficient matching then reduces to finding a maximum cardinality matching in this undirected graph. They use results from graph theory to optimally solve this problem and give the structure of the optimal policy.

## 1.6 Conclusions

Organ allocation is one of the most active areas in medical optimization. Unlike many other optimization applications in medicine, it has multiple perspectives. The individual patient's perspective typically considers the patient's health and how he or she should behave when offered choices, e.g., whether or not to accept a particular cadaveric organ or when to transplant a living-donor organ. The societal perspective designs an allocation mechanism to optimize at least one of several possible objectives. One possible objective is to maximize the total societal health benefit. Another is to minimize some measure of inequity in allocation.

   Given the rapid changes in organ allocation policy, it seems likely that new optimization issues will arise in organ allocation. A critical issue in future research is modeling disease progression as it relates to allocation systems. The national allocation systems are increasingly using physiology and laboratory values in the allocation system (e.g., the MELD system described in Section 1.3). Furthermore, new technologies may mean more choices to be optimized for patients in the future. For example, artificial organs and organ assist devices are becoming more common. Given the intense emotion that arises in organ allocation, more explicit modeling of the political considerations of various parties will yield more interesting and more applicable societal-perspective optimization models.

## Acknowledgments

The authors wish to thank Scott Eshkenazi for assistance in preparing this chapter. This work was supported in part by grants DMI-0223084, DMI-0355433, and CMMI-0700094 from the National Science Foundation, grant R01-HS09694 from the Agency for Healthcare Research and Quality, and grant LM 8273-01A1 from the National Library of Medicine of the National Institutes of Health.

## References

[1] J.H. Ahn and J.C. Hornberger. Involving patients in the cadaveric kidney transplant allocation process: A decision-theoretic perspective. *Management Science*, 42(5):629–641, 1996.

[2] O. Alagoz, L.M. Maillart, A.J. Schaefer, and M.S. Roberts. The optimal timing of living-donor liver transplantation. *Management Science*, 50(10):1420–1430, 2004.

[3] O. Alagoz, L.M. Maillart, A.J. Schaefer, and M.S. Roberts. Determining the acceptance of cadaveric livers using an implicit model of the waiting list. *Operations Research*, 55(1):24–36, 2007.

[4] R.E. Barlow and F. Proschan. *Mathematical Theory of Reliability*. John Wiley and Sons, New York, NY, 1965.

[5] CORE, 2003. Available from http://www.core.org, information and data accessed on January 22, 2003.

[6] I. David and U. Yechiali. A time-dependent stopping problem with application to live organ transplants. *Operations Research*, 33(3):491–504, 1985.

[7] I. David and U. Yechiali. Sequential assignment match processes with arrivals of candidates and offers. *Probability in the Engineering and Informational Sciences*, 4:413–430, 1990.

[8] I. David and U. Yechiali. One-attribute sequential assignment match processes in discrete time. *Operations Research*, 43(5):879–884, 1995.

[9] V.S. Elliott. Transplant scoring system called more fair. *American Medical News*, May 13, 2002. Available from http://www.ama-assn.org/amednews/2002/05/13/hll20513.htm.

[10] K. Garber. Controversial allocation rules for liver transplants. *Nature Medicine*, 8(2):97, 2002.

[11] J.C. Hornberger and J.H. Ahn. Deciding eligibility for transplantation when a donor kidney becomes available. *Medical Decision Making*, 17:160–170, 1997.

[12] D.H. Howard. Why do transplant surgeons turn down organs?: A model of the accept/reject decision. *Journal of Health Economics*, 21(6):957–969, 2002.

[13] HowStuffWorks, 2004. Available from http://health.howstuffworks.com/organ-transplant2.htm, information and data accessed on June 14, 2004.

[14] IOM. *Organ Procurement and Transplantation*. National Academy Press, Washington, D.C., 1999. Available from Institute of Medicine (IOM) website, http://www.iom.edu/.

[15] P.S. Kamath, R.H. Wiesner, M. Malinchoc, W. Kremers, T.M. Therneau, C.L. Kosberg, G.D'Amico, E.R. Dickson, and W.R. Kim. A model to predict survival in patients with end-stage liver disease. *Hepatology*, 33(2):464–470, 2001.

[16] M. Malinchoc, P.S. Kamath, F.D. Gordon, C.J. Peine, J. Rank, and P.C. ter Borg. A model to predict poor survival in patients undergoing transjugular intrahepatic portosystemic shunts. *Hepatology*, 31:864–871, 2000.

[17] Government Accounting Office, 2003. Available from http://www.gao.gov/special.pubs/organ/chapter2.pdf, information and data accessed on January 21, 2003.

[18] M.L. Puterman. *Markov Decision Processes*. John Wiley and Sons, New York, NY, 1994.

[19] R. Righter. A resource allocation problem in a random environment. *Operations Research*, 37(2):329–338, 1989.

[20] A. Roth, T. Sonmez, and U. Unver. Kidney exchange. *Quarterly Journal of Economics*, 119(2):457–488, 2004.

[21] A. Roth, T. Sonmez, and U. Unver. Pairwise kidney exchange. *Journal of Economic Theory*, 125(2):151–188, 2005.

[22] SRTR. Transplant primer: Liver transplant, 2004. Available from http://www.ustransplant.org/, information and data accessed on September 9, 2004.

[23] J.E. Stahl, N. Kong, S. Shechter, A.J. Schaefer, and M.S. Roberts. A method-ological framework for optimally reorganizing liver transplant regions. *Medical Decision Making*, 25(1):35–46, 2005.

[24] J.E. Stahl, J.E. Kreke, F. Abdullah, and A.J. Schaefer. The effect of cold-ischemia time on primary nonfunction, patient and graft survival in liver trans-plantation: a systematic review. Forthcoming in *PLoS ONE*.

[25] X. Su and S. Zenios. Patient choice in kidney allocation: a sequential stochastic assignment model. *Operations Research*, 53(3):443–455, 2005.

[26] J.F. Trotter and M.J. Osgood. MELD scores of liver transplant recipients according to size of waiting list: Impact of organ allocation and patient out-comes. *Journal of American Medical Association*, 291(15):1871–1874, 2004.

[27] P.A. Ubel and A.L. Caplan. Geographic favoritism in liver transplantation–unfortunate or unfair? *New England Journal of Medicine*, 339(18):1322–1325, 1998.

[28] UNOS. Organ distribution: Allocation of livers, 2004. Available from http://www.unos.org/resources/, information and data accessed on Septem-ber 9, 2004.

[29] UNOS. View data sources, 2004. Available from http://www.unos.org/data/, information and data accessed on September 9, 2004.

[30] R.H. Wiesner, S.V. McDiarmid, P.S. Kamath, E.B. Edwards, M. Malinchoc, W.K. Kremers, R.A.F. Krom, and W.R. Kim. MELD and PELD: Application of survival models to liver allocation. *Liver Transplantation*, 7(7):567–580, 2001.

[31] S.A. Zenios, G.M. Chertow, and L.M. Wein. Dynamic allocation of kidneys to candidates on the transplant waiting list. *Operations Research*, 48(4):549–569, 2000.

# 2

# Can We Do Better? Optimization Models for Breast Cancer Screening

Julie Simmons Ivy

Edward P. Fitts Department of Industrial and Systems Engineering, North Carolina State University, Raleigh, North Carolina 27695-7906
jsivy@ncsu.edu

**Abstract.** "An ounce of prevention is worth a pound of cure." In healthcare, this well-known proverb has many implications. For several of the most common cancers, the identification of individuals who have early-stage disease enables early and more effective treatment. Historically, however, the effectiveness of and the frequency with which to perform these screening tests have been questioned. This is particularly true for breast cancer where survival is highly correlated with the stage of disease at detection. Breast cancer is the most common noncutaneous cancer in American women, with an estimated 240,510 new cases and 40,460 deaths in 2007 (http://www.cancer.gov). Mammography is currently the only screening method recommended by American Cancer Society (ACS) Guidelines for Breast Cancer Screening: Update 2003 for community-based screening (Smith et al. [33]). Mammography seeks to detect cancers too small to be felt during a clinical breast examination by using ionizing radiation to create an image of the breast tissue. However, screening mammography detects noncancerous lesions as well as in situ and invasive breast cancers that are smaller than those detected by other means. The ACS suggests that establishing the relative value between screening and non-screening factors is complex and can be only indirectly estimated. Operations researchers have a unique opportunity to determine the optimal future for breast cancer screening and treatment by developing models and mechanisms that can accurately describe the dynamic nature of quality costs as well as the interaction between such costs, resulting activities, and system improvement. This is not a new question; in addition to a rich body of empirical breast cancer research, there is more than 30 years of mathematical-modeling-based breast cancer screening research. In this chapter, we present a critical analysis of optimization-based models of the breast cancer screening decision problem in the literature and provide guidance for future research directions.

## 2.1 Introduction

"An ounce of prevention is worth a pound of cure" — a well-known proverb — is a simple description of the advantages of proactive health maintenance: personal activities intended to enhance health or prevent disease and disability.

P.M. Pardalos, H.E. Romeijn (eds.), *Handbook of Optimization in Medicine*,
Springer Optimization and Its Applications 26, DOI: 10.1007/978-0-387-09770-1_2,
© Springer Science+Business Media LLC 2009

For several of the most common cancers, cardiovascular diseases, and other illnesses, examining seemingly healthy individuals to detect a disease before the surfacing of clinical symptoms enables early and more effective treatment (Parmigiani [28]). Screening, in particular, is one of the most important areas both in clinical practice and research in medicine today (United States Preventive Services Task Force, 2003). Historically, however, the effectiveness of and the frequency with which to perform these screening tests have been questioned. In fact, medical economist Louise Russell (Russell [30]) contests this well-known proverb in her text of the same name, which challenges the conventional wisdom that more frequent screening is necessarily better. Cancer screening, in particular, has received much of this attention. Russell, for example, argues that standard recommendations such as annual Pap smears for women and prostate tests for men over 40 are, in fact, simply rules of thumb that ignore the complexities of individual cases and the trade-offs between escalating costs and early detection (Russell [31]). Rising healthcare costs exacerbate these concerns. Healthcare costs account for a high fraction of the gross domestic product in industrial countries, ranging from approximately 7% in the United Kingdom to 14% in the United States. In the United States, this percentage is projected to exceed 16% by 2010. These growing costs are especially noteworthy in today's economy, as policymakers are forced to trim healthcare benefits or other social services, and healthcare systems are under significant pressure to control expenditures and improve performance (Baily and Garber [2]). Both public and private payers are demanding increased efficiency and "value for money" in the provision of healthcare services (Earle et al. [10]).

In addressing these concerns, some medical experts question the value of screening tests for cancers, including breast and ovarian cancers in women, prostate cancer in men, and lung cancer in both sexes. A key issue in determining the effectiveness of testing is whether the tests can adequately distinguish between nonmalignant and malignant tumors so that patients with nonmalignant tumors are not subjected to the risks of surgery, radiation, or chemotherapy. This debate is further complicated because the ability of screening tests to detect very tiny tumors in the breast, prostate, and other organs has far outpaced scientists' understanding of how to interpret and respond to the findings (The New York Times, April 14, 2002).

Olsen and Gotzsche [26] suggested that breast cancer screening might be ineffective in terms of outcomes (also refer to Gotzsche and Olsen [15]). The authors evaluated the randomized trials of breast cancer screening through meta-analysis, concluding that five of the seven trials were flawed and should not be regarded as providing reliable scientific evidence. Further, they concluded that there is no reliable evidence that screening reduces breast cancer mortality. Although numerous guideline groups, national health boards, and authors dispute Olsen and Gotzsche's methodology and conclusions, the Olsen and Gotzsche article reignited a debate about the value of breast cancer screening and put into question some of the evidence-based support for the cost-effectiveness of breast cancer screening. In fact, according to the American Cancer Society (ACS), the inherent limitations of the breast cancer screening randomized control trials (RCTs) in estimating mammography benefits have led to increased interest in evaluating community-based screening (Smith et al. [33]). The question is whether *routine* mammograms should be recommended and, if so, for whom (The Wall Street Journal, February 26, 2002). Mammograms obviously aid in the detection and diagnosis of breast cancer. At issue is whether the breast cancer screening test makes any difference in preventing breast cancer deaths (The New York Times, February 1, 2002).

There seems to be a consensus that we could do better, but the question is how. By improving the management and/or treatment of diseases, decision makers can alleviate spending pressure on their systems while maintaining outcomes, or even improve outcomes without increasing spending (Baily and Garber [2]). To provide these policymakers with an informed understanding of system and health maintenance, operations researchers have a unique opportunity to determine the optimal future for breast cancer screening and treatment by developing models and mechanisms that can accurately describe the dynamic nature of quality costs as well as the interaction between such costs, resulting activities, and system improvement.

### 2.1.1 Background on breast cancer and mammography techniques

#### Breast cancer

Breast cancer is a disease in which malignant cancer cells form in the tissues of the breast. Breast cancer is a progressive disease that is classified into a variety of histological types. The standard taxonomy for categorizing breast cancer is given by the American Joint Committee on Cancer staging system based on tumor size and spread of the disease. According to this taxonomy, patients with smaller tumors are more likely to be in the early stage of the disease, have a better prognosis, and are more successfully treated.

It is the most common noncutaneous cancer in American women, with an estimated 267,000 new cases and 39,800 deaths in 2003 (http://www.cancer.gov). The average lifetime cumulative risk of developing breast cancer is 1 in 8. Breast cancer incidence, however, increases with age. For the average 40-year-old woman, the risk of developing breast cancer in the next 10 years is less than 1 in 60. The 10-year risk of developing breast cancer for the average 70-year-old woman, however, is 1 in 25.

#### Screening by mammography

Mammography is currently the only screening method recommended by ACS Guidelines for Breast Cancer Screening: Update 2003 for community-based screening (Smith et al. [33]). Mammography seeks to detect cancers too small to be felt during a clinical breast examination by using ionizing radiation to create an image of the breast tissue. The examination is performed by compressing the breast firmly between a plastic plate and an x-ray cassette which contains special x-ray film. Screening mammography detects noncancerous lesions as well as *in situ* and invasive breast cancers that are smaller than those detected by other means. Mammography screening reduces the risk of mortality by increasing the likelihood of detecting cancer in its preclinical state, thus allowing earlier treatment and more favorable prognoses associated with early-stage cancers (Szeto and Devlin [34]).

Currently, mammography is the best way available to detect breast cancer in its earliest, most treatable stage, on average 1.7 years before a woman can feel the lump (The National Breast and Cervical Cancer Early Detection Program, 1995). The remaining unanswered question, however, is whether "routine" mammograms should be recommended and, if so, beginning and ending at what ages (The Wall Street Journal, February 26, 2002). The reason this question remains open is because mammograms do not achieve perfect sensitivity (a true positive) or specificity (a true negative). As a result, the issue of adverse consequences of screening for women who do not have breast cancer, as well as women who have early-stage breast cancer that will not progress, has become one of the core issues in recent debates about mammography (Smith et al. [33]).

## Mammography efficacy

Sensitivity refers to the likelihood that a mammogram correctly detects the presence of breast cancer when breast cancer is indeed present (a true positive). Sensitivity depends on several factors, including lesion size, lesion conspicuousness, breast tissue density, patient age, hormone status of the tumor, overall image quality, and the interpretive skill of the radiologist. Retrospective correlation between mammogram results with population-based cancer registries shows that sensitivity ranges from 54% to 58% in women under 40 and from 81% to 94% in those over 65 (http://www.cancer.gov).

Specificity refers to the likelihood that a mammogram correctly reports no presence of breast cancer when breast cancer is indeed not present (a true negative). Mammography specificity directly affects the number of "unnecessary" interventions performed due to false-positive results, including additional mammographic imaging (e.g., magnification of the area of concern), ultrasound, and tissue sampling (by fine-needle aspiration, core biopsy, or excision biopsy). It is interesting to note that the emotional effects of false positives are often assumed to be negligible in most mathematical models for mammography, although scarring from surgical biopsy can mimic a malignancy on subsequent physical or mammographic examinations. Patient characteristics associated with an increased chance of a false-positive result include younger age, increased number of previous breast biopsies, family history of breast cancer, and current estrogen use. Radiologist characteristics associated with an increased chance of a false-positive result include longer time between screenings, failure to compare the current image with prior images, and the individual radiologist's tendency to interpret mammograms as abnormal (cancer.gov). Although the average specificity of mammography exceeds 90%, this rate varies with patient age; the specificity rate for women ages 40 through 49 is 85% to 87%, and for women over age 50 it is 88% to 94% (http://www.womenssurgerygroup.com).

### 2.1.2 Mammogram screening recommendations

The 2003 American Cancer Society guidelines recommend that women at average risk should begin annual mammography screening at age 40. Although the potential for mammography screening to reduce the risk of breast cancer mortality is generally accepted for women older than 50, some authors argue that the benefits for younger women are less certain because the incidence of the disease, as well as the efficacy of the screening test, are lower in younger women. That is, younger women are less likely to develop breast cancer and more likely to receive false test results.

On the other hand, the ACS Guidelines for Breast Cancer Screening: Update 2003 (Smith et al. [33]) states that the importance of annual screening is clearly greater in premenopausal women ($<55$) compared with postmenopausal women. The guidelines speculate that younger women might benefit from more frequent screening because the disease is more aggressive in younger women (Althuis et al. [1]; Brenner and Hakulinen [6]; Foxcroft et al. [13]; Kroman et al. [19]; Mathew et al. [24]). Peer et al. report that in patients younger than 50, tumors have a median doubling time of 80 days, whereas the tumors in patients ages 50 to 70 have a median doubling time of 157 days, and tumors in patients older than 70 have a median doubling time of 188 days (Michaelson et al. [25]).

### Screening for women with increased risk

The ACS guidelines admit that the age at which screening should be initiated for women at high risk is not well established and speculate that women at increased risk might benefit from additional screening strategies beyond those offered to women of average risk, such as earlier initiation of screening, shorter screening intervals, or the addition of screening modalities (such as ultrasound or magnetic resonance imaging (MRI)). A woman's risk of breast cancer is higher if her mother, sister, or daughter has had breast cancer, especially if the family member developed the disease before age 40 (http://www.cancer.gov). A woman's risk can increase by as much as 50% if her mother has had breast cancer and by 25% if her sister has had breast cancer (Collaborative Group on Hormonal Factors in Breast Cancer, 2001).

**Screening the elderly**

Similar uncertainty surrounds recommendations for elderly women and women with high comorbidity, i.e., high risk of contracting and dying from other diseases. Postmenopausal women frequently have one or more preexisting comorbid conditions (e.g., heart disease, chronic obstructive pulmonary disease, diabetes, hypertension, and arthritis) at the time of breast cancer diagnosis (Yancik et al. [35]). Yancik et al. [35] found that comorbidity in older patients also may limit the ability to obtain prognostic information (i.e., axillary lymph node dissection) and tend to minimize treatment options (e.g., breast-conserving therapy). In light of these facts, the 2003 American Cancer Society guidelines state that screening decisions in "older women" should be individualized by considering the potential benefits and risks of mammography in the context of current health status and estimated life expectancy but offer no concrete recommendations.

The ACS suggests that establishing the relative value between screening and non-screening factors is complex and can be only indirectly estimated. "Insofar as additional RCTs of breast cancer screening are unlikely, the evaluation of service screening represents an important new development for several reasons, specifically by measuring the value of modern mammography in the community and measuring the benefit from mammography screening to women who actually get screened (Smith et al. [33])." These statements demonstrate the need for mathematical modeling and simulation to evaluate the impact of screening.

## 2.2 Optimization Models for Mammography Screening

To address this concern, a number of studies have examined the relationship between screening and mortality reduction (see, e.g., Szeto and Devlin [34]; Beemsterboer [4]; de Koning [9]). In addition to the empirical breast cancer research, there exists a body of mathematical-modeling-based breast cancer research. This research has important implications for physician and patient decision making as well as for insurance pricing. It seeks to address the current mammography screening debate and shows great promise as a means for developing and evaluating cost-effective screening and treatment policies for breast cancer as well as other health screening problems. There are several research issues to be addressed to further this field, e.g., risk estimation, defining patient costs, dealing with multiple decision makers, and so forth. Although there are many models in the literature focusing on modeling the disease progression (Retsky et al. [29]), with some based on Markov processes (e.g., Chen et al. [8]), our discussion will highlight those optimization-based models that also incorporate the decision process.

Kirch and Klein [18] developed an early model for determining examination schedules for the detection of age-dependent diseases such as breast cancer. Their model minimizes the expected detection delay for a given number of examinations. They specify the age span for which the examination schedule is to be developed and assume that it consists of $n$ equal-length periods, specifically five-year periods. Further, they assume that each period starts with an examination and that all examinations within a period are at equal intervals. For this model, they demonstrate that an optimal examination schedule is nonperiodic; specifically, the frequency of examination will be either approximately or exactly proportional to the square root of the age-specific incidence probability of the disease. For breast cancer examinations, they demonstrate that optimal nonperiodic schedules will result in savings of 2% to 3% in the expected number of examinations when periodic and nonperiodic schedules have the same expected detection delay. One major limitation of their model is that they assume examinations are perfect or error-free.

Shwartz [32] developed one of the earliest and most complex mathematical models to address the breast cancer screening problem. He defines 21 disease states consisting of seven tumor-size categories and, for each size category, three lymph-node involvement levels. He hypothesizes that tumors grow exponentially with a random growth rate and uses data from tumor-doubling times to estimate the distribution of tumor growth rates in the population of women who have been treated for breast cancer. Shwartz acknowledges that there is no "unarbitrary" method for determining lymph-node involvement levels for tumors detected by screening due to a lack of data. Shwartz [32] incorporates non-stationarity, i.e., age-dependency, for incidence and mortality rates, and assumes stationary test-efficacy and disease aggression. Shwartz uses his model to predict the benefits of alternative screening strategies in terms of life expectancy gain by considering both the probability of lymph node involvement at detection and the probability of recurrence. However, Shwartz's model does not incorporate the possibility of false-positive test results and assumes the decision maker knows disease state information.

Ozekici and Pliska [27] model disease progression as a delayed Markov process in which the sojourn time in the good state is a general, non-negative random variable, and the progression through the deterioration states follows a Markov process. Their model determines the optimal inspection schedule by minimizing the total expected value of the following costs: inspection costs, false-positive (supertest or biopsy) costs, true positive-costs as a function of disease state, and false-negative costs as a function of disease state. They acknowledge that these costs, particularly the true-positive and false-negative costs, are quite subjective and difficult to estimate in practice. Some other limitations of this model are that it does not include death as a possible state and uses stationary transition probabilities and test efficacy parameters. Further, their model assumes that once the disease is "apparent to the individual," no diagnosis is necessary.

Parmigiani [28] developed a Bayesian framework for examining the effect of age on the optimal schedule of examinations. He presented a general, non-Markovian stochastic process to model disease progression that includes a death state, although he did not distinguish breast cancer death and non–breast cancer death. Parmigiani acknowledged one of the significant limitations of modeling for breast cancer: that data required to determine the form and parameters of general sojourn time distributions is often not available. He assumed that the sensitivity of the screening test is a function of patient age and the sojourn time in the detectable preclinical disease state. Screening schedules are chosen to minimize total expected cost, including examination costs (financial charges, undesired side effects, psychological stress, etc.), mortality/morbidity costs, and treatment costs that depend on the timing of detection. Parmigiani suggested generating trade-off curves as a function of these costs and quality of life judgments. Parmigiani examined the dynamics of test-efficacy, incidence rate, disease aggression, and mortality rate; however, he did so for each factor in isolation.

Zelen [36] developed a weighted utility function for determining the optimal screening schedule. The utility function is equivalent to a fixed budget that allows for a fixed number of examinations. This utility function is linear in the probabilities of finding a case at examination and of being clinically incident between examinations. For this model, Zelen proves that if the disease incidence is independent of time, then a necessary and sufficient condition for the screening intervals to be equally spaced is that the sensitivity of the examination be equal to one. Further he derives the equations for the optimal intervals and shows they depend on the distribution of the preclinical sojourn times and the sensitivity of the test. Zelen proves if the sojourn distribution is exponential, the optimal intervals are equal except for the first and the last intervals. Zelen acknowledges that selection of the weights for the utility function is both critical and fairly difficult given the available data. Another limitation of this model is that it does not involve penalties or discounting for false-positive diagnoses and it does not address the question of when screening should begin.

To address some of these limitations, Lee and Zelen [21, 22] and Lee et al. [20] developed a stochastic model that predicts breast cancer mortality for different examination schedules and screening modalities as a function of the stage shift distribution, examination schedules, population age distribution, follow-up time, and survival conditional on stage at diagnosis. Their model consists of three health states: disease-free or nondetectable state; a preclinical state, in which an individual has disease but is asymptomatic and detectable by examination; and a clinical state in which the disease has been diagnosed by routine methods. Their model assumes the natural history of the disease is progressive, and any benefit from earlier diagnosis is due to change in the distribution of disease stages at diagnosis (stage shift). They applied their model to eight randomized clinical trials and found the model predicted the reduction in mortality for seven of the eight trials within the reported confidence levels. They explore the relationship between reduction in disease-specific mortality and schedule sensitivity and lifetime schedule sensitivity.

The main features of their schedules are the initial age to begin a scheduled examination program, the intervals between subsequent examinations, and the number of examinations. Their evaluation of schedules compares examination schedules with equal intervals between examinations as well as staggered schedules using a threshold method. Their threshold method constructs examination schedules so that the probability of an individual being in the preclinical state is always bounded by a preselected value. They also introduce the concept of schedule sensitivity, the ratio of the expected number of cases diagnosed on scheduled examinations to the expectation of the total number of cases. By combining the threshold and schedule sensitivity methods, they assess the trade-offs between the initial age at examination and the cost per case found. Calculation of the schedule sensitivities requires data on age-specific incidence of disease, sensitivity of the screening modalities, and distribution of sojourn times in the preclinical state. Lee and Zelen (1998) acknowledge one limitation is that sojourn time distribution tends to be less readily available.

Mangasarian et al. [23] did not focus on screening schedule development but developed linear programming–based machine-learning techniques to increase the accuracy and objectivity of breast cancer diagnosis and prognosis. Their techniques use characteristics of individual cells, obtained from minimally invasive fine-needle aspirate, to discriminate benign from malignant breast lumps. Given two matrices, one of malignant vectors and the other of benign vectors, in general, the two sets are not strictly linearly separable. Hence they solve a linear program that minimizes the average sum of their violations. Their linear program will generate a strict separating plan if it exists or it will minimize the average sum of the violations. In addition, they have developed a method that constructs a surface that predicts when breast cancer is likely to recur in patients who have had their cancers excised. The linear-program-based recurrence surface approximation is based on the idea of constructing a surface that bounds from above the disease-free survival times for the nonrecurring training cases and closely bounds from below the time to recur times of the recurrent training cases. The method is unique in its ability to handle cases for which cancer has not recurred (censored data) as well as cases for which cancer has recurred at a specific time.

Baker [3] developed a model of breast cancer screening in which the processes of tumor development and growth, detection of tumors at screening, presentation of women with cancers to physicians (non–screen detected cancers), and survival after diagnosis are modeled parametrically using maximum likelihood estimation. The model is fit to data from the Northwest of the United Kingdom for 413 women who screened positive and for 761 women who developed interval cancers. Her model does not require a Markov assumption and does not discretize state variables such as tumor size; they are modeled as continuous variables. The model is an evaluation tool used to assess different screening policies. Baker defines the cost of breast cancer to be the sum of the cost of screening and the cost of person years of life lost (PYLL) due to cancer. To evaluate the cost of a policy requires the specification of the number of screens that could be paid for with the cost of one month of life. Baker derives a range of optimal policies by varying the value placed on human life. As the cost of screening decreases, the optimal policy requires screening most intensively in the 40 to 48 age range, in which at higher screening cost, no screens at all would have been done. The model recommends most screening effort going to where it reduces PYLL due to cancer only slightly, given sufficient money devoted to screening. Some limitations of this model include: the sample data used was small, the model's parameterization does not allow it to fit the data well, the model does not include non-invasive breast cancers, and the cost associated with person years of life lost is quite subjective and difficult to estimate.

Michaelson et al. [25] developed a computer simulation method, based on the rates of breast cancer growth and spread, to evaluate the ability of different screening intervals to detect breast cancer prior to distant metastatic spread. In their model, breast cancer incidence, mortality rate, and disease aggression are assumed to be constant with respect to patient age.

Gunes et al. [16] developed a system dynamics model for breast cancer screening incorporating dynamics of healthcare states, program outreach, and the screening volume quality relationship. Their model is unique in that it takes a more inclusive view of screening that addresses the public health concerns for screening implementation. They view breast cancer screening provision as a problem of matching supply (screening service) and demand (participation in the program, screening frequency) while ensuring sufficient quality (high sensitivity and specificity of the test). Their focus is not the screening schedule but to reduce breast cancer deaths, keeping system cost in mind. Their results indicate a strong relation between screening quality and the cost of screening and treatment and emphasize the importance of accounting for service dynamics when assessing the performance of healthcare intervention.

The aforementioned models, with the exception of Mangasarian et al. [23], focus on the screening decision and address the treatment decision as a separate problem (refer to Shwartz [32]; Ozekici and Pliska [27]; Parmigiani [28]; and Zelen [36]). Most of these models also assume a fixed screening interval and solve for the optimal interval length. These decision models assume either the disease state is observable or that perfect information through biopsy will be available, e.g., Ozekici and Pliska [27].

Ivy [17] uses the partially observable Markov decision process (POMDP) structure to develop a medical decision-making tool for determining a "cost-effective" methodology for mammography screening and breast cancer treatment. Specifically, this model explores mammography screening from the payer and patient perspectives. The model identifies those breast cancer detection and treatment actions that minimize the total expected cost over the lifetime of a patient and identifies screening policies that maximize the total expected patient utility over the patient's lifetime. Ivy [17] defines cost-effective screening strategies according to efficient frontiers developed using a constrained POMDP-based model for balancing patient and payer costs. This model incorporates the diagnosis of the disease, its treatment, and is unique in that it incorporates the uncertainty associated with the partial observability of the disease process by the decision maker and the uncertainty associated with the outcome of the treatment in determining the effectiveness of screening for a given patient. Another unique feature of this model is that it integrates the screening and treatment decisions. However, this model does not address the age dynamics associated with screening efficacy and death rates. Ivy acknowledges it can be difficult to estimate disease costs and patient utilities.

In the following section, we will discuss in chronological order a representative sample of several of the aforementioned models in greater detail.

## 2.3 Models for Scheduling Screening Examinations

### 2.3.1 Kirch and Klein [18]

As mentioned in Section 2.2, Kirch and Klein developed a general model for minimizing the expected detection delay for a given number of examinations. A class of optimal schedules was found by varying the allowed number of examinations. To do so, the age span for which a schedule is to be developed was specified, consisting of $n$ equal-length periods. They assumed each period starts with an examination, all examinations within a period are at equal intervals, and examinations are error-free.

In their model, Kirch and Klein defined $T$ as the disease detection point, the earliest time that the disease could be detected by an examination. Then the detection delay is the time between $T$ and the time of actual detection, which occurs only if an examination is scheduled. $x_i$ is defined as the number of examinations scheduled for the $i^{th}$ period, then the length of the inter-examination interval is $60/x_i$ and $0 < T < 60/x_i$, where periods are assumed to be five years in length because age-specific disease incidence rates are usually tabulated for five-year periods.

Kirch and Klein's model is based on the following logic: If disease is detected in the $i^{th}$ period, this is the result of either (1) patient compliance resulting in an examination or (2) a scheduled examination. If disease is detected due to patient compliant, they defined $D$ as the patient detection delay, the length of time between the detectability point and the examination. In their paper, Kirch and Klein considered two cases: (1) if $D$ is constant and (2) if $D$ is a random variable with a known distribution. We will discuss the first case in detail.

If $D$ is constant, they defined $Q$ as the detection delay if the disease becomes detectable in the $i^{th}$ period: $Q$ is a function of the length of the interval between scheduled examinations in the $i^{th}$ period, the location of the detectability point $T$, and $D$, i.e.,

$$Q(x_i; D, T) = \begin{cases} D & \text{if } T + D \leq 60/x_i \\ 60/x_i - T & \text{if } T + D > 60/x_i \end{cases}$$

Assuming $T$ is a continuous random variable over $(0, 60/x_i)$, they defined the conditional expected delay, given that the disease becomes detectable during the $i^{\text{th}}$ period, as

$$E\left(Q(x_i; D)\right) = D \int_0^{60/x_i - D} + \int_{60/x_i - D}^{60/x_i} (60/x_i - t) f_T(t)\, dt.$$

They assumed $T$ is uniformly distributed, so this reduces to

$$E\left(Q(x_i; D)\right) = \begin{cases} D\left(1 - Dx_i/120\right) & \text{if } 1 \leq x_i \leq 60/D \\ 30/x_i & \text{if } x_i > 60/D \end{cases}$$

which is a convex decreasing function of $x_i$.

Kirch and Klein proposed a constrained optimization model that minimizes the expected detection delay for a given schedule $(x_1, \ldots, x_n)$ and given constant detection delay $D$, as

$$G(x_1, \ldots, x_n; D) = \sum_{i=1}^{n} p_i E\left(Q(x_i; D)\right)$$

where $p_i$ is the conditional probability that the detectability point occurs in period $i$, given that it will occur sometime within the $n$ periods of interest. They minimized $G$ subject to a bound, $K$, on the expected number of examinations for patients who do not get the disease, i.e.,

$$\sum_{i=1}^{n} \sum_{j=1}^{i} x_j q_i \leq K$$

where $q_i$ is the probability that the patient dies in the $i^{\text{th}}$ period and $x_i \geq 1$, $i = 1, \ldots, n$. Because $E\left(Q(x_i; D)\right)$ is convex, $G$ is a convex combination of convex functions. Therefore, Kirch and Klein used the Kuhn–Tucker theorem to determine the necessary and sufficient conditions for a feasible schedule to be optimal, i.e.,

$$\lambda(K) = E\left(Q'(x_i; D)\right) \frac{p_i}{s_i} \qquad \text{for all } x_i > 60/D$$

and

$$E\left(Q'(x_i; D)\right) \frac{p_k}{s_k} > \lambda(K) \qquad \text{for } x_k = 1$$

where $s_i$ is the probability of surviving to the start of period $i$. Similarly, Kirch and Klein determined the corresponding optimality conditions for the case when $D$ is a random variable.

In the application of this model to breast cancer, assuming $D$ is constant and equal to 18 months, Kirch and Klein calculated the expected number of examinations per patient for a given detection delay and determined that for the same detection delay, the optimal non-periodic schedules involve 2% to 3% fewer expected examinations than do the periodic schedules. They used 1963–1965 breast cancer age-specific incidence rates ($r_i$) in Connecticut, for women ages 25 to 79, to estimate the conditional incidence probabilities ($p_i$), where

$$p_i = \frac{s_i r_i}{\sum_{j=1}^{n} s_j r_j} \qquad \text{for } i = 1, \ldots, n.$$

They used the 1967 United States estimated survival probabilities for white females for $s_i$.

### 2.3.2 Shwartz [32]

As mentioned in Section 2.2, for his model of breast cancer development, Shwartz defined 21 disease states, consisting of seven tumor sizes and, for each size category, three lymph-node involvement levels. He defined the following state and transition rate variables:

- $S(t)$ = tumor volume at tumor age $t$
- $S(0)$ = tumor volume at time 0
- $i(A)$ = rate at which a tumor develops at age $A$
- $n(t)$ = rate at which a group of lymph nodes becomes involved at tumor age $t$
- $c(t)$ = rate at which a tumor surfaces clinically at tumor age $t$
- $hp(T)$ = rate at which death from breast cancer occurs at $T$ years after treatment, given that the tumor was detected in prognostic class $p$
- $p$ is a function of tumor size and number of lymph nodes involved
- $d(A)$ = rate at which death from causes other than breast cancer occurs at age $A$

Shwartz hypothesized that tumors grow at an exponential rate, $\Lambda$. He used data on tumor-doubling times to estimate the distribution of tumor growth rates in the population of women who have been treated for breast cancer. He considered two distributions to model the growth rate, the hyperexponential (which predicts slower-growing tumors) and lognormal distributions.

Shwartz evaluated screening strategies for a woman as a function of her current age, at-risk category, and compliance level (i.e., the probability that the woman complies with any future planned screen), incorporating the false-negative rate of a screen by tumor size category, the probability a tumor is missed on $v$ screens, and the amount of radiation exposure per screen. A screening strategy (policy) is defined by the number of screens to be given to a woman over her lifetime and the age at which the $m_{\text{th}}$ screen will be given, $m = 1, \ldots, n$. He considered independent and dependent false negatives on successive screens. He defined dependence by the following rule: If the tumor is in the same size category in screens $v - 1$ and $v$, then probability of false positive in $v$ screens is equal to the probability of false negative in $v-1$ screens, otherwise the screens are independent.

Shwartz used this disease model to determine the disease state of a woman at the time of detection (incorporating detection by screening or clinical surfacing). For a woman in risk category $R$ with compliance probability $g$, Shwartz defined:

- $\phi_{R,g}(u, j, A|\lambda_k) =$ the probability that a woman of current age $A_c$ develops breast cancer at age $u$ ($u$ may be $\leq A_c$) and that at age $A$, $A \geq A_c$, she is alive, the tumor has not surfaced clinically, and the tumor is in lymph-node category $j$, $j = 1, 2, 3$, given that when she develops the disease, her tumor has growth rate $\lambda_k$.
- $P_{R,g}(i, j, A|\lambda_k)\Delta t =$ the probability that a woman of current age $A_c$ develops breast cancer and that at age $A$, $A \geq A_c$, she is alive and the disease is detected in tumor size category $i$, $i = 1, \ldots, 7$ and lymph-node category $j$, $j = 1, 2, 3$ between age $A$ and age $A + \Delta t$, given that when she develops the disease, her tumor has growth rate $\lambda_k$.
- $O_{R,g}(A|\lambda_k)\Delta t =$ the probability that a woman of current age $A_c$ develops breast cancer and that she will die from causes other than breast cancer between $A$ and $A + \Delta t$, $A \geq A_c$, given that when she develops the disease, her tumor has growth rate $\lambda_k$.

If no screens are given,

$$\phi(u, j, A|\lambda_k) = R_i(u)e^{-R_i u}L_j(A-u)e^{-C(A-u)}\frac{e^{-D(A)}}{e^{-D(A_c)}}$$

$$P\left(S^*(A-u), j, A|\lambda_k\right)\Delta t = \int_0^A \phi(u, j, A|\lambda_k)c(A-u)\Delta t\, du$$

where $S^*(t) =$ size category of the tumor at tumor age $t$ and

$$O(A|\lambda_k)\Delta t = \int \phi(u, j, A|\lambda_k)d(A)\Delta t\, du.$$

Assuming screening strategy $E_m$, $m = 1, \ldots, n$, where $E_0 = 0$ and $E_{n+1} = 110$. Then for $1 \leq e \leq n$, and $E_e < A < E_{e+1}$, the probability of an interval cancer, i.e., that the disease surfaces in some disease state $(i, j)$ between screen $e$ and $e + 1$, is

$$P\left(S^*(A-u), j, A|\lambda_k\right) =$$
$$\sum_{y=1}^{e+1}\int_{E_{y-1}}^{\min(E_y, A)} \phi(u, j, A|\lambda_k c(A-u)\theta\Delta t X(e+1-y)\, du.$$

Shwartz proposed calculating this for all $e$, $e = 1, \ldots, n$ to determine the total probability that the disease will surface between planned screens; then the probability of being detected at some screen $E_e$ is

$$P\left(S^*(E_e - u), j, E_e | \lambda_k\right) =$$
$$\sum_{y=1}^{e} \int_{E_{y-1}}^{E_y} \phi(u, j, E_e | \lambda_k X(e - y) \left(1 - q\left(S^*(E - u)\right)\right) g \, du.$$

This can be calculated for all screens, and by appropriate summing he can determine $P(i, j, A | \lambda_k)$ and $O(A | \lambda_k)$ for any screening strategy. In addition, he adjusted $i(A)$ to account for the possibility of breast cancer induced from the radiological exposure associated with a screen.

In order to calculate $P(i, j, A)$ and $O(A)$ from $P(i, j, A | \lambda_k)$ and $O(A | \lambda_k)$, Shwartz must estimate the probability that a woman develops a tumor that has a growth rate of $\lambda_k$, for $k = 1, \ldots, 16$. Shwartz estimated the tumor growth rate parameters and lymph node involvement parameters based on the Bross et al. [7] model using a pattern search procedure.

Shwartz's model is a policy evaluation model rather than an optimization model. He evaluated various screening policies (those presented all assume uniform screening intervals from yearly screening to screening every ten years) and evaluated various potential decision metrics: number of screens, life expectancy, percentage of possible gain realized, life expectancy if breast cancer surfaces, probability of no lymph node involvement at detection, and probability of no recurrence. Though this model has the potential to incorporate various tumor growth rates, compliance levels, and dependent false negatives, the results presented assume prognosis is independent of tumor growth rate, perfect compliance, and independent false negatives. In addition, Shwartz assumed that the threat of death from breast cancer is constant. Further, Shwartz acknowledged that there is no "unarbitrary" method for determining lymph-node involvement levels for tumors detected by screening (as the data are only for tumors that clinically surface).

### 2.3.3 Ozekici and Pliska [27]

Ozekici and Pliska presented a stochastic model using dynamic programming for their optimization. Their model of disease progression followed a Markov process with state space $E = \{0, 1, \ldots, k, \Delta\}$, where state 0 is the good or no tumor state and state $\Delta$ is absorbing and represents the failure state defined as sickness that is apparent to the individual. The remaining states represent increasing levels of defectiveness (e.g., increasing tumor size, but the condition is not recognized by the individual). They defined $X_t$ as the state of the deterioration process at time $t$ and $T_n$ as the time of the $n^{th}$ transition where $T_0 = 0$. $\{X_t; t \geq 0\}$ is an increasing process with the Markov transition matrix

$$P(i, j) = P\left(X_{T_{n+1}} = j | X_{T_n} = i\right)$$

where $P(i, j) = 0$ if $j \leq i$. Uniquely, Ozekici and Pliska defined a delayed Markov process with $G$, the sojourn time in state 0 with the corresponding distribution function $F$ where $\alpha \equiv F(\infty) < 1$, i.e., with probability $1 - \alpha$ the individual never contracts the disease. For all other states, they assumed the sojourn time in state $i$ is exponentially distributed whenever $1 \leq i \leq k$.

Ozekici and Pliska defined an inspection schedule in which inspections are binary and imperfect. If an inspection indicates that disease is present, the patient is treated and is assumed to leave the model, i.e., once the deterioration is detected through a true-positive outcome, the decision process ends. If the underlying state is $i$, they defined $u_i$ as the probability the corrective action will be unsuccessful and failure will occur (in the context, however, this is a bit unclear because they do not include a death state); and they assumed $u_1 < u_2 < \cdots < u_k$, i.e., the earlier the disease state, the more likely the medical treatment will be successful. Though they assumed inspections are imperfect, they assumed it is possible to identify false positives with a supertest such as a biopsy, which does not affect the deterioration process. If an inspection yields a positive outcome in a defective state, then corrective action is taken, the deterioration process is affected, and no more tests are performed. As soon as the processes reaches state $\Delta$, the failure is known to the inspector, and no further tests are performed.

Their model selects the inspection schedule that minimizes the expected cost, where an inspection schedule defines when to inspect based on the observed history. The observed history is defined by $t_i$, the time of the $i^{\text{th}}$ inspection, and $Y_i$ the corresponding inspection outcome. $F_n$ is the observed history just after the $n^{\text{th}}$ inspection, $F_n = \{t_1, Y_1, t_2, Y_2, \ldots, t_n, Y_n\}$. Ozekici and Pliska identify the potential outcomes of an inspection. If $Y_n = 1$ and this is a true positive, the problem ends at time $t_n$. If $Y_n = 1$ and this is a false positive or if $Y_n = 0$, then the problem continues and the inspector must chose an action, $\tau$, from the set $\{-1, \mathbb{R}_+\}$ where the action $\tau = -1$ means no more inspections and $\tau = \mathbb{R}_+ \equiv [0, \infty)$ means perform the next inspection after $\tau$ more time units.

Because of the delayed Markov process structure of their model, Ozekici and Pliska are able to transform the intractable Markov decision chain $F_n$ into a simpler one by using the sufficient statistic $(t_n, p)$ where $p \equiv (p_0, p_1, \ldots, p_k)$ is the conditional probability distribution of the state of the underlying deterioration process immediately after the time $t_n$ inspection given the history $F_n$. This facilitates their ability to model and solve for the optimal inspection schedule using dynamic programming. They defined the following dynamic program for determining the inspection schedule that minimizes expected cost, where the minimum expected cost during $(t, \infty)$ given $P(X_t = i) = p$ for $i \in E_\Delta$, is

$$v(t, p) =$$
$$\min \begin{cases} a(t, p), \\ \inf_{\tau \geq 0} \left[ c(t, \tau, p) + b(t, \tau, p) v\left(t + \tau, h(t, \tau, p)\right) + d(t, \tau, p) v(t + \tau, \hat{I}) \right], \end{cases}$$

with $a(t, p)$ defined as the expected cost of failure if no more inspections are scheduled, $c(t, \tau, p)$ as the expected value of costs incurred during $(t, t + \tau]$,

$b(t, \tau, p)$ as the probability that inspection occurs at time $t + \tau$ with a negative outcome, i.e., $Y_{t+\tau} = 0$, and $d(t, \tau, p)$ as the probability that inspection occurs at time $t + \tau$ with a false-positive outcome.

In applying their model to breast cancer, Ozekici and Pliska discretized time into six-month periods with period $t = 0$ corresponding with 20 years of age and $T = 140$ corresponding with 70 years at age 90, then they determined the distribution $F$ by distributing the mass $\alpha$ on the discrete set $\{0, 1, \ldots, 140\}$ according to the age distribution of breast cancer given in Eddy [11] where the distribution $F$ is assumed to be uniform over each five-year interval. The age distribution data of breast cancer are based on age of detection rather than age at which the cancer is first detectable. They assumed a single preclinical stage and estimated the value for the probability of curing the disease based on Shwartz's [32] model. They derived their cost parameters from Eddy et al. [12] and varied the cost of failure as they assumed it includes the value of the loss of life, which they acknowledged does not have a uniformly agreed upon value.

### 2.3.4 Parmigiani [28]

Parmigiani developed a four-state stochastic model natural where state I is disease is absent or too early to be detectable, state II is detectable preclinical disease, state III is clinical or symptomatic disease, and state IV is death. Transitions are from I to II, II to III, and any state to IV. Transitions from I and II to IV represent death from other causes, and transitions from III to IV may have any cause. He assumed the patient is disease-free at the start of the problem and defined $Y$ and $U$ as the sojourn times in states I and II, respectively, where $Y + U$ is the age of the patient when clinical symptoms surface. $f_{II}(y)$ and $f_{IV}(y)$ are defined as the transition densities from I to II and I to IV; $h_{III}(u|y)$ and $h_{IV}(u|y)$ are the conditional transition densities from II to III and IV, given arrival in II at time $y$. All densities are assumed to be continuous. Parmigiani defined the state transition probabilities as follows:

- $\xi = \int_0^\infty f_{II}(y)\,dy < 1$ is the probability of transition from I to II;
- $\theta_{III}(y) = \int_0^\infty h_{III}(u|y)\,du$ is the probability the disease reaches III given an arrival in II at time $y$;
- $\theta_{III} = \int_0^\infty \theta_{III}(y)f_{II}(y)\,dy$ is the marginal probability of contracting the disease and reaching the clinical stage.

Parmigiani defined $\beta(x, u)$, the sensitivity of the screening test or the probability the test detects the disease if the patient is in state II, as a function of the patient age, $x$, and the sojourn time, $u$, in II at examination time. Further he assumed that for cancer, $\beta$ is increasing in $u$.

Parmigiani defined an examination schedule as a sequence $\tau = \{\tau_i\}_{i=1,2,\ldots}$, where $\tau_i$ is the time of the $i^{\text{th}}$ examination and $\tau_0 = 0$. $n = \sup\{i : \tau_i < \infty\}$ is the number of planned examinations, finite or infinite, and if $n$ is infinite, $F_{II}(\lim_{i\to\infty} \tau_i) = 1$. He assumed that screening examinations occur until the disease is detected in state II, or the individual reaches state III, IV, or age $\tau_n$, and if the exam is positive, treatment follows and screening terminates. Parmigiani also assumed that an unplanned examination is necessary to identify disease that has reached the clinical stage.

Examination schedules were chosen based on expected losses, for which Parmigiani proposed a general function, $L_s(y, u)$, for losses associated with disease-associated factors, such as mortality, morbidity, and treatment, where $y$ and $u$ are the sojourn times in I and II and where $s$ is II for screen detection, III for clinical detection, and IV for death from other causes. The schedule affects losses through s and may enter directly when $s = II$. Parmigiani required that $L$ must be continuous and differentiable in $y$ and $u$. Further, Parmigiani made assumptions about the structure of $L$. He assumed: (i) a longer sojourn time in II increases losses so $\partial L_s/\partial u > 0$ for $s = II, III$; (ii) $L_{II}(y, u) \leq L_{III}(y, u)$ for every $(y, u)$ so early detection is always advantageous; and (iii) $L_{II}(y, u) \leq L_{III}(y, u)$ for every $(y, u)$ survival is preferred to death.

The optimal schedule was chosen to minimize the total expected loss or risk: $R(\tau) \equiv kI(\tau) + c\mathcal{L}(\tau)$. Parmigiani defined $I(\tau)$ as the expected number of examinations (which is a function of the expected number of false negatives and the examination sensitivity) and $\mathcal{L}(\tau)$ as the expected value of the function of $L$ for a fixed schedule $\tau$; the expectations are taken with respect to the joint distribution of $Y$ and $U$ and are assumed finite. The optimal $\tau$ depends on $k$ and $c$ only through the $k/c$; however, Parmigiani acknowledged that it can be difficult to specify this ratio.

Parmigiani considered the two components of this loss function in determining the optimal examination schedule. Specifically, given a transition from I to II at age $y$, the loss has a deterministic component, depending on the number of examinations already performed, and a stochastic component, depending on $U$ and on the number of false negatives. He defined $\lambda_i(y)$ as the expected value of the stochastic component, conditional on $y \in (\tau_{i-1}, \tau_i]$. Parmigiani determined that the optimal examination schedule $\tau$ must satisfy

$$\lambda_{i+1}(\tau_i) - \lambda_i(\tau_i) =$$

$$\sum_{k=1}^{i} \int_{\tau_{k-1}}^{\tau_k} \frac{\partial \lambda_k(y)}{\partial \tau_i} \frac{f_{II}(y)}{f_{II}(\tau_i)} \, dy - \frac{k}{c} \left( \frac{f_{IV}(\tau_i)}{f_{II}(\tau_i)} + 1 \right) \qquad i = 1, 2, \ldots.$$

This gives the optimal increment in the $\lambda$'s as a function of the previous examination ages so it is possible to utilize this recursive structure for numerical solution.

### 2.3.5 Zelen [36]

Zelen presented a model of a person's health consisting of three possible states: $S_0$, a health state where an individual is free of disease or has disease that cannot be detected by any specific diagnostic examination; $S_p$, preclinical disease state where an individual unknowingly has disease that can be detected by a specific examination; and $S_c$, the state where the disease is clinically diagnosed. The disease is progressive transitioning from $S_0$ to $S_p$ to $S_c$. $\beta$ is the test sensitivity, the probability of the examination detecting an individual in $S_p$ conditional on being in $S_p$. $P(t)$ is the prevalence of preclinical disease, the probability of being in $S_p$ at time $t$; $w(t)\Delta t$ is the probability of making the transition from $S_0$ to $S_p$ during $(t, t + \Delta t)$. $I(t)$ is the point incidence function of the disease, where $I(t)\Delta t$ is the probability of making the transition from $S_p$ to $S_c$ during $(t, t + \Delta t)$. Define $q(t)$ as the probability density function of the sojourn time in $S_p$ where $Q(t) = \int_t^\infty q(x)\,dx$ where $t$ is time relative to when the individual entered $S_p$. For $P(t)$, $w(t)$, and $I(t)$, $t$ is time relative to a time origin.

For the interval $[0, T]$ within which are $n + 1$ examinations at the ordered time points $t_0 < t_1 < t_2 < \cdots < t_n$, Zelen denotes the $i^{\text{th}}$ interval by $(t_{i-1}, t_i)$ and its length by $\Delta_i = t_i - t_{i-1}$ for $I = 1, 2, \ldots, n$ where $t_0 = 0$ and $t_n = T$. The comparison of different screening programs is made based on the following utility function, which allows for exactly $n + 1$ examinations:

$$U_{n+1} = U_{n+1}(\beta, T) = A_0 D_0(\beta) + A \sum_{r=1}^{n} D_r(\beta) - B \sum_{r=1}^{n} I_r(\beta).$$

The weights $A_0$ and $A$ represent the probability of a cure when disease is found through screening examination, and $B$ represents the probability of cure for an interval case. $D_r(\beta)$ is the probability that the disease is detected at the $r^{\text{th}}$ screening examination, when the sensitivity is $\beta$ where $r = 0$ corresponds with the first screening examination. $D_r(\beta) = \beta P(t_{r-}|r)$ where $t_{r-} = \lim_{\delta \downarrow 0}(t_r - \delta)$ and $P(t|r)$ is the probability of being in state $S_p$ at time $t$ $(t_{r-1} \le t \le t_r)$ after having $r$ examinations at time $t_0 = 0, t_1, t_2, \ldots, t_{r-1}$, i.e., the probability of having undiagnosed preclinical disease at time $t$. If the weights represent the probability of a cure, then $U_{n+1}$ represents the difference in cure rates between those found on examination compared with interval cases. The optimal spacing of the examinations can be found by determining the values of $\{t_r\}$ that maximize $U_{n+1}$. The solutions for systems of equations defining the optimal intervals require knowledge of the preclinical sojourn time distribution.

To apply this model to breast cancer mammography screening requires the estimation of $q(t)$, $\beta$, and $\theta = B/A$. Zelen used data from the Swedish two-county trial and the Health Insurance Plan of Greater New York (HIP) studies. Based on the finding from the HIP study that the mean age of women diagnosed on the first examination was identical to that of the control group (the no examination group) clinically diagnosed with breast cancer, Zelen applied the proof of Zelen and Feinleib [37] to justify the assumption of exponentially distributed preclinical sojourn times. His estimations of $\beta$, examination sensitivity, and $m$, the mean of the preclinical sojourn time, are based on the Swedish two-county trial; however, he acknowledges that the confidence intervals for both are wide and so the values of $\beta$ are varied from 0.85 to 1 in increments of 0.05. He defined the weights $A_0$, $A$, and $B$ as the probabilities of having no axillary nodal involvement at the time of diagnosis where no distinction is made between $A_0$ and $A$ due to the limitations of the existing data.

Zelen determined the optimal equal-interval screening program and compared it with the recommended annual screening program using comparable sensitivities, screening horizons, and number of examinations. The major assumption of this model was that of a stable disease model. Under the stable disease model, the transition into $S_p$ is assumed to be independent of time. If the incidence of the disease is dependent on age, then Zelen indicated that a constant interval between examinations is not optimal even when the sensitivity is one.

In the next section, we will discuss the disease structure and the decision process for most of the aforementioned models using Ivy [17] as a guide.

## 2.4 Optimization Models for Breast Cancer Screening (and Treatment)

### 2.4.1 Modeling disease development and progression

Optimization models for breast cancer typically define a finite number of disease states to represent the progression of breast cancer. The typical state definitions are disease-free, preclinical disease (the individual has the disease but is asymptomatic and unaware of it (Lee and Zelen [21]), and clinical disease. Some models also include death as a possible state with a few distinguishing death from breast cancer from non–breast cancer death. It should be noted that Shwartz [32] presents a significantly more detailed representation for the disease progression including 21 disease states. However, his model requires several assumptions about the state transition rates. For each of these models, the disease progression is modeled based on a Markov assumption with a few exceptions. Ozekici and Pliska [27] model disease progression as a delayed Markov process in which the transition from no cancer to the preclinical disease state (the sojourn time) is a general, non-negative random variable. Parmigiani [28] and Baker [3] assume a general, non-Markovian stochastic process to model disease progression. Zelen [36] and Lee and Zelen [22] assume models of a more general disease progression with exponential transition rates presented as a possible and reasonable assumption for the development of breast cancer.

As shown in Figure 2.1, in applying her decision-making model for breast cancer screening, Ivy [17] defines the patient's condition to be in one of three states: "no disease (NC)," "non-invasive (*in situ*) breast cancer," and "invasive breast cancer." *In situ* breast cancer refers to breast cancer that "remains in place" and has not spread through the wall of the breast cell. Ductal carcinoma

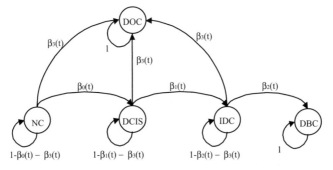

**Fig. 2.1.** An example of a breast cancer state transition diagram from Ivy [17].

*in situ* (DCIS) is the most common type of *in situ* cancer (accounting for approximately 87% of the *in situ* breast cancer cases diagnosed among women) and is breast cancer at its earliest and most curable stage — still confined to the ducts. DCIS often occurs at several points along a duct and appears as a cluster of calcifications, or white flecks, on a mammogram. Most cases of DCIS are detectable *only* by mammography. Because of its potential to recur or to become invasive, DCIS is treated with excision (or lumpectomy and radiotherapy) if the area of DCIS is small or with mastectomy if the disease is more extensive. Invasive ductal carcinoma (IDC) accounts for 70% to 80% of invasive breast cancer cases. It begins in a duct, breaks through the duct wall, and invades the supporting tissue (stroma) in the breast. From there, it can metastasize to other parts of the body. IDC is usually detected as a mass on a mammogram or as a palpable lump during a breast exam.

The state transition diagram in Figure 2.1 illustrates the patient's progression across the three states of breast cancer. The state transition rates, $\beta_0(t)$ and $\beta_1(t)$, are typically time dependent to reflect the impact of patient age on the disease progression. Once the patient enters the *IDC* state, she remains there until some exogenous action is taken, or until she dies from breast cancer (DBC) or from another cause (DOC). Notice a patient can die from other causes from each of the patient condition states. The Ivy model differs from other models of breast cancer in the following ways: Typically, disease is defined as preclinical (which may include both non-invasive disease and early invasive disease) and clinical rather by non-invasive versus invasive disease; and distinguishing death from breast cancer from death from other causes.

### 2.4.2 Monitoring the patient condition

In the modeling of breast cancer monitoring and decision making, there are many types of available information; two of the most common information sources are

- annual clinical breast exam (CBE) with the outcome: lump or no lump;

- mammogram available only for a fee with the outcome: "abnormal" or "normal" in the simplest case. (Note that it is possible to generalize this outcome to follow a continuous distribution.)

The information available through mammography is typically assumed to be superior to the information available from a CBE. Mammography locates cancers too small to be felt during a clinical breast examination. It is the best way to detect breast cancer in its earliest, most treatable stage, an average of 1.7 years before a woman can feel the lump. As this suggests, the Type I information does not distinguish between the NC and DCIS states. For both types of monitoring observations, in Ivy [17] the parameters for the Bernoulli distributions represent probabilities of a true positive (i.e., sensitivity), a false positive, a true negative (i.e., specificity), and a false negative for a CBE and a mammogram. Some authors also consider self-breast exams, and many do not specify a particular screening (or examination) modality.

### 2.4.3 Decision process

Applying the Ivy [17] model to this situation, the decision maker must first decide whether to pay for a mammogram and then select the appropriate treatment action. It is assumed that if the decision to have a mammogram is made and the mammogram is abnormal, a biopsy will be performed. It is assumed that a second action may be selected within the same time period that the mammography is performed. If a mammogram is selected, the decision maker has the following options: *do nothing*, perform a *lumpectomy* (*with radiation*), or perform a *mastectomy* (*with reconstruction*). The decision tree in Figure 2.2 summarizes the sequence of decisions. Notice, treatment decisions such as a lumpectomy and mastectomy are made only with a prior mammogram. In this decision model, the patient's condition is known with

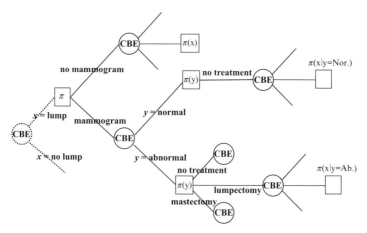

**Fig. 2.2.** An example of a breast cancer decision tree from Ivy [17].

certainty only immediately after a mastectomy, when the patient is assumed to be in a "treated" cancer-free (TrNC) state.

### 2.4.4 The optimization model

The decision maker's objective is to select the course of action (i.e., when to have a mammogram and, given the information provided by the mammogram, what course of action to take) that minimizes the total expected cost over the lifetime of the patient or population of patients with similar risk characteristics. In many of the breast cancer screening optimization models, the models evaluate specified screening schedules, and the optimal screening policy is one that results in the smallest cost. These policies are not fully dynamic in nature, and the optimization model does not drive the selection of the optimal policy in the traditional sense. Frequently, the structure of the screening policy is predefined (e.g., a threshold policy; Lee and Zelen [21]), and then the "best" policy is selected among various predefined alternatives.

Ivy [17] extends a cost-minimization model to model the patient's perspective by defining utilities for patient conditions and treatment and screening actions with the objective of minimizing the total expected utility-loss (where utility-loss $= 1 -$ utility). The patient's objective is to maximize the effectiveness of screening and treatment in terms of survival and quality of life with the goal of determining when to have a mammogram and the appropriate ("best") treatment given the results. The screening and treatment effectiveness is expressed in terms of the patient's utility, where the patient's objective is to maximize the total expected utility over their screening lifetime.

In order to define a cost-effective screening and treatment strategy, Ivy [17] balances both the payer's and the patient's objectives. Ivy [17] develops an efficient frontier in order to explore the relationship between patient and payer preferences and to determine conditions for the cost-effectiveness of mammography screening. Ivy [17] presents a constrained cost model that minimizes total expected cost subject to constraints on total expected utility-loss.

## 2.5 Areas for Future Research

Although there has been more than 30 years of research devoted to developing optimization models for breast cancer screening, most of these models have not had their desired impact. Our models are currently not influencing breast cancer screening policy or physician-patient decision making. In general, these are all good, general models; the devil is in the details. The question is how can we best implement, specifically and accurately parameterize these models, so that they may be applied effectively for improved breast cancer decision making.

The need for this research is especially evident now, particularly for the uninsured population in the 40 to 49 age group. For example, in 2002, the state of Michigan changed the age to qualify for screening mammograms from 40 to 50 for low-income women with a normal clinical breast exam who participate in the Breast and Cervical Cancer Control Program. Although this may result in budget savings in the short-term, the impact of later detection of possibly more advanced stages of breast cancer could result in significantly greater infrastructure costs.

As operations researchers and mathematical modelers, we have a rich opportunity to influence the direction of screening policy and improve the quality of screening outcomes. Through mathematical modeling and optimization, it is possible to determine the impact of screening policies on various populations without costly and invasive clinical trials. It is possible to answer questions that it would never be possible to answer in a clinical trial either due to the expense or the infeasibility of such studies. Through optimization and mathematical modeling, it is possible for truly patient-centered care to become a reality.

However, the application of optimization models to mammography screening requires several considerations.

a. *Ensuring the models are realistic while retaining enough simplicity to make the model useful.* There are several simplifying assumptions in the application of optimization models to breast cancer detection and treatment: The cancer stages are reduced to three states, the disease is assumed to progress from non-invasive to invasive cancer, and the treatment options are simplified or excluded.

b. *Simplifying the user interface.* In Ivy [17], for example, the model requires the decision maker to estimate the risk of each stage of the disease and express these risks probabilistically. Other models require similar complex computations. A user interface must be developed to simplify these tasks as risk factors have important implications for the screening and treatment decisions.

c. *Incorporating the preferences of multiple decision makers.* The physician, patient, and payer are all decision makers. Each has a different objective function, e.g., maximizing the probability of survival (or quality of life), minimizing cost, or some combination of these objectives.

d. *Changing the focus of screening from simple death avoidance to disease incidence reduction.* Because the goal of screening is to reduce the incidence of advanced disease, the screening interval should be set for a period of time in which adherence to routine screening is likely to result in the detection of the majority of cancers while they are still occult and localized (Smith et al. [33]).

e. *Defining cost.* An accurate cost estimate (i.e., the dollar cost associated with performing a procedure, the patient cost in terms of the effect on the patient's quality of life, and the cost of each disease state) is necessary for determining the cost-effectiveness of mammography. This is one of the most challenging issues associated with the cost-minimization and cost-effectiveness components of this research. Cost-effectiveness for medical applications has a very different meaning than cost-effectiveness in machine maintenance and other manufacturing fields. It is much easier to quantify the cost associated with deteriorated equipment than with a deteriorated health state. But understanding and quantifying this cost is critical to understanding and quantifying the true value of screening.

f. *Accurately estimating risk.* The current standard for estimating breast cancer risk is the Gail model (Gail et al. [14]). The Gail model is a logistic regression model based on sample data from white American females who are assumed to participate in regular screenings. The applicability of this model to other populations is not clear. In fact, although African-American women are at increased risk of breast cancer mortality compared with white American women, the Gail model is known to underestimate breast cancer risk in African-American women. Further, Bondy and Newman [5] find that African-American ethnicity is an independent predictor of a worse breast cancer outcome. The higher breast cancer mortality rate among African-American women is related to the fact that, relative to white women, a larger percentage of their breast cancers are diagnosed at a later, less treatable stage. This is due in part to early incidence and possibly more aggressive cancers.

It is possible to define and address the implications of this disparity with the optimization models such as the ones described here. In addition, this research can be used to investigate the impact of biologically aggressive breast cancers on younger women in relation to screening and treatment policies. As mentioned earlier, according to the American Cancer Society Guidelines for Breast Cancer Screening: Update 2003, though annual screening likely is beneficial for all women, the importance of annual screening clearly is greater in pre- versus post-menopausal women. However, this benefit is not reflected in existing screening policy.

Optimization models stand to offer substantial benefits to society in terms of improved public policy and healthcare delivery. The outcomes of this research can provide breast cancer screening *policy insights*, useful at the physician/patient level, but with public-policy-level implications as well. These models have great potential to help resolve currently unanswered questions concerning the relationship between screening policy and mortality risk for average-risk women, as well as for women at high risk and/or with existing comorbid conditions. The outcomes of this research also could provide information that may impact ACS policy recommendations for breast cancer screening intervals and/or technologies. Further, the framework of these models also is easily extendable to other types of disease screening, such as cervical cancer screening, colorectal cancer screening, and pregnancy-based HIV screening.

## Acknowledgments

The author would like to thank the anonymous referees and editor for their helpful comments; they have greatly improved the quality of the chapter. During much of the preparation of this chapter, the author was on the faculty at the Stephen M. Ross School of Business at the University of Michigan.

## References

[1] M. Althuis, D. Borgan, R. Coates, J. Daling, M. Gammon, K. Malone, J. Schoenberg, and L. Brenton. Breast cancers among very young pre-menopausal women (united states). *Cancer Causes and Control*, 14:151–160, 2003.

[2] M. Bailey and A. Garber. Health care productivity. In M.N. Bailey, P.C. Reiss, and C. Winston, editors, *Brookings Papers on Economic Activity: Microeconomics*, pages 143–214. The Brookings Institution, Washington, D.C., 1997.

[3] R. Baker. Use of a mathematical model to evaluate breast cancer screening policy. *Health Care Management Science*, 1:103–113, 1998.

[4] P. Beemsterboer. *Evaluation of Screening Programmes*. PhD thesis, Erasmus University Rotterdam, Rotterdam, The Netherlands, 1999.

[5] M. Bondy and L. Newman. Breast cancer risk assessment models applicability to African-American women. *Cancer Supplement*, 97(1):230–235, 2003.

[6] H. Brenner and T. Hakulinen. Are patients diagnosed with breast cancer before age 50 years ever cured? *Journal of Clinical Oncology*, 22(3):432–438, 2004.

[7] I. Bross, L. Blumenson, N. Slack, and R. Priore. A two disease model for breast cancer. In A.P.M. Forrest and P.B. Kunkler, editors, *Prognostic Factors in Breast Cancer*, pages 288–300. Williams & Wilkins, Baltimore, Maryland, 1968.

[8] H. Chen, S. Duffy, and L. Tabar. A Markov chain method to estimate the tumour progression rate from preclinical to clinical phase, sensitivity and positive value for mammography in breast cancer screening. *The Statistician*, 45(3):307–317, 1996.

[9] H. de Koning. *The Effects and Costs of Breast Cancer Screening*. PhD thesis, Erasmus University Rotterdam, Rotterdam, The Netherlands, 1993.

[10] C. Earle, D. Coyle, and W. Evans. Cost-effectiveness analysis in oncology. *Annals of Oncology*, 9:475–482, 1998.

[11] D. Eddy. *Screening for Cancer: Theory, Analysis, and Design*. Prentice-Hall, Englewood Cliffs, New Jersey, 1980.

[12] D. Eddy, V. Hasselblad, W. Hendee, and W. McGiveney. The value of mammography screening in women under 50 years. *Journal of the American Medical Association*, 259:1512–1519, 1988.

[13] L. Foxcroft, E. Evans, and A. Porter. The diagnosis of breast cancer in women younger than 40. *The Breast*, 13:297–306, 2004.

[14] M. Gail, L. Brinton, D. Byar, D. Corle, S. Green, C. Schairer, and J. Mulvihill. Projecting individualized probabilities of developing breast cancer for white females who are being examined annually. *Journal of the National Cancer Institute*, 81(24):1879–1886, 1989.

[15] P. Gotzsche and O. Olsen. Is screening for breast cancer with mammography justifiable? *Lancet*, 355:129–134, 2000.

[16] E. Gunes, S. Chick, and O. Aksin. Breast cancer screening services: Trade-offs in quality, capacity, outreach, and centralization. *Health Care Management Science*, 7:291–303, 2004.

[17] J. Ivy. A maintenance model for breast cancer treatment and detection. Working paper, 2007.

[18] R. Kirch and M. Klein. Surveillance schedules for medical examinations. *Management Science*, 20(10):1403–1409, 1974.

[19] N. Kroman, M. Jensen, J. Wohlfart, J. Mouridsen, P. Andersen, and M. Melbye. Factors influencing the effect of age on prognosis in breast cancer: Population based study. *British Medical Journal*, 320:474–479, 2000.

[20] S. Lee, H. Huang, and M. Zelen. Early detection of disease and scheduling of screening examinations. *Statistical Methods in Medical Research*, 13:443–456, 2004.

[21] S. Lee and M. Zelen. Scheduling periodic examinations for the early detection of disease: Applications to breast cancer. *Journal of the American Statistical Association*, 93(444):1271–1281, 1998.

[22] S. Lee and M. Zelen. Modelling the early detection of breast cancer. *Annals of Oncology*, 14:1199–1202, 2003.

[23] O. Mangasarian, W. Street, and W. Wolberg. Breast cancer diagnosis and prognosis via linear programming. *Operations Research*, 43(4):570–577, 1995.

[24] A. Mathew, B. Rajan, and M. Pandey. Do younger women with non-metastatic and non-inflammatory breast carcinoma have poor prognosis? *World Journal of Surgical Oncology*, 2:1–7, 2004.

[25] J. Michaelson, E. Halpern, and D. Kopans. Breast cancer: Computer simulation method for estimating optimal intervals for screening. *Radiology*, 212:551–560, 1999.

[26] O. Olsen and P. Gotzsche. Cochrane review on screening for breast cancer with mammography. *Lancet*, 358:1340–1342, 2001.

[27] S. Ozekici and S. Pliska. Optimal scheduling of inspections: A delayed markov model with false positive and negatives. *Operations Research*, 39(2):261–273, 1991.

[28] G. Parmigiani. On optimal screening ages. *Journal of the American Statistical Association*, 88(422):622–628, 1993.

[29] M. Retsky, R. Demicheli, D. Swartzendruber, P. Bame, R. Wardwell, G. Bonadonna, J. Speer, and P. Valagussa. Computer simulation of breast cancer metastasis model. *Breast Cancer Research and Treatment*, 45:193–202, 1997.

[30] L. Russell. *Is Prevention Better Than Cure?* Brookings Institution Press, Washington, D.C., 1986.

[31] L. Russell. *Educated Guesses: Making Policy About Medical Screening Tests.* University of California Press, Berkeley, California, 1994.

[32] M. Shwartz. A mathematical model used to analyze breast cancer screening strategies. *Operations Research*, 26(6):937–955, 1998.

[33] R. Smith, D. Saslow, K. Sawyer, W. Burke, M. Constanza, W. Evans, R. Foster, E. Hendrick, H. Eyre, and S. Sener. American Cancer Society guidelines for breast cancer screening: Update 2003. *CA A Cancer Journal for Clinicians*, 53(3):141–169, 2003.

[34] K. Szeto and N. Devlin. The cost-effectiveness of mammography screening: Evidence from a microsimulation model for New Zealand. *Health Policy*, 38:101–115, 1996.

[35] R. Yancik, M. Wesley, L. Ries, R. Havlik, B. Edwards, and J. Yates. Effect of age and comorbidity in postmenopausal breast cancer patients aged 55 years and older. *Journal of the American Medical Association*, 285(7):885–892, 2001.

[36] M. Zelen. Optimal scheduling of examinations for the early detection of disease. *Biometrika*, 80(2):279–293, 1993.

[37] M. Zelen and M. Feinleib. On the theory of screening for chronic disease. *Biometrika*, 56:601–614, 1969.

# 3

# Optimization Models and Computational Approaches for Three-dimensional Conformal Radiation Treatment Planning

Gino J. Lim

Department of Industrial Engineering, University of Houston, 4800 Calhoun Road, Houston, Texas 77204
ginolim@uh.edu

**Abstract.** This chapter describes recent advances in optimization methods for three-dimensional conformal radiation treatment (3DCRT) planning. A series of optimization models are discussed for optimizing various treatment parameters: *beam weight optimization, beam angle optimization,* and *wedge orientation.* It is well-known that solving such optimization models in a clinical setting is extremely difficult. Therefore, we discuss solution time reduction methods that are easy to use in practice. Techniques for controlling dose-volume histograms (DVHs) are described to meet the treatment planner's preference. Finally, we present a clinical case study to demonstrate the computational performance and effectiveness of such approaches.

## 3.1 Introduction

### 3.1.1 Background

Cancer is the second leading cause of death in the United States [1]. Treatment options are determined by the type and the stage of the cancer and include surgery, *radiation therapy,* chemotherapy, and so forth. Physicians often use a combination of those treatments to obtain the best results. Our aim is to describe techniques to improve the delivery of radiation to cancer patients. We will focus on using optimization approaches to improve the treatment planning process. The objective of treatment planning problems is to control the local tumor (target) volume by delivering a uniform (homogeneous) dose of radiation while sparing the surrounding normal and healthy tissue. A major challenge in treatment planning is the presence of organs-at-risk (OARs). An OAR is a critical structure located very close to the target for which the dose of radiation must be severely constrained. This is because an overdose of radiation within the critical structure may lead to medical complications. OAR is also termed "sensitive structure" or "critical structure" in the literature.

P.M. Pardalos, H.E. Romeijn (eds.), *Handbook of Optimization in Medicine,*
Springer Optimization and Its Applications 26, DOI: 10.1007/978-0-387-09770-1_3,
© Springer Science+Business Media LLC 2009

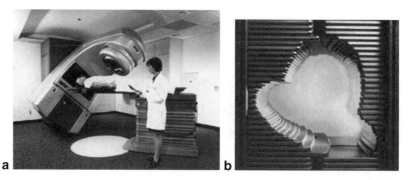

**Fig. 3.1.** External beam therapy machine: (a) a linear accelerator and (b) a multileaf collimator.

External-beam radiation treatments are typically delivered using a linear accelerator (see Figure 3.1(a)) with a multileaf collimator (see Figure 3.1(b)) housed in the head of the treatment unit. The shape of the aperture through which the beam passes can be varied by moving the computer-controlled leaves of the collimator. There are two types of radiation treatment planning: forward planning and inverse planning. In forward planning, treatment plans are typically generated by a trial and error approach. An improved treatment plan is produced by a sequence of experiments with different radiation beam configurations in external beam therapy. Because of the complexity of the treatment planning problem, this process, in general, is very tedious and time-consuming and does not necessarily produce "high-quality" treatment plans. Better strategies for obtaining treatment plans are therefore desired. Because of significant advances in modern technologies such as imaging technologies and computer control to aid the delivery of radiation, there has been a significant move toward inverse treatment planning (it is also called computer-based treatment planning). In inverse treatment planning, an objective function is defined to measure the goodness (quality) of a treatment plan. Two types of objective functions are often used: dose-based models and biological (radiobiological) models. The biological model argues that optimization should be based on the biological effects resulting from the underlying radiation dose distributions. The treatment objective is usually to maximize the tumor control probability (TCP) while keeping the normal tissue complication probability (NTCP) within acceptable levels. The type of objective functions we use in this chapter is based solely on dose, meaning that achieving accurate dose distributions is the main concern. The biological aspect is implicitly given by the physician's prescription. The inverse treatment planning procedure allows modeling highly complex treatment planning problems from brachytherapy to external beam therapy. Examples of these more complex plans include conformal radiotherapy, intensity modulated radiation therapy (IMRT) [10, 15, 27, 30, 54], and tomotherapy [16, 26].

### 3.1.2 Use of optimization techniques

Radiation treatment planning for cancer patients has emerged as a challenging application for optimization [3, 4, 5, 8, 24, 25, 47, 52, 53, 55]. Two major goals in treatment planning are speed and quality. Solution quality of a treatment plan can be measured by *homogeneity, conformity,* and *avoidance* [18, 19, 31, 33, 32]. Fast solution determination in a simple manner is another essential part of a clinically useful treatment planning procedure. Acceptable dose levels of these requirements are established by various professional and advisory groups.

It is important for a treatment plan to have uniform dose distributions on the target so that *cold* and *hot spots* can be minimized. A cold spot is a portion of an organ that receives below its required dose level. On the other hand, a *hot spot* is a portion of an organ that receives more than the desired dose level. The homogeneity requirement ensures that radiation delivered to tumor volume has a minimum number of *hot spots* and *cold spots* on the target. This requirement can be enforced using lower and upper bounds on the dose or approximated using penalization. The conformity requirement is used to achieve the target dose control while minimizing the damage to OARs or healthy normal tissue. This is generally expressed as a ratio of cumulative dose on the target over total dose prescribed for the entire treatment. This ratio can be used to control conformity in optimization models. As we mentioned earlier, a great difficulty of producing radiation treatment plans is the proximity of the target to the OARs. An avoidance requirement can be used to limit the dose delivered to OARs. Finally, simplicity requirements state that a treatment plan should be as simple as possible. Simple treatment plans typically reduce the treatment time as well as implementation error.

In this chapter, we introduce a few optimization models and solution techniques that are practically useful to automate an external-beam radiation treatment planning process. Potential benefits of the automated treatment planning process include the reduction in planning time and improved quality of dose distributions of treatment plans. Such planning systems should depend less on the experience of the treatment planner. In other words, treatment planners will expedite their learning curve much faster in this automated system than in the conventional forward planning system. However, it should be noted that the treatment goals may vary from one planner to another. Therefore, an automated treatment planning system must be able to self-adjust to these changes and accommodate different treatment goals.

## 3.2 Three-dimensional Conformal Radiation Therapy

Although IMRT delivers superb quality treatment plans, optimizing IMRT plans within a clinically acceptable time frame still remains a challenging task. Therefore, we focus on the conventional three-dimensional conformal radiotherapy (3DCRT) techniques [31, 33, 32]. This approach has several advantages over an IMRT plan optimization. First, the optimization procedure is much simpler because we do not consider each pencil beam in the optimization model. Second, both fluence map optimization and leaf-sequencing optimization are not required. Beam shapes and their uniform monitor units are determined as a result of beam weight optimization. Therefore, much faster solution determination can be achieved. Third, monitor units contain real values. Note that an IMRT plan optimization requires discretized fluence maps for leaf-sequencing, which can easily introduce discretization error to the model. Finally, significantly less beams are used for the treatment, which has a practical advantage over IMRT.

One of the main strategies for minimizing morbidity in 3DCRT is to reduce the dose delivered to normal tissues that are spatially well separated from the tumor. This can be done by using multiple beams from different angles.

### 3.2.1 Effect of multiple beams

A single radiation beam leads to a higher dose delivered to the tissues in front of the tumor than to the tumor itself. In consequence, if one were to give a dose sufficient to control the tumor with a reasonably high probability, the dose to the upstream tissues would likely lead to unacceptable morbidity. A single beam would only be used for very superficial tumors, where there is little upstream normal tissue to damage. For deeper tumors, one uses multiple cross-firing beams delivered within minutes of one another: All encompass the tumor, but successive beams are directed toward the patient from different directions to traverse different tissues outside the target volume. The delivery of cross-firing beams is greatly facilitated by mounting the radiation-producing equipment on a gantry, as illustrated in Figure 3.1(a).

Several directed beams noticeably change the distribution of dose, as is illustrated in Figure 3.2. As a result, dose outside the target volume can often be quite tolerable even when dose levels within the target volume are high enough to provide a substantial probability of tumor control.

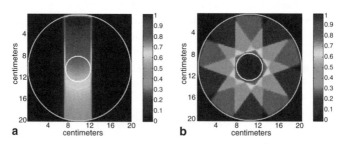

**Fig. 3.2.** Effect of multiple beams. (a) Single beam: tissue on top receives significant dose; (b) five beams: a hot spot is formed by five beams.

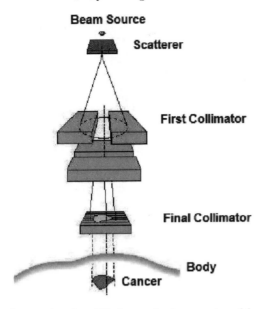

**Fig. 3.3.** A beam's-eye view is a 2D shape of a tumor viewed by the beam source at a fixed angle.

### 3.2.2 Beam shape generation and collimator

The leaves of the multileaf collimator are computer controlled and can be moved to the appropriate positions to create the desired beam shape. From each beam angle, three-dimensional anatomical information is used to shape the beam of radiation to match the shape of the tumor. Given a gantry angle, the view on the tumor that the beam source can see through the multileaf collimator is called the *beam's-eye view* of the target (see Figure 3.3); [20]. This beam's-eye view (BEV) approach ensures adequate irradiation of the tumor while reducing the dose to normal tissue.

### 3.2.3 Wedge filters

A wedge (also called a "wedge filter") is a tapered metallic block with a thick side (the *heel*) and a thin edge (the *toe*) (see Figure 3.4). This metallic wedge varies the intensity of the radiation in a linear fashion from one side of the radiation field to the other. When the wedge is placed in front of the aperture, less radiation is transmitted through the heel of the wedge than through the toe. Figure 3.4(b) shows an external 45° wedge, so named because it produces isodose lines that are oriented at approximately 45°. The quality of the dose distribution can be improved by incorporating a wedge filter into one or more of the treatment beams. Wedge filters are particularly useful in compensating for a curved patient surface, which is common in breast cancer treatments.

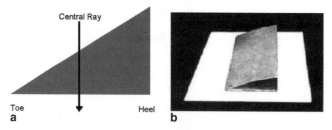

Central Ray

Toe                              Heel
a                                    b

**Fig. 3.4.** Wedges: (a) a wedge filter; (b) an external wedge.

Two different wedge systems are used in clinical practice. In the first system, four different wedges with angles 15°, 30°, 45°, and 60° are available, and the therapist is responsible for selecting one of these wedges and inserting it with the correct orientation. In the second system, a single 60° wedge (the *universal wedge*) is permanently located on a motorized mount located within the head of the treatment unit. This wedge can be rotated to the desired orientation or removed altogether, as required by the treatment plan.

Lim et al. [33] show in the following theorem that a treatment plan that requires the use of a wedge is in some cases equivalent to one that uses a wedge with different properties in combination with an open (unwedged) beam of the same shape. This result implies that a single "universal" wedge suffices in designing a wide range of treatment plans; not much is to be gained by using a range of wedges with different properties.

**Theorem 1.** *When a universal wedge is appropriately used for radiation therapy, all plans deliverable by the four-wedge systems can be reproduced.*

### 3.2.4 Radiation treatment procedure

1. The patient is immobilized in an individual cast so that the location of the treatment region remains the same for the rest of the treatment process.
2. A CT scan is performed with the patient in the cast to identify the three-dimensional shapes of organs of interest.
3. Conformal treatment plans are generated using the organ geometries.
4. Treatments are performed 5 times a week for 5 to 7 weeks.

### 3.2.5 Treatment planning process

A typical treatment planning process of a 3DCRT includes the following tasks:

1. Beam's-eye view generation for each beam (gantry) angle.
2. Dose matrix calculation for each beam angle.
3. Optimal treatment parameter generation.
4. Treatment plan validation and implementation.

There are several input data required for optimization models in radiation treatment planning. The first input describes the machine that delivers radiation. The second and troublesome input is the dose distribution of a particular treatment problem. A dose distribution consists of radiation dose contribution to each voxel of the region of interest when unit radiation intensity is exposed from a fixed gantry angle. It can be expressed as a functional form or a set of data. However, difficulties of using such distributions include high nonlinearity of the functional form, or the large amount of data that specifies the dose distribution. This problem needs to be overcome in a desirable automated treatment planning tool. The third common input is the set of organ geometries that are of interest to the physician. Further common inputs are the desired dose levels for each organ of interest. These are typically provided by physicians. Other types of inputs can also be specified depending on the treatment planning problems. However, a desirable treatment planning system should be able to generate high-quality treatment plans with minimum additional inputs and human guidance.

## 3.3 Formulating the Optimization Problems

### 3.3.1 Optimizing beam weights

We start with the simplest model, in which the angles from which beams are to be delivered are selected in advance, wedges are not used, and the apertures are chosen to be the beam's-eye view from each respective angle. All that remains is to determine the beam weights for each angle.

We now introduce notation that is used below and in later sections. The set of beam angles is denoted by $\mathcal{A}$. We let $\mathcal{T}$ denote the set of all voxels that comprise the PTV (Planning Target Volume), $\mathcal{S}$ denote the voxels in the OAR (Organ-At-Risk: typically this is a collection of organs $\mathcal{S}^p$, $p = 1, \ldots, m$), and $\mathcal{N}$ be the voxels in the normal tissue. We use $\Delta$ to denote the prescribed dose level for each PTV voxel, and the hot spot control parameter $\phi$ defines a dose level for each voxel in the critical structure that we would prefer not to exceed. The beam weight delivered from angle $A$ is denoted by $w_A$, and the dose contribution to voxel $(i,j,k)$ from a beam of unit weight from angle $A$ is denoted by $\mathcal{D}_{A,(i,j,k)}$. (It follows that a beam of weight $w_A$ produces a dose of $w_A \mathcal{D}_{A,(i,j,k)}$ in voxel $(i,j,k)$.) We obtain the total dose $D_{(i,j,k)}$ to voxel $(i,j,k)$ by summing the contributions from all angles $A \in \mathcal{A}$. We use $\mathcal{D}_{A,\Omega}$ (and $D_\Omega$) to denote the submatrices consisting of the elements $\mathcal{D}_{A,(i,j,k)}$ (and $D_{(i,j,k)}$) for all $(i,j,k)$ in a given set of voxels $\Omega$.

The beam weights $w_A$, for $A \in \mathcal{A}$, are nonnegative and are the unknowns in the optimization problem. The general form of this problem is

$$\min_{w} f(D_\Omega) \quad \text{s.t.}$$
$$D_\Omega = \sum_{A \in \mathcal{A}} w_A \mathcal{D}_{A,\Omega}, \ \Omega = \mathcal{T} \cup \mathcal{S} \cup \mathcal{N}, \tag{3.1}$$
$$w_A \geq 0, \qquad \qquad \forall A \in \mathcal{A}.$$

The choice of objective function $f(D_\Omega)$ in (3.1) depends on the specific goal of the treatment planner. In general, the objective function measures the mismatch between the prescription and the delivered dose. For voxels in the PTV region $\mathcal{T}$, there may be terms that penalize any difference between the delivered dose and the prescribed dose. For the voxels in each OAR $\mathcal{S}^p$ ($p = 1, \ldots, m$), there may be terms that penalize the amount of dose in excess of $\phi_p$, the desired upper bound on the dose to voxels in $\mathcal{S}^p$. For simplicity of exposition, we consider only a single OAR from now on. The objective often includes terms that penalize any dose to voxels in the normal region $\mathcal{N}$.

The $L_1$-norm (sum of absolute values) and squared $L_2$ norm (sum of squares; see [11]) are both used to penalize difference between delivered and desired doses in the objective $f(D_\Omega)$. Two possible definitions of $f$ based on these norms are

$$f(D_\Omega) = \lambda_t \frac{\|D_\mathcal{T} - \Delta e_\mathcal{T}\|_1}{|\mathcal{T}|} + \lambda_s \frac{\|(D_\mathcal{S} - \phi \Delta e_\mathcal{S})_+\|_1}{|\mathcal{S}|} + \lambda_n \frac{\|D_\mathcal{N}\|_1}{|\mathcal{N}|}, \tag{3.2}$$

$$f(D_\Omega) = \lambda_t \frac{\|D_\mathcal{T} - \Delta e_\mathcal{T}\|_2^2}{|\mathcal{T}|} + \lambda_s \frac{\|(D_\mathcal{S} - \phi \Delta e_\mathcal{S})_+\|_2^2}{|\mathcal{S}|} + \lambda_n \frac{\|D_\mathcal{N}\|_2^2}{|\mathcal{N}|}. \tag{3.3}$$

The notation $(\cdot)_+ := \max(\cdot, 0)$ in the second term defines the overdose to voxels in the OAR, and $e_\mathcal{T}$ is the vector whose components are all 1 and whose dimension is the same as the cardinality of $\mathcal{T}$ (similarly for $e_\mathcal{S}$). The parameters $\lambda_t$, $\lambda_s$, and $\lambda_n$ are nonnegative weighting factors applied to the objective terms for the PTV, OAR, and normal tissue voxels, respectively, and $|\mathcal{T}|$, $|\mathcal{S}|$, and $|\mathcal{N}|$ denote the number of voxels in these respective regions.

An objective function based on $L_\infty$-norm terms (3.4) allows effective penalization of hot spots in the OAR and of cold spots in the PTV. We define such a function by

$$\lambda_t \|(D_\mathcal{T} - \Delta e_\mathcal{T})\|_\infty + \lambda_s \|(D_\mathcal{S} - \phi \Delta e_\mathcal{S})_+\|_\infty + \lambda_n \|D_\mathcal{N}\|_\infty. \tag{3.4}$$

Combinations of these objective functions can be used to achieve specific treatment goals, as described later.

Problems of the form (3.1) in which $f$ is defined by (3.2) or (3.4) can be formulated as linear programs using standard techniques. For example, the term $\lambda_s \|(D_\mathcal{S} - \phi de_\mathcal{S})_+\|_1 / |\mathcal{S}|$ in (3.2) can be modeled by introducing a vector $V_\mathcal{S}$ into the formulation, along with the constraints $V_\mathcal{S} \geq D_\mathcal{S} - \phi de_\mathcal{S}$ and $V_\mathcal{S} \geq 0$, and including the term $(\lambda_s / |\mathcal{S}|) e_\mathcal{S}^T V_\mathcal{S}$ in the objective. Problems in which $f$ is defined by (3.3) can be formulated as convex quadratic programs.

The treatment planner's goals are often case specific. For example, the planner may wish to keep the maximum dose violation on the PTV low and also to control the integral dose violation on the OAR and the normal tissue. (Note that $L_\infty$-norm is recommended on the OAR only if it is a serial organ that must limit a maximum radiation dose in order to avoid medical complications.) These goals can be met by defining the objective to be a weighted sum of the relevant terms. For the given example, we might obtain the following definition of $f(D_\Omega)$ in (3.1):

$$\lambda_t \| D_\mathcal{T} - \Delta e_\mathcal{T} \|_\infty + \lambda_s \frac{\|(D_\mathcal{S} - \phi \Delta e_\mathcal{S})_+\|_1}{|\mathcal{S}|} + \lambda_n \frac{\|D_\mathcal{N}\|_1}{|\mathcal{N}|}. \tag{3.5}$$

In practice, voxels in the PTV that receive a dose within specified limits may be acceptable as a treatment plan. Furthermore, voxels that receive below the lower dose specification (cold spots) may be penalized more severely than hot spots in the PTV. Therefore, we consider the following definition of $f$:

$$f(D_\Omega) = \lambda_t^+ \|(D_\mathcal{T} - \theta_u \Delta e_\mathcal{T})_+\|_\infty + \lambda_t^- \|(\theta_L \Delta e_\mathcal{T} - D_\mathcal{T})_+\|_\infty \tag{3.6}$$
$$+ \lambda_s \frac{\|(D_\mathcal{S} - \phi \Delta e_\mathcal{S})_+\|_1}{|\mathcal{S}|} + \lambda_n \frac{\|D_\mathcal{N}\|_1}{|\mathcal{N}|}.$$

In this objective, $\theta_L$ is the PTV cold-spot control parameter. If the dosage delivered to a voxel in $\mathcal{T}$ falls below $\theta_L \Delta$, a penalty term for the violation is added to the objective. Likewise, a voxel in the PTV incurs a penalty if the dose exceeds $\theta_u \Delta$.

All the models described in this paper can accommodate this separation of *hot* and *cold* spots. However, we simplify the exposition throughout by using a combined objective function. Alternative objectives have been discussed elsewhere. For example, the papers [40, 44] use score functions to evaluate and compare different plans, whereas [23] use a multi-objective approach.

Building on the beam-weight optimization formulations described above, we now consider extended models in which beam angles and wedges are included in the optimization problem.

### 3.3.2 Optimizing beam angles

We now consider the problem of selecting a subset of at most $K$ beam angles from a set $\mathcal{A}$ of candidates while simultaneously choosing optimal weights for the selected beams. In the model, the binary variables $\psi_A$, $A \in \mathcal{A}$ indicate whether or not angle $A$ is selected to be one of the treatment beam orientations. The constraint $w_A \leq M\psi_A$ (for some large $M$) ensures that weight $w_A$ is nonzero only if $\psi_A = 1$. The resulting mixed integer programming formulation is as follows:

$$\min_{w, \psi} \quad f(D_\Omega) \quad \text{s.t.}$$
$$D_\Omega = \sum_{A \in \mathcal{A}} w_A \mathcal{D}_{A, \Omega}, \ \Omega = \{\mathcal{T} \cup \mathcal{S} \cup \mathcal{N}\}$$
$$0 \leq w_A \leq M\psi_A, \quad \forall A \in \mathcal{A}, \tag{3.7}$$
$$\sum_{A \in \mathcal{A}} \psi_A \leq K,$$
$$\psi_A \in \{0, 1\}, \qquad \forall A \in \mathcal{A}.$$

Some theoretical considerations of optimizing beam orientations are also discussed in [3]. A treatment plan involving few beams (say, 3 to 5) generally is preferable to one of similar quality that uses more beams because it requires less time and effort to deliver. Furthermore, it has been shown that, when many beams are used (say $\geq 5$), beam orientation becomes less important in the overall optimization [9, 12, 15, 46]. In many cited cases, the objective is to find a minimum number of beams that satisfy the treatment goals.

The beam angles and the weights can be selected either sequentially or simultaneously. Most of the earlier work in the literature uses sequential schemes [7, 21, 35, 42, 43], in which a certain number of beam angles are decided first, and their weights are subsequently determined. Rowbottom et al. [41] optimizes both variables simultaneously. To reduce the initial search space, a heuristic approach to remove some beam orientations *a priori* is used, while the overall optimization problem is solved with the simplex method and simulated annealing. Prior information is included in the simultaneous optimization scheme outlined in [39].

A different approach has been proposed by [22]. They address a geometric formulation of the coplanar beam orientation problem by means of a hybrid multi-objective genetic algorithm, which attempts to replicate the approach of a (human) treatment planner while reducing the amount of computation required. When the approach is applied without constraining the number of beams, the solution produces an indication of the minimum number of required beams. Webb [48] applies simulated annealing to a two-dimensional treatment planning problem. Three-dimensional problems using simulated annealing approach are described in [41, 49, 50, 51], and column generation approaches are discussed in [37].

### 3.3.3 Optimizing wedge orientations

Several researchers have studied the treatment planning problem with wedges. Xing et al. [56] optimize the beam weights for an open field and two orthogonal wedged fields. Li et al. [29] describe an algorithm for selecting both wedge orientation and beam weights, and [45] describes a mathematical basis for selection of wedge angle and orientation. It is noted in [57] that including wedge angle selection in the optimization makes for excessive computation time. Design of treatment plans involving wedges is also discussed in [13].

Suppose that four possible wedge orientations are considered at each beam angle: "north," "south," "east," and "west." At each angle $A$, we calculate dose matrices for the beam's-eye view aperture and for each of these four wedge settings, along with the dose matrix for the open beam, as used in the formulations above. We let $\mathcal{F}$ denote the set of wedge settings; $\mathcal{F}$ contains 5 elements in this case. Extending our previous notation, the dose contribution to voxel $(i, j, k)$ from a beam delivered from angle $A$ with wedge setting $F$ is denoted by $\mathcal{D}_{A,F,(i,j,k)}$, and we use $\mathcal{D}_{A,F,\Omega}$ to denote the collection of doses for all $(i, j, k)$ in some set $\Omega$. The weight assigned to a beam from angle $A$ with wedge setting $F$ is denoted by $w_{A,F}$.

To include wedges in the optimization problem, we do not simply replace $\mathcal{A}$ by $\mathcal{A} \times \mathcal{F}$ in (3.7); there are some additional considerations. First, in selecting beams, we do not wish to place a limit on the total number of beams delivered, as in Section 3.3.2, but rather on the total number of distinct angles used. (In the clinical situation, changing the wedge orientation takes relatively little time.) It follows that a single binary variable suffices for each angle $A$, so we can state the MIP model that includes beam orientation selection as follows:

$$
\begin{aligned}
\min_{w, \psi} \; & f(D_\Omega) \qquad \text{s.t.} \\
D_\Omega \;&=\; \sum_{A \in \mathcal{A}, F \in \mathcal{F}} w_{A,F} \mathcal{D}_{A,F,\Omega}, \; \Omega \in \mathcal{T} \cup \mathcal{S} \cup \mathcal{N}, \\
0 \le w_{A,F} \;&\le\; M\psi_A, \qquad\qquad \forall A \in \mathcal{A}, \; \forall F \in \mathcal{F}, \\
\textstyle\sum_{A \in \mathcal{A}} \psi_A \;&\le\; K, \qquad\qquad\qquad \psi_A \in \{0, 1\}, \forall A \in \mathcal{A}.
\end{aligned}
\tag{3.8}
$$

A second consideration is that we do not wish to deliver two beams from the same angle for two diametrically opposite wedge settings. We can accommodate this restriction by introducing separate binary variables $\pi_{A,F}$ for each pair of angle $A$ and orientation $F$. A less expensive approach is to postprocess the solution whenever

$$
\{w_{A,\text{south}} > 0 \text{ and } w_{A,\text{north}} > 0 \} \text{ or } \{w_{A,\text{west}} > 0 \text{ and } w_{A,\text{east}} > 0\},
$$

for any $A$, to zero out one of the weights for each pair.

To illustrate the postprocessing technique, consider the "west" and "east" wedge orientations. We introduce a wedge transmission factor $\tau$ that defines the reduction in dose caused by the wedge. Wedges are characterized by $\tau_0$ and $\tau_1$, with $0 \le \tau_0 < \tau_1 \le 1$ which indicate the smallest and largest transmission factors for the wedge among all pencil beams in the field. Specifically, $\tau_0$ indicates the factor by which the dose is decreased for pencil beams along the heel of the wedge, and $\tau_1$ is the transmission factor along the opposite (toe) edge. Suppose now that we have a treatment plan in which for some $A$ the weight corresponding with the open beam (no wedge) is $w_{A,\text{open}} \ge 0$, and the weights corresponding with the west and east beams are $w_{A,\text{west}} > 0$ and $w_{A,\text{east}} > 0$, respectively. Suppose also for the moment that $w_{A,\text{west}} \ge w_{A,\text{east}}$. Lim et al. [33] show that an identical dose could be delivered to each affected voxel $(i, j, k)$ by using weight $w_{A,\text{open}} + w_{A,\text{east}}(\tau_1 - \tau_0)$ for the open beam, $(w_{A,\text{west}} - w_{A,\text{east}})$ for the west wedge, and 0 for the east wedge. A similar result holds for the case of $w_{A,\text{west}} \le w_{A,\text{east}}$.

Note that if there are other constraints on the number of wedges being used, we need to replace (3.8) by a formulation with additional binary variables $\pi_{A,F}$.

### 3.3.4 Computing tight upper bounds on the beam weights

If the upper bound $M$ on the beam weights $w_{A,F}$ is too large (as is usually the case), the feasible set is larger and the algorithm often takes longer to solve the problem. A key preprocessing technique to overcome this problem is to calculate a stringent bound on the continuous decision variables ([36]) that allows $M$ to be chosen sufficiently large to produce an optimal solution, but not larger than necessary. We now describe a technique of this type for problem (3.8).

Let $\mu_A$ be the maximum dose deliverable to the PTV by a beam angle $A$ with a unit beam intensity. Because the open beam delivers more radiation to a voxel (per unit beam weight) than any wedged beam, we have

$$\mu_A := \max_{F\in\mathcal{F},\,(i,j,k)\in\mathcal{T}} \mathcal{D}_{A,F,(i,j,k)}$$
$$= \max_{(i,j,k)\in\mathcal{T}} \mathcal{D}_{A,(i,j,k)}, \quad A = 1,2,\cdots,|\mathcal{A}|, \tag{3.9}$$

where, as before, $\mathcal{D}_{A,(i,j,k)}$ denotes the dose delivered to voxel $(i,j,k)$ from a unit weight of the open beam at angle $A$. For a given angle $A$, the maximum dose deliverable to a PTV voxel using wedge filters is given as follows:

$$\mu_A \left( w_{A,0} + \tau_1 \sum_{F\in\mathcal{F}\backslash\{0\}} w_{A,F} \right), \tag{3.10}$$

where $0 \in \mathcal{F}$ denotes the open beam. Suppose now that we modify the model in (3.8) to include explicit control of hot spots by introducing an upper bound $u$ on the dose allowed in any PTV voxel. We add the constraint

$$D_{\mathcal{T}} \leq u e_{\mathcal{T}} \tag{3.11}$$

to (3.8). By combining (3.11) with (3.10), we deduce that

$$w_{A,0} + \tau_1 \sum_{F\in\mathcal{F}\backslash\{0\}} w_{A,F} \leq \frac{u}{\mu_A}, \quad \forall A \in \mathcal{A}.$$

Accordingly, we can replace the constraint $M\psi_A \geq w_{A,F}$ in (3.8) by

$$w_{A,0} + \tau_1 \sum_{F\in\mathcal{F}\backslash\{0\}} w_{A,F} \leq \left(\frac{u}{\mu_A}\right)\psi_A, \quad \forall A \in \mathcal{A}, \tag{3.12}$$

where $\psi_A$ is the binary variable that indicates whether or not the angle $A$ is selected. The resulting optimization problem becomes

$$\min_{w,\psi} \quad f(D_\Omega) \quad \text{s.t.}$$

$$D_\Omega = \sum_{A \in \mathcal{A}, F \in \mathcal{F}} w_{A,F} \mathcal{D}_{A,F,\Omega}, \ \Omega \in \mathcal{T} \cup \mathcal{S} \cup \mathcal{N},$$

$$(u/\mu_A)\psi_A \geq w_{A,0} + \tau_1 \sum_{F \in \mathcal{F} \backslash 0} w_{A,F} \tag{3.13}$$

$$K \geq \sum_{A \in \mathcal{A}} \psi_A,$$
$$w_{A,F} \geq 0, \qquad\qquad \forall A \in \mathcal{A}, \forall F \in \mathcal{F},$$
$$\psi_A \in \{0,1\}, \qquad\qquad \forall A \in \mathcal{A}, \forall F \in \mathcal{F}.$$

Note that if we also impose an upper bound on dose level to normal-tissue voxels, we can trivially derive additional bounds on the beam weights using the same approach.

## 3.4 Solution Quality in Clinical Perspective

As we mentioned in Section 3.1, the solution quality of a treatment plan must meet at least three basic requirements to be practically useful: conformity, uniformity, and homogeneity. Researchers proposed several different visualizations to help the treatment planner in assessing the quality of a treatment plan: tumor control probability (TCP), normal tissue complication probabilities (NTCPs), dose-volume histogram (DVH), and dose distribution (dose plot). Typically, the DVH and the three-dimensional radiation dose distribution are used as the means of evaluating the treatment quality.

### 3.4.1 Dose-volume histogram

Dose-volume histograms are a compact way to represent dose distribution information for subsets of the treatment region. By placing simple constraints on the shape of the DVH for a particular region, radiation oncologists attempt to control the fundamental aspects of the treatment plan. For instance, the oncologist is often willing to sacrifice some specified portion of an OAR (such as the lung) in order to provide an adequate probability of tumor control (especially if the OAR lies near the tumor). This aim is realized by requiring that at least a specified percentage of the OAR must receive a dose less than a specified level. DVH constraints are used to control uniformity of the dose to the PTV and to avoid cold spots. For example, the planner may require all voxels in the PTV to receive doses of between 95% and 107% of the prescribed dose $\Delta$.

Figure 3.5(a) shows a DVH example of a treatment plan based on a prostate tumor data. There are three lines: one for the PTV, one for the OAR, and the third for the normal tissue. The point $p$ can be interpreted as 50% of the entire volume of target region receiving 100% or less of the prescribed dose level. Ideally, we aim to achieve a solution such that the DVH of the

66      G.J. Lim

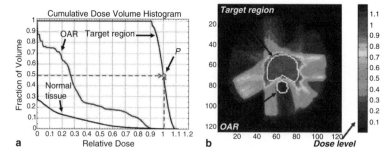

**Fig. 3.5.** Solution quality is typically assessed by (a) DVH and (b) radiation dose plot.

target region is perpendicular at relative dose 1.0 (i.e., all PTV voxels receive the exact amount of the prescribed dose) and the DVHs of the OARs and the normal tissues are perpendicular at relative dose 0 (i.e., they receive no dose at all). A dose plot is another useful visualization tool to measure the solution quality. A series of dose plots can provide positional information of the organs and the dose distribution to verify if a treatment plan meets the treatment goals. Figure 3.5(b) shows an axial slice of the dose distribution. This plot shows that the high-dose-radiation region conforms to the PTV whereas the OAR receives a very low radiation dose. Overall, this solution was designed such that more than 90% of the OAR receives a radiation dose below 30% of the target prescribed dose level.

### 3.4.2 DVH control techniques

Suppose that our aim is to control the DVH such that no more than $\alpha\%$ of the PTV receives $\beta$ $Gy$ or higher. Mathematically such constraints can be written as follows:

$$\frac{\sum_{(i,j,k)\in\mathcal{T}} I_{D(i,j,k)>\beta}}{|\mathcal{T}|} \le \alpha, \tag{3.14}$$

where $I_{D(i,j,k)>\beta}$ is an indicator variable that takes 1 if the total dose on the $voxel(i,j,k)$ receives higher than $\beta$ $Gy$, otherwise 0, and $\alpha \in [0,1]$. Similar constraints can be defined for all organs of interest. Therefore, it is not difficult to see that adding such constraints to any optimization models can lead to an *NP-hard* problem. Therefore, researchers have proposed algorithms that converge to local solutions. Such algorithms include *Simulated Annealing*, *Implicit DVH Control* by optimization model parameters, and *Column Generation Approach*. In this subsection, we will discuss an approach of [33] to implicitly control DVH using optimization control parameters. This approach is very simple and easy to use and can deliver a desired DVH most of the time.

**Implicit DVH control for 3DCRT**

Modelers usually are advised to update the weights $(\lambda_t^+, \lambda_t^-, \lambda_s, \lambda_n)$ to achieve DVH control. However, as pointed out in [14], understanding the relationship between the $\lambda$ values and their intended consequences is far from straightforward. Rather than focusing the tuning efforts on these weights, we can manipulate other parameters in the model; specifically, the PTV control parameters $\theta_U$ and $\theta_L$ and the hot-spot control parameter $\phi$ in (3.6). We describe these techniques with reference to the problem in (3.7).

In this approach, homogeneity is controlled by $\theta_L$ and $\theta_U$, which define the lower and upper bounds on the dose to PTV voxels (we have $\theta_L \leq 1 \leq \theta_U$). The conformity constraints, which require the dose to the normal tissue to be as small as possible, can be implemented by increasing the weight $\lambda_n$ on the normal-tissue term in the objective. Avoidance constraints, which take the form of DVH constraints on the OAR, are implemented via the hot-spot control parameter $\phi$.

**Choice of norms in the objective functions**

One can use infinity-norms to control hot and cold spots in the treatment region, and $L_1$-norm penalty terms are useful for controlling the integral dose over a region. Here we illustrate the effectiveness of using *both* types of terms in the objective by comparing results obtained from an objective with only $L_1$ terms with results for an objective with both $L_1$ and infinity-norm terms. Specifically, we compare the function in (3.6) (with $\lambda_t^+ = \lambda_t^- = \lambda_s = \lambda_n = 1$) against a function in which the infinity norms in the first two terms are replaced by $L_1$ norms, scaled by the cardinality of the target set.

We use data from a pancreatic tumor that includes four critical structures (two kidneys, the spinal cord, and the liver) in the vicinity of the tumor for this illustration. The optimization parameters are set as follows: $\theta_L = 0.95$, $\theta_U = 1.07$, $\phi = 0.2$, and $K = 4$.

As might be expected, (3.6) has better control on the PTV as shown in Figure 3.6; the infinity-norm yielded a stricter enforcement of the constraints on the PTV. The two objective functions can produce a similar solution if the values of $\lambda_t$'s are chosen appropriately. It is noted in [33] that it is easier to choose an appropriate value of $\lambda_t$ for the $L_\infty$ penalty than it is to tune this parameter for the $L_1$ norm. (In the normal and OAR regions, the difference in quality of the solutions obtained from these two alternative objectives was insignificant.)

**DVH control on the PTV**

Here we consider the optimization problem (3.7) with objective function $f(D_\Omega)$ defined by (3.6). We aim to attain homogeneity of the dose on $\mathcal{T}$ without sacrificing too much quality in the dose profile for the normal region

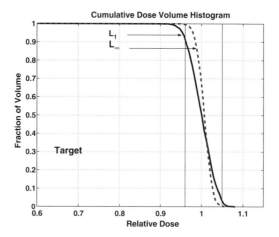

**Fig. 3.6.** Dose-volume Histogram on the PTV.

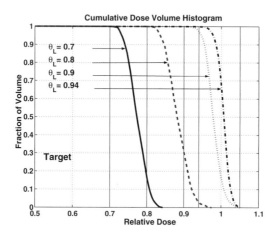

**Fig. 3.7.** DVH control for different choices of parameter $\theta_L$.

and OAR. As discussed above, the key parameters in (3.6) with respect to this goal are $\theta_U$ and $\theta_L$. In this experiment, we fix $\theta_U = 1.07$, and try the values 0.7, 0.8, 0.9, 0.94 for the lower-bound fraction $\theta_L$. Figure 3.7 shows four DVH plots based on the four different values of $\theta_L$. For each value, we find that 100% of the PTV receives more than the desired lower bound $\theta_L$; we manage to completely avoid PTV cold spots in this example. We might expect that larger values of $\theta_L$ (which confine the target dose to a tighter range) would result in a less attractive solution in the OAR and the normal tissue, but it turns out that the loss of treatment quality is not significant. Therefore, the use of $\theta_U$ and $\theta_L$ to implement homogeneity constraints appears to be effective.

## DVH control on the OAR

We show here that the dose to the OAR can be controlled by means of the parameter $\phi$ in (3.6), assuming that the weights $\lambda_t$, $\lambda_s$, and $\lambda_n$ have been fixed appropriately. As shown in Figure 3.8(a), we set $\phi$ to various values in the range $[0, 1]$. For $\phi = 0.5$, almost all of the OAR receives dose less than 50% of the prescribed target dose. Similar results hold for the values $\phi = 0.2$ and $\phi = 0.1$. (For $\phi = 0.1$, about 20% of the OAR receives more than 10% of the prescribed dose, but only about 5% receives more than 20% of the prescribed dose.) Better control of the dose to OAR causes loss of treatment quality on the PTV and the normal tissue, but Figure 3.8 shows that the degradation is not significant.

Note that if our goal is to control hot spots in the OAR rather than the integral dose, we could replace the term $\|(D_{\mathcal{S}} - \phi\Delta e_{\mathcal{S}})_+\|_1$ in the objective (3.6) by its infinity-norm counterpart $\|(D_{\mathcal{S}} - \phi\Delta e_{\mathcal{S}})_+\|_\infty$.

The parameter $\phi$ can be updated on a per-organ basis if the DVH requirement for a given OAR is not satisfied. Furthermore, there can be some conflict between the goals of controlling DVH on target and non-target regions, as the proximity of PTV to normal regions and OAR makes it inevitable that some nontarget voxels will receive high doses. If the PTV dose control is most important (as is usually the case), the control parameters $\theta_L$, $\theta_U$, $\phi$ should be chosen with $(\theta_U - \theta_L)$ small and $\phi$ as a fairly large (but smaller than 1) fraction of the prescribed target dose $\Delta$. However, if the OAR dose control is most important, a smaller value of $\phi$ should be used in conjunction with $L_1$-norm penalties for the OAR terms in the objective. In addition, a larger value of $(\theta_U - \theta_L)$ is appropriate in this case.

## DVH control via wedges

In general, the use of wedges gives more flexibility in achieving adequate coverage of the tumor volume while sparing normal tissues. To show the effect of wedges, we test the optimization models on a different set of data from the one used in the subsections above, from a prostate cancer patient. Figure 3.9 shows DVH graphs obtained for a treatment plan using wedges (3.13) and one using no wedges (3.15). Conventionally, 4 or 6 beams are usually used to treat cases of this type. However, we use three beam angles ($K = 3$) to emphasize the effect of wedges. Figure 3.9(a) shows that a significant improvement on the OAR is achieved by adding wedges. In Figure 3.9(b), we see that there is also a slight improvement in the DVH for the PTV and little difference between the wedge and no-wedge cases for the normal tissue. Note that it took 3 minutes 23 seconds to solve the optimization problem without the wedge whereas it took 5 minutes 45 seconds with the wedge parameter.

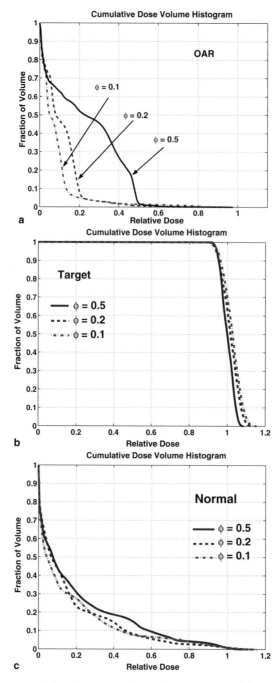

**Fig. 3.8.** DVH control for different values of parameter $\phi$: (a) OAR; (b) PTV; (c) normal.

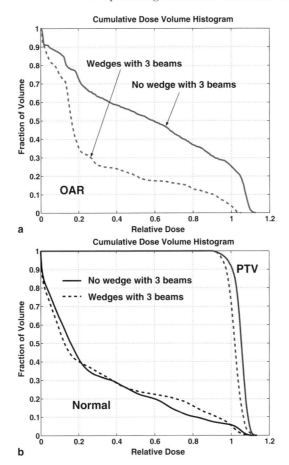

**Fig. 3.9.** Effect of wedges on the DVHs in a prostate cancer case with 3 beam angles: (a) Organ at risk; (b) target and normal.

## 3.5 Solution Time Reduction Techniques

Most optimization models discussed in this chapter involve numerous variables (some of them discrete) and a large amount of data, mostly in the form of dose matrices. Therefore, the optimization problem is time-consuming to construct and solve. In this section, we describe techniques to reduce the solution time. First, optimization problem size can be reduced by carefully selecting voxels that have direct impact on the final solution. For example, a substantial amount of voxels on the normal tissue can be easily removed (by 50% or more). This is because a typical treatment volume contains a vast amount of voxels that may not receive any radiation.

Second, solving the mixed-integer programming (MIP) problems in a clinical setting becomes extremely difficult. Therefore, there is a need to speed up the solution process. One of such methods solves a lower-resolution problem first to identify the most promising beam angles, then considers only these angles in solving the full-resolution problem. Legras et al. [28] describe a multiphase approach in which linear programs are solved to determine the most promising beam angles, with a refined solution being obtained from a final nonlinear program. Lim et al. [33] propose a similar method, *three-phase approach*, where each of the phases involves the solution of MIPs in which the angles are selected explicitly. The phases differ from each other in the reduced sets of voxels that are used as the basis of the problem formulation.

### 3.5.1 Normal tissue voxel reduction

*Preprocessing*

In practice, a fixed number of equi-spaced beam angles are often considered in the treatment planning. Some voxels $\bar{N}$ between two beam angles may never receive any radiation or they may receive an insignificant amount of radiation for the treatment planning, i.e.,

$$\bar{N} := \left\{ (i, j, k) \in \mathcal{N} \mid D_{(i,j,k)} \le \epsilon \right\}.$$

We can simply exclude such voxels from the optimization models *a priori*. For example, when 36 beam angles are considered for a prostate case example, the total number of normal voxels is reduced from 136,000 voxels to 60,000 voxels (about 56% reduction with $\epsilon = 10^{-5}$). Further voxel reduction can be achieved by reducing the grid resolution on the normal tissues.

*Reducing resolution in the normal tissue*

Because the main focus of the planning problem is to deliver enough dose to the PTV while avoiding organs at risk, the dosage to normal regions that are some distance away from the PTV need not be resolved to high precision. It suffices to compute the dose only on a representative subset of these normal-region voxels and use this subset to enforce constraints and to formulate their contribution to the objective.

Given some parameter $\rho$, we define a neighborhood of the PTV as follows:

$$\mathcal{R}_\rho(\mathcal{T}) := \{ (i, j, k) \in \mathcal{N} \mid \text{dist}\,((i, j, k), \mathcal{T}) \le \rho, \},$$

where $\text{dist}\,((i, j, k), \mathcal{T})$ denotes the Euclidean distance of the center of the voxel $(i, j, k)$ to the PTV. We also define a reduced version $\mathcal{N}_1$ of the normal region, consisting only of the voxels $(i, j, k)$ for which $i$, $j$, and $k$ are all even; that is

$$\mathcal{N}_1 := \{ (i, j, k) \in \mathcal{N} \mid i \bmod 2 = j \bmod 2 = k \bmod 2 = 0 \}.$$

Finally, we include in the optimization problem only those voxels that are close to the PTV, or that lie in an OAR; or that lie in the reduced normal region; that is,

$$(i, j, k) \in \mathcal{T} \cup \mathcal{S} \cup \mathcal{R}_\rho(\mathcal{T}) \cup \mathcal{N}_1,$$

(see related work in [2, 34]). Because each of the voxels $(i, j, k) \in \mathcal{N}_1$ effectively represents itself and seven neighboring voxels, the weights applied to the voxels $(i, j, k) \in \mathcal{N}_1$ in the objective functions (3.2) and (3.3) should be increased correspondingly. An appropriate replacement for the term $\|D_\mathcal{N}\|_1 / |\mathcal{N}|$ in (3.2) could then be

$$\frac{\|\mathcal{D}_{\mathcal{R}_\rho(\mathcal{T})}\|_1 + \|\mathcal{D}_{\mathcal{N}_1}\|_1 \, (|\mathcal{N} \setminus \mathcal{R}_\rho(\mathcal{T})| / |\mathcal{N}_1|)}{|\mathcal{N}|}.$$

### 3.5.2 A three-phase approach

This is a multiphase approach that "ramps up" to the solution of the full problem via a sequence of models. Essentially, the models are solved in increasing order of difficulty, with the solution of one model providing a good starting point for the next. The models differ from each other in the selection of voxels included in the formulation and in the number of beam angles allowed.

If the most promising beam angles can be identified in advance, the full problem can be solved with a small number of discrete variables. One simple approach for removing unpromising beam angles is to remove from consideration those that pass directly through any OAR [41]. A more elaborate approach [38] introduces a score function for each candidate angle, based on the ability of that angle to deliver a high dose to the PTV without exceeding the prescribed dose tolerance to OAR or to normal tissue located along its path. Only beam angles with the best scores are included in the model.

These heuristics can reduce solution time appreciably, but their effect on the quality of the final solution cannot be determined *a priori*. We propose instead the following incremental modeling scheme, which obtains a near-optimal solution within a small fraction of the time required to solve the original formulation directly. Our scheme proceeds as follows.

*Phase 1: Selection of promising beam angles*

The aim in this phase is to construct a subset of beam angles $\mathcal{A}_1$ that are likely to appear in the final solution of (3.8). (A similar technique can be applied to (3.13).) We solve a collection of $r$ MIPs, where each MIP is constructed from a reduced set of voxels consisting of the voxels in the PTV, a randomly sampled 10% of the OAR voxels ($\mathcal{S}'$), and the voxels in $\mathcal{R}_\rho(\mathcal{T})$; that is,

$$\Omega_1 = \{\mathcal{T} \cup \mathcal{S}' \cup \mathcal{R}_\rho(\mathcal{T})\}.$$

We define $\mathcal{A}_1$ as the set of all angles $A \in \mathcal{A}$ for which $w_A > 0$ for at least one of these $r$ sampled problems.

*Phase 2: Treatment beam angle determination*

In the next phase, we select $K$ or fewer treatment beam angles from $\mathcal{A}_1$. We solve a version of (3.8) using $\mathcal{A}_1$ in place of $\mathcal{A}$ and a reduced set of voxels defined as follows:

$$\Omega_2 = \{\mathcal{T} \cup \mathcal{S} \cup \mathcal{R}_\rho(\mathcal{T}) \cup \mathcal{N}_1\}.$$

Note that $|\mathcal{A}_1|$ is typically greater than or equal to $K$, so the binary variables play a nontrivial role in this phase.

*Phase 3: Final approximation*

In the final phase, we fix the $K$ beam angles (by fixing $\psi_{A_1} = 1$ for the angles selected in Phase 2 and $\psi_A = 0$ otherwise) and solve the resulting simplified optimization problem over the complete set of voxels. This final approximation typically takes much less time to solve than does the full-scale model, because of both the smaller amount of data (due to fewer beam angles) and the absence of binary variables.

   Although there is no guarantee that this technique will produce the same solution as the original full-scale model (3.8), Lim et al. [33] have found that the quality of its approximate solution is close to optimal based on several numerical experiments.

## 3.6 Case Study

In this section, we present the computational performance of the *three-phase approach* introduced in Section 3.5.2 coupled with the sampling strategy. Our test data is the pancreatic data set introduced in Section 3.4.2. This data consists of 1,244 voxels in the PTV, 69,270 voxels in the OAR, and 747,667 voxels in the normal region.

   The specific optimization model considered in this section is as follows:

$$
\begin{aligned}
\min_{w,\psi} \quad & f(D_\Omega) \quad \text{s.t.} \\
& D_\Omega = \sum_{A \in \mathcal{A}} w_A D_{A,\Omega}, \ \Omega = \mathcal{T} \cup \mathcal{S} \cup \mathcal{N}, \\
& D_\mathcal{T} \le u e_\mathcal{T}, \\
& 0 \le w_A \le M\psi_A, \quad \forall A \in \mathcal{A}, \\
& \sum_{A \in \mathcal{A}} \psi_A \le K, \\
& \psi_A \in \{0,1\}, \qquad \forall A \in \mathcal{A},
\end{aligned}
\tag{3.15}
$$

where $f(D_\Omega)$ is defined by (3.6). Optimization model parameters in (3.15) are as follows: $\theta_L = 0.95$, $\theta_u = 1.07$, $\phi = 0.2$, $K = 4$, $\lambda_t^+ = \lambda_t^- = \lambda_s = \lambda_n = 1$, $u = 1.15$, and $|\mathcal{A}| = 36$. The set of angles $\mathcal{A}$ consists of angles equally spaced by $10°$ in a full $360°$ circumference. Dose matrices are calculated based on a BEV for each beam angle.

*Computational performance of the three-phase approach*

First, the optimization model (3.15) was solved using the full set of voxels. The MIP solver was set to terminate when the gap between upper and lower bound of the objective value falls below 1% (in relative terms). This calculation and the others in this section were performed on a Pentium 4, 1.8 GHz PC running Linux. The problems were modeled in the GAMS modeling language [6], and CPLEX 7.1 was used as the linear programming (LP) and MIP solver.

Figure 3.10 shows changes of upper and lower bounds on the optimal objective value as the iteration number increases, where iteration count is the total number of branch-and-bound nodes explored. Only slight improvements to the upper bound (which represents the best integer solutions found to date) occur after the first 220,000 iterations, and the lower bound of the objective value increases slowly beyond this point. We set the "big M" value to 2 for this experiment. The total computation time of over 112 hours is shown in column I of Table 3.1. This table also shows the effects of the computational speedups described in Section 3.5. In columns II, III, and IV we use the tight bound (3.12) on $w_A$, specialized to the case in which no wedges are used. Note that the constraint $w_A \leq M\psi_A$ in (3.15) was replaced by $w_A \leq (u/\mu_A)\psi_A$. In addition, column III shows the effects of using the reduced-voxel version of the problem discussed in Section 3.5.1. Finally, column IV shows results obtained with the three-phase approach of Section 3.5.2 using $r = 10$.

For purposes of comparing the quality of the computational results obtained with these four approaches, the final objective values are calculated on the full set of voxels. To three significant figures, these values were the same. The next rows in Table 3.1 show the CPU times required (in hours) for each of the four experiments and the savings in comparison with

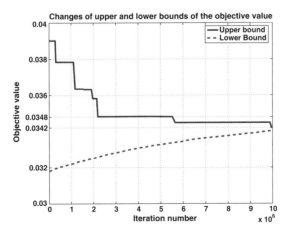

**Fig. 3.10.** Progress of upper and lower bounds during MIP algorithm.

**Table 3.1.** Comparisons among different solution schemes.

|  | I | II | III | IV |
|---|---|---|---|---|
| Approach | Single Solve | Single Solve | Reduced Model | Three-Phase |
| Bound ($M$) | 2 | $u/\mu_A$ | $u/\mu_A$ | $u/\mu_A$ |
| Final objective | 0.0342 | 0.0342 | 0.0342 | 0.0342 |
| Time (hours) | 112.3 | 93.5 | 29.9 | 0.5 |
| Time saved (%) | - | 16.8 | 73.3 | 99.5 |

the time in column I. By comparing columns I and II, we can see that a modest reduction is obtained by using the tighter bound. Column III shows a computational savings of almost three quarters, without degradation of solution quality, when a reduced model is used. Note that the reduced model has 1,244 voxels in the PTV, 14,973 voxels in the OAR, and 96,154 voxels in the normal tissue (i.e., 86% total voxel reduction). The most dramatic savings, however, were for the three-phase scheme, which yielded a savings of 99.5% over the direct solution scheme with no appreciable effect on the quality of the solution.

The difficulty of the full problem arises in large part from the hot-spot and cold-spot control terms. Using looser values for these parameter values speeds up the the solution time considerably.

*Solution quality*

Let us examine the quality of a solution that 3DCRT can produce. We use the three-phase approach in the treatment planning to speed up the solution generation. Wedges are included in the formulation. The specific goals of the treatment plan were defined as follows:

1. Four beam angles.
2. As the highest priority, the target volume should receive a dose of between 95% and 107% of the prescribed dose.
3. 90% of each OAR should receive less than 20% of the target prescribed dose level.
4. The integral dose delivered to the normal tissue should be kept as small as possible.

Figure 3.11 shows DVH plots of this experiment. The homogeneity constraints are satisfied for the PTV; every voxel in the PTV receives between 95% and 107% of the prescribed dose. It is also clear that approximately 90% of each OAR receives at most 20% of the target prescribed dose, as specified; the DVH plot for each OAR passes very close to the point $(0.2, 0.1)$ that corresponds with the aforementioned treatment goal.

Figure 3.12 shows isodose lines on the slices through the treatment region obtained by computed tomography. The PTV is outlined within four isodose lines. The outermost line is 20% isodose line, which encloses a region in

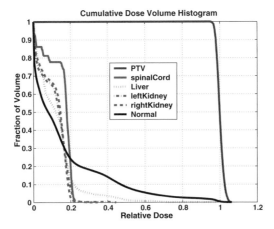

Fig. 3.11. Dose-volume Histogram at optimum.

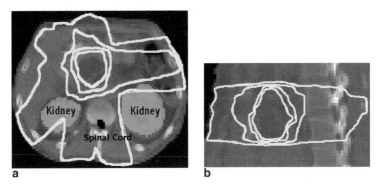

Fig. 3.12. Isodose plots: (a) axial; (b) sagittal. Lines represent 20%, 50%, 80%, and 95% isodoses (20% line outermost).

which the voxels receive a dose of at least 20% of the PTV prescribed dose. Moving inwards toward the PTV, we see 50%, 80%, and 95% isodose lines. Figure 3.12(a) shows an axial slice. The kidneys are outlined as two circles directly below the PTV. As can be seen, the PTV lies well inside the 95% isodose line, and the dose to the organs at risk remains reasonable. Figure 3.12(b) shows a sagittal view of the PTV with those four isodose lines also.

The three-phase approach outlined here has been used in a number of other studies. Examples of the benefits of this procedure on breast, pancreatic, head and neck cases, for example, can be found in [17].

## 3.7 Discussion

Three-dimensional conformal radiation therapy (3DCRT) is widely used in practice due to its simplicity when compared with other commercially available radiation delivery techniques and its ability to generate good-quality solutions. We have introduced optimization models and computational approaches for 3DCRT planning. Most of the models discussed in this chapter are based on mixed integer programming (MIP). The strength of MIP is that it guarantees global optimality. However, it is extremely difficult to solve such optimization models on real patient data because the problem size becomes very large, say over 500,000 constraints, 500,000 variables (including integer variables, especially when the DVH constraints (3.14) are imposed), and it requires significant computational efforts. Therefore, many researchers have developed various solution techniques that can solve the problem quickly and be easily used in a clinical setting. Some of them are based on entirely heuristic methods whereas others are a combination of optimization techniques and heuristic approaches. As we showed in this chapter, the three-phase approach appears to be an excellent choice for 3DCRT planning when it is coupled with a sequential sampling for reducing the beam angle set. More studies need to be done for imposing DVH constraints while solving the problem in a reasonable time.

## References

[1] American Cancer Society. Cancer facts and figures. *www.cancer.org*, 2005.
[2] G.K. Bahr, J.G. Kereiakes, H. Horwitz, R. Finney, J. Galvin, and K. Goode. The method of linear programming applied to radiation treatment planning. *Radiology*, 91:686–693, 1968.
[3] T. Bortfeld and W. Schlegel. Optimization of beam orientations in radiation therapy: some theoretical considerations. *Physics in Medicine and Biology*, 38(2):291–304, 1993.
[4] T.R. Bortfeld, A.L. Boyer, W. Schlegel, D.L. Kahler, and T.J. Waldron. Realization and verification of three-dimensional conformal radiotherapy with modulated fields. *International Journal of Radiation Oncology, Biology and Physics*, 30(4):899–908, 1994.
[5] T.R. Bortfeld, J. Burkelbach, R. Boesecke, and W. Schlegel. Methods of image reconstruction from projections applied to conformation radiotherapy. *Physics in Medicine and Biology*, 25(10):1423–1434, 1990.
[6] A. Brooke, D. Kendrick, and A. Meeraus. *GAMS: A User's Guide*. The Scientific Press, South San Francisco, California, 1988.
[7] G.T.Y. Chen, D.R. Spelbring, C.A. Pelizzari, J.M. Balter, L.C. Myrianthopoulous, S. Vijayakumar, and H. Halpern. The use of beam eye view volumetrics in the selection of noncoplanar radiation portals. *International Journal of Radiation Oncology: Biology, Physics*, 23:153–163, 1992.
[8] Y. Chen, D. Michalski, C. Houser, and J.M. Galvin. A deterministic iterative least-squares algorithm for beam weight optimization in conformal radiotherapy. *Physics in Medicine and Biology.*, 47:1647–1658, 2002.

[9] B.C.J. Cho, W.H. Roa, D. Robinson, and B. Murray. The development of target-eye-view maps for selection of coplanar or noncoplanar beams in conformal radiotherapy treatment planning. *Medical Physics*, 26(11):2367–2372, 1999.

[10] P.S. Cho, S. Lee, R.J. Marks, S. Oh, S. Sutlief, and H. Phillips. Optimization of intensity modulated beams with volume constraints using two methods: cost function minimization and projection onto convex sets. *Medical Physics*, 25(4):435–443, 1998.

[11] A. Cormack and E. Quinto. The mathematics and physics of radiation dose planning using X-rays. *Contemporary Mathematics*, 113:41–55, 1990.

[12] S.M. Crooks, A. Pugachev, C. King, and L. Xing. Examination of the effect of increasing the number of radiation beams on a radiation treatment plan. *Physics in Medicine and Biology*, 47:3485–3501, 2002.

[13] J. Dai, Y. Zhu, and Q. Ji. Optimizing beam weights and wedge filters with the concept of the super-omni wedge. *Medical Physics*, 27(12):2757–2762, 2000.

[14] J. Dennis and I. Das. A closer look at drawbacks of minimizing weighted sums of objectives for Pareto set generation in multicriteria optimization problems. *Structural Optimization*, 14:63–69, 1997.

[15] M. Ehrgott and R. Johnston. Optimisation of beam directions in intensity modulated radiation therapy planning. *OR Spectrum*, 25(2):251–264, 2003.

[16] G. Fang, B. Geiser, and T.R. Mackie. Software system for UW/GE tomotherapy prototype. In D.D. Leavitt and G. Starkshall, editors, *Proceedings of the 12th International Conference on the Use of Computers in Radiation Therapy, Salt Lake City*, pages 332–334, St. Louis, Missouri, 1997. Medical Physics Publishing.

[17] M.C. Ferris, R. Einarsson, Z. Jiang, and D. Shepard. Sampling issues for optimization in radiotherapy. Mathematical programming technical report, Computer Sciences Department, University of Wisconsin, Madison, Wisconsin, 2004.

[18] M.C. Ferris, J.-H. Lim, and D.M. Shepard. Optimization approaches for treatment planning on a Gamma Knife. *SIAM Journal on Optimization*, 13:921–937, 2003.

[19] M.C. Ferris, J.-H. Lim, and D.M. Shepard. Radiosurgery treatment planning via nonlinear programming. *Annals of Operations Research*, 119:247–260, 2003.

[20] M. Goitein, M. Abrams, S. Rowell, H. Pollari, and J. Wiles. Multi-dimensional treatment planning: II. Beam's eye-view, back projection, and projection through CT sections. *International Journal of Radiation Oncology: Biology, Physics*, 9:789–797, 1983.

[21] P. Gokhale, E.M.A. Hussein, and N. Kulkarni. Determination of beam orientation in radiotherapy planning. *Medical Physics*, 21(3):393–400, 1994.

[22] O.C.L. Haas, K.J. Burnham, and J.A. Mills. Optimization of beam orientation in radiotherapy using planar geometry. *Physics in Medicine and Biology*, 43(8):2179–2193, 1998.

[23] H.W. Hamacher and K.-H. Küfer. Inverse radiation therapy planning — a multiple objective optimization approach. *Discrete Applied Mathematics*, 118:145–161, 2002.

[24] Intensity Modulated Radiation Therapy Collaborative Working Group. Intensity-modulated radiotherapy: Current status and issues of interest. *International Journal of Radiation Oncology: Biology, Physics*, 51(4):880–914, 2001.

[25] T.J. Jordan and P.C. Williams. The design and performance characteristics of a multileaf collimator. *Physics in Medicine and Biology*, 39:231–251, 1994.

[26] J.M. Kapatoes, G.H. Olivera, J.P. Balog, H. Keller, P.J. Reckwerdt, and T.R. Mackie. On the accuracy and effectiveness of dose reconstruction for tomotherapy. *Physics in Medicine and Biology*, 46:943–966, 2001.

[27] E.K. Lee, T. Fox, and I. Crocker. Beam geometry and intensity map optimization in intensity-modulated radiation therapy. *International Journal of Radiation Oncology, Biology and Physics*, 64(1): 301–320, 2006.

[28] J. Legras, B. Legras, and J. Lambert. Software for linear and non-linear optimization in external radiotherapy. *Computer Programs in Biomedicine*, 15:233–242, 1982.

[29] J.G. Li, A.L. Boyer, and L. Xing. Clinical implementation of wedge filter optimization in three-dimensional radiotherapy treatment planning. *Radiotherapy and Oncology*, 53:257–264, 1999.

[30] G.J. Lim, J. Choi, and R. Mohan. Iterative solution methods for beam angle and fluence map optimization in IMRT. *OR Spectrum*, 30(2):289–309, 2008.

[31] J.-H. Lim. *Optimization in Radiation Treatment Planning*. PhD thesis, University of Wisconsin, Madison, Wisconsin, December 2002.

[32] J.-H. Lim, M.C. Ferris, and D.M. Shepard. Optimization tools for radiation treatment planning in matlab. In M.L. Brandeau, F. Sainfort, and W.P. Pierskalla, editors, *Operations Research and Health Care: A Handbook of Methods and Applications*, pages 775–806. Kluwer Academic Publishers, Boston, 2004.

[33] J.-H. Lim, M.C. Ferris, S.J. Wright, D.M. Shepard, and M.A. Earl. An optimization framework for conformal radiation treatment planning. INFORMS Journal on Computing, 19(3):366–380, 2007.

[34] S. Morrill, I. Rosen, R. Lane, and J. Belli. The influence of dose constraint point placement on optimized radiation therapy treatment planning. *International Journal of Radiation Oncology, Biology and Physics*, 19:129–141, 1990.

[35] L.C. Myrianthopoulos, G.T.Y. Chen, S. Vijayakumar, H. Halpern, D.R. Spelbring, and C.A. Pelizzari. Beams eye view volumetrics — an aid in rapid treatment plan development and evaluation. *International Journal of Radiation Oncology: Biology, Physics*, 23:367–375, 1992.

[36] G.L. Nemhauser and L.A. Wolsey. *Integer and Combinatorial Optimization*. John Wiley & Sons, 1988.

[37] F. Preciado-Walters, R. Rardin, M. Langer, and V. Thai. A coupled column generation, mixed-integer approach to optimal planning of intensity modulated radiation therapy for cancer. *Mathematical Programming*, 101:319–338, 2004.

[38] A. Pugachev and L. Xing. Pseudo beam's-eye-view as applied to beam orientation selection in intensity-modulated radiation therapy. *International Journal of Radiation Oncology, Biology, Physics*, 51(5):1361–1370, 2001.

[39] A. Pugachev and L. Xing. Incorporating prior knowledge into beam orientation optimization in IMRT. *International Journal of Radiation Oncology, Biology and Physics*, 54:1565–1574, 2002.

[40] A. Pugachev and L. Xing. Computer-assisted selection of coplanar beam orientations in intensity-modulated radiation therapy. *Physics in Medicine and Biology*, 46(9):2467–2476, 2001.

[41] C.G. Rowbottom, V.S. Khoo, and S. Webb. Simultaneous optimization of beam orientations and beam weights in conformal radiotherapy. *Medical Physics*, 28(8):1696–1702, 2001.

[42] C.G. Rowbottom, S. Webb, and M. Oldham. Improvements in prostate radiotherapy from the customization of beam directions. *Medical Physics*, 25:1171–1179, 1998.

[43] C.G. Rowbottom, S. Webb, and M. Oldham. Beam-orientation customization using an artificial neural network. *Physics in Medicine and Biology*, 44:2251–2262, 1999.

[44] S. Shalev, D. Viggars, M. Carey, and P. Hahn. The objective evaluation of alternative treatment plans. II. Score functions. *International Journal of Radiation Oncology: Biology, Physics*, 20(5):1067–1073, 1991.

[45] G.W. Sherouse. A mathematical basis for selection of wedge angle and orientation. *Medical Physics*, 20(4):1211–1218, 1993.

[46] S. Soderstrom, A. Gustafsson, and A. Brahme. Few-field radiation-therapy optimization in the phase-space of complication-free tumor central. *International Journal of Imaging Systems and Technology*, 6(1):91–103, 1995.

[47] J. Tervo and P. Kolmonen. A model for the control of a multileaf collimator in radiation therapy treatment planning. *Inverse Problems*, 16:1875–1895, 2000.

[48] S. Webb. Optimisation of conformal radiotherapy dose distributions by simulated annealing. *Physics in Medicine and Biology*, 34(10):1349–1370, 1989.

[49] S. Webb. Optimization by simulated annealing of three-dimensional, conformal treatment planning for radiation fields defined by a multileaf collimator. *Physics in Medicine and Biology*, 36(9):1201–1226, 1991.

[50] S. Webb. Optimization by simulated annealing of three-dimensional, conformal treatment planning for radiation fields defined by a multileaf collimator: II. Inclusion of the two-dimensional modulation of the x-ray intensity. *Physics in Medicine and Biology*, 37(8):1992, 1992.

[51] S. Webb. *The Physics of Conformal Radiotherapy: Advances in Technology*. Taylor & Francis, London, U.K., 1997.

[52] S. Webb. Configuration options for intensity-modulated radiation therapy using multiple static fields shaped by a multileaf collimator. *Physics in Medicine and Biology*, 43:241–260, 1998.

[53] X. Wu and Y. Zhu. A global optimization method for three-dimensional conformal radiotherapy treatment planning. *Physics in Medicine and Biology*, 46:109–119, 2001.

[54] P. Xia and L.J. Verhey. Multileaf collimator leaf sequencing algorithm for intensity modulated beams with multiple static segments. *Medical Physics*, 25(8):1424–1434, 1998.

[55] Y. Xiao, Y. Censor, D. Michalski, and J.M. Galvin. The least-intensity feasible solution for aperture-based inverse planning in radiation therapy. *Annals of Operations Research*, 119:183–203, 2003.

[56] L. Xing, R.J. Hamilton, C. Pelizzari, and G.T.Y. Chen. A three-dimensional algorithm for optimizing beam weights and wedge filters. *Medical Physics*, 25(10):1858–1865, 1998.

[57] L. Xing, C. Pelizzari, F.T. Kuchnir, and G.T.Y. Chen. Optimization of relative weights and wedge angles in treatment planning. *Medical Physics*, 24(2):215–221, 1997.

# 4

# Continuous Optimization of Beamlet Intensities for Intensity Modulated Photon and Proton Radiotherapy

Rembert Reemtsen[1] and Markus Alber[2]

[1] Institut für Mathematik, Brandenburgische Technische Universität Cottbus, Universitätsplatz 3–4, D-03044 Cottbus, Germany
reemtsen@math.tu-cottbus.de

[2] Radioonkologische Klinik, Universitätsklinikum Tübingen, Hoppe-Seyler-Strasse 3, D-72076 Tübingen, Germany
msalber@med.uni-tuebingen.de

**Abstract.** Inverse approaches and, in particular, intensity modulated radiotherapy (IMRT), in combination with the development of new technologies such as multileaf collimators (MLCs), have enabled new potentialities of radiotherapy for cancer treatment. The main mathematical tool needed in this connection is numerical optimization. In this article, the continuous optimization approaches that have been proposed for the computation of optimal or locally optimal beam and beamlet intensities respectively are surveyed, and an approach of the authors is described in detail. Also, the use of optimization in connection with intensity modulated proton therapy (IMPT) and, in particular, with the IMPT spot-scanning technique is discussed.

## 4.1 Introduction

### 4.1.1 Radiotherapy treatment planning

*Radiation therapy* is an essential medical tool for cancer treatment. About $500,000$ patients in the United States and $150,000$ patients in Germany are treated yearly by radiation therapy. The hazard with radiotherapy, however, is that it does not only destroy tumor cells, but similarly also affects healthy tissue. Therefore, based on the images of *computed tomography*, for each patient a compromise has to be found between the two conflicting goals: to deposit a sufficiently high dose into the *planning target volume(s)* (PTVs), i.e., the tumor(s) and/or the possibly involved tissue, and to simultaneously spare, as much as possible, the *organs at risk* (OARs) and the other healthy tissue. As a consequence, radiotherapy treatment planning involves the selection of several suitable directions for the incident beams and the determination

P.M. Pardalos, H.E. Romeijn (eds.), *Handbook of Optimization in Medicine*, 83
Springer Optimization and Its Applications 26, DOI: 10.1007/978-0-387-09770-1_4,
© Springer Science+Business Media LLC 2009

of beam intensities or, if these are modulated, beamlet intensities so that, through superposition of the doses delivered by the single beams or beamlets respectively, a desired dose is deposited in the PTVs and simultaneously no critical doses are administered to the normal-tissue volumes. (Introductions into the field are found, e.g., in [16, 24, 54, 62, 108, 124, 125].)

Conventionally in radiotherapy, the radiation is produced by beams of highly energetic photons delivered by a *linear accelerator*. The treatment itself is standardized in most hospitals. Depending on the position and the type of the tumor(s), the number of *radiation fields* or *beams* respectively is prescribed (typically between 2 and 5), the field or beam angles are essentially predetermined, and the beam intensities are homogeneous or have a constant gradient. The radiation fields are rectangular, and often custom-made apertures or *multileaf collimators* (MLCs) are used to cover parts of the fields and thereby protect portions of the patient's body. (A MLC consists of typically 25–60 tungsten slabs that can be shifted from each of two opposite sides by computer control.)

In case of such a conventional approach, an individual treatment plan is normally obtained by a trial-and-error procedure, where the radiation effects of a few differing arrangements are considered with respect to their dose distributions. In contrast with this *forward approach*, an *inverse approach* starts from the definition of treatment goals, defined by requirements on the doses for the PTVs and the OARs, and it results in the problem of finding beam or beamlet intensities for a certain number of well-positioned radiation fields such that the delivered doses meet these requirements or are close to them (e.g., [18, 19, 27]). Hence, in an inverse approach, restrictions on doses are often established in form of inequalities or equalities, and goals are described by one or, as in case of a multicriteria approach, by several objective functions. Thus an inverse approach naturally is connected with numerical optimization.

Many articles over the past 20 years have dealt with the improvement of conventional arrangements by inverse approaches, and the work in this direction still continues. Simultaneously and starting with the seminal works of Brahme et al. [23, 26] and planning techniques by Censor et al. [11, 31, 32], research on the more complex inverse approach of *intensity modulated radiation therapy* (IMRT) emerged and has attracted a rapidly growing interest. This approach, first employed clinically around 1994, "is regarded by many in the field as a quantum leap forward in treatment delivery capability" [62]. In IMRT, the photon beams are split into thousands of *beamlets* or *pencil beams*, which enables the creation of much more sophisticated and precise dose distributions and thereby renders possible the treatment of cancer patients by radiotherapy who could not be treated adequately before. Mathematically, IMRT leads to large-scale optimization problems.

### 4.1.2 Optimization models

For the optimization of an IMRT treatment plan, a variety of parameters may be considered. Besides the *beamlet weights* determining the beamlet intensities, the main degrees of freedom are the number of beams used, the beam angles, and parameters connected with the realization of an intensity profile by a MLC. Ideally, all of these parameters should enter an optimization model, and experiments in this direction also have been performed. However, in particular, if integer variables are included in a model to find, for example, an optimal set of $r$ beams from a given set of $s \geq r$ beams with prescribed angles (e.g., [45, 47, 79]), the size of the resulting problems and the state-of-the-art of *mixed-integer programming* exclude nonlinear functions in the model. Therefore, currently, the optimization of IMRT treatment plans requires the *a priori* decision whether integer variables are allowed in the model, in which case only linear functions should be used for the beamlet weights optimization, or whether certain parameters as beam angles and beam directions are fixed so that nonlinear constraints can be explored.

The relevance of biological treatment goals for radiotherapy, leading to nonlinear constraints corresponding with *equivalent uniform dose* (EUD) or *partial volume* (PV) *constraints*, has been generally acknowledged during the past years (see Section 4.3.8). On the other hand, nonlinear programming is associated with the risk that local minimizers are computed, having an objective function value far away from the global minimum value. For this reason, some researchers have substituted or approximated intrinsically nonlinear conditions by (a typically much larger number of) linear or convex constraints. For example, the nonlinear convex EUD function of [93] has been replaced by an expression that results in a large number of linear constraints ([121]), or a convex objective function defined for each volume element of the irradiated volume has been approximated by a finite number of linear constraints ([103]). Several authors have implemented *dose-volume constraints* (see Section 4.3.8) by means of binary variables and have used *mixed-integer linear programming* (MILP) in order to take partial volume effects into account, which naturally are described by nonconvex functions (e.g., [13, 76, 79, 100]). Such treatment of dose-volume effects, however, can lead to tens or hundreds of thousands of additional binary variables, which increases the complexity of the problem considerably.

In this connection, it is important to note that the process of finding an optimal IMRT treatment plan cannot be fully automated, as it requires the participation of an expert, who has to set up the treatment goals, to evaluate the computed treatment plan, and to modify, quite commonly, the original goals, assessing simultaneously the related risks for the patient. By their experience, experts often have a good feeling for reasonable beam numbers and beam directions in a particular case. Also the avoidance of selecting beam angles in an optimal way and hence of binary variables, as described above, may be partially compensated for by the use of a slightly increased number of beams ([116]).

Therefore we prescribe the number of beams and beam angles (as demand most clinical software packages) and give preference to the improvement of the model for beamlet weights optimization in the framework of continuous optimization by respecting biological considerations ([10]). Our model may be supplemented by a heuristic procedure to "optimize" the beam angles ([85]). Also, in order to translate an obtained intensity profile into a sequence of MLC openings, an iterative procedure can be executed that normally leads only to small loss concerning the optimality of the goals ([8]).

Other, partly MILP, optimization procedures for *leaf sequencing* were proposed in ([13, 15, 42, 46, 58, 70, 71, 72, 74, 77, 84, 102, 110, 111, 118, 130]). Most procedures require the solution of a beamlet weights optimization program in a first phase, so that this program should yield relatively smooth weight profiles that can be converted into MLC openings efficiently. In this connection, proper measures have recently been studied to cope with the ill-posedness of beamlet weights optimization problems and, thereby, to avoid the computation of strongly oscillating weight profiles ([5, 7, 29, 39, 97]).

The problem of finding an IMRT treatment plan necessitates compromises between competing goals that may be rated differently. Accordingly, some authors recently have investigated this problem in the framework of *multicriteria optimization*, with the aim to produce a set of treatment plans that relate to different weightings of the objectives ([22, 41, 57, 75, 104]). However, multicriteria optimization requires a multiple of the computation time needed for ordinary optimization of similar type, so that, at the current stage of development of the field, clinical practicability forces the number of objectives to be small and the involved functions to be linear or convex.

The authors handle the IMRT treatment planning problem as an ordinary optimization problem in continuous variables, and they combine the solution of the problem with a sensitivity analysis in order to detect those constraints of the problem for which small changes in bounds have the largest effect on the EUD in the target. In this way normally only one or few constraints of the problem have to be changed if original treatment goals have to be relaxed.

The problems themselves resulting from our approach are convex or nonconvex optimization problems, which have the beamlet weights as variables and typically contain only 10–25 constraints apart from the simple bounds for the beamlet weights. This is distinguished, for example, from linear models that include at least one inequality constraint and, by that, one additional slack variable for each volume element. Moreover, it is shown in this paper that our optimization model and algorithm for its solution, both presented in [10], can also be extended to the much larger problems of *intensity modulated proton therapy* (IMPT) treatment planning and can yield optimal solutions for these within a few minutes of computing times.

### 4.1.3 Organization of the chapter

The general tools and our notation for the description of IMRT treatment planning problems are given in Section 4.2. In Section 4.3, we review the most prominent approaches to continuous beamlet weights optimization for inverse treatment planning, where we distinguish between linear programming, linear approximation, piecewise linear approximation, and multicriteria models and models that include or attempt to simulate nonlinear conditions on the doses representing probability measures, partial volume or equivalent uniform dose requirements. The description of the latter models encompasses a detailed discussion of our approach from [10]. A sensitivity analysis used in combination with this is presented in Section 4.4. Finally, in Section 4.5, we consider treatment planning in connection with the 3D *spot scanning technique* of IMPT. The paper concludes in Section 4.6 with a clinical case example for both IMRT and IMPT, for which the optimization was performed with the barrier-penalty multiplier method from [10].

## 4.2 Preliminaries

Radiotherapy and IMRT in particular require the selection of a number $p$ of *radiation fields*, also called *incident beams*, and, associated with that, $p$ *beam angles*, where for practical reasons normally $p$ is a number between 3 and 6 and is smaller than 12. As we have argued in the introduction, we assume here that the fields and beam angles are predetermined either by experience of an expert, referring to the type and position of the tumor in relation to surrounding OARs, or by trial-and-error. Clearly, for a fixed number of beams, the continuous optimization of both doses and beam directions would be desirable but is impeded by the computationally expensive dependence of the dose absorbed in the patient's body on the orientation of the radiation fields and by the combinatorial nature of the problem.

The IMRT treatment problem is fully discretized according to techniques that have been suggested first in ([11, 31, 32]). Each of the $p$ *radiation fields* is a 2D region with a polygonal boundary, normally originating from a projection of the PTVs onto a plane at the position of the collimator. Each (remaining) field $j$ is partitioned into $n_j$ rectangular *field elements* of equal size, also denoted as *bixels* (see Figure 4.1), where typically the number $n_j$

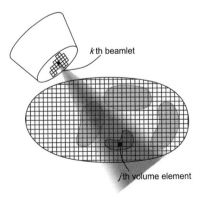

*k*th beamlet

*j*th volume element

**Fig. 4.1.** Discretization of radiation field and body.

varies between 100 and 2,000. Accordingly, each of the $p$ beams is divided into $n_j$ *beamlets* or *pencil beams* respectively so that the total number of beamlets over all fields amounts to $n = \sum_{j=1}^{p} n_j$.

The portion of the human body to be irradiated is considered to be divided into $q$ not necessarily disjoint $3D$ *volumes* that represent the PTVs and the regions of normal tissue as, e.g., OARs. Furthermore, the $\ell$th of these $q$ volumes is partitioned into $m_\ell$ cubic *volume elements* or *voxels* of equal size, having a side length of normally $\geq 2$ mm. Typically $q$ is smaller than 15, and the total number $m = \sum_{\ell=1}^{q} m_\ell$ of voxels is of order $10^5$ or $10^6$. We number all volume elements consecutively from 1 to $m$ and let $V_\ell$ ($\ell = 1, \ldots, q$) be the index set of all elements belonging to the $\ell$th volume, having a cardinality $|V_\ell|$. For convenience, we also identify $V_\ell$ with the $\ell$th volume itself.

Let now $d_{jk} \geq 0$ be the dose deposited in the $j$th volume element by the $k$-th beamlet at unit beam intensity and let $D = (d_{jk})$ be the resulting $m \times n$ *dose matrix*. This matrix $D$ needs to be determined for each individual patient, which can be done by a Monte Carlo simulation of the radiation transport through the patient ([78]) or with sufficient accuracy by a method that adapts a dose distribution computed for a homogeneous medium, so-called *pencil beam kernels*, to the geometry and density distribution of the patient ([2]). The dose matrix $D$ is sparse because the $k$th beamlet predominantly affects volume elements only in proximity of its line of propagation. Typically, at a reasonable cutoff for the minimal dose, less than 3–8% of the coefficients of $D$ are nonzero so that $D$ can be stored in a closed form.

For the optimization process, the matrix $D$ is assumed to be known. Then the goal of IMRT is to find, for each beamlet and according to the optimization goals of the respective model, a suitable nonnegative *beamlet weight* defining its radiation intensity. The total dose absorbed by the $j$th volume element is linearly dependent on the vector $\phi \geq 0$ of beamlet weights, $\phi = (\phi_1, \ldots, \phi_n)^\top$, and is given by

$$D_j^\top \phi = \sum_{k=1}^{n} d_{jk}\phi_k \geq 0, \qquad (4.1)$$

where $D_j^\top$ contains the entries of the $j$th row of the dose matrix $D$. The $n$ beamlet weights $\phi_k \geq 0$ are unknowns of an optimization model for IMRT treatment planning.

The technical realization of a set of *beamlet weights* or an *intensity profile*, which nowadays is typically performed by a MLC, is a difficult problem in itself which is not discussed here (see the references given in Section 4.1.2). A MLC is part of the treatment machine and can expose a polygonal geometry formed by automatically shifted tungsten leaves. Hence, following the dose optimization, an intensity pattern has to be found for each field, which is close to the optimal profile determined by the optimization process and which can be generated by a relatively small number (typically 10–30) of MLC openings (see Figure 4.2). Clearly, the *a priori* inclusion of a comprehensive set of constraints into an optimization model, which would guarantee that the optimal

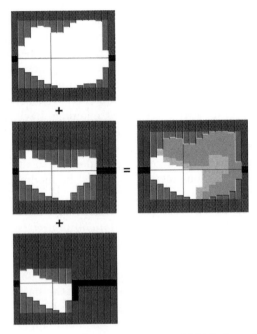

**Fig. 4.2.** Intensity modulation by superposition of MLC-shaped fields. Dark gray bars from top and bottom symbolize the tungsten leaves, white area in the centers symbolizes the exposed area of the field. On the right, total intensity levels are symbolized by gray values.

dose obtained by the model is realizable by a MLC, would be desirable (see [37, 107, 117, 119] for approaches in this direction). However, for the optimization model used by the authors and by a heuristic reoptimization of the MLC field shapes, the loss in dose quality caused by the translation of an optimal intensity profile to deliverable MLC field segments can be kept small and amounts to 0–5% of the target EUD, depending on the case complexity ([8]).

## 4.3 Optimization Models for IMRT Treatment Planning

### 4.3.1 Introduction

In this section, we discuss the main continuous optimization models related to inverse approaches for radiotherapy treatment planning. Considering the huge number of papers existing in this regard, we do not intend here to provide a complete review on the topic, but rather to survey the most prevalent ideas and problem types and point out their differences in terms of gains and drawbacks. In addition we present our own approach in detail.

In our review, we do not distinguish between inverse radiotherapy treatment planning with and without intensity modulation, as the models used for intensity optimization of unmodulated beams have likewise or similarly been applied to IMRT or could in principle be applied to that. Often, and naturally before IMRT had been invented, the number of incident beams and hence continuous variables in the optimization problem were generally less than 12 and rarely more than 36. In contrast with that, the number of beamlets and hence continuous variables for IMRT typically amounts to 3,000–8,000, whereas an optimization problem for IMPT treatment planning, having the same mathematical appearance as a model for IMRT, may possess 40,000 variables and more (see Section 4.5). Also, for IMRT and IMPT, the resolution in regard to volume elements has to be increased considerably so that the responses of tissues to the inhomogeneous intensities caused by modulation of the beams can be traced appropriately.

Three special treatment techniques of radiotherapy are *tomotherapy, intensity modulated arc therapy* (IMAT), and *radiosurgery* (see [16, 108], [135, 136], and [49, 62], respectively, for descriptions). Tomotherapy employs a specifically designed treatment machine that can deliver a narrow intensity modulated fan beam from a large number of fixed beam directions. While the radiation source rotates around the patient, the patient couch position is stepped forward so that the radiation source follows a helical trajectory relative to the patient. IMAT, on the other hand, employs a standard linear accelerator to deliver a constant beam while the radiation source rotates around the patient. During the delivery of such an arc, the field shape can change by virtue of a MLC. By repeating the rotation several times with various field shapes, a modulated fluence profile per arc angle results. Finally, radiosurgery is a quite specialized treatment technique, which has been primarily designed to destroy malignancies in the brain. The basic mathematical ideas used for treatment planning in case of these three techniques are similar to those for IMRT and are therefore included in our discussion (see, e.g., [44, 48, 49, 80, 109] for some recent developments concerning these topics).

For $x \in \mathbb{R}^r$, we employ the $\ell_p$-norm

$$\|x\|_p = \left( \sum_{i=1}^r |x_i|^p \right)^{1/p} \quad (1 \le p < \infty), \qquad \|x\|_\infty = \max_{i=1,\ldots,r} |x_i| \,,$$

where the dimension $r$ of the space is assumed to be clear from the circumstances. The nonnegative vector $[x]_+$ is defined by

$$[x]_+ = (\max\{0,\, x_i\})_{i=1,\ldots,r} \,,$$

and $e \in \mathbb{R}^r$ is the vector with all elements being 1. Furthermore, the $|V_\ell| \times n$ matrix with lines $D_j^\top$, $j \in V_\ell$, for some $\ell$ is denoted by $D_{(\ell)}$. Concerning standard concepts and algorithms of optimization used in our presentation, we refer to textbooks on optimization as, e.g., [14, 51, 95].

*Remark 1.* If an optimization problem in $\mathbb{R}^n$ is a convex problem, each stationary point, i.e., each point that satisfies the first-order necessary optimality conditions of the problem, is a local minimizer, and each local minimizer also is a global minimizer, i.e., a "solution" of the problem. Furthermore, the solution set of a convex problem is a convex set, and therefore a convex problem either has no solution (consider the problem $\min_{x \in \mathbb{R}} e^x$), a unique solution, or infinitely many solutions (e.g., [51]).

For standard descent algorithms in optimization, convergence is proven only, under suitable assumptions, to a stationary point, and for different starting points such an algorithm may converge to distinct stationary points, if there exists more than one such point. Thus, in case of a convex problem and under proper assumptions, a descent algorithm always converges to a global minimizer, but applied to a nonconvex problem it may get trapped in a point that is not a local minimizer as, e.g., a saddle point if the problem is unconstrained. The latter event has to be taken into account in case of nonconvex radiotherapy treatment planning models.

There is some confusion in the area concerning convexity. For example, a local minimizer of a strongly quasiconvex function (see the much cited paper [43]), if such exists, also is its unique global minimizer, but a strongly quasiconvex function can have saddle points ([12, p. 113]) and no local minimizer at all. (Consider the strongly quasiconvex functions $f(x) = x^3$ and $f(x) = x^3(x + 1)$.) Moreover, the existence of nonglobal local minimizers sometimes is erroneously thought to be connected with the use of certain gradient algorithms. Thus, an algorithm that finds "multiple solutions" with distinct objective function values in case of a convex optimization problem simply does not properly converge.

### 4.3.2 Linear programming models with dose bound constraints

Surveys on inverse approaches are found, e.g., in [54, 62, 108, 124, 125]. In the early approaches, one treatment goal for each volume $V_\ell$ was to not exceed an upper dose bound of $\Delta_\ell^u$ Gy, i.e., to satisfy the linear constraints

$$D_j^\top \phi \leq \Delta_\ell^u, \quad j \in V_\ell. \tag{4.2}$$

Typically, for each PTV $V_\ell$, this was combined with the requirement to not fall short of a lower dose bound of $\Delta_\ell^l$ Gy with $\Delta_\ell^l < \Delta_\ell^u$ and hence to fulfill the constraints

$$D_j^\top \phi \geq \Delta_\ell^l, \quad j \in V_\ell. \tag{4.3}$$

The purpose of such lower bound constraints is to guarantee a specified dose and, in combination with upper bounds as in (4.2), a nearly homogeneous dose in the targets. Sometimes, for the sake of a uniform description for all volumes, a lower dose bound is added also for each normal-tissue volume, where this can be set to zero. In this way, a large system of linear inequalities

$$A_\ell \phi \leq b_\ell \quad (\ell = 1, \ldots, q), \qquad \phi \geq 0, \tag{4.4}$$

is obtained, with $A_\ell \in \mathbb{R}^{s_\ell \times n}$, $b_\ell \in \mathbb{R}^{s_\ell}$, $\phi \in \mathbb{R}^n$, and $s_1 + \cdots + s_q \geq m \gg n$, where different actions described in the following have been taken to deal with such system.

Some authors have been of the opinion that each vector $\phi$ of the (often relatively small) feasible set of the system in (4.4) would be of equal clinical value and have proposed algorithms to find such a vector, where special measures have to be considered in case the feasible set is empty (see, e.g., [34] and, for a more recent development, [131]). A feasible point of a linear system of inequalities can be computed by phase 1 of the *simplex algorithm*. Moreover, the inequalities satisfied with equality or almost equality for a solution of phase 1 give information about the constraints that should be relaxed in case of infeasibility.

Most authors, however, sought a feasible vector for the system in (4.4) that minimizes or maximizes some objective function, where different views have been taken concerning a suitable goal to be reached. For that we let $\mathcal{P} \subseteq \{1, \ldots, q\}$ be some index set, $\Pi = \sum_{\ell \in \mathcal{P}} |V_\ell|$ be the total number of elements in volumes $V_\ell$ ($\ell \in \mathcal{P}$), and

$$f_\mathcal{P}(\phi) = \frac{1}{\Pi} \sum_{\ell \in \mathcal{P}} \sum_{j \in V_\ell} D_j^\top \phi = \frac{1}{\Pi} \sum_{\ell \in \mathcal{P}} \left\| D_{(\ell)} \phi \right\|_1 \tag{4.5}$$

be the *integral dose* over these volumes. Then, if $\mathcal{Q} = \{1, \ldots, q\}$ is the index set of all volumes, $\mathcal{N}$ that of all normal-tissue volumes including OARs, and $\mathcal{T}$ that of all PTVs, typical goals have been the minimization of $f_\mathcal{Q}(\phi)$ or $f_\mathcal{N}(\phi)$ and the maximization of $f_\mathcal{T}(\phi)$ or $f_\mathcal{T}(\phi) - f_\mathcal{N}(\phi)$ (see [105] for the latter). In these functions, the factor related to $1/\Pi$ can be ignored for the minimization, and each sum $\left\| D_{(\ell)} \phi \right\|_1$ may be weighted differently, according to its presumed importance.

In addition, maximization of the minimal dose in the PTVs was suggested (e.g., [81, 86, 100]), which is equivalent to maximization of the variable $\tau$ over all vectors $(\phi, \tau)$ under the additional constraints $\tau \leq D_j^\top \phi$ ($j \in V_\ell$, $\ell \in \mathcal{T}$). Finally, the minimization of a linear combination of some linear functions has been suggested, including the integral dose over all volumes and the maximum beamlet weight ([59]). The latter goal can be expressed by a new variable $\phi_{\max}$ and the inclusion of the additional constraints

$$\phi_k \leq \phi_{\max} \quad (k = 1, \ldots, n). \tag{4.6}$$

Also, in [81], an objective function containing linear penalties on critical beamlet weights was investigated.

The resulting *linear programming* (LP) problems include at least one inequality constraint for each voxel. Therefore, in case of IMRT, these problems comprise tens or hundreds of thousands of inequality constraints and as many slack variables in addition to the $n$ unknown beamlet weights, as most codes start from the standard form of a LP problem, which requires the introduction of such variables. If new variables $d_j$ with

$$d_j = D_j^\top \phi \quad (j = 1, \ldots, m) \tag{4.7}$$

are introduced in the problem for the doses and the inequalities are written in terms of these $d_j$'s as several authors do, the problem is even enlarged by $m$ variables and equality constraints. Thus LP treatment planning problems typically are large-scale problems in regard to the number of variables and constraints, even for conventional radiotherapy with unmodulated beams. Such problems usually have been solved by the simplex algorithm and, more recently, also by software packages as CPLEX (e.g., [86, 103, 108]), which includes a LP barrier *interior-point method* in addition to the simplex algorithm.

### 4.3.3 Linear programming models with elastic constraints

In clinical routine, initial treatment goals often turn out to be too restrictive. Consequently, a natural shortcoming of any optimization problem including both upper and lower dose bounds is that the related inequalities may be inconsistent. For this reason, *elastic constraints* have been introduced, which include parameters that allow some over- and underdosage of volume elements and thereby avoid the possible infeasibility of the inequality system.

In the quite general framework of [61] and [62], a system $A_\ell \phi \leq b_\ell$ is replaced, for example, by

$$A_\ell \phi \leq b_\ell + \theta_\ell u_\ell, \tag{4.8}$$

where $u_\ell > 0$ is a given vector from $\mathbb{R}^{s_\ell}$, whose components weight the allowed amount of violation for the constraints in (4.8), and $\theta_\ell \geq 0$ is a variable that controls the (weighted) maximum violation of this system. (More generally $u_\ell \theta_\ell$ can be a matrix-vector product, for example with $u_\ell = I$, for an unknown vector $\theta_\ell \geq 0$.) Then, e.g., the objective function $\sum_{\ell=1}^{q} w_\ell \theta_\ell$ with *importance weights* $w_\ell > 0$ is minimized with respect to $(\phi, \theta_\ell) \geq 0$ and the constraints in (4.8) for all $\ell$. Note that the feasible set of such a problem is nonempty because, for given $\phi$, each vector $(\phi, \theta_\ell) \geq 0$ with sufficiently large $\theta_\ell$ satisfies the inequality system in (4.8). Elastic constraints were similarly employed for a LP problem in [59] and for a multicriteria weighted sum approach in [57], in which the weights $w_\ell$ of the objective function $\sum_\ell w_\ell \theta_\ell$ are varied (see Section 4.3.7).

Note that, for $u_\ell = e$ in (4.8), the problem of minimizing the term $w_\ell \theta_\ell$ alone over all vectors $(\phi, \theta_\ell) \geq 0$ under the constraints in (4.8) is equivalent with the problem of minimizing, over all $\phi \geq 0$, the function

$$F_\ell(\phi) = \left\| [A_\ell \phi - b_\ell]_+ \right\|_\infty, \tag{4.9}$$

i.e., the maximum violation of the system $A_\ell \phi \leq b_\ell$. Moreover, if $V_\ell$ is a target volume and the system in (4.8) stands for

$$\Delta_\ell e - \theta_\ell e \leq D_{(\ell)} \phi \leq \Delta_\ell e + \theta_\ell e \tag{4.10}$$

with some dose $\Delta_\ell > 0$, it is equivalent with the *linear Chebyshev approxima-tion problem* of minimizing

$$F_\ell(\phi) = \|A_\ell\phi - b_\ell\|_\infty \tag{4.11}$$

with respect to $\phi \geq 0$ ([40]). Thus, problems with elastic constraints are closely related to certain minimum norm problems discussed in Sections 4.3.5 and 4.3.6.

### 4.3.4 Further linear programming related results

An attempt to overcome some limitations while still remaining in the frame-work of LP is provided in [103], where a convex voxel-based objective function, as it is given, for instance, in Sections 4.3.6 and 4.3.8 below, is approximated by a piecewise linear function. In this way, however, at least $K \cdot m$ inequality constraints and hence slack variables are added to the problem, where, for the numerical results, $K$ was a number between 2 and 4. Also, several authors, including those of [103], suggest LP approaches to deal with *partial-volume constraints*. The latter approaches are discussed in Section 4.3.8.

Robust LP (and second-order cone programming) approaches respecting uncertainties in regard to the patient positioning or the dose matrix were recently studied in [38] and [98]. In connection with LP models for radio-therapy treatment planning, the results of [99] show that the choices of the problem formulation and the algorithm for its solution are quite relevant in order to solve the large-scale LP problems within clinically acceptable compu-tation times. (See also the work in [104] on equivalent problem formulations in this context.)

### 4.3.5 Linear approximation models

The possible inconsistency of the constraints in a LP approach to the treat-ment planning problem has stimulated the study of various constrained linear approximation problems, with the aim of finding an intensity weight vector that is nearest to the desired goals in some sense. Some authors have consid-ered the (squared) simple-bound constrained *linear least-squares approxima-tion problem*

$$\min_{\phi \geq 0} \sum_{\ell=1}^{q} w_\ell \frac{1}{|V_\ell|} \|A_\ell\phi - b_\ell\|_2^2 \tag{4.12}$$

(e.g., [63, 132, 133]). Alternatively, the simple-bound constrained *Chebyshev approximation problem*

$$\min_{\phi \geq 0} \max_{1 \leq \ell \leq q} \left\{ w_\ell \frac{1}{|V_\ell|} \|A_\ell\phi - b_\ell\|_\infty \right\} \tag{4.13}$$

was investigated ([60]). Both problems always have a solution (cf. Remark 2 below). However, minimum norm problems of this type can be interpreted as an attempt to find an approximate solution of an overdetermined system of equations and hence force all normal tissue volumes to receive doses closely below or above the allowed maximum doses, which usually is not desirable.

The latter drawback is remedied if, for all normal-tissue volumes, one approximates zero doses with respect to the (squared) weighted $\ell_2$-norms, under homogeneity constraints on the targets. Such a way of proceeding results in the solution of a constrained linear least-squares approximation problem of the type

$$\text{minimize} \sum_{\ell \in \mathcal{N}} w_\ell \frac{1}{|V_\ell|} \left\| D_{(\ell)} \phi \right\|_2^2$$
$$\text{s.t.} \quad A_\ell \phi \leq b_\ell (\ell \in \mathcal{T}), \tag{4.14}$$
$$\phi \geq 0,$$

where the inequality system stands for lower and upper dose bounds ([67]). The problem in (4.14) resembles the aforementioned simpler LP problem for the objective function (4.5) with $\mathcal{P} = \mathcal{N}$ and additional importance weights.

Interchange of the roles of $\mathcal{T}$ and $\mathcal{N}$ in (4.14) yields the alternative problem

$$\text{minimize} \sum_{\ell \in \mathcal{T}} \left\| A_\ell \phi - b_\ell \right\|_2^2$$
$$\text{s.t.} \quad A_\ell \phi \leq b_\ell (\ell \in \mathcal{N}), \tag{4.15}$$
$$\phi \geq 0,$$

which was investigated, e.g., in [81] (see also the references in [62]). The matrix inequality constraints in (4.15) typically result from upper dose bounds for healthy volumes so that $\phi = 0$ is feasible for the problem. Simultaneous minimization with respect to a given set of normal tissue dose bounds $b_\ell$ ($\ell \in \mathcal{N}$) was recently studied in [138].

Instead of the squared $\ell_2$-norm in (4.14) and (4.15), one may exploit the maximum norm, which for problem (4.15) was done in [28]. Other meaningful variations of linear minimum norm problems can be found, for example, in [62] and [108]. In particular, the linear least-squares problems with linear constraints can be written as ordinary *quadratic programming* (QP) problems, and (linearly constrained) problems involving the $\ell_1$- or $\ell_\infty$-norm typically can be transformed straightforwardly into LP problems ([40]). The latter is true for all $l_\infty$-problems given here. Thus the $l_2$-problems can be solved by an algorithm for QP or some *nonlinear programming* (NLP) method like a *penalty type method* ([67]) or a *gradient projection method* ([16, 19]), and (linearly constrained) linear Chebyshev approximation problems can be solved by the Simplex algorithm or an interior-point method.

### 4.3.6 Piecewise linear approximation models and extensions

A very popular modification of the least-squares approach in (4.12), which avoids its drawbacks and includes only simple-bound constraints, is to let only those constraints of the system in (4.4) enter the linear approximation problem, at least for the normal-tissue volumes, which are violated for $\phi$ (e.g., [16, 17, 19, 47, 66, 115, 128]). The resulting convex simple-bound constrained *piecewise linear least-squares problem* has the form

$$\min_{\phi \geq 0} \sum_{\ell=1}^{q} w_\ell \frac{1}{|V_\ell|} F_\ell(\phi) \qquad (4.16)$$

where $F_\ell$ equals either the quadratic function

$$F_\ell(\phi) = \|A_\ell\phi - b_\ell\|_2^2 \qquad (4.17)$$

or the piecewise quadratic function

$$F_\ell(\phi) = \|[A_\ell\phi - b_\ell]_+\|_2^2 . \qquad (4.18)$$

The *importance weights* $w_\ell \geq 0$, not all being zero, may be normalized such that

$$w = (w_1, \ldots, w_\ell)^\top, \quad \|w\|_1 = \sum_{\ell=1}^{q} w_\ell = 1. \qquad (4.19)$$

Typically, the quadratic function in (4.17) is used for a PTV and the piecewise quadratic function in (4.18) for each other volume (e.g., [16, 19, 47, 66, 115]). This approach has been realized in most clinical software, e.g., in the package KonRad of the German Cancer Research Center in Heidelberg ([21, 96]).

The least-squares type problem in (4.16) has been solved, for example, by a scaled gradient projection algorithm ([16]), a variant of a Newton projection method ([47]), and an active set method ([66]). Some authors also heuristically adapt gradient type methods for unconstrained problems, like conjugate gradient methods, to problems with constraints. Others consider a piecewise linear least-squares problem including functions of type (4.18) as an ordinary QP problem, which, however, can lead to failures. Note that the function in (4.18) possesses a continuous first derivative on $\mathbb{R}^n$ but typically is not twice continuously differentiable everywhere. (If existence of second derivatives is required for an algorithm, the power 2 in (4.18) has to be increased by at least 1.)

Observe that, if $F_\ell$ in problem (4.16) is defined through (4.18) for all $\ell \in \{1, \ldots, q\}$ as in [16], each feasible point of the related linear inequality system is a minimizer of (4.16) with objective function value zero. Hence, in this case, the least-squares type approach in (4.16) is distinguished from the approach of searching for a feasible point of a linear inequality system, mentioned in Section 4.3.2, only insofar as the types of these problems motivate the use of different algorithms and different measures in case the system is inconsistent.

For the linear feasible-point approach, in [131] an algorithm is discussed that always finds the unique feasible point for which $\|\phi\|_2$ becomes minimal. This latter approach may be viewed as an attempt to find a feasible point that produces a small integral dose over the irradiated volume. The feasible point of a linear system having minimal Euclidean norm could also be found by solution of a linearly constrained QP problem with objective function $\|\phi\|_2^2$ ([30]). If the squared Euclidean norm $\|\phi\|_2^2$ in this problem would be exchanged for the maximum norm $\phi_{\max} = \|\phi\|_\infty$, serving the same goal, the problem could be solved as a LP problem with the additional constraints from (4.6). Also observe in this connection that, if the maximum norm is employed rather than the squared $\ell_2$-norm (see (4.9)–(4.11)), then problem (4.16) is equivalent to the LP problem

$$minimize \sum_{\ell=1}^{q} w_\ell \frac{1}{|V_\ell|} \theta_\ell$$
$$s.t. \quad A_\ell \phi \le b_\ell + \theta_\ell e (\ell = 1, \ldots, q),$$
$$(\phi, \theta) \ge 0,$$

for $\theta = (\theta_1, \ldots, \theta_\ell)^\top$, which just is a prominent case of the LP elastic constraints approach from [61] and [62].

Our discussion reveals that there exist close relations between many of the LP, the feasible-point, and the (piecewise) linear approximation models to the IMRT treatment planning problem. By their nature, all of these problems are linear in the sense that, for each volume $V_\ell$, they relate to the linear system $A_\ell \phi \le b_\ell$ representing a dose bound and that they are distinguished only by measuring possible constraint violations in different ways.

From the computational point of view, it may also be desirable to deal with such linear systems only. But, as is well-known, the response of a complex organ to radiation does depend on the absorbed dose in a nonlinear way and not only on the amount of dose violations in the individual volume elements (see Section 4.3.8). Also, the linearity of an approach normally requires the presence of at least one constraint for each voxel, but it is by no means clear that a LP problem with a very large number of inequality constraints is preferable to, for example, a nonlinear convex problem with few if any inequality constraints apart from the bounds $\phi \ge 0$. Furthermore, large numbers of quite similar linear constraints generated by some discretization process (concerning, e.g., the volumes) typically lead to very ill-conditioned constraint matrices and hence may be liable to numerical difficulties.

A first natural extension of the model in (4.16)–(4.19) would be to consider, for each volume $V_\ell$, a constraint

$$G_\ell(\phi) \le 0$$

with a sufficiently smooth goal function $G_\ell$ defined on a proper subset of $\mathbb{R}^n$, where for simplicity we assume here the presence of only one goal for each volume. Then, in generalization of problem (4.16), we arrive at the problem

$$\min_{\phi \ge 0} \sum_{\ell=1}^{q} w_\ell \frac{1}{|V_\ell|} [G_\ell(\phi)]_{(+)}^2, \tag{4.20}$$

where $[\cdot]^2_{(+)}$ stands for either $[\cdot]^2$ or $[\cdot]^2_+$. This is a convex optimization problem if, for example, $G_\ell$ in a term $[G_\ell(\phi)]^2_+$ is a convex and in a term $[G_\ell(\phi)]^2$ a linear function.

*Remark 2.* All minimization problems, studied up to this point in Section 4.3 and preceding problem (4.20), are convex optimization problems. For the LP and the linearly constrained QP problems of this and the previous subsections, existence of a solution is guaranteed if the set of feasible points is nonempty, as their objective functions are bounded below by zero on the respective feasible sets (e.g., [126, p. 130]). For problem (4.16)–(4.19), the existence of a solution can be proved along the lines of the proof given for the example case in [10]. A sufficient condition for the existence of only one solution is that the objective function is strictly convex (e.g., [14, 51]). In particular, the objective function related to a linear least-squares problem is strictly convex if the matrix $A$, associated with such problem, has full column rank or, equivalently, if $A^\top A$ is nonsingular (e.g., [95]).

### 4.3.7 Multicriteria optimization models

The choice of the weights $w_\ell$ in (4.16) and (4.20) respectively is quite arbitrary. For a prescribed selection of these weights, the maximum amount of a possible constraint violation for a particular volume at a solution of the problem is not predictable and may turn out to be not acceptable clinically. In fact, it has been reported that computed doses are extremely sensitive to the selection of weights (e.g., [57, 86]). Therefore, by trying different settings of weights, one may end up in a very time-consuming trial-and-error process.

From its nature, the problem of finding a radiotherapy treatment plan is a *multicriteria optimization problem* (e.g., [69]) with a finite number of well-defined objective functions. Such a problem is associated with a manifold of solutions, the *(Edgeworth–) Pareto minimal points*, which refer to the differing importance that may be given to the single objectives. These Pareto minimizers are closely related to minimizers of the *scalar optimization problem* in (4.16) that are obtained for different weight vectors $w > 0$. If the $F_\ell$ $(\ell = 1, \ldots, q)$ are any convex functions, a solution of problem (4.16) for a given vector $w > 0$ is a *properly (Edgeworth–) Pareto minimal point* of the problem associated with the $q$ objectives $F_\ell$, and, conversely, each properly (Edgeworth–) Pareto minimal point of that problem solves problem (4.16) for some weights $w > 0$ ([69, p. 299]).

The determination of Pareto minimizers via such scalar optimization problem is known in the framework of multicriteria optimization as the *weighted sum approach*. In practice, usually a finite set of optimization problems as in (4.16) is solved for a proper discrete set of weight vectors $w \geq 0$, where typically all vectors $w$ with $\|w\|_1 = 1$ from a uniform grid in $[0, 1]^q$ are chosen. Then either solutions for all of these problems are offered to the decision maker or the solution of these scalar problems is accompanied by some decision process, according to which irrelevant solutions are ignored and a suitable solution is extracted for use.

Several authors have recently studied multicriteria weighted sum approaches for radiotherapy treatment planning. In [134], the problem in (4.12) is studied in a multicriteria setting, and the obtained plans are evaluated by a *dose-volume histogram* function. Another approach of this type is discussed in [75] (and similarly in [41]) for

$$\mathcal{F}(\phi) = w_1 \sum_{\ell \in \mathcal{T}} \frac{1}{|V_\ell|} \left\| D_{(\ell)}\phi - \Delta_\ell e \right\|_2^2 + w_2 \sum_{\ell \in \mathcal{NT}} \frac{1}{|V_\ell|} \left\| D_{(\ell)}\phi \right\|_2^2$$
$$+ \sum_{\ell \in \mathcal{O}} w_{3,\ell} \frac{1}{|V_\ell|} \left\| \left[ D_{(\ell)}\phi - \Delta_\ell e \right]_+ \right\|_2^2, \qquad (4.21)$$

where the $\Delta_\ell$ are reference doses and $\mathcal{T}$, $\mathcal{O}$, and $\mathcal{NT}$ are the index sets of all volumes representing PTVs, OARs, and the remaining normal tissues respectively. Solutions are computed for all weights $w \geq 0$ on a uniform mesh in the cube $[0, 1]^{|\mathcal{T}|+|\mathcal{NT}|+|\mathcal{O}|}$ so that the number of structures (at most 6 in ([75])) and the width of this mesh determines the total number of problems to be solved. Naturally, especially for high-dimensional problems, this number needs to be kept small. (Compare, e.g., the example case of Section 4.6 and those in [10], which include up to 25 goals for head-and-neck cancer cases.)

The scalar problem of minimizing the convex function in (4.21) subject to the simple bounds $\phi \geq 0$ could be solved, for example, by some gradient projection method. In order to arrive at an unconstrained optimization problem, the authors of [41] and [75] recommend instead replacing the weights $\phi_k$ by weights $\psi_k^2 = \phi_k$. However, this transformation may have consequences concerning the convergence of the algorithm ([51, p. 147]) and, what is not mentioned, transforms the convex problem into a nonconvex one so that nonglobal local minimizers have to be discussed. (The function $f(x) = (x - a)^2$ with some $a > 0$ is convex, but $g(y) = (y^2 - a)^2$ is not.) Then the nonconvex problems are solved by a conjugate gradient method ([41]) and the limited memory BFGS method (e.g., [95]) respectively. Note at this point that several authors use (quasi-) Newton type methods that directly or indirectly need second derivatives, though these do not exist in all points, for example, when functions including expressions of type $\| [\cdot]_+ \|_2^2$ are used. But such action is known to possibly lead to very slow convergence ([114]), as the iteration numbers reported in [75] also seem to indicate.

In [57], a linear multicriteria weighted sum approach is studied using elastic constraints (see Section 4.3.3), where the allowed maximum violations of the constraints for the various structures form the objectives. The approach is combined with a strategy to find a certain representative subset of Pareto solutions rather than to compute solutions for all weights vectors of a given discrete set, and some numerical experiments are presented.

A more sophisticated multicriteria optimization approach, which requires the solution of optimization problems including constraints on the various goals, is developed in [22] and [120]. The aim again is to find suitable representatives of the set of Pareto minimizers, where the total number of problems to be solved does not depend on the fineness of some mesh, but only on the number $q$ of goals (which should not exceed about 6, as is said). This approach also makes use of the EUD model (see (4.25) in Section 4.3.8) where, for the numerical realization, the $\ell_p$-norm, $1 < p < \infty$, on the dose in the EUD function ([93]) is replaced by a suitable convex combination of the $\ell_1$- and $\ell_\infty$-norm ([121]). This replacement has the advantage of leading to LP problems, but means that a single nonlinear convex constraint for a volume $V_\ell$ is exchanged for $|V_\ell|$ linear constraints. Strategies to reduce the large number of linear constraints are implemented, and numerical experience with the total approach is reported.

The authors of [104] discuss a unifying framework providing conditions under which multicriteria optimization problems including well-known nonconvex treatment planning criteria can be transformed into problems with convex criteria, having the same set of Pareto minimizers.

### 4.3.8 Nonlinear conditions

### General discussion

The LP and similarly the linear and piecewise linear approximation models for IMRT considered up to this point are merely based on *physical criteria*, i.e., on measurable physical quantities such as volumes and doses. It has been observed by a number of authors that such approaches have serious limitations (see, e.g., the discussions and references in [24, 88, 127, 129]). They take the biology of radiation into account only insofar as they try to avoid critical structures, but they do not adequately model the responses of healthy and tumorous tissues to radiation, which behave neither linearly nor quadratically. The sensitivity of a healthy organ to radiation does not simply depend on the maximum dose absorbed by some of its volume elements, but rather on the total dose distribution in the organ. Moreover, for example, a *cold spot*, i.e., a small underdosed volume, in a target may not greatly influence a quadratic objective formed by the differences of desired and actual doses, but may significantly reduce the tumor control probability.

Therefore the insertion of biological considerations for both dose prescriptions and the rules for control of their violation have been proposed (e.g., [20, 25, 53, 88, 89, 101]), and alternative biological optimization models, which respect the dose responses of the different tissues and the response to inhomogeneous dose distributions, have been developed (e.g., [3, 6, 9, 10, 55, 73, 123]). Biological conditions are inherently nonlinear so that their direct implementation necessarily leads to large-scale nonlinear convex or nonconvex optimization problems. Naturally, the problems may have multiple local minimizers, but in general clinically usable solutions seem to have been found (see [17, 43, 106] for studies in this connection).

## Probability functions and an overdosage penalty constraint

Several authors have studied objective functions in an optimization model, representing *normal tissue complication probability* (NTCP) and *tumor control probability* (TCP). The authors of [123] optimize, by a gradient technique, an objective function including both probabilities and dose-volume criteria in addition. In [54], which integrates earlier results from [55, 56, 112, 113], various formulations of optimization problems with biologically motivated linear constraints and a nonlinear objective function have been studied, including the *probability of uncomplicated tumor control* $P_+ = P_B - P_{B \cap I}$ as an objective function, where $P_B$ is the probability of tumor control and $P_{B \cap I}$ is the probability of simultaneous tumor control and severe normal-tissue complications. For the solution of these (by today's standards relatively small) problems, several algorithms based on an augmented Lagrangian approach have been compared with a *Sequential Quadratic Programming* (SQP) method, where it was found that the augmented Lagrangian approach, combined with a limited memory BFGS method, was the most favorable one. In [66], a probability function $P_- = 1 - P_{++}$ was minimized under the constraints $\phi \geq 0$ by an active set method, where $P_{++}$ is taken from [1] and similar to $P_+$ (see the discussion in [54]). It has been remarked, however, that these types of probability functions "are simplistic, and the data they rely on are sparse and of questionable quality" ([123]).

The authors of this paper favor the use of the *logarithmic tumor control probability* (LTCP)

$$\text{LTCP}(\phi; V, \Delta, \alpha) = \frac{1}{|V|} \sum_{j \in V} \exp(-\alpha \, (D_j^\top \phi - \Delta)) \qquad (4.22)$$

for each PTV $V$ as objective function ([10]), where $\Delta > 0$ is the dose value requested for $V$ and $\alpha > 0$ is a constant related to cell survival and the only biological constant needed. Minimization of this convex function is easily seen to be equivalent to the maximization of the TCP function (see [91])

$$\prod_{j \in V} \exp \left\{ -\frac{1}{|V|} \exp(-\alpha \, (D_j^\top \phi - \Delta)) \right\}.$$

In the absence of, e.g., adequate dose bounds on the normal tissue volumes, minimization of the LTCP could result in a prohibitively high dose in the targets (as is observed in [62, p. 4–25]). Therefore, for each PTV $V$, the authors use a *quadratic overdosage penalty* (QOP) constraint of the type

$$\text{QOP}(\phi; V, \Delta, \delta) = \frac{1}{|V|} \sum_{j \in V} [D_j^\top \phi - \Delta]_+^2 - \delta^2 \leq 0, \qquad (4.23)$$

where $\delta > 0$ is a given bound. Such a constraint prevents an excessively high dose in $V$ and simultaneously allows a mild mean violation of the acceptable dose value $\Delta$ in $V$ by some $\delta$. A constraint of this type is also applied to permit a certain overdosing of some volume $V$ neighboring a PTV, as a sharp dose drop from the PTV to $V$ is not realizable physically. Note that the function $QOP(\,\cdot\,; V, \Delta, \delta)$ is once but not twice continuously differentiable everywhere, so that the power 2 in $QOP$ has to be increased by at least one if second derivatives of functions are needed in an algorithm.

In this connection, observe that an underdosage in some voxels of $V$ for a solution of an optimization problem, involving a term as in (4.22) in its objective function, would lead to positive powers in the exponential function and hence tends to affect the objective function value considerably more than in case of a quadratic (type) function. This observation implies intuitively, though not rigorously, that using an additional minimum dose constraint for $V$ as in (4.3) would not significantly increase the actual minimum dose attained by such a program. Therefore, like other authors, we avoid the implementation of lower dose bounds as they may cause infeasibility of the program and hence difficulties for algorithms. In either case, when cold spots are detected in a target or if a system of constraints turns out to be inconsistent, the original treatment goals need to be reconsidered and modified.

**Partial volume conditions**

It has been generally accepted that, for each involved critical *parallel organ* (lung, parotid gland, kidney, etc.), an optimization model should reflect the property that a certain percentage of such an organ can be sacrificed without serious consequences for the patient, if this is of advantage for the overall treatment. Thus, instead of merely pursuing the goal for a particular parallel OAR to stay below an upper dose bound, the model should provide a solution exhibiting an acceptable dose distribution for this organ in regard to the dose versus the percentage-of-volume. Such a relationship can be depicted in a *cumulative dose-volume histogram* (DVH) and is typically considered, in combination with other criteria, to evaluate the quality of a treatment plan.

In this connection, many authors start from an ideal clinical DVH curve and seek dose distributions that match these inherently nonlinear curves at one or multiple points. Constraints in a model that are designed for this purpose are often denoted as *dose-volume* (DV) *constraints* (see Figure 4.3 and, e.g., [54, p. 32] for a summary of the application of such constraints).

Though we intend to concentrate on continuous optimization models, we would like to point out that a mathematically rigorous description of pointwise DV constraints can be given by mixed-integer linear constraints. For that, a binary variable $y_j \in \{0, 1\}$ is assigned to each element of the respective volume $V_\ell$, depending on its dose level, and dose-bound constraints for $V_\ell$ including these new variables as, e.g.,

$$D_j^\top \phi \le \Delta_\ell^u + 100 * y_j, \quad j \in V_\ell,$$

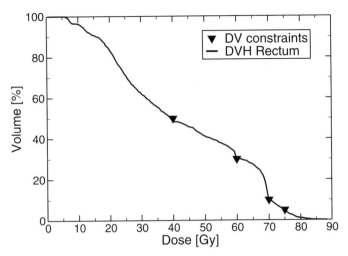

**Fig. 4.3.** Cumulative DVH of a rectum in a prostate example case. Four DV constraints were set for the optimzation: (40 Gy/50%), (60 Gy/30%), (70 Gy/10%), (75 Gy/5%). Each constraint ensures that no more than $y$% of the organ volume receives more than $x$ Gy dose. The treatment dose of the prostate was 84 Gy.

are combined with an additional constraint that only allows a desired portion $p_\ell \in (0, 1]$ of these binary variables to be 1, e.g.,

$$\sum_{j \in V_\ell} y_j \leq p_\ell \, |V_\ell| \,.$$

The solution of related large-scale MILP programs has been studied, e.g., in [13, 76, 79, 86].

Some authors suggest the inclusion of certain continuous linear DV constraints to remain within a LP framework. In [90] several "collars" around a target are formed and an upper dose-bound constraint of type (4.2) is used for each such neighborhood, where the dose bound is decreasing with increasing distance from the target, and the thickness of the collars is determined by the percentage of volume elements that shall be below the given bound. This procedure is modified for IMRT in [59] where, in addition to the distance from the target, a heuristic concerning the expected number of beamlets meeting a structure is used in order to select the voxels related to a certain dose bound. In [103], a new type of linear constraints, derived from a technique used in finance, bounds the tail averages of DVHs, but entails a number of artificial variables proportional to the number of voxels of the respective structures.

While in these approaches the LP program remains unchanged during the iteration process, the authors of [86] employ dose bounds as in (4.2) for subvolumes of a particular structure and make use of sensitivity information to adapt the respective voxel sets in each iteration, in case DVH requirements are not satisfied for the current solution. A dynamic adaptation of such linear bounds is also applied in [65] in combination with a least-squares objective function for the PTVs. However, it is not clear whether these latter procedures always converge to a desired solution.

Another technique for respecting DV conditions, which is applied by some authors and was developed in [21] in connection with the least-squares type problem (4.16)–(4.19), is to check at each iteration whether the current solution meets a particular DVH specification, and, if not, to add a "penalty" $w_\ell [D_j^\top \phi - \Delta]_+^2$ for certain $j \in V_\ell$ to the objective function as, e.g., for all or some of those voxels that exceed a desired dose $\Delta$, assuming that the number of these voxels is greater than a permitted number (e.g., [109, 115]). Differing from that, in [36], a continuous, though not everywhere differentiable, linear-quadratic "penalty" function defined on the total respective volume is added to a least-squares function for the PTVs, where this penalty is multiplied by a factor depending on the current fraction of the structure surpassing a required dose. Similarly, in [35], a least-squares error function is adapted properly in each iteration so that a sequence of least-squares approximation problems is solved. Hence, in these approaches the objective functions of the optimization models are redefined during the iteration process, and it is not clear to what point such procedure converges, in case it converges at all.

The authors of [87] extend the approach of searching for a feasible point of a certain linear system (see Section 4.3.2) to include a new type of (nonconvex) quasi-convex DV constraints and report satisfying results for an algorithm, which has originally been designed for the solution of the convex feasibility problem only.

Objections to pointwise DV conditions are that it is unclear how many of such conditions are needed to obtain an acceptable DVH curve and that the precise fulfillment of such conditions is a somewhat artificial goal and not justified medically. Usually also several DVHs have to be considered simultaneously so that their *a priori* fixing by pointwise DV conditions may entail a significant loss of freedom in the search space for the beamlet weights, while modified conditions would still be medically tolerable and could result in an overall improvement for the patient. On the other hand, trial-and-error procedures in this respect are very time consuming. Furthermore, some approaches that alter definitions of objective functions or constraint sets during the performance of an algorithm lack a rigorous mathematical convergence analysis and are therefore uncertain concerning their outcomes.

The direct translation of a dose-versus-percentage constraint into a continuous mathematical condition, however, is known to lead to a nonlinear constraint. The authors apply the *partial volume* (PV) constraint

$$\mathrm{PV}(\phi; V, \Delta, p, \zeta) = \frac{1}{|V|} \sum_{j \in V} \frac{(\frac{1}{\Delta} D_j^\top \phi)^p}{1 + (\frac{1}{\Delta} D_j^\top \phi)^p} - \zeta \leq 0 \qquad (4.24)$$

with some constant $\zeta \in (0,1)$ for each parallel OAR $V$ (see [3, 6, 68] for details). For example, in relation to our example case in Section 4.6, the data

$\Delta = 20$, $p = 3$, and $\zeta = 0.1$ for the right parotid gland express that, at a dose of 20 Gy, a volume element of this organ loses 50% of its function (e.g., the production of saliva) and that at most 10% of the total function of the organ may be lost.

The constraint in (4.24) is formed by the sigmoidal function $\sigma(x) = x^p/(1 + x^p)$, which has a relatively smoothly increasing step and hence offers some freedom concerning the dose distribution in $V$. (Alternative experiments with the sigmoidal error function can be found in [101, 108].) Note in this connection that constraints of type (4.24) can also be utilized to obtain continuous pointwise DV constraints as discussed above, when the step of the sigmoidal function is contracted ([123]).

**Equivalent uniform dose conditions**

In contrast with conventional radiation therapy, IMRT normally leads to nonuniform dose distributions in organs. Niemierko ([93]) has introduced the *(generalized) equivalent uniform dose* (EUD)

$$\left\{ \frac{1}{|V|} \sum_{j \in V} (D_j^\top \phi)^p \right\}^{1/p} \tag{4.25}$$

as a model for a biologically permissible nonuniform dose distribution in a volume $V$ that, in regard to the irradiation response, is comparable with a uniform dose distribution of $\Delta$ Gy. In this function, $p \in \mathbb{Z}$ is some tissue-specific power, which is negative for PTVs and positive for OARs. Note that for $p = 1$, the function in (4.25) becomes the mean dose and for $p = \infty$ the maximum dose for $V$, both used above. The EUD concept has by now been widely accepted, in particular for *serial organs*, i.e., the spinal cord, nerves, and all other structures that can be seriously damaged by a high dose in a small spot. It has been observed that the use of the EUD model can lead to greater normal tissue sparing, compared with merely dose-based optimization [129]. "Inverse planning based on the probabilities of tumor control and normal tissue complication remains the ultimate goal, and the equivalent uniform dose is a step in this direction" [122].

The EUD concept was applied for optimization in [10, 22, 97, 103, 120, 122, 121, 127, 129]. In particular, the authors of [129] investigate an objective function that makes use of the EUD model for tumors as well as normal tissues. The resulting nonconvex function is minimized by a gradient technique. In [127], the EUD-based model is combined with a dose-volume approach to further improve the treatment plans, and in [137] various gradient algorithms are compared for dose-volume-based and EUD-based objective functions. The recent convex approach from [120, 122] employs an upper bound on the EUD as an optimization constraint for all OARs and PTVs and a lower bound on the EUD of the PTVs. In this way, a convex constraint set is obtained and a quadratic least-squares error function, which is adapted in each iteration similarly as it has been suggested in [21] for DV constraints (see above), is minimized over this set by a componentwise Newton method that is combined with a projection technique. In [97], several variants of a gradient projection algorithm are investigated to solve nonlinear optimization problems with an EUD-based objective function and nonnegativity constraints on the weights.

The authors themselves employ an EUD constraint of the type

$$\mathrm{EUD}(\phi; V, \Delta, p, \varepsilon) = \frac{1}{|V|} \sum_{j \in V} \left( \frac{1}{\Delta} D_j^\top \phi \right)^p - \varepsilon^p \leq 0 \qquad (4.26)$$

for each serial OAR $V$ only, where $\Delta > 0$ is some given dose value, $\varepsilon > 0$ a given constant, and $p \geq 1$ some tissue-dependent power ([10]). For instance, in the optimization problem of the example case in Section 4.6, we include an EUD constraint for the spinal cord with the settings $\Delta = 28$, $p = 12$, and $\varepsilon = 1$. This constraint allows a tiny excess of 28 Gy for fractions of this organ, with the extent of overdosage depending on the size of the volume in which it occurs. Note that a single convex constraint as in (4.26) normally can replace the $|V|$ linear constraints entering a program if an upper dose bound as in (4.2) is used for $V$.

### Remarks on the model of the authors

The functions $LTCP$, $QOP$, and $EUD$ are nonquadratic convex, and the function $PV$ is nonconvex. Moreover, in this ideal description concerning the beamlet weights (see [10] for this), the zero vector is feasible for the respective constraints. Therefore, use of these functions leads to a feasible convex or, if the irradiation of, e.g., parotid glands and lungs is to be controlled, nonconvex optimization problem with sufficiently smooth functions in $n$ variables. Technical limitations of a MLC may enforce additional constraints on the weights that have to be included in the program (see [7, 8] for examples). However, in contrast with, for example, LP models, the model of the authors involves rarely more than 15–20 constraints besides the constraints $\phi \geq 0$.

The nonlinear optimization problems resulting from this model are solved by a barrier-penalty multiplier method ([10]) and combined with a sensitivity analysis discussed in the following section. For the solution of the subproblems in the algorithm, a conjugate gradient method is used, as such a method is well suited to deal with the typical ill-conditionedness of beamlet weight optimization problems mentioned in the introduction (see [5, 10] and the recent results in [52]). For ill-conditioned problems of the occurring type, a conjugate gradient method finds a good approximate solution with respect to the optimal objective function value in relatively few iterations (but normally not with respect to the variables) and can be used in a regularizing manner in the sense that it can be stopped before serious ill-conditioning starts (see also [29, 97] in this connection).

In view of the possible nonconvexity and hence the existence of nonglobal local minimizers of our optimization problems, we would like to mention that, in all our experiments, different starting points for the barrier-penalty algorithm have led to "solutions" with objective function values of equal orders of magnitudes. Our algorithm needed typically 3–5 minutes of execution time by a Xeon 2.66 GHz processor for standard cases and up to 30 minutes for complex cases of a set of several hundred clinical case examples. Both, the algorithm and the sensitivity analysis, were implemented in the software package HYPERION, which was developed at the University Hospital in Tübingen and is used already in daily clinical routine in several hospitals in Germany and the United States.

## 4.4 Sensitivity Analysis

In clinical routine, the initially provided dose distribution framework, which is needed for the development of a treatment plan, often defines constraints for the OARs that are not compatible with the desired PTV dose, so that an obtained solution is unacceptable and the constraints need to be relaxed in a controlled and sensible manner. The relaxation of bounds on the other hand can result in serious consequences for the patient and therefore has to include the considerations of physicians.

In order to come to proper decisions in this regard and to simultaneously avoid a time-consuming trial-and-error process, the physicians can be supported for the IMRT treatment optimization model of the authors by a *sensitivity analysis*, which was introduced in [4] and is developed in this section. Sensitivity analysis is a standard tool in optimization ([14, 50]) and was used for linear models in radiotherapy treatment planning already in, e.g., [33, 34, 86].

Let $f : \mathbb{R}^n \to \mathbb{R}$ be the objective function and $g_i(\phi) \le c$ be some constraint of the problem. Furthermore, let $\phi^*(0)$ be the solution of the problem for $c = 0$ and let $\lambda_i^* \ge 0$ be the related Lagrange multiplier. Next consider $f$ as a function depending on $c$, i.e., as $f(\phi(c))$. Then a standard result of sensitivity analysis in optimization says that, under suitable assumptions and for $|c|$ sufficiently small, the problem has a local minimizer $\phi^*(c)$ and

$$\frac{\partial f}{\partial c}(\phi^*(c))_{c=0} = -\lambda_i^*$$

(see [14, p. 315] and [50]). Thus, for some small perturbations $c$, one arrives at

$$\frac{f(\phi^*(c)) - f(\phi^*(0))}{c} \approx -\lambda_i^*, \tag{4.27}$$

saying that a relaxation of the inequality constraints with the largest multipliers causes the largest local changes of the optimal objective function value.

It is relevant that, in case the objective function $f$ equals the LTCP of a single target $V$, i.e., if $f(\phi) =$ LTCP$(\phi; V, \Delta, \alpha)$ for some prescribed dose value $\Delta$, the change in the optimal value of the problem by a small relaxation of an inequality constraint can also be translated into a change of the EUD in $V$, which is a more significant number for the physicians. The EUD for the dose distribution in $V$, differing from the function in (4.25) that sometimes is denoted as the generalized EUD, is defined by

$$\mathcal{E}(\phi; V, \alpha) = -\frac{1}{\alpha} \log \left\{ \frac{1}{|V|} \sum_{j \in V} \exp(-\alpha D_j^\top \phi) \right\}$$

([92]) and, for $f$ as assumed, can be written in the form

$$\mathcal{E}(\phi; V, \alpha) = -\frac{1}{\alpha} \log \left\{ f(\phi) \exp(-\alpha \Delta) \right\}.$$

Thus, in this case, the change of the EUD in the target

$$\delta_{EUD} = \mathcal{E}(\phi^*(c); V, \alpha) - \mathcal{E}(\phi^*(0); V, \alpha)$$

affected by a constraint perturbation $c$ is given, with (4.27), approximately by

$$\delta_{EUD} \approx -\frac{1}{\alpha} \left[ \log \left\{ f(\phi^*(0)) - \lambda_i^* c \right\} - \log \left\{ f(\phi^*(0)) \right\} \right]$$

$$= -\frac{1}{\alpha} \log \left\{ 1 - \frac{\lambda_i^* c}{f(\phi^*(0))} \right\}. \tag{4.28}$$

Note in this connection that, ideally, $f(\phi^*(0))$ would equal 1 and that one has $\delta_{EUD} \approx 0$ in case $\lambda_i^* = 0$, which in particular is true if $g_i$ is inactive at the solution, i.e., if $g_i(\phi^*(0)) < 0$ (e.g., [95]).

Thus, via the size of the Lagrange multipliers, a sensitivity analysis as described guides the decision making of the expert in regard to those bounds for which small enlargements have the largest effects with respect to the desired target dose, where the quantitative information given by (4.28) is reasonably accurate if these enlargements are small (see [4] and the numerical results in Section 4.6). Typically, in the clinical practice, only few dose-limiting objectives for sensitive structures conflict seriously with the target objectives so that, in general, only 1 to 4 bounds in a program need to be relaxed, whereas all others can be kept unchanged. This reduces the amount of user interaction to a small number of well-directed trials.

For the treatment planning model of the authors, the biological interpretation of a change of bounds is immediate. For example, if $\zeta$ in a PV constraint of type (4.24) is increased from 0.3 to 0.4, this means that the percentage of a volume to be sacrificed in the worst case is raised from 30% to 40%. In contrast with that, for least-squares type or multicriteria approaches, the effect for a single volume by a change of desired dose bounds, percentages of a volume, or importance weights is not known, and typically such changes lead to alterations in the optimal doses for all volumes. For multicriteria approaches in particular, it was observed that computed doses are very sensitive to the selection of weights ([57, 86]).

## 4.5 Intensity Modulated Proton Therapy

About 60 years ago, it was proposed that irradiation with beams of *protons* or *heavy ions* would often be a better tool for cancer treatment than would be conventional irradiation with photons. In contrast with photons, which deposit the maximum dose near the beginning of their path through the body, protons deliver the maximum dose briefly before they stop and only relatively little before and almost none behind this point (see Figure 4.4).

The depth of the *Bragg peak*, i.e., the position of maximum dose deposition, is directly correlated with the energy of the incident particles and can be tuned precisely. Hence, by modulating the kinetic energy of the particles and the beam intensities, i.e., the exposure times of the beams, one can generate a nearly homogeneous *spread-out Bragg peak* (SOBP) in the direction of the beam. This can be performed with *passive scattering techniques*, where the proton beam passes through rotating devices of angularly variable thickness that reduce the particle energy for an appropriate, fixed fraction of the rotation time. In contrast, in *intensity modulated proton therapy* (IMPT), the exposure time of the proton beam for every scanning position and every beam energy is a free variable. This technique is also called *spot scanning* (SC), which highlights the fact that the irradiated volume is covered by Bragg peaks of narrow beams that are scanned in 3 dimensions (two lateral deflections and the depth via the particle energy).

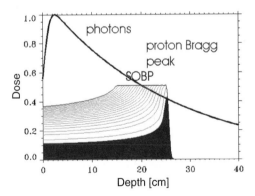

**Fig. 4.4.** Schematic depth dependence of 6 MV photons (solid line), a proton beam at fixed energy (solid filled curve), and a superposition of proton beams of various energies (thin lines) yielding a SOBP (from http://p-therapie.web.psi.ch/wirkung1.html).

Experience and small-scale case studies show that, compared with conventional photon radiotherapy, proton therapy normally leads to similar results in terms of the targets but may yield some or much improvement concerning the OARs and considerable improvement in regard to the total dose administered to the patient. (In a case study of [94], the total dose was reduced, relative to the dose obtained by IMRT, with the SC technique by about 46%.) The latter fact is relevant especially when children have to be irradiated. However, in contrast with the case of photons, the sharpness of the proton profiles requires a highly precise set-up of the planning problem and the treatment by IMPT, since any error in these can be fatal if OARs are very close to the tumor.

For a long time, the technical problems and costs to perform irradiation with protons and other heavy charged particles have been prohibitive, at least for application on a large scale. In particular, the Paul Scherrer Institute (PSI) in Villigen, Switzerland, has played a pioneering role in overcoming some of these problems ([82, 83]) so that, during the past years, the spot scanning has attracted considerable interest and is currently being implemented in many places. The recent success with proton therapy also has led to the development of cyclotrons exclusively for proton therapy, whereas, in the past, the cyclotrons needed for the proton acceleration had been constructed primarily for research in atomic physics and not for medical applications. Several dedicated proton sites will go into operation in the near future.

Treatment planning tools for proton therapy are not as far developed yet as for conventional therapy with photons. However, the optimization models discussed earlier in this paper can be straightforwardly transferred to the SC technique. In contrast with IMRT, where the beamlet intensities, determining the number of variables in an optimization model, depend on 2 parameters (the position in the fields), in case of IMPT with the SC technique these are specified by 3 parameters (position in the fields and particle energy). On the other hand, due to the favorable properties of protons, irradiation of a patient by IMPT is performed normally only from 2 or 3 directions. In total, the optimization model has the same appearance as for IMRT. However, the dose matrix $D = (d_{jk})$ is computed differently and has a considerably larger number of columns because of the third dimension of beamlet variability. In this context, it is remarkable that an obtained proton beam intensity profile can be realized directly up to some negligible deviations so that, in contrast with IMRT, no translation into MLC openings is needed.

Optimization problems for IMPT with the SC technique can have 40,000 variables and more. This fact and the newness of the method explain why only very few references can be found commenting on an optimization model and algorithms for IMPT treatment planning. In [82], application of a least-squares approach is reported, which is said to be similar to the one used in [64] and [19] (which is not purely least-squares) and is known as a method for image reconstruction, e.g., in computed tomography. The authors of [94] employ the least-squares type approach from [17] and [96], which has also been implemented in the software package KonRad (see Section 4.3.6). In the following section, it is shown by a clinical example case that our approach and algorithm from [10] can also be successfully applied to IMPT.

## 4.6 Example Case

The patient of our example case was an 11-year-old boy having a rhabdomyosarcoma, which reached from the interior of the lower jaw to the base of the skull. Irradiation of the patient by conventional radiotherapy was impossible as the tumor had infiltrated the second vertebra and because of the proximity of the tumor to the optic chiasm, optical nerves, and the brain stem. The decision was made to irradiate the vertebra with 36 Gy to avoid unilateral growth inhibition and simultaneously spare the spinal cord. The volume of gross tumor was treated to 57.6 Gy, the volume of suspected microscopic expansion to 48.6 Gy. The organs at risk (chiasm, optical nerves, eyes, spinal cord, brain stem) were defined with a 3-mm margin for setup errors, and a dose reduction in the overlap of the optical chiasm and the PTV was accepted.

The constraints of the optimization model are of the type introduced and explained in Section 4.3.8. In total, constraints for 12 volumes entered the optimization model. In particular, $V_1$ is the *gross tumor volume* (GTV), i.e., the solid tumor, $V_2 \supseteq V_1$ is the *clinical target volume* (CTV), which is the GTV together with a margin in which tumor cells are suspected, and $V_3 \supseteq V_2$ is the *planning target volume* (PTV), which adds a safety margin to the CTV, in order to respect small movements of the patient and other inaccuracies. The optimization model had the following form, where "$V_\ell \pm 5$ mm" means that an area of 5 mm width was added or subtracted respectively from $V_\ell$. In particular, the remaining volume "$V_{12} - 5$ mm" consists of the entire head and neck not otherwise classified as organ at risk or target volume, with an additional margin of 5 mm around all targets.

$$\text{minimize } \text{LTCP}(\phi; V_1, 57.6, 0.25) + \text{LTCP}(\phi; V_2, 48.6, 0.25)$$
$$+\text{LTCP}(\phi; V_3, 36, 0.25)$$

$$\text{s.t.} \quad \text{QOP}(\phi; V_1, 57.6, 1) \leq 0 \ (GTV),$$
$$\text{QOP}(\phi; V_2 \setminus V_1, 57.6, 0.2) \leq 0 \ (CTV),$$
$$\text{QOP}(\phi; V_2 \setminus (V_1 + 5\,mm), 48.6, 1) \leq 0 \ (CTV),$$
$$\text{QOP}(\phi; V_3 \setminus V_2, 48.6, 0.3) \leq 0 \ (PTV),$$
$$\text{QOP}(\phi; V_3 \setminus (V_2 + 5\,mm), 36, 1) \leq 0 \ (PTV),$$
$$\text{EUD}(\phi; V_4, 8, 12, 1) \leq 0 \ (right \ eye),$$
$$\text{EUD}(\phi; V_5, 14, 12, 1) \leq 0 \ (left \ eye),$$
$$\text{EUD}(\phi; V_6, 40, 12, 1) \leq 0 \ (optic \ chiasm),$$
$$\text{EUD}(\phi; V_7, 28, 12, 1) \leq 0 \ (r. \ optical \ nerve),$$
$$\text{EUD}(\phi; V_8, 40, 12, 1) \leq 0 \ (l. \ optical \ nerve),$$

$$\text{EUD}(\phi; V_9, 28, 12, 1) \leq 0 \quad \textit{(spinal cord)},$$
$$\text{EUD}(\phi; V_{10}, 28, 12, 1) \leq 0 \quad \textit{(brain stem)},$$
$$\text{PV}(\phi; V_{11}, 3, 20, 0.1) \leq 0 \quad \textit{(right parotid)},$$
$$\text{QOP}(\phi; V_{12} - 5\,mm, 30, 0.1) \leq 0 \quad \textit{(remaining vol.)},$$
$$\phi \geq 0.$$

For image processing, 112 computed tomographic slices with 3 mm spacing had been generated. A $18 \times 21 \times 33$ cm$^3$ box was irradiated, and the size of a volume element was $2 \times 2 \times 2$ mm$^3$ so that the total number of volume elements amounted to about $m = 1,600,000$, of which about 50% belonged to the patient's body. For the application of IMRT, 7 radiation fields were used, partitioned in field elements of $10 \times 2$ mm$^2$ size. The number of field elements and beamlets respectively totaled $n = 3,727$ so that each field had about 532 elements on average. In contrast with that, for IMPT, only 2 beam directions were chosen. The beams were scanned over a $3 \times 3 \times 2.4$ mm$^3$ grid ($x \times y \times$ energy), resulting in $n = 47,043$ proton spots, where the number of the spots equals the product of the number of beam directions and of those grid points of the *scanning grid* that belong to the PTV. The maximum proton energy needed to cover the PTV was 138 MeV.

Consequently, the optimization problem had $n = 3,727$ (IMRT) and $n = 47,043$ (IMPT) variables respectively and contained 14 inequality constraints apart from the $n$ nonnegativity constraints $\phi \geq 0$. In both cases, the problem was solved with the algorithm introduced in [10]. Some characteristic numbers for its performance are listed in Table 4.1, where the CPU times refer to a Xeon 2.66 GHz processor. The given results show the typical behavior of the algorithm. (The average sizes of a set of 127 clinical problems and the average iteration numbers for their solution in case of IMRT can be found in [10].) As for most nonlinear optimization algorithms, the computed solution of our algorithm normally is not a feasible, but only an almost feasible point, where the maximum amount of constraint violations naturally depends on the size of the stopping threshold used and, for our settings, typically corresponds with much less than 0.5% of the EUD of the respective organ.

Both plans were optimized with the same set of OAR constraints. Given that the obtained solutions have to satisfy these constraints, noticeable differences could only be found in the target volume dose distributions and the total

**Table 4.1.** Performance of algorithm.

| Results | IMRT | IMPT |
|---|---|---|
| No. outer iterations | 3 | 6 |
| No. inner iterations | 148 | 274 |
| Average no. inner iterations per outer iteration | 49 | 46 |
| No. objective function evaluations for step size | 1,413 | 2,595 |
| Average no. objective function evaluations | 10 | 9 |
| CPU time (minutes:seconds) | 5:41 | 10:51 |

dose delivered to the entire normal tissue (see the isodose lines in Figure 4.5).
Because of the superior properties of protons, the total normal tissue dose is
clearly much lower, especially in the brain. However, the target coverage of
both plans is comparable. This is a consequence of the comparatively shallow
lateral gradient and large diameter of scanned proton beams that partially off-
sets the advantages of protons compared with photons in cases where OARs
are extremely close to target volumes. Still, for pediatric cases in particular,
IMPT is the superior method.

Finally, Table 4.2 shows sensitivity results for the obtained IMRT solution.
Each figure signifies the predicted amount by how many Gy the EUD in the
GTV, CTV, and PTV respectively would increase if the respective constraint
$i$ were relaxed in such a way that, if it is a QOP constraint (see (4.23)), $\delta$ is
replaced by $\delta+1$, if it is a partial volume constraint (see (4.24)), $\zeta$ is replaced by
$1.01\zeta$, and if it is an EUD constraint (see (4.26)), $\Delta^p \varepsilon^p$ is replaced by $(\Delta+1)^p \varepsilon^p$
after the constraint has been multiplied by $\Delta^p$. Also the change $\delta_{EUD}$ in
(4.28) for the objective function obtained by such a relaxation is partitioned
into three portions for the terms relating to the GTV, the CTV, and the

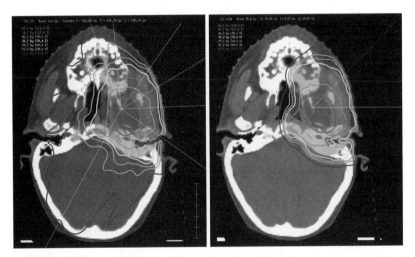

**Fig. 4.5.** Transversal section close to the base of skull of the example case. Left,
IMRT; right, IMPT. The isodose lines correspond with 25%, 50%, 60%, 70%, 95%,
112.5% of the prescription dose to the CTV of 48.6 Gy.

**Table 4.2.** Predicted changes of EUD.

| No. constr. | 1 | 2 | 3 | 4 | 5 | 6–10 | 11 | 12 | 13 | 14 |
|---|---|---|---|---|---|---|---|---|---|---|
| GTV | 0.3 | 0.3 | 0.3 | 0.1 | 0.4 | 0.0 | 0.1 | 0.4 | 0.0 | 0.4 |
| CTV | 0.1 | 0.1 | 0.3 | 0.0 | 0.4 | 0.0 | 0.3 | 0.4 | 0.2 | 0.4 |
| PTV | 0.1 | 0.1 | 0.2 | 0.0 | 0.5 | 0.0 | 0.3 | 0.3 | 0.2 | 0.4 |

**Table 4.3.** Resulting changes of EUD.

|       | Prescribed | Obtained | Pred. change | Resulting |
|-------|------------|----------|--------------|-----------|
| GTV   | 57.6       | 55.0     | 0.3          | 55.3      |
| CTV   | 48.6       | 45.5     | 0.3          | 46.0      |
| PTV   | 36.0       | 34.6     | 0.2          | 34.8      |

PTV by proper division of the Lagrange multiplier $\lambda_i^*$ into three portions $\lambda_{i,j}^*$ ($j = 1, 2, 3$). If $\nabla f_1^*$, $\nabla f_2^*$, and $\nabla f_3^*$ are the gradients of these terms and $\nabla g_i^*$ is that of the constraint $i$ in the solution, then $\lambda_{i,j}^*$ is taken as $\lambda_i^*$ times the weight factor $\left|\nabla f_j^{*T} \nabla g_i^* / (\nabla f_1^* + \nabla f_2^* + \nabla f_3^*)^\top \nabla g_i^*\right|$. The latter weight is straightforwardly motivated by consideration of the gradient condition of the first-order necessary optimality conditions for the problem (e.g., [95]).

Results that actually are obtained are recorded in Table 4.3 for the case that the QOP constraint 3 is relaxed in the aforesaid way and the problem is solved again with this altered constraint. Column 2 in this table gives the prescribed EUD, column 3 the obtained one in the solution, column 4 the predicted change of the EUD (see Table 4.2), and column 5 the resulting EUD for the solution of the modified problem. Concerning the resulting EUD for the CTV, observe that a relaxation by 1 for this QOP constraint is rather large and that the estimate in (4.28) only holds for sufficiently small changes of the bounds. Thus, predictions are not always correct, but, in any case, consideration of the multipliers guides the way to the most dominant constraints for a solution in regard to a requested change of the EUD in the targets.

## Acknowledgments

The authors are grateful to two anonymous referees who have helped to improve the original version of this paper by many valuable suggestions and remarks.

## References

[1] A. Agren, A. Brahme, and I. Turesson. Optimization of uncomplicated control for head and neck tumors. *International Journal of Radiation Oncology Biology Physics*, 19:1077–1085, 1990.

[2] A. Ahnesjö and M.M. Aspradakis. Dose calculations for external photon beams in radiotherapy. *Physics in Medicine and Biology*, 44:R99–R156, 1999.

[3] M. Alber. *A Concept for the Optimization of Radiotherapy*. PhD thesis, Universität Tübingen, Tübingen, Germany, 2000.

[4] M. Alber, M. Birkner, and F. Nüsslin. Tools for the analysis of dose optimization: II. sensitivity analysis. *Physics in Medicine and Biology*, 47:1–6, 2002.

[5] M. Alber, G. Meedt, F. Nüsslin, and R. Reemtsen. On the degeneracy of the IMRT optimisation problem. *Medical Physics*, 29:2584–2589, 2002.

[6] M. Alber and F. Nüsslin. An objective function for radiation treatment optimization based on local biological measures. *Physics in Medicine and Biology*, 44(2):479–493, 1999.

[7] M. Alber and F. Nüsslin. Intensity modulated photon beams subject to a minimal surface smoothing constraint. *Physics in Medicine and Biology*, 45:N49–N52, 2000.

[8] M. Alber and F. Nüsslin. Optimization of intensity modulated radiotherapy under constraints for static and dynamic MLC delivery. *Physics in Medicine and Biology*, 46:3229–3239, 2001.

[9] M. Alber and F. Nüsslin. Ein Konzept zur Optimierung von klinischer IMRT. *Zeitschrift für Medizinische Physik*, 12:109–113, 2002.

[10] M. Alber and R. Reemtsen. Intensity modulated radiotherapy treatment planning by use of a barrier-penalty multiplier method. *Optimization Methods and Software*, 22:391–411, 2007.

[11] M.D. Altschuler and Y. Censor. Feasibility solutions in radiation therapy treatment planning. In *Proceedings of the Eighth International Conference on the Use of Computers in Radiation Therapy*, pages 220–224, Silver Spring, Maryland, USA, 1984. IEEE Computer Society Press.

[12] M.S. Bazaraa, H.D. Sherali, and C.M. Shetty. *Nonlinear Programming — Theory and Algorithms*. Wiley, New York, 1993.

[13] G. Bednarz, D. Michalski, C. Houser, M.S. Huq, Y. Xiao, P. R. Anne, and J.M. Galvin. The use of mixed-integer programming for inverse treatment planning with predefined field segments. *Physics in Medicine and Biology*, 47:2235–2245, 2002.

[14] D.P. Bertsekas. *Nonlinear Programming*. Athena Scientific, Belmont, Massachusetts, 2nd edition, 1999.

[15] N. Boland, H.W. Hamacher, and F. Lenzen. Minimizing beam-on time in cancer radiation treatment using multileaf collimators. *Networks*, 43:226–240, 2004.

[16] T. Bortfeld. Dosiskonfirmation in der Tumortherapie mit externer ionisierender Strahlung: Physikalische Möglichkeiten und Grenzen. Habilitationsschrift, Universität Heidelberg, Heidelberg, Germany, 1995.

[17] T. Bortfeld. Optimized planning using physical objectives and constraints. *Seminars in Radiation Oncology*, 9:20–34, 1999.

[18] T. Bortfeld, A.L. Boyer, W. Schlegel, D.L. Kahler, and T.J. Waldron. Realization and verification of three-dimensional conformal radiotherapy with modulated fields. *International Journal of Radiation Oncology Biology Physics*, 30:899, 1994.

[19] T. Bortfeld, J. Bürkelbach, R. Boesecke, and W. Schlegel. Methods of image reconstruction from projections applied to conformation radiotherapy. *Physics in Medicine and Biology*, 35:1423–1434, 1990.

[20] T. Bortfeld, W. Schlegel, C. Dykstra, S. Levegrün, and K. Preiser. Physical vs biological objectives for treatment plan optimization. *Radiotherapy & Oncology*, 40(2):185, 1996.

[21] T. Bortfeld, J. Stein, and K. Preiser. Clinically relevant intensity modulation optimization using physical criteria. In D.D. Leavitt and G. Starkschall, editors, *XIIth International Conference on the Use of Computers in Radiation Therapy*, pages 1–4. Medical Physics Publishing, Madison, WI, 1997.

[22] T. Bortfeld, C. Thieke, K.-H. Küfer, M. Monz, A. Scherrer, and H. Trinkhaus. Intensity-modulated radiotherapy — a large scale multi-criteria programming problem. Technical Report ITWM, Nr. 43, Fraunhofer Institut für Techno- und Wirtschaftsmathematik, Kaiserslautern, Germany, 2003.

[23] A. Brahme. Optimisation of stationary and moving beam radiation therapy techniques. *Radiotherapy & Oncology*, 12:129–140, 1988.

[24] A. Brahme. Treatment optimization using physical and radiobiological objective functions. In A.R. Smith, editor, *Radiation Therapy Physics*, pages 209–246. Springer, Berlin, 1995.

[25] A. Brahme and B.K. Lind. The importance of biological modeling in intensity modulated radiotherapy optimization. In D.D. Leavitt and G. Starkschall, editors, *XIIth International Conference on the Use of Computers in Radiation Therapy*, pages 5–8. Medical Physics Publishing, Madison, WI, 1997.

[26] A. Brahme, J.E. Roos, and I. Lax. Solution of an integral equation encountered in rotation therapy. *Physics in Medicine and Biology*, 27:1221–1229, 1982.

[27] A. Brahme, J.E. Roos, and I. Lax. Solution of an integral equation in rotation therapy. *Medical Physics*, 27:1221, 1982.

[28] R.E. Burkard, H. Leitner, R. Rudolf, T. Siegl, and E. Tabbert. Discrete optimization models for treatment planning in radiation therapy. In H. Hutten, editor, *Science and Technology for Medicine: Biomedical Engineering in Graz*, pages 237–249. Pabst Science Publishers, Lengerich, 1995.

[29] F. Carlsson and A. Forsgren. Iterative regularization in intensity modulated radiation therapy optimization. *Medical Physics*, 33:225–234, 206.

[30] Y. Censor. Mathematical optimization for the inverse problem of intensity-modulated radiation therapy. In J.R. Palta and T.R. Mackie, editors, *Intensity-Modulated Radiation Therapy: The State of the Art*, pages 25–49. Medical Physics Publ., Madison, WI, 2003.

[31] Y. Censor, M. Altschuler, and W. Powlis. A computational solution of the inverse problem in radiation-therapy treatment planning. *Applied Mathematics and Computation*, 25:57–87, 1988.

[32] Y. Censor, W.D. Powlis, and M.D. Altschuler. On the fully discretized model for the inverse problem of radiation therapy treatment planning. In K.R. Foster, editor, *Proceedings of the Thirteenth Annual Northeast Bioengineering Conference, Vol. 1*, pages 211–214, New York, NY, USA, 1987. IEEE.

[33] Y. Censor and S.C. Shwartz. An iterative approach to plan combination in radiotherapy. *International Journal of Bio-medical Computing*, 24:191–205, 1989.

[34] Y. Censor and S.A. Zenios. *Parallel Optimization: Theory, Algorithms, and Applications*. Oxford University Press, Oxford, 1997.

[35] Y. Chen, D. Michalski, C. Houser, and J.M. Galvin. A deterministic iterative least-squares algorithm for beam weight optimization in conformal radiotherapy. *Physics in Medicine and Biology*, 47:1647–1658, 2002.

[36] P.S. Cho, S. Lee, R.J. Marks II, S. Oh, S.G. Sutlief, and M.H. Phillips. Optimization of intensity modulated beams with volume constraints using two methods: Cost function minimization and projections onto convex sets. *Medical Physics*, 25:435–443, 1998.

[37] P.S. Cho and R.J. Marks. Hardware-sensitive optimization for intensity modulated radiotherapy. *Physics in Medicine and Biology*, 45:429–440, 2000.

[38] M. Chu, Y. Zinchenko, S.G. Henderson, and M.B. Sharpe. Robust optimization for intensity modulated radiation therapy treatment planning under uncertainty. *Physics in Medicine and Biology*, 50:5463–5477, 2005.

[39] A.V. Chvetsov, D. Calvetti, J.W. Sohn, and T.J. Kinsella. Regularization of inverse planning for intensity-modulated radiotherapy. *Medical Physics*, 32:501–514, 2005.

[40] L. Collatz and W. Wetterling. *Optimization Problems*. Springer, New York, 1975.

[41] C. Cotrutz, M. Lahanas, C. Kappas, and D. Baltas. A multiobjective gradient based dose optimization algorithm for conformal radiotherapy. *Physics in Medicine and Biology*, 46:2161–2175, 2001.

[42] W. De Gersem, F. Claus, C. De Wagter, B. Van Duyse, and W. De Neve. Leaf position optimization for step-and-shoot IMRT. *International Journal of Radiation Oncology Biology Physics*, 51:1371–1388, 2001.

[43] J.O. Deasy. Multiple local minima in radiotherapy optimization problems with dose-volume constraints. *Medical Physics*, 24:1157–1161, 1997.

[44] M.A. Earl, D.M. Shepard, S. Naqvi, X.A. Li, and C.X. Yu. Inverse planning for intensity-modulated arc therapy using direct aperture optimization. *Physics in Medicine and Biology*, 48:1075–1089, 2003.

[45] M. Ehrgott and R. Johnston. Optimisation of beam directions in intensity modulated radiation therapy planning. *OR Spectrum*, 25:251–264, 2003.

[46] K. Engel. A new algorithm for optimal multileaf collimator field segmentation. *Discrete Applied Mathematics*, 152:35–51, 2005.

[47] K. Engel and E. Tabbert. Fast simultaneous angle, wedge, and beam intensity optimization in inverse radiotherapy planning. *Optimization and Engineering*, 6:393–419, 2005.

[48] M.C. Ferris, J. Lim, and D.M. Shepard. An optimization approach for radiosurgery treatment planning. *SIAM Journal on Optimization*, 13:921–937, 2003.

[49] M.C. Ferris, J. Lim, and D.M. Shepard. Radiosurgery treatment planning via nonlinear programming. *Annals of Operations Research*, 119:247–260, 2003.

[50] A.V. Fiacco. *Introduction to Sensitivity and Stability Analysis in Nonlinear Programming*. Academic Press, New York, 1983.

[51] R. Fletcher. *Practical Methods of Optimization*. John Wiley & Sons, Chichester, 2nd edition, 1991.

[52] A. Forsgren. On the behavior of the conjugate-gradient method on ill-conditioned problems. Technical Report TRITA-MAT-2006-OS1, Dept. of Math., KTH Stockholm, 2006.

[53] M. Goitein and A. Niemierko. Intensity modulated therapy and inhomogeneous dose to the tumor: A note of caution. *International Journal of Radiation Oncology Biology Physics*, 36:519–522, 1996.

[54] A. Gustaffson. *Development of a Versatile Algorithm for Optimization of Radiation Therapy*. PhD thesis, University of Stockholm, Stockholm, Sweden, 1996.

[55] A. Gustafsson, B.K. Lind, and A. Brahme. A generalized pencil beam algorithm for optimization of radiation therapy. *Medical Physics*, 21:343–356, 1994.

[56] A. Gustafsson, B.K. Lind, R. Svensson, and A. Brahme. Simultaneous optimization of dynamic multileaf collimation and scanning patterns or compensation filters using a generalized pencil beam algorithm. *Medical Physics*, 22:1141–1156, 1995.

[57] H.W. Hamacher and K.-H. Küfer. Inverse radiation therapy planning — a multiple objective optimization approach. *Discrete Applied Mathematics*, 118:145–161, 2002.

[58] H.W. Hamacher and F. Lenzen. A mixed-integer programming approach to the multileaf collimator problem. In W. Schlegel and T. Bortfeld, editors, *The Use of Computers in Radiation Therapy*, pages 210–212. Springer, Berlin-Heidelberg-New York, 2000.

[59] M. Hilbig. *Inverse Bestrahlungsplanung für intensitätsmodulierte Strahlenfelder mit Linearer Programmierung als Optimierungsmethode*. PhD thesis, Technische Universität München, München, Germany, 2003.

[60] J. Höffner. *New Methods for Solving the Inverse Problem in Radiation Therapy Planning*. PhD thesis, Universität Kaiserslautern, Kaiserslautern, Germany, 1996.

[61] A. Holder. Designing radiotherapy plans with elastic constraints and interior point methods. *Health Care Management Science*, 6:5–16, 2003.

[62] A. Holder and B. Salter. A tutorial on radiation oncology and optimization. In H. Greenberg, editor, *Tutorials on Emerging Methodologies and Applications in Operations Research*, pages 4.1–4.47. 2004.

[63] T. Holmes and T.R. Mackie. A comparison of three inverse treatment planning algorithms. *Physics in Medicine and Biology*, 39:91–106, 1994.

[64] T. Holmes and T.R. Mackie. A filtered backprojection dose calculation method for inverse treatment planning. *Medical Physics*, 21:303–313, 1994.

[65] Q. Hou, J. Wang, Y. Chen, and J.M. Galvin. An optimization algorithm for intensity modulated radiotherapy - the simulated dynamics with dose-volume constraints. *Medical Physics*, 30:61–68, 2003.

[66] D.H. Hristov and B.G. Fallone. An active set algorithm for treatment planning optimization. *Medical Physics*, 24:1455–1464, 1997.

[67] D.H. Hristov and B.G. Fallone. A continuous penalty function method for inverse treatment planning. *Medical Physics*, 25:208–223, 1998.

[68] A. Jackson, G.J. Kutcher, and E.D. Yorke. Probability of radiation-induced complications for normal tissues with parallel architecture subject to non-uniform irradiation. *Medical Physics*, 20:613–625, 1993.

[69] J. Jahn. *Vector Optimization*. Springer, Berlin-Heidelberg, 2004.

[70] T. Kalinowski. A duality based algorithm for multileaf collimator field segmentation with interleaf collision constraint. *Discrete Applied Mathematics*, 152:52–88, 2005.

[71] T. Kalinowski. *Optimal Multileaf Collimator Field Segmentation*. PhD thesis, Universität Rostock, Germany, 2005.

[72] T. Kalinowski. Reducing the number of monitor units in multileaf collimator field segmentation. *Physics in Medicine and Biology*, 50:1147–1161, 2005.

[73] P. Källman, B.K. Lind, and A. Brahme. An algorithm for maximizing the probability of complication free tumor control in radiation therapy. *Physics in Medicine and Biology*, 37:871–890, 1992.

[74] S. Kamath, S. Sahni, J. Palta, S. Ranka, and J. Li. Optimal leaf sequencing with elimination of tongue-and-groove. *Physics in Medicine and Biology*, 2004:N7–N19, 2004.

[75] M. Lahanas, E. Schreibmann, and D. Baltas. Multiobjective inverse planning for intensity modulated radiotherapy with constraint-free gradient-based optimization algorithms. *Physics in Medicine and Biology*, 48:2843–2871, 2003.

[76] M. Langer, R. Brown, M. Urie, J. Leong, M. Stracher, and J. Shapiro. Large scale optimization of beam weights under dose-volume restrictions. *International Journal of Radiation Oncology Biology Physics*, 18:887–893, 1990.

[77] M. Langer, V. Thai, and L. Papiez. Improved leaf sequencing reduces segments or monitor units needed to deliver IMRT using multileaf collimators. *Medical Physics*, 28:2450–2458, 2001.

[78] W. Laub, M. Alber, M. Birkner, and F. Nüsslin. Monte Carlo dose computation for IMRT optimization. *Physics in Medicine and Biology*, 45:1741–1754, 2000.

[79] E.K. Lee, T. Fox, and I. Crocker. Integer programming applied to intensity-modulated radiation therapy treatment planning. *Annals of Operations Research*, 119:165–181, 2003.

[80] J. Lim. *Optimization in Radiation Treatment Planning*. PhD thesis, University of Wisconsin, Madison, 2002.

[81] W.A. Lodwick, S. McCourt, F. Newman, and S. Humphries. Optimization methods for radiation therapy plans. In C. Börgers and F. Natterer, editors, *Computational Radiology and Imaging: Therapy and Diagnostics*, pages 229–249. Springer, Berlin, 1999.

[82] A. Lomax. Intensity modulated methods for proton radiotherapy. *Physics in Medicine and Biology*, 44:185–205, 1999.

[83] A.J. Lomax, T. Boehringer, A. Coray, E. Egger, G. Goitein, M. Grossmann, P. Juelke, S. Lin, E. Pedroni, B. Rohrer, W. Roser, B. Rossi, B. Siegenthaler, O. Stadelmann, H. Stauble, C. Vetter, and L. Wisser. Intensity modulated proton therapy: A clinical example. *Medical Physics*, 28:317–324, 2001.

[84] L. Ma, A. Boyer, L. Xing, and C.-M. Ma. An optimized leaf-setting algorithm for beam intensity modulation using dynamic multileaf collimators. *Physics in Medicine and Biology*, 43:1629–1643, 1998.

[85] G. Meedt, M. Alber, and F. Nüsslin. Non-coplanar beam direction optimization for intensity modulated radiotherapy. *Physics in Medicine and Biology*, 48(18):2999–3019, 2003.

[86] M. Merritt, Y. Zhang, H. Liu, and R. Mohan. A successive linear programming approach to IMRT optimization problem. Technical Report TR02-16, The Department of Computational & Applied Mathematics, Rice University, Houston, Texas, 2002.

[87] D. Michalski, Y. Xiao, Y. Censor, and J.M. Galvin. The dose-volume constraint satisfaction problem for inverse treatment planning with field segments. *Physics in Medicine and Biology*, 49:601–616, 2004.

[88] R. Mohan, X. Wang, A. Jackson, T. Bortfeld, A.L. Boyer, G.J. Kutcher, S.A. Leibel, Z. Fuks, and C.C. Ling. The potential and limitations of the inverse radiotherapy technique. *Radiotherapy & Oncology*, 32:232–248, 1994.

[89] R. Mohan and X.-H. Wang. Response to Bortfeld et al. re physical vs biological objectives for treatment plan optimization. *Radiotherapy & Oncology*, 40(2):186–187, 1996.

[90] S.M. Morill, R.G. Lane, J.A. Wong, and I.I. Rosen. Dose-volume considerations with linear programming optimization. *Medical Physics*, 18:1201–1210, 1991.

[91] T.R. Munro and C.W. Gilbert. The relation between tumor lethal doses and the radiosensitivity of tumor cells. *The British Journal of Radiology*, 34:246–251, 1961.

[92] A. Niemierko. Reporting and analyzing dose distributions: A concept of equivalent uniform dose. *Medical Physics*, 24:103–110, 1997.

[93] A. Niemierko. A generalized concept of equivalent uniform dose (EUD) (abstract). *Medical Physics*, 26:1100, 1999.

[94] S. Nill, T. Bortfeld, and U. Oelfke. Inverse planning of intensity modulated proton therapy. *Zeitschrift für Medizinische Physik*, 14:35–40, 2004.

[95] J. Nocedal and S.J. Wright. *Numerical Optimization*. Springer, New York-Berlin-Heidelberg, 1999.

[96] U. Oelfke and T. Bortfeld. Inverse planning for photon and proton beams. *Medical Dosimetry*, 26:113–124, 2001.

[97] A. Ólafsson, R. Jeraj, and S.J. Wright. Optimization of intensity-modulated radiation therapy with biological objectives. *Physics in Medicine and Biology*, 50:5357–5379, 2005.

[98] A. Ólafsson and S.J. Wright. Efficient schemes for robust IMRT treatment planning. Technical Report Optimization TR 06-01, Department of Computer Science, University of Wisconsin, Madison, 2006.

[99] A. Ólafsson and S.J. Wright. Linear programming formulations and algorithms for radiotherapy treatment planning. *Optimization Methods and Software*, 21:201–231, 2006.

[100] F. Preciado-Walters, R. Rardin, M. Langer, and V. Thai. A coupled column generation, mixed integer approach to optimal planning of intensity modulated radiation therapy for cancer. *Mathematical Programming*, 101:319–338, 2004.

[101] C. Raphael. Mathematical modelling of objectives in radiation therapy treatment planning. *Physics in Medicine and Biology*, 37:1293–1311, 1992.

[102] H.E. Romeijn, R.K. Ahuja, J.F. Dempsey, and A. Kumar. A column generation approach to radiation therapy treatment planning using aperture modulation. *SIAM Journal on Optimization*, 15:838–862, 2005.

[103] H.E. Romeijn, R.K. Ahuja, J.F. Dempsey, A. Kumar, and J.G. Li. A novel linear programming approach to fluence map optimization for intensity modulated radiation therapy treatment planning. *Physics in Medicine and Biology*, 48:3521–3542, 2003.

[104] H.E. Romeijn, J.F. Dempsey, and J.G. Li. A unifying framework for multicriteria fluence map optimization models. *Physics in Medicine and Biology*, 49:1991–2013, 2004.

[105] I.I. Rosen, R.G. Lane, S.M. Morrill, and J.A. Belli. Treatment plan optimization using linear programming. *Medical Physics*, 18:141–152, 1991.

[106] G.R. Rowbottom and S. Webb. Configuration space analysis of common cost functions in radiotherapy beam-weight optimization algorithms. *Physics in Medicine and Biology*, 47:65–77, 2002.

[107] J. Seco, P.M. Evans, and S. Webb. Modelling the effects of IMRT delivery: constraints and incorporation of beam smoothing into inverse planning. In W. Schlegel and T. Bortfeld, editors, *Proceedings of the XIII International Conference on the Use of Computers in Radiation Therapy*, pages 542–544, Heidelberg, 2000. Springer.

[108] D.M. Shepard, M.C. Ferris, G.H. Olivera, and T.R. Mackie. Optimizing the delivery of radiation therapy to cancer patients. *SIAM Review*, 41:721–744, 1999.

[109] D.M. Shepard, G.H. Olivera, P.J. Reckwerdt, and T.R. Mackie. Iterative approaches to dose optimization in tomotherapy. *Physics in Medicine and Biology*, 45:69–90, 2000.

[110] J.V. Siebers, M. Lauterbach, P.J. Keall, and R. Mohan. Incorporating multileaf collimator leaf sequencing into iterative IMRT optimization. *Medical Physics*, 29:952–959, 2002.

[111] R.A.C. Siochi. Minimizing static intensity modulation delivery time using an intensity solid paradigm. *International Journal of Radiation Oncology Biology Physics*, 42:671–680, 1999.

[112] S. Söderström, A. Gustafsson, and A. Brahme. The clinical value of different treatment objectives and degrees of freedom in radiation therapy optimization. *Radiotherapy & Oncology*, 29:148–163, 1993.

[113] S. Söderström, A. Gustafsson, and A. Brahme. Few-field radiation therapy optimization in the phase space of complication-free tumor control. *International Journal of Imaging Systems and Technology*, 6:91–103, 1995.

[114] P. Spellucci. *Numerische Verfahren der nichtlinearen Optimierung.* Birkhäuser, Basel-Boston-Berlin, 1993.

[115] S.V. Spirou and C.-S. Chui. A gradient inverse planning algorithm with dose-volume constraints. *Medical Physics*, 25:321–333, 1998.

[116] J. Stein, R. Mohan, X.-H. Wang, T. Bortfeld, Q. Wu, K. Preiser, C.C. Ling, and W. Schlegel. Number and orientations of beams in intensity-modulated radiation treatments. *Medical Physics*, 24:149–160, 1997.

[117] J. Tervo and P. Kolmonen. A model for the control of a multileaf collimator in radiation therapy treatment planning. *Inverse Problems*, 16:1875–1895, 2000.

[118] J. Tervo, P. Kolmonen, T. Lyyra-Laitinen, J.D. Pintér, and T. Lahtinen. An optimization-based approach to the multiple static delivery technique in radiation therapy. *Annals of Operations Research*, 119:205–227, 2003.

[119] J. Tervo, T. Lyyra-Laitinen, P. Kolmonen, and E. Boman. An inverse treatment planning model for intensity modulated radiation therapy with dynamic MLC. *Applied Mathematics and Computation*, 135:227–250, 2003.

[120] C. Thieke. *Multicriteria Optimization in Inverse Radiotherapy Planning.* PhD thesis, University of Heidelberg, Heidelberg, Germany, 2003.

[121] C. Thieke, T.R. Bortfeld, and K.-H. Küfer. Characterization of dose distributions through the max and mean dose concept. *Acta Oncologica*, 41:158–161, 2002.

[122] C. Thieke, T. Bortfeld, A. Niemierko, and S. Nill. From physical dose constraints to equivalent uniform dose constraints in inverse radiotherapy planning. *Medical Physics*, 30:2332–2339, 2003.

[123] X-H. Wang, R. Mohan, A. Jackson, S.A. Leibel, Z. Fuchs, and C.C. Ling. Optimization of intensity-modulated 3d conformal treatment plans based on biological indices. *Radiotheraphy & Oncology*, 37:140–152, 1995.

[124] S. Webb. *The Physics of Conformal Radiotherapy: Advances in Technology.* Medical Science Series. IOP Publishing, Bristol, 1997.

[125] S. Webb. *Intensity-Modulated Radiation Therapy.* Medical Science Series. IOP Publishing, Bristol, 2000.

[126] J. Werner. *Optimization Theory and Applications.* Vieweg, Braunschweig, Germany, 1984.

[127] Q. Wu, D. Djajaputra, Y. Wu, J. Zhou, H.H. Liu, and R. Mohan. Intensity-modulated radiotherapy optimization with gEUD-guided dose-volume objectives. *Physics in Medicine and Biology*, 48:279–291, 2003.

[128] Q. Wu and R. Mohan. Algorithms and functionality of an intensity modulated radiotherapy optimization system. *Medical Physics*, 27:701–711, 2000.

[129] Q. Wu, R. Mohan, A. Niemierko, and R. Schmidt-Ullrich. Optimization of intensity-modulated radiotherapy plans based on the equivalent uniform dose. *International Journal of Radiation Oncology Biology Physics*, 52:224–235, 2002.

[130] P. Xia and L.J. Verhey. Multileaf collimator leaf sequencing algorithm for intensity modulated beams with multiple static segments. *Medical Physics*, 25:1424–1434, 1998.

[131] Y. Xiao, D. Michalski, J.M. Galvin, and Y. Censor. The least-intensity feasible solution for aperture-based inverse planning in radiation therapy. *Annals of Operations Research*, 119:183–203, 2003.

[132] L. Xing and G.T.Y. Chen. Iterative methods for inverse treatment planning. *Physics in Medicine and Biology*, 41:2107–2123, 1996.

[133] L. Xing, R.J. Hamilton, D. Spelbring, C.A. Pelizzari, G.T.Y. Chen, and A.L. Boyer. Fast iterative algorithms for three-dimensional inverse treatment planning. *Medical Physics*, 25:1845–1849, 1998.

[134] L. Xing, J.G. Li, S. Donaldson, Q.T. Le, and A.L. Boyer. Optimization of importance factors in inverse planning. *Physics in Medicine and Biology*, 44:2525–2536, 1999.

[135] C.X. Yu. Intensity-modulated arc therapy with dynamic multileaf collimation: An alternative to tomotherapy. *Physics in Medicine and Biology*, 40:1435–1449, 1995.

[136] C.X. Yu, X.A. Li, L. Ma, D. Chen, S. Naqvi, D. Shepard, M. Sarfaraz, T.W. Holmes, M. Suntharalingam, and C.M. Mansfield. Clinical implementation of intensity-modulated arc therapy. *International Journal of Radiation Oncology Biology Physics*, 53:453–463, 2002.

[137] X. Zhang, H. Liu, X. Wang, L. Dong, Q. Wu, and R. Mohan. Speed and convergence properties of gradient algorithms for optimization of IMRT. *Medical Physics*, 31:1141–1152, 2004.

[138] Y. Zhang and M. Merritt. Fluence map optimization in IMRT cancer treatment planning and a geometric approach. In W.W. Hager, S.-J. Huang, P.M. Pardalos, and O.A. Prokopyev, editors, *Multiscale Optimization Methods and Applications*, pages 205–228. Springer, Berlin-Heidelberg-New York, 2006.

# 5

# Multicriteria Optimization in Intensity Modulated Radiotherapy Planning

Karl-Heinz Küfer[1], Michael Monz[1], Alexander Scherrer[1], Philipp Süss[1], Fernando Alonso[1], Ahmad Saher Azizi Sultan[1], Thomas Bortfeld[2], and Christian Thieke[3]

[1] Department of Optimization, Fraunhofer Institut for Industrial Mathematics (ITWM), Gottlieb-Daimler-Straße 49, D-67663 Kaiserslautern, Germany
`{kuefer,monz,scherrer,suess}@itwm.fhg.de`
[2] Department of Radiation Oncology, Massachusetts General Hospital and Harvard Medical School, 30 Fruit Street, Boston, Massachusetts 02114
`tbortfeld@partners.org`
[3] Clinical Cooperation Unit Radiation Oncology, German Cancer Research Center (DKFZ), Im Neuenheimer Feld 280, D-69120 Heidelberg, Germany
`c.thieke@dkfz-heidelberg.de`

**Abstract.** The inverse treatment planning problem of IMRT is formulated as a multicriteria optimization problem. The problem is embedded in the more general family of *design problems*. The concept of *virtual engineering*, when interpreted as an optimization paradigm for design problems, reveals favorable structural properties. The numerical complexity of large-scale instances can then be significantly reduced by an appropriate exploitation of a structural property called *asymmetry*.

Methods to treat the multicriteria problem appropriately are developed. The methods proposed serve as ingredients for a system that incorporates (a) calculations of efficient IMRT plans possessing high clinical quality and (b) an interactive decision-making framework to select solutions. The plan calculations are done fast even for relatively large dimensions by exploiting the asymmetry property. They result in a database of plans that delimit a large set of clinically relevant plans. A sophisticated navigation scheme allows one to obtain plans that conform to the preference of the decision-maker while conveying the chances and limitations of each interaction with the system. The resulting workflow can be embedded into the clinical decision-making process to truly address the multicriteria setting inherent to IMRT planning problems.

## 5.1 The IMRT Treatment Planning Problem

Radiotherapy is, besides surgery, the most important treatment option in clinical oncology. It is used with both curative and palliative intention, either solely or in combination with surgery and chemotherapy. The vast majority of all radiotherapy patients is treated with high-energy photon beams. Hereby,

P.M. Pardalos, H.E. Romeijn (eds.), *Handbook of Optimization in Medicine*, Springer Optimization and Its Applications 26, DOI: 10.1007/978-0-387-09770-1_5, © Springer Science+Business Media LLC 2009

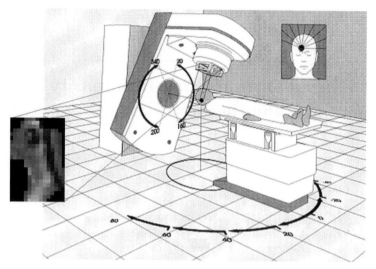

**Fig. 5.1.** The gantry moves around the couch on which the patient lies. The couch position may also be changed to alter the beam directions.

the radiation is produced by a linear accelerator and delivered to the patient by several beams coming from different directions (see Figure 5.1).

In conventional conformal radiation therapy, only the outer shape of each beam can be smoothly adapted to the individual target volume. The intensity of the radiation throughout the beam's cross section is uniform or only modified by the use of pre-fabricated wedge filters. This, however, limits the possibilities to fit the shape of the resulting dose distribution in the tissue to the shape of the tumor, especially in the case of irregularly shaped non-convex targets like para-spinal tumors.

This limitation is overcome by a technique called *intensity modulated radiation therapy* (IMRT) [77]. Using multileaf collimators (MLCs) (see Figure 5.2), the intensity is modulated by uncovering parts of the beam only for individually chosen opening times (monitor units) and covering the rest of the beam opening by the collimator leafs. This allows for a more precise therapy by tightly conforming the high dose area to the tumor volume.

An IMRT treatment plan is physically characterized by the beam arrangement given by the angle of the couch relative to the gantry and the rotation angle of the gantry itself and by the intensities on each beam (see Figure 5.1). The treatment aim is to deliver sufficient radiation to the tumor while sparing as much of the healthy tissue as possible. Finding ideal balances between these inherently contradictory goals challenges dosimetrists and physicians in their daily practice.

The *treatment planning problem* is to find an optimal set of parameters describing the patient treatment. Although the choice of a particular delivery

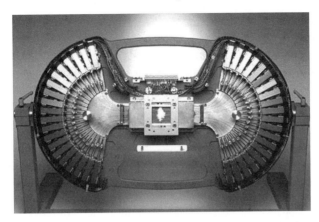

**Fig. 5.2.** A multileaf collimator (MLC). The square opening in the center of the machine is partially covered by leafs, each of which can be individually moved (picture from [66]).

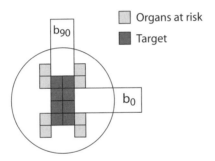

**Fig. 5.3.** Optimization of the setup geometry is highly non-convex even for the single-beam case.

scheme (radiation modality, energies, fractionation, etc.) is by far not a trivial decision to make, it will be considered given in our context.

Finding an optimal setup geometry and an optimal set of intensity maps using a global optimization model [13, 40] has the disadvantage that the resulting problem is highly non-convex, as the following single-beam example demonstrates. Assume the target is given by the rectangular shape in the middle of Figure 5.3 and the smaller rectangular structures at the corners of the target are critical volumes.

The beam may be arranged anywhere on the depicted circle. Then the optimal beam direction is given by beam $b_{90}$. A search for this optimum may, however, get stuck in a local minimum like beam $b_0$. Notice that the search would have to make a very big change in the current solution to leave this local optimum. Coupling beam orientation search with the optimization of intensity maps results in a non-convex objective function, and traditional search methods designed for convex situations are prone to get stuck in local optima [8]. Consequently, empirical and heuristic search methods with no quality guarantees are used to decide on the geometry setup. Most studies that have addressed the problem of beam orientation in IMRT employ stochastic optimization approaches including evolutionary or simulated annealing algorithms in which intensity map optimization is performed for every individual selection of beam orientations [8, 54, 69].

Beam geometry optimization is still an interesting problem in its own right and it has been addressed in many publications (see, e.g., [46, 53, 54] and the references listed there). Brahme [11] mentions some techniques useful to handle setups with only a few beams. However, in this chapter we will not discuss any such solution approaches but merely assume that the irradiation geometry is given.

On the other hand, a very detailed search on the potential beam positions is in many cases not necessary. If the critical part of the body is covered by the beam's irradiation, the optimization of the intensity maps will mitigate the errors due to non-optimal beam directions substantially. However, in some cases like head-and-neck treatments, computer-based setup optimization might be of considerable value.

Aside from the dose distribution resulting from the beam geometry, its complexity also has to be considered when evaluating the quality of a treatment plan. In most cases (see, e.g., [7, 9, 11, 25]), an *isocentric* model is used for the choice of the setup geometry, i.e., the central rays of the irradiation beams meet in one single point, the *isocenter of irradiation* (see Figure 5.4).

To further facilitate an automated treatment delivery, usually a coplanar beam setting is used. Then the treatment can be delivered completely by

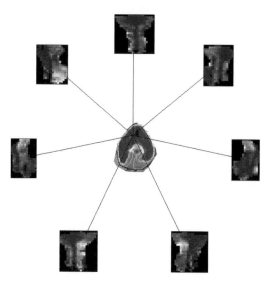

**Fig. 5.4.** Intensity maps for different beam directions intersecting at the tumor.

just rotating the gantry around the patient without the need to rotate or translate the treatment couch between different beams. This leads to shorter overall treatment times, which are desirable both for stressing the patient as little as possible and for minimizing the treatment costs.

In the classic approach to solve the treatment planning problem, "forward treatment planning" has been the method of choice. The parameters that characterize a plan are set manually by the planner, and then the dose distribution in the patient is calculated by the computer. If the result is not satisfactory, the plan parameters are changed again manually, and the process starts over. As the possibilities of radiotherapy become more sophisticated (namely with intensity modulation), Webb argues [76, Chapter 1.3] that *"it becomes quite impossible to create treatment plans by forward treatment planning because:*

- *there are just too many possibilities to explore and not enough human time to do this task*
- *there is little chance of arriving at the optimum treatment plan by trial-and-error*
- *if an acceptable plan could be found, there is no guarantee it is the best, nor any criteria to specify its precision in relation to an optimum plan."*

Furthermore, there is no unified understanding of what constitutes an optimal treatment plan, as the possibilities and limitations of treatment planning are case-specific.

Therapy planning problems have in the past two decades been modeled using an *inverse* or *backward* strategy (see the survey paper [11]): given desired dose bounds, parameters for the treatment setup are found using computerized models of computation.

The approach to work on the problem from a description of a desired solution to a configuration of parameters that characterizes it is an established approach in product design called *virtual engineering.*

## 5.2 Optimization as a Virtual Engineering Process

In this section, we introduce our understanding of the concept of virtual engineering, together with the involved mathematical structures. We then argue that multicriteria optimization is the appropriate tool to realize virtual engineering. Then we discuss existing strategies to cope with multicriteria optimization problems and introduce our method.

Virtual engineering has been used in various disciplines, and there are many ways to interpret its meaning. In software development, for example, the functionality of a program is specified before the first line of code is written. Another typical example of virtual engineering is a company using an envisioned or existing product as a model for development. It is possible to extract the same concept of all formulations of virtual engineering: one of an inverse approach to solve a problem. This means that the solution is obtained from

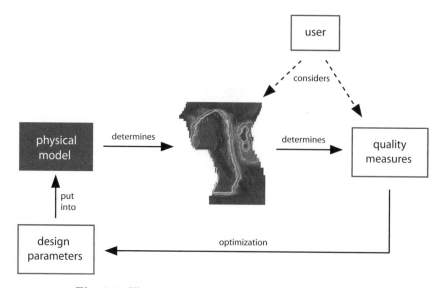

**Fig. 5.5.** Illustration of the virtual engineering concept.

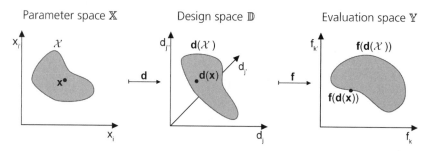

**Fig. 5.6.** Illustration of the spaces involved in the general design problem.

the specification of ideal properties, rather than from a syntactical description of parameters and some recipe. The setting we find in IMRT planning is typical of a large-scale design problem. Refer to Figures 5.5 and 5.6 for an illustration of the following discussion. In order to maintain an abstract setting here, we will only assume that there exists the need to tailor a complex design $\mathbf{d}(\mathbf{x})$ contained in the design space $\mathbb{D}$ (a physical product, a treatment plan, a financial portfolio, etc.) depending on an element $\mathbf{x}$ of the space of parameters $\mathbb{X}$ (part configurations, beamlet intensities, investments, etc.) that fulfills certain constraints. Given design parameters that distinguish a solution, it can be simulated with the help of a "virtual design tool." This aid possesses the capability to virtually assemble the final solution when given the setup parameters and relevant constraints, effectively constructing an element of $\mathbb{D}$.

This element is then evaluated using functions $f_k : \mathbb{D} \to \mathbb{R}$. Their combination yields a vector valued mapping $\mathbf{f} = (f_k)_{k \in \mathcal{K}} : \mathbb{D} \to \mathbb{Y}$ from $\mathbb{D}$ into an evaluation space $\mathbb{Y}$. The elements $\mathbf{f}(\mathbf{d}) \in \mathbb{Y}$ provide a condensed information on the design quality and thus direct the design process. Formulation of restrictions and desirable goals on the evaluations yields an optimization problem based on the criterion functions $F_k = f_k \circ \mathbf{d}$.

Focusing on the design process, the main difference to forward engineering is that the reverse approach is a more systematic solution procedure. Although the iterations in forward engineering resemble a trial-and-error approach, it is the mathematical optimization that is characteristic for our virtual engineering concept.

Because a complex design usually cannot be assigned a single "quality score" accepted by any decision-maker, there are typically several criterion functions $F_k, k \in \mathcal{K}$. Thus, the problem is recognized as a *multicriteria optimization problem*

$$\mathbf{F}(\mathbf{x}) \to \min \quad \text{subject to} \tag{5.1}$$
$$\mathbf{x} \in \mathcal{X} \subseteq \mathbb{X},$$

where $\mathbf{F}(\mathbf{x}) = (F_k(\mathbf{x}))_{k \in \mathcal{K}}$, and $\mathcal{X}$ is the set of all $\mathbf{x}$ that fulfill the problem-specific constraints.

The quest for a solution that has a single maximal quality score should be transformed to one for solutions that are *Pareto optimal* [24] or *efficient*. A solution satisfying this criterion has the property that none of the individual criteria can be improved while at least maintaining the others. The affirmative description is: if a Pareto optimal solution is improved in one criterion, it will worsen in at least one other criterion. A solution is *weakly Pareto optimal* or *weakly efficient* if there is no solution for which all the criterion functions can be improved simultaneously. Or, put differently: if the solution is improved in one criterion, there is at least one criterion that cannot be improved. The set of all Pareto optimal solutions in $\mathcal{X}$ is called the *Pareto set* and denoted by $\mathcal{X}_{\text{Par}}$. The set of all evaluations of the Pareto set is called the *Pareto boundary* and denoted by $\mathbf{F}(\mathcal{X}_{\text{Par}})$.

### 5.2.1 Multicriteria optimization strategies

A realization of the conventional forward strategies mentioned before is a method we label the "Human Iteration Loop": this is depicted in Figure 5.7. This strategy has several pitfalls that are all avoidable. First, the decision-maker is forced to transform the several criteria into a scalarized objective function. A *scalarization* of a multicriteria optimization problem is a transformation of the original problem into a single or a family of scalar optimization problems to create single solutions that are at least weakly efficient. A standard scalarization, for example, is the weighted sum approach (see Section 5.3.3). Weights in this scalarization approach are nothing but an

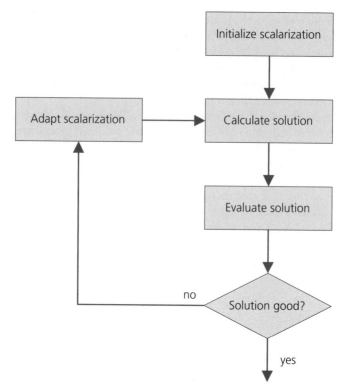

**Fig. 5.7.** Human iteration loop described as method of *successive improvements*.

attempt to translate the decision-maker's ideal into artificial weights. This artificial nature results from having to quantify global trade-off rates between different criteria that are often of very different nature. We will address some issues with scalarizations in Section 5.3.

Further, given a large dimension of $\mathbb{X}$, it is impossible to ask the decision-maker for optimization parameters like weights that directly lead to an ideal solution. An iterative adjustment of the parameters converges to a solution that hardly incorporates any wishes that are not explicitly modeled. It is perhaps even presumptuous to expect a decision-maker to specify an ideal solution in terms of the criteria exactly, let alone to ask for exact global trade-offs between objectives.

Moreover, initial demands on the solution properties might very well be reverted when the outcome is seen as a whole. This may be a result of the realization that the initial conviction of an ideal was blemished or simply the realization that the description was not complete. In any case, any model of an ideal that is "cast in stone" and inflexible is detrimental to the design process. A truly multicriteria decision-making framework would allow for flexible modeling because it is able to depict more information.

With a large number of criteria comes a high-dimensional Pareto boundary and with this a large number of directions to look for potential solutions for our problem. For this reason, there exist methods that attempt to convey the shape of the boundary to the decision-maker. This is most often done implicitly, either by giving the decision maker a myriad of solutions that approximate the boundary or by interactively calculating representative solutions that are Pareto optimal. Miettinen [47, Part II] adopts the following classification of methods to attain practical solutions:

1. **no-preference methods**
   methods where no preference information is used, i.e., methods that work without specification of any preference parameters
2. ***a posteriori* methods**
   preference is used *a posteriori*, i.e., the system generates a set of Pareto optimal solutions of which the decision maker selects one
3. ***a priori* methods**
   where the decision maker has to specify his preferences through some parameters prior to the calculation and
4. **interactive methods**
   where preferences are not necessarily stated prior to calculations, but are revealed throughout.

The large-scale situation in treatment planning forbids the application of purely interactive methods, and pure *a priori* methods are not flexible enough for our demands. With *a posteriori* methods in the strict sense usually comes a complex re-evaluation after the actual optimization, whereas no-preference methods do not allow any goal-directed planning.

We therefore develop in the later sections a hybrid method where some information is given *a priori* and used to automatically attain a set of Pareto optimal points. The final plan is selected using a real-time interactive method that works on pre-computed plans obtained via an *a posteriori* method. In a sense, the methodology described here incorporates advantages of all the methods classified above. Unlike the Human Iteration Loop, our method does not require the decision-maker to formulate an individual scalarization problem. Rather, after specifying aspired values and bounds, the user will be presented with a database of Pareto optimal solutions calculated offline, which can be navigated in real-time in a clever way to choose a plan that is in accordance with the preferences of the decision-maker.

In Section 5.3, we will describe how the solutions are obtained, and in Section 5.5, we will address the issues faced when selecting from a range of solutions. Note that the pre-computation of plans is done without any human interaction, thus taking the *Human* out of the mundane *Iteration Loop*. But even if the user is left out of the optimization, with high dimensions of the spaces involved comes a costly computation of a candidate solution set. Thus, the problems that need to be solved have to be manageable by an optimization routine.

Specifying properties a solution should have, to return to the idea of virtual engineering, implicitly places restrictions on the parameters. As the parameters are not evaluated directly, the mapping $\mathbf{d} : \mathbb{X} \to \mathbb{D}$ necessarily becomes a "subroutine" in the iterations of the optimization process – only a design can be qualitatively judged in a meaningful way. Many descent algorithms used in optimization evaluate the objective function rather often during run-time, making this subroutine a significant driver for the complexity.

In applications, this subroutine often corresponds with time-consuming simulations. In the IMRT case it is the dose mapping, which is a costly calculation if the degree of discretization is rather fine. A method to cope with this computational difficulty has to be established. Fortunately, typical design problems have a numerical property that, if exploited, makes the many necessary computations possible: asymmetry.

### 5.2.2 Asymmetry in linear programming

Typical design problems are *asymmetric* in the sense that the number of parameters is rather small compared with the description of the corresponding design like, e.g., its characterization by the criterion values or its discrete representation in the design space with $dim(\mathbb{X}) \ll dim(\mathbb{D})$.

The latter case is well-known in linear optimization: according to [6], the number of pivot steps that a simplex method needs to reach the solution of a linear optimization problem strongly depends on the surface structure of the polyhedral feasible region in the vicinity of the solution, i.e., on the number of linear constraints defining facets close to the optimum.

Furthermore, in absence of strong degeneracy, the number of linear constraints that characterize a feasible element $\mathbf{x} \in \mathbb{X}$ of the parameter space is about $dim(\mathbb{X})$. If, for example, these constraints arise from bounds on the different components of the corresponding design $\mathbf{d}(\mathbf{x}) \in \mathbb{D}$ obtained under the linear mapping $\mathbf{d} : \mathbb{X} \to \mathbb{D}$, there are rather few active constraints in comparison with the other roughly $dim(\mathbb{D}) - dim(\mathbb{X})$ inactive constraints, which play no role in this particular $\mathbf{x}$.

This asymmetry is often exploited by aggregation methods, see [22]. Consider the linear problem

$$\mathbf{c}^T \mathbf{x} \to \min \quad \text{subject to} \qquad (5.2)$$
$$\mathbf{Ax} \leq \mathbf{b}$$

where $\mathbf{c} \in \mathbb{X}$ and $\mathbf{A} \in \mathbb{R}^{\dim(\mathbb{D}) \times \dim(\mathbb{X})}$ is a matrix with the row vectors $\mathbf{a}_j$, i.e., $\mathbf{d}(\mathbf{x}) = \mathbf{Ax}$.

If there were comparably few inequalities $\mathbf{a}_j \cdot \mathbf{x} \leq b_j$ that are fulfilled with equality in a neighborhood of the solution $\mathbf{x}^*$ and thus require exact knowledge of the values $\mathbf{a}_j \cdot \mathbf{x}$ to characterize the solution, using more or less exact approximations of the other values would not affect the solution at all. In other words, one could get away with an approximate $\mathbf{A}'$, which is ideally of a much simpler form and thus allows a faster evaluation of the approximate mapping $\mathbf{d}' : \mathbf{x} \mapsto \mathbf{A}'\mathbf{x}$.

Aggregation methods would then construct such an $\mathbf{A}'$ by replacing families of similar inequalities by single surrogate ones that form an approximation of the surface structure of the feasible region with a moderate approximation error. An aggregation method called the *adaptive clustering method*, which was invented in the context of IMRT plan optimization, is presented in Section 5.4.

To summarize,

1. virtual engineering problems naturally lend themselves to formulations of multicriteria optimization problems,
2. the multicriteria setting can be coped with by appropriate optimization methods and clever schemes for selecting a solution (presented in later sections), and
3. in order to manage the computations, the asymmetry inherent to many design problems can be exploited.

In the following section we describe the *a posteriori* part of the multicriteria framework for the treatment planning problem. The interactive component is the subject of Section 5.5.

## 5.3 Multicriteria Optimization

In IMRT, the multicriteria setting stems from the separate evaluations of a dose distribution in the various volumes of interest (VOIs). As there is typically no solution that simultaneously optimizes all criterion functions, there exist trade-offs in changing from one treatment plan to another.

There are many examples in which some organs are natural "opponents." In cases of prostate cancer, the rectum and the bladder are such opponents (see Figure 5.8). In head-and-neck cases, sparing more of the spinal cord typically means inflicting more damage on the brain stem or the parotid glands.

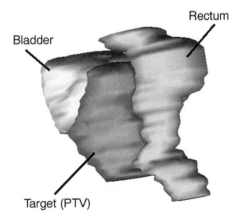

**Fig. 5.8.** Exemplary prostate case where the target volume is situated between two critical structures.

Another example of conflicting goals is the aspired homogeneity of the dose distribution in the target versus the overdosage of some risk VOI. Also, choosing more than one criterion for a volume renders the problem multicriterial.

IMRT planning using multicriteria optimization formulations and techniques has been a fruitful area of research in recent years. Among the earliest approaches is the weighted sum method of Haas [27] who employed a genetic algorithm to search for "good" scalarization weights. Some ideas to include the decision-maker further into the solution generation process was developed by a subset of the authors of [27] in [26]. In essence, this was another form of a Human Iteration Loop. Cotrutz et al. [15] first applied multicriteria optimization to inverse IMRT planning. However, they could only achieve reasonable computation times for the case of two to three criteria.

Multiobjective linear programming formulations were also proposed, for example in [29] or [28]. Holder [29] applied some results from interior point methods to attain solutions of a multiobjective linear programming formulation with different solution characteristics. Hamacher and Küfer [28] put more focus on "attractive" dose distributions by first formulating a (mixed integer, linear) inequality system to specify allowable ranges for dose values in volume parts and then minimizing maximal deviations from this range of "ideal" dose. They proposed a continuous relaxation to be solved by standard linear programming (LP) techniques. Küfer et al. [37] use a linear model based on the equivalent uniform dose concept presented in [71] by a subset of the authors. Yet, the restriction to linear modeling limits the possibilities to formulate clinically meaningful objective functions.

There have also been several suggestions of global nonlinear optimization models. Solutions to these problems are most often found by some evolutionary scheme or a randomized algorithm. In both cases, it is very difficult to make statements about the quality of the solutions. Lahanas et al. [38] describe an approach to a global formulation, together with some decision support using projections of the Pareto front on 2 dimensions. Their methodology suffers from the non-convexity and the resulting complexity in solving the problem, as well as from the limited flexibility in the decision-making process: the set of solutions they calculate is a static set, unaltered once created.

We use a convex nonlinear modeling that covers the shortcomings of the approaches mentioned above. Its ingredients will be described in the next two sections. In Section 5.3.1, we specify the criterion functions we use, and in Section 5.3.2, we introduce the constraints of our model. In the remainder of the section, we discuss the applicability of different multicriteria optimization approaches.

### 5.3.1 Modeling the inverse treatment planning problem

An oncologist assesses the quality of a treatment plan predominantly based on the shape of the dose distribution. Because there does not exist an accepted notion of how to judge the quality of a dose distribution even for individual

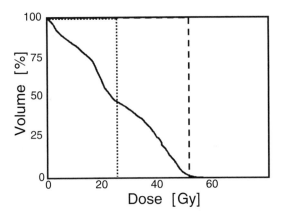

**Fig. 5.9.** Exemplary DVH curve with the resulting EUDs for a parallel and a serial organ.

organs, we will discuss some common choices. For a more complete survey, see [67] and the references therein.

A popular choice are so called DVH constraints. The *dose-volume histogram* (DVH) depicts for each VOI the volume percentage that receives at least a certain dose as a function of the dose (see Figure 5.9). It thus condenses the information present in the dose distribution by neglecting geometric information.

A DVH constraint enforces one of the curves to pass above or below a specified dose-volume point. So either the percentage of volume that receives less than the specified dose or the volume that receives more than a specified dose is restricted for the chosen VOI. DVH constraints are widely used, in particular some clinical protocols are formulated using DVH constraints.

Unfortunately, incorporating DVH constraints into the optimization results in a nonconvex feasible region and thus a global optimization problem. Hence, given a local optimum of the problem, there is no guarantee for global optimality. Therefore, either an enormous computational effort has to be spent for finding all local optima or a suboptimal solution, whose deficiency in quality compared with the global optimum is unknown, has to be accepted.

Hence, convex evaluation functions of the dose distribution in a VOI have been devised that try to control the DVH. For a numerical treatment of the planning problem, the relevant part of the patient's body is partitioned into small cubes called *voxels* $V_j$. Then the dose distribution can be expressed as a vector of values $d(V_j)$ with one dose value per voxel. Using this notation, one such evaluation function is

$$f_k(\mathbf{d}) := \sum_{V_j \subseteq R_k} (\max\{d(V_j) - U_k, 0\})^q, \quad q \in [1, \infty), \tag{5.3}$$

where $R_k$ is some risk VOI. This function penalizes the parts of the volume, where the dose distribution exceeds a specified threshold $U_k$. In terms of the DVH curve, this function penalizes nonzero values beyond the threshold of $U_k$.

In [57], Romeijn et al. propose a different type of dose-volume constraint approximation, which yields a piece-wise linear convex model analogous to the well-known Conditional Value-at-Risk measure in financial engineering.

Another approach to quantify the quality of a dose distribution in a VOI considers the biological impact. The biological impact is assessed using statistical data on the tumor control probability (TCP) and the normal tissue complication probability (NTCP) [76, Chapter 5]. These statistics are gained from experiences with thousands of treated patients, see, e.g., [21].

The concept of equivalent uniform dose (EUD) was first introduced by Brahme in [10]. The EUD is the uniform dose that is supposed to have the same biological impact in a VOI as a given non-uniform dose distribution and depends on the type of the VOI.

The most well-known is Niemierko's EUD concept [51], which uses the $L_a$-norm to compute the EUD:

$$f_k(\mathbf{d}) = \left( \frac{1}{\#\{V_j \subseteq R_k\}} \cdot \sum_{V_j \subseteq R_k} d(V_j)^a \right)^{\frac{1}{a}}, \quad a \in (-\infty, 0) \cup (1, \infty). \quad (5.4)$$

Figure 5.9 illustrates EUD evaluations of a given DVH for two different $a$-parameters. The dotted and the dashed lines are EUD measures with $a$ about 1 and $a$ close to $\infty$, respectively. Organs that work in *parallel*, i.e., organs such as lungs or kidneys that are viable even after part is impaired, are evaluated with an $a$ close to 1, whereas *serial* organs, i.e., structures that depend on working as an entity like the spinal cord, are evaluated with high $a$ values.

Romeijn et al. [59] show that for multicriteria optimization, many common evaluation functions can be expressed by convex functions yielding the same Pareto set.

The numerical solutions presented later are calculated using Niemierko's EUD concept for the risk VOIs and a variant of (5.3) for the upper and lower deviations in tumor volumes. However, the methods described in this paper are valid for any set of convex evaluation functions.

## 5.3.2 Pareto solutions and the planning domain

Our method is based on gathering preference information from the decision maker after the automatic calculation of some Pareto solutions. It is neither possible nor meaningful to calculate all efficient solutions. It is impossible because in the case of convex evaluation functions, the Pareto set is a connected subset of the set of feasible solutions [75] and therefore uncountable. It is also not meaningful as there are many Pareto solutions that are clinically irrelevant.

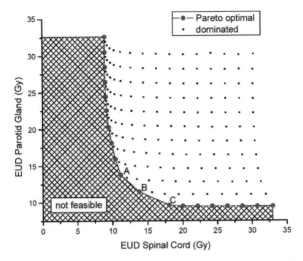

**Fig. 5.10.** Exploration of the Pareto set for a head-and-neck case by enumeration methods. Every dot represents a treatment plan. A total of $16 \times 16 = 256$ plans were generated. The large round dots represent the Pareto set for this case, i.e., the set of efficient treatment plans.

For instance, in the example given in Figure 5.10, one would not go from point A with dose levels of 11 Gy in the spinal cord and 13 Gy in the parotid gland to the upper left efficient solution with dose levels of 9 Gy (spinal cord) and 33 Gy (parotid gland). In other words, the 2 Gy dose reduction in the spinal cord at this low dose level is not worth the "price" of a 20 Gy dose increase in the parotid gland, which may cause xerostomia.

To avoid unnecessary computations, we focus on parts of the Pareto boundary that contain clinically meaningful plans. Because it is easier to classify clinical irrelevance than relevance, we try to exclude as many irrelevant plans as possible and call the remaining set of possible plans the *planning domain*.

To exclude plans that exceed the clinically acceptable values in the criteria, hard constraints are added. Let $\mathbf{F}$ be the vector of criteria and $\mathbf{x}$ be the vector of beamlet intensities, see Section 5.4, the so called *intensity map*. Then, these *box constraints*

$$\mathbf{F}(\mathbf{x}) \leq \mathbf{u}$$

for upper bounds $\mathbf{u}$ should be set rather generously in order to allow a flexible range of solution outcomes. Of course, the more flexible this range is chosen, the more calculations will be necessary.

In exceptional cases, i.e., if they are chosen too strict, they may lead to infeasibility. This serves as an indication to the decision-maker that the initial appraisal of the situation was utopian. If after a relaxation of the box constraints there are still no feasible solutions, the oncologist may realize that more irradiation directions, i.e., more degrees of freedom, are needed to find a clinically acceptable solution and alter the geometry accordingly.

### 5.3.3 Solution strategies

Usually, multicriteria problems are solved by formulating multiple scalarized versions of the problem. There are several standard methods along with their variations that can be used to scalarize the multicriteria problem and that exhibit different characteristics. In this section, we introduce some standard scalarizations and the one used in this work.

Once a planning domain is fixed, the problem to solve is given by

$$\mathbf{F}(\mathbf{x}) \to \min \quad \text{subject to} \tag{5.5}$$
$$\mathbf{x} \in \mathcal{X}_u,$$

where

$$\mathcal{X}_u := \{\mathbf{x} \geq 0 \mid \mathbf{F}(\mathbf{x}) \leq \mathbf{u}\}$$

is the set of feasible intensity maps.

In the *weighted sum approach*, weights $w_k > 0$ are chosen for each evaluation function $F_k, k \in \mathcal{K}$ and the weighted sum of the function values is minimized.

$$\sum_{k \in \mathcal{K}} w_k F_k(\mathbf{x}) \to \min \quad \text{subject to} \tag{5.6}$$
$$\mathbf{x} \in \mathcal{X}_u.$$

For convex multicriteria problems, every set of positive weights yields a Pareto optimal plan, and every Pareto optimal plan is an optimum of (5.6) for an appropriate set of non-negative weights (see [47, Chapter 3.1] for more details).

Another standard approach is the *ε-constraint* method.

$$F_l(\mathbf{x}) \to \min \quad \text{subject to} \tag{5.7}$$
$$F_k(\mathbf{x}) \leq \varepsilon_k \text{ for all } k \in \mathcal{K}$$
$$\mathbf{x} \in \mathcal{X}_u,$$

where all $l \in \mathcal{K}$ must be minimized successively to ensure Pareto optimality. The bounds $\varepsilon_k$ are varied to obtain different results. If chosen appropriately, every Pareto optimal plan can be found [47]. The ε-constraint method is typically used to compute a fine grid of solutions on the Pareto boundary.

A further approach is the *compromise programming* or *weighted metric* approach [47, 85, 88]. Here, a reference point is chosen, and the distance to it is minimized using a suitable metric. The reference point must be outside the feasible region to ensure (weak) Pareto optimality. The ideal point (the point given by the minima of the individual criteria) or some utopia point (a point that is smaller than the ideal point in each component) can be used as reference points.

The different components of $\mathbf{F}(\mathbf{x})$ are scaled to obtain different solutions. Alternatively, the metric can be varied, or both. The solutions obtained are guaranteed to be Pareto optimal if the metric is chosen appropriately and the scaling parameters are positive.

A popular choice is the *Tchebycheff problem*

$$\max_{k \in \mathcal{K}} \{\sigma_k F_k(\mathbf{x})\} \rightarrow \min \quad \text{subject to} \qquad (5.8)$$

$$\mathbf{F}(\mathbf{x}) \leq \mathbf{u}$$

$$\mathbf{x} \in \mathcal{X}_u.$$

Solutions to (5.8) are in general weakly efficient. For that reason, the objective is often augmented by

$$\epsilon \sum_{k \in \mathcal{K}} F_k(\mathbf{x}),$$

with $\epsilon > 0$ arbitrarily small, resulting in *augmented Tchebycheff problems* that produce properly efficient solutions.

The scaling can be derived from ideal values for the criteria as in [37]. Still, the choice of the scaling factors $\sigma_k$ is difficult for the same reasons that it is difficult to formulate a relevant planning domain: the decision-maker may not know enough about the case *a priori*.

Note that scaling is not the same as choosing weights for a problem like the weighted scalarization (5.6) above. The coefficients $\sigma_k$ contain information about the willingness to deteriorate relative to the specified reference point. Thus, deviations from the treatment goals can be much better controlled by reference point methods than by (5.6), as the solutions obtained by varying weights provide no information about the trade-offs of the criteria, see [17].

The concept of *achievement scalarization functions* introduced in [78, 79] and discussed in [47, 80] generalizes the weighted metrics approach. It allows improvements that exceed the specified aspiration levels and thus does not require *a priori* knowledge about the ranges of the different criteria.

The scalar problems our approach utilizes are so-called *extreme compromises*. The extreme compromises successively minimize the maximum values occurring in subsets of the criteria. They partition the set of criteria into the subsets of *active* and inactive ones, then first care for the successive maxima in the active criteria and thereafter treat the *inactive* criteria likewise.

Let $\emptyset \neq \mathcal{M} \subseteq \mathcal{K}$ be the set of indices of the active criteria. Define

$$\pi_{\mathcal{M}} : \mathbb{Y} \times \mathcal{K} \rightarrow \mathcal{K}$$

such that

$$y_{\pi_{\mathcal{M}}(y,k)} \geq y_{\pi_{\mathcal{M}}(y,k')} \quad \text{for} \quad k, k' \in \mathcal{M}, k \leq k'$$
$$y_{\pi_{\mathcal{M}}(y,k)} \geq y_{\pi_{\mathcal{M}}(y,k')} \quad \text{for} \quad k, k' \notin \mathcal{M}, k \leq k'$$
$$\pi_{\mathcal{M}}(y, k) \leq \pi_{\mathcal{M}}(y, k') \quad \text{for} \quad k \in \mathcal{M}, k' \notin \mathcal{M}.$$

in analogy to [19, Chapter 6.3], let

$$sort(y) := \left( \left( y_{\pi_{\mathcal{M}}(y,k)} \right)_{k \in \mathcal{M}}, \left( y_{\pi_{\mathcal{M}}(y,k)} \right)_{k \notin \mathcal{M}} \right).$$

The solution of

$$sort\,(\mathbf{F}(\mathbf{x})) \to \min \quad \text{subject to} \tag{5.9}$$
$$\mathbf{x} \in \mathcal{X}_u$$

is called extreme compromise for the active criteria $\mathcal{M}$. It can easily be seen that the extreme compromises are Pareto optimal for every non-empty set $\mathcal{M}$. The resulting criterion vector $\mathbf{F}(\mathbf{x}^*)$ will consist of several groups of indices with decreasing function value. Here, the groups lying in $\mathcal{M}$ and in $\mathcal{K} \setminus \mathcal{M}$ by construction form independent scales.

The extreme compromise with all criteria active, i.e., $\mathcal{M} = \mathcal{K}$, is known as *lexicographic max-ordering* problem [19], *variant lexicographic* optimization problem optimization problem [61], or as *nucleolar solution* [44] and *nucleolus* in game theory (see references in [44, 61]). The latter articles also describe methods for computing it. Sankaran [61] is able to compute it solving $|\mathcal{K}|$ optimization problems using $|\mathcal{K}|$ additional variables and constraints. We call this particular extreme compromise the *equibalanced solution*.

The general extreme compromises can be determined by applying the above method lexicographically to the two subsets. Alternatively, if upper bounds and lower bounds for the criterion functions are known, the functions can be scaled and shifted such that the largest values in the inactive criteria are always smaller than the smallest values in the active criteria. Sankaran's algorithm will then directly yield the corresponding extreme compromise.

The interactive method presented in Section 5.5 works with a precomputed approximation of the relevant part of the Pareto boundary. To construct such approximations, the scalarizations presented in this section are repeatedly used by higher level routines yielding the approximation. The applicability of several common approximation schemes for our problem is discussed in the following section.

### 5.3.4 Approximation of the Pareto boundary

Following the classification used in [60], we briefly discuss the applicability of point-based approximations of the efficient set in $\mathcal{X}$ and point-based, outer, inner, and sandwich approximations in $\mathcal{Y} := \mathbf{F}(\mathcal{X})$.

"Since the dimension of $\mathcal{Y}_{\text{Par}} := \mathbf{F}(\mathcal{X}_{\text{Par}})$ is often significantly smaller than the dimension of $\mathcal{X}_{\text{Par}}$, and since the structure of $\mathcal{Y}_{\text{Par}}$ is often much simpler than the structure of $\mathcal{X}_{\text{Par}}$," it is worthwhile to search for a solution in the outcome space $\mathcal{Y}_{\text{Par}}$ [5]. Thus, in IMRT planning, where the dimensions are 4–10 for $\mathcal{Y}$, and 400–2000 for $\mathcal{X}$, it is futile to work with a point-based approximation in $\mathcal{X}$.

The methods for point-based approximations of $\mathbf{F}(\mathcal{X}_{\text{Par}})$ usually fall into one of the following categories:

1. they use fixed grids for the scalarization parameters [12, 55, 70],
2. state relations between approximation quality and distance of scalarization parameters for arbitrary grids [1, 50] or
3. try to create a fine grid directly on the Pareto boundary [23, 62].

The methods in (1) cover the scalarization parameter set whose dimension is at least $|\mathcal{K}| - 1$ with regular grids of maximum distance $\varepsilon$. The methods in (2) and (3) in turn cover the Pareto boundary — which is in general a $|\mathcal{K}| - 1$ dimensional manifold — with grids of maximum distance $\varepsilon$. In any case, the number of points needed is at least

$$O\left( (1/\varepsilon)^{|\mathcal{K}|-1} \right).$$

Such a number of points is neither tractable nor actually needed for our problem. We use interpolation yielding a continuous approximation of the Pareto boundary to overcome the need for a fine grid.

The continuous approximation using convex combination is insensitive to large gaps as long as the interpolates stay close to the Pareto boundary. Hence, the distance between grid points is not an appropriate quality measure. Because of convexity, the interpolated solution is at least as good as the interpolated $\mathbf{F}$-vectors of the interpolation points (Figure 5.11).

Thus, if the distance between the convex hull of the pre-computed criterion values and the Pareto boundary is small, so is the distance for the interpolated plans. Note though, that the approximation attained by the convex hull of the pre-computed plans will probably contain notably suboptimal points and possibly convex combinations that approximate non-Pareto optimal parts of the boundary of the feasible set.

The above reasoning motivates the use of distance-based methods for approximating the Pareto boundary in the first phase of our hybrid procedure. There are three types of distance-based methods:

- outer approximation methods
- inner approximation methods and
- sandwich approximation methods.

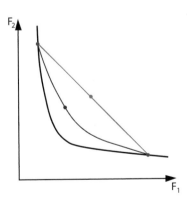

**Fig. 5.11.** The $\mathbf{F}$-vector of the convex combination of solutions is in general better than the convex combination of the original $\mathbf{F}$-vectors.

Outer approximation methods successively find new supporting hyperplanes and approximate the Pareto boundary by the intersection of the corresponding half-spaces. The methods proposed by Benson [3] and Voinalovich [74] for linear multicriteria problems are clearly not directly applicable to our nonlinear case, although some of the ideas can be combined with inner approximation methods to form sandwich approximation schemes.

Inner approximation methods [14, 16, 64] create successively more Pareto optimal points and approximate the Pareto boundary with the close-by facets of the convex hull of the computed points. Sandwich approximation methods [35, 68] determine supporting hyperplanes for every computed Pareto optimal point and use the corresponding half-spaces to simultaneously update the outer approximation. Having an inner and outer approximation, the sandwich approximation schemes are able to give worst-case estimates for the approximation error.

All methods mentioned try to choose the scalar subproblems such that the maximal distance between the Pareto boundary and the approximation is systematically reduced. The construction of the inner approximation is conducted in all five methods by variations of the same basic idea:

1. create some starting approximation that consists of one face, i.e., a $|\mathcal{K}| - 1$ dimensional facet;
2. find the Pareto point that is farthest from the chosen face by solving a weighted scalarization problem with a weight vector that is perpendicular to the face;
3. add the point to the inner approximation and update the convex hull;
4. if the approximation is not yet satisfactory, choose a face from the inner approximation and go to 2, otherwise stop.

The methods are perfectly suited for our needs as they control the distance to the Pareto boundary and systematically reduce it. Unfortunately, all of them use convex hull computations as a subroutine in step 3 — a method with immense computational expense and memory needs in higher dimensions. The best available algorithms for convex hull computations usually work for dimensions up to 9 [2, 14]. But the trade-off between computational and memory expense for the convex hull subroutine against computational savings due to well-chosen scalar problems reaches its breakeven point much earlier.

In multicriteria IMRT planning, often two to three nested tumor volumes are considered, each having separate criterion functions for the lower and upper dose deviation. Furthermore, several risk VOIs are within reach of the tumor so that we can easily exceed the dimensionality where applying the distance-based approximation methods is still reasonable.

Therefore, we either have to apply heuristic or stochastic approximation approaches. The covered range plays a crucial role in the interactive selection process, and there is no guarantee that a reasonable range is achieved with stochastic procedures. Hence, we propose to use a heuristic to supply the appropriate ranges and a stochastic procedure to improve the approximation with further points.

To achieve the ranges, we compute the extreme compromises for every non-empty subset of $\mathcal{K}$. The rationale behind the definition of the extreme compromises is to fathom the possibilities for simultaneously minimizing a group of criteria for all possible such groups (see Figure 5.12(a)).

Note that the solutions minimizing individual criteria, the so-called *individual minima*, are contained in the extreme compromises. Thus, also the convex hull of individual minima (CHIM) – the starting point for the inner approximation in [16, 68] and a possible starting point in [35, 64] – is contained in the convex hull of the extreme compromises. Figure 5.12(a) shows that the extreme compromises cover substantially more than the CHIM, which is in this case even sub-dimensional. This is due to the fact that two of the three criteria share a common minimum.

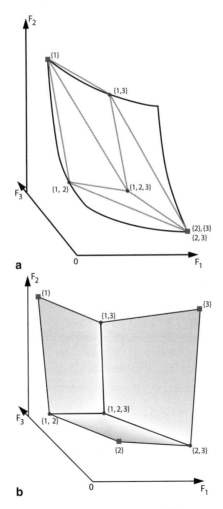

**Fig. 5.12.** Extreme compromises in 3 dimensions. The integer sets state the active set for the corresponding extreme compromise. The squares are individual minima.

The extreme compromises are often not of high clinical relevance as the inactive criteria may reach their upper bounds and are thus close to the plans that were *a priori* characterized as clinically irrelevant.

The complexity of calculating the extreme compromises is exponential, as the number of optimization problems of type (5.9) that have to be solved is equal to the number of non-empty subsets of $\mathcal{K}$ which is $2^{|\mathcal{K}|} - 1$.

Figure 5.12(b) shows the position of the extreme compromises for $\mathcal{Y}$ being a bent open cube. Again the squares depict the individual minima. As one can see, we need the full number of extreme compromises to cover the Pareto boundary for this case. In the general case, the "grid" given by the extreme compromises is distorted, with occasional degeneracies (see Figure 5.12(a)).

One method to reduce the number of computations is to group related VOIs and treat them as one component. They are turned active or inactive as a group and hence only count as one criterion in the exponential complexity term. In a head-and-neck case, one could for example group the eyes with their respective optical nerve as it is meaningless to spare one while applying critical doses to the other.

To improve the approximation of the Pareto boundary, we add further points to the approximation by a stochastic procedure. These points will most likely not change the range of the approximation but improve the distance between the approximation and the Pareto boundary. This allows us to better convey the shape of the Pareto boundary to the planner in the navigation process (see Section 5.5).

For the stochastic procedure, the scaling in an augmented Tchebycheff problem (5.8) is chosen randomly from a uniform distribution. It is so far an open question whether it is worthwhile to use non-uniform distributions, which make it less likely that a new parameter set is chosen that is close to an already used one. The distribution of the solutions for the non-uniform distribution would clearly be better, but to update the distribution after the calculation of a solution and the evaluation of the distribution requires additional computational effort that could thwart the effect of the improved distribution.

As it was mentioned in Section 5.2, the computation of the extreme compromises and the intermediate points are technical and done without any human interaction, so the calculations could for example run overnight. Nonetheless, the overall number of solutions to be calculated can be large making it essential to improve the speed of the individual calculations.

## 5.4 The Numerical Realization

In this section, we introduce the dose calculation used by our optimization method. We then explain our approach to deal with the high dimensionality of our problems that exploit the asymmetry introduced in Section 5.2.

The width of the leaves of the MLC used to apply the treatment implies natural slices of each beam. A further dissection of each slice into rectangular

**Fig. 5.13.** Schematic form of an intensity map for a head-and-neck case. Different gray levels correspond with different intensities.

areas leads to a partition of the beam into *beamlets*. The intensity modulation of a beam is now given by the intensity values of each beamlet (see Figure 5.13).

The discretization in the body volume is typically based on small cuboid-shaped voxels $V_j$. The dose distribution on the volume can now be represented by voxel-related dose values. As there are typically up to a few thousand beamlets and several hundred thousand voxels, we are dealing with a truly large-scale optimization problem. As a consequence of the superposition principle of dose deposits in the volume in the case of photon therapy, the dose distribution for an intensity vector then follows as

$$\mathbf{d} : \mathbb{X} \to \mathbb{D},$$
$$\mathbf{x} \mapsto \mathbf{P} \cdot \mathbf{x},$$

with the matrix $\mathbf{P}$ being the *dose information matrix*. The entry $p_{ji}$ of this matrix represents the contribution of the $i$-th beamlet to the absorbed dose in voxel $V_j$ under unit intensity. There are several methods to attain these values. They might be calculated using the pencil beam approach, a superposition algorithm, or some Monte Carlo method. In this chapter we do not discuss this important issue; we assume $\mathbf{P}$ to be given in some satisfactory way.

### 5.4.1 The adaptive clustering method

Plan quality is measured by evaluating a dose distribution in the involved clinical structures. Typically, during plan optimization the dose distribution will attain an acceptable shape in most of the volume, such that the final quality of a treatment plan strongly depends on the distribution in some small volume parts. Often this effect is observed in regions where the descent of radiation from the cancerous to the healthy tissue implies undesirable dose-volume effects.

Based on this problem characteristic, several approaches to reduce the computational complexity of the problem by manipulations in the volume have been tried (see the listing in [86]). These manipulations are done by heuristic means prior to the optimization routine and incorporate, for example, the use of large voxels in less critical volume parts, a restriction to the voxels located in pre-defined regions of interest, or a physically motivated selection of voxels. However, such heuristic problem modifications may lead to an insufficient control on the dose distribution during the optimization and thus result in a plan with inferior quality.

The *adaptive clustering method* overcomes these defects by an sequential adaptation in the volume. It was introduced in [37] and it is discussed in detail in [65]. We will thus only briefly explain it here and provide a small example in Section 5.4.2.

In a preprocessing step, voxels with their corresponding dose information are aggregated to *clusters*. This process is repeated to form a *cluster hierarchy* (see Figure 5.14). This *hierarchical discretization process* is independent of how dose distributions are evaluated — the same cluster hierarchy may be used in several models with different criterion functions. Figure 5.15 shows for a clinical head-and-neck case the progress of the hierarchical clustering process in a transversal voxel layer.

Created only once for an IMRT planning problem, the resulting cluster hierarchy then serves as a "construction kit" to generate adapted volume

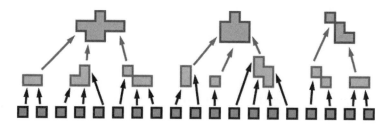

**Fig. 5.14.** Illustration of the cluster hierarchy.

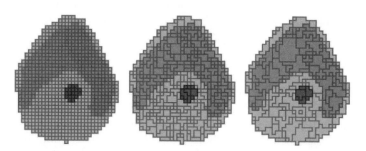

**Fig. 5.15.** The progress of the hierarchical clustering process in time in a transversal voxel layer. In this layer, the clinical target volume is located on the right side, the planning target volume on the left, and the spinal cord in the center. The remainder is unclassified tissue. The voxels (left) are iteratively merged to clusters of increasing size (center and right).

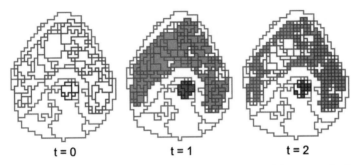

$t = 0$        $t = 1$        $t = 2$

**Fig. 5.16.** The adapted clusterings for a transversal voxel layer at the beginning of the local refinement process ($t = 0$) and after the first ($t = 1$) and second ($t = 2$) local refinement step. The filled clusters are the ones that were refined in the previous step.

discretizations for the scalarized multicriteria planning problems. Each optimization starts on a coarse clustering that consists of large clusters. The method formulates a series of optimization problems, each of which solves the planning problem on adapted volume discretization. While the optimization runs, the algorithm gradually detects those volume parts responsible for large contributions to the evaluation functions and replaces the corresponding clusters in *local refinement* steps by smaller ones to improve the local control on the dose distribution. Discretizations with clusters of different resolution are called *adapted clusterings*. Transversal slices of such adapted clusterings are shown in Figure 5.16.

Because of the individual adaptation of the volume structure during the optimization by the *local refinement process*, the result is numerically optimal with respect to the original problem but can be obtained with a significantly smaller expense than a computation on the original voxel-based volume structure would have required. Numerical experiments on several sets of real clinical data typically show a reduction in computation time by a factor of about 10 compared with an optimization on the original volume structure, where both computations yield plans with almost identical evaluation function values.

### 5.4.2 Asymmetry in the inverse treatment planning problem

The decisively reduced computational expense obtained by the adaptive clustering method traces back to the asymmetry of the inverse treatment planning problem. Consider the treatment planning problem as a general design problem, see Section 5.2.2. The quality of a dose distribution in a VOI is measured by an evaluation function $f$.

Assume that voxels with similar dose values that result from similar row vectors $\mathbf{p}(V)$ of the dose information have similar influence on the evaluation functions. In case of $f_{R_k}$ for a VOI $R_k$, $\frac{\partial}{\partial d(V)} f_{R_k}(\mathbf{d})$ is continuous in $\mathbf{d}$. Then the asymmetry is manifested in the following sense.

Let $\mathbf{x}^*$ be a solution to a scalarization of the planning problem like the Tchebycheff problem (5.8) and $\mathbf{P}' = (\mathbf{p}'(V)) \in \mathbb{R}^{\dim(\mathbb{D}) \times \dim(\mathbb{X})}$ be an approximation of $\mathbf{P}$. Then

$$f_{R_k}(\mathbf{P} \cdot \mathbf{x}^*) \approx f_{R_k}(\mathbf{P}' \cdot \mathbf{x}^*) + \sum_{V \subseteq R_k} \frac{\partial f_{R_k}}{\partial d(V)}(\mathbf{P}' \cdot \mathbf{x}^*) \cdot (\mathbf{p}(V) - \mathbf{p}'(V))\mathbf{x}^*,$$

and a reasonable choice of $\mathbf{p}'(V')$ for each family of voxels $V'$ with similar partial derivatives and dose values yields $f_{R_k}(\mathbf{P}' \cdot \mathbf{x}) \approx f_{R_k}(\mathbf{P} \cdot \mathbf{x})$ with a very moderate error. This means that the problem using $\mathbf{P}'$ instead of $\mathbf{P}$ is then a good approximation of the original problem.

The soundness of this approach follows from standard results of sensitivity analysis. As our approximations of the row vectors become better, i.e., $\max_V \|\mathbf{p}(V) - \mathbf{p}'(V)\| \to 0$, the optimal solutions of the approximate problems converge to the optimal solutions of the original problem. However, even larger $\|\mathbf{p}(V) - \mathbf{p}'(V)\|$ could be easily accepted, provided the resulting dose differences $(\mathbf{p}'(V) - \mathbf{p}'(V)) \cdot \mathbf{x}^*$ only marginally affect the quality of the dose distribution as measured by the criterion functions.

This implies the following conclusion: if one had an approximation $\mathbf{P}'$ of $\mathbf{P}$, for which $\mathbf{P}' \cdot \mathbf{x}$ could be cheaply computed, then the optima of the corresponding approximate problem would also be (almost) optimal for the original problem but could be obtained with a much smaller computational expense.

In contrast with many other optimization problems, the continuous background of this problem provides a possibility to exploit the asymmetry by constructive means. Voxels lying in the vicinity of each other are irradiated similarly by most of the beamlets and thus play a similar role in the optimization. Critical voxels with a strong influence on the quality of the dose distribution $\mathbf{d}$ and its evaluation will thus concentrate in local volume parts that depend more or less continuously on $\mathbf{d}$. Hence, a thorough examination of the voxels to detect the critical ones also reveals the subspace of $\mathbb{D}$ that requires a mapping with the original row vectors and the ones for which even large gaps $\|\mathbf{p}(V) - \mathbf{p}'(V)\|$ can be accepted. This allows a highly efficient construction of an approximate $\mathbf{P}'$ as done in the adaptive clustering method.

To summarize, the concept of asymmetry provides a strategy to tackle the large-scale optimization problems that have to be solved to generate a database of plans. We are able to calculate efficient solutions of the original multicriteria problem comparably fast.

## 5.5 Navigating the Database

When the plan database computation is finished, a vital part of the planning is not yet accomplished. The planner still has to decide on a plan. Because inspecting a plan usually involves sifting through all its slices, a manual inspection of all plans contained in the database is infeasible. In particular, when we additionally consider convex combination of plans and thus an infinite number of plans, manual inspection is not an option.

We use an interactive multicriteria optimization method that works on the convex hull $\hat{\mathcal{X}} := conv\left\{\mathbf{x}^{(l)}, l \in \mathcal{L}\right\}$ of pre-computed plans $\mathbf{x}^{(l)}, l \in \mathcal{L}$. User actions are transformed into optimization problems on this restricted domain. The restriction of domain together with the structural information gained during the calculation of the database allows the problems to be solved in real-time. Therefore, the execution of the interactive method feels more like navigation than optimization.

The feeling of direct control is strengthened by constantly providing the user with a visualization of the current solution and up-to-date estimates of the ideal and the nadir point. The former is the combination of the individual minima of the criteria over the current domain. The latter combines the maxima of the individual criteria over the Pareto optimal plans in the current domain. Having thus a clear picture of the possibilities and limitations, the planner's decisions are based on a much firmer ground.

There are two basic mechanisms in our interactive method (patented for radiotherapy planning by the Fraunhofer ITWM [72]):

- the *restriction* mechanism that changes the feasible region and
- a search mechanism called *selection* that changes the current solution.

The former updates the ideal and nadir point estimates when the user changes the box constraint for a criterion. The latter searches for a plan that best adheres to a planner's wish.

A special variant of the restriction mechanism is the use of a *lock*, which is a shortcut for restricting the chosen criterion to the current or better values. Furthermore, the whole database can be re-normalized, i.e., all plans can be scaled to a new mean dose in the target.

### 5.5.1 The restriction mechanism

The restriction mechanism allows the planner to set feasible hard constraints on the criterion values *a posteriori*. He can thus exclude unwanted parts of the Pareto boundary of $\hat{\mathcal{Y}} := \mathbf{F}(\hat{\mathcal{X}})$, the image of the restricted domain under the vector of evaluation functions. Let $\hat{\mathcal{X}}_{\mathbf{u}} := \left\{\mathbf{x} \in \hat{\mathcal{X}} : \mathbf{F}(\mathbf{x}) \le \mathbf{u}\right\}$ be the set of solutions that are feasible for the current upper bounds $\mathbf{u}$ and $\hat{\mathcal{Y}}_{\mathbf{u}} := \mathbf{F}(\hat{\mathcal{X}}_{\mathbf{u}})$ be the set of corresponding criterion vectors.

Every change in the right-hand side of the box constraints $\mathbf{F}(\mathbf{x}) \le \mathbf{u}$ causes the system to update its estimate of the ideal and nadir point, thus providing the planner with an update of the so-called *planning horizon* – the multidimensional interval between the ideal and nadir point estimate.

This interval is important information for the decision maker, as *"(t)he ideal criterion values are the most optimistic aspiration levels which are possible to set for criteria, and the nadir criterion values are the most pessimistic reservation values that are necessary to accept for the criteria"* (see [36]).

**Minimum values: the ideal point**

Let us now introduce some notation. The intensity maps of the pre-calculated plans are combined into a matrix $\mathbf{X} := \left(\mathbf{x}^{(l)}\right)_{l \in \mathcal{L}}$ with columns consisting of the intensity maps. Likewise, the criterion vectors $\mathbf{y}^{(l)} := \mathbf{F}(\mathbf{x}^{(l)}), l \in \mathcal{L}$ are combined into a matrix $\mathbf{Y} = \left(\mathbf{y}^{(l)}\right)_{l \in \mathcal{L}}$. Thus, the entry $(k, l)$ of $\mathbf{Y}$ represents the $k^{th}$ criterion value of the $l^{th}$ solution.

As the change of an upper bound in some criterion changes the feasible region, it may alter the minima of the criteria as well. If the upper bound $\mathbf{u}$ changes, the new ideal point can be found by solving the following problem for each criterion function $F_k \in \mathcal{K}$:

$$
\begin{aligned}
F_k(\mathbf{X\lambda}) &\to \min \quad \text{subject to} \qquad\qquad (5.10)\\
\mathbf{F}(\mathbf{X\lambda}) &\le \mathbf{u}\\
\boldsymbol{\lambda} &\in \Sigma,
\end{aligned}
$$

where $\Sigma := \left\{ \boldsymbol{\lambda} \in \mathbb{R}_{+}^{|\mathcal{L}|} : \mathbf{e}^T \boldsymbol{\lambda} = 1 \right\}$ is the simplex of convex combination coefficients.

The optimization problem (5.10) finds a convex combination, minimizing the $k^{th}$ criterion, while observing the upper bounds $\mathbf{u}$.

Problem (5.10) is convex because the functions $F_k$ are convex and can therefore be efficiently solved. However, it is not clear *a priori* how fast these problems can be solved — at least it is not known if they can be solved fast enough to allow a real-time navigation. Hence, a linear approximation of (5.10) is formulated as:

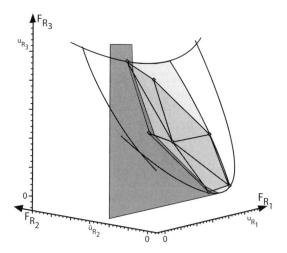

**Fig. 5.17.** A new upper bound for $F_{R_2}$ is introduced.

$$(\mathbf{Y}\boldsymbol{\lambda})_k \to \min \quad \text{subject to} \tag{5.11}$$
$$\mathbf{Y}\boldsymbol{\lambda} \leq \mathbf{u}$$
$$\boldsymbol{\lambda} \in \Sigma.$$

The linear problem (5.11) overestimates the true criterion values

$$\mathbf{Y}\boldsymbol{\lambda} = \left(\mathbf{F}(\mathbf{x}^{(l)})\right)_{l \in \mathcal{L}} \boldsymbol{\lambda} \geq \mathbf{F}(\mathbf{X}\boldsymbol{\lambda}) \tag{5.12}$$

due to convexity. Therefore, the feasible region of problem (5.11) $\Lambda' := \{\boldsymbol{\lambda} \in \Sigma \mid \mathbf{Y}\boldsymbol{\lambda} \leq \mathbf{u}\}$ is contained in the feasible region of the original problem (5.10):

$$\Lambda' \subseteq \Lambda := \{\boldsymbol{\lambda} \in \Sigma \mid \mathbf{F}(\mathbf{X}\boldsymbol{\lambda}) \leq \mathbf{u}\} \tag{5.13}$$

Hence, the LP works on a subset of the original domain, and due to convexity, its objective function is larger than the original convex objective function. Therefore, the result gives an upper bound for the true minimum.

Depending on the computational complexity of the original formulation, one can either solve the $|\mathcal{K}|$ original problems for the individual minima or use the linear estimates.

## Maximum values: the nadir point

A different problem is faced in finding the maximum values of the criteria. For this, let $\hat{\mathbf{u}}$ be the vector of the individual maxima contained in the database, i.e., $\hat{u}_k := \max_{l \in \mathcal{L}} \left\{ y_k^{(l)} \right\}$. It can easily be shown that this is the nadir point of the multicriteria problem restricted to $\hat{\mathcal{X}}$. From the definitions, it directly follows that $\hat{\mathcal{X}} = \hat{\mathcal{X}}_{\hat{\mathbf{u}}}$.

However, for general $\mathbf{u}$ the situation is not as straightforward. The problem formulation to obtain the $k^{th}$ coordinate of the nadir point reads:

$$F_k(\mathbf{X}\boldsymbol{\lambda}) \to \max \quad \text{subject to} \tag{5.14}$$
$$\mathbf{F}(\mathbf{X}\boldsymbol{\lambda}) \leq \mathbf{u}$$
$$\mathbf{F}(\mathbf{X}\boldsymbol{\lambda}) \quad \text{Pareto optimal}$$
$$\boldsymbol{\lambda} \in \Sigma.$$

Computing the exact nadir point components is a convex *maximization* problem over a non-convex domain — the Pareto boundary — and thus a global optimization problem, which is difficult to solve in three or more dimensions (see, e.g., the abstract of [4]). In [81] an overview of methods for optimization over the Pareto optimal set is given — a class of algorithms that is more general, but can be used for the nadir point detection. More such algorithms are proposed in [18, 30, 31, 42], all of which involve global optimization subroutines and at best converge in finitely many iteration but are inappropriate for a real-time procedure. An exception is the algorithm proposed in [20] that is less computationally involved, but as it heavily relies on bicriteria subproblems, it only works for up to three criteria.

Because exact methods are intractable, heuristic estimates for the nadir point have to be used. Estimates using the so-called payoff table (see, e.g., [47]) are problematic, because they can be too large or too small and arbitrarily far away from the true value (see [32]). But in [20], small algorithmic changes to the payoff table heuristic are proposed that make it either a lower or upper bound for the true value. Applying these small changes to the problems solved when looking for the ideal point, the improved payoff table entries can be computed with almost no additional effort.

In [36], a heuristic to approximate the nadir point for linear multicriteria optimization based on the simplex algorithm is proposed. It uses its objective function to enforce Pareto optimality and successively changes the right-hand side to maximize the currently considered criterion. Furthermore, a cutting plane is used to cut off the part of the polyhedron that contains smaller values than the most current estimate. The heuristic yields a lower bound for the true nadir value, as in general it only detects local maxima. It involves no global optimization subroutines and is thus eligible for our purposes. Additionally, it can be stopped any time still yielding a lower bound for the nadir point; although the estimate is less accurate then.

The heuristic works on the fully linearized problem and thus we still have to calculate the true $F_k(\mathbf{X}\boldsymbol{\lambda}^{(k)})$ values for the optimal convex combination coefficients $\boldsymbol{\lambda}^{(k)}$ of problem (5.14) for all $k \in \mathcal{K}$ to get an estimate for the nadir point of the convex problem.

Depending on the time restrictions and the problem complexity, one can either evaluate the payoff tables in conjunction with the ideal point detection or use the more sophisticated nadir point heuristic above. Furthermore, the payoff table heuristic can be used while the upper bound is changed, and the simplex-based nadir point heuristic is used to correct the values when the changes have taken place.

### 5.5.2 The selection mechanism

Thus far, the user can only manipulate the planning horizon but cannot change the current solution. This is done with the selection mechanism.

The first solution shown can be any plan from the database. Usually, the equibalanced solution is presented first. Now, the user can change one of the criterion values of this solution within the bounds given by the ideal and nadir point estimates. The system searches for a solution that attains the modified value in the chosen criterion and degrades the other criterion values least possible.

This search is accomplished by solving an achievement scalarization problem for a specifically chosen reference point. Let $\mu$ be the value chosen for $\mathbf{F}_{k'}$,

$\bar{\mathbf{y}}$ be the criterion vector of the former solution, and $\mathcal{K}' := \mathcal{K} \setminus \{k'\}$. Then the selection mechanism problem is formulated as

$$\max_{k \in \mathcal{K}'} \{F_k(\mathbf{X}\boldsymbol{\lambda}) - \bar{y}_k\} + \epsilon \sum_{k \in \mathcal{K}'} F_k(\mathbf{X}\boldsymbol{\lambda}) \to \min \quad \text{subject to}$$

$$\mathbf{F}(\mathbf{X}\boldsymbol{\lambda}) \leq \mathbf{u} \tag{5.15}$$
$$F_{k'}(\mathbf{X}\boldsymbol{\lambda}) = \mu$$
$$\boldsymbol{\lambda} \in \Sigma,$$

for a small $\epsilon > 0$. Approximating again $\mathbf{F}(\mathbf{X}\boldsymbol{\lambda})$ by $\mathbf{Y}\boldsymbol{\lambda}$, we attain the linear approximation

$$\max_{k \in \mathcal{K}'} \{(\mathbf{Y}\boldsymbol{\lambda})_k - \bar{y}_k\} + \epsilon \sum_{k \in \mathcal{K}'} (\mathbf{Y}\boldsymbol{\lambda})_k \to \min \quad \text{subject to}$$

$$\mathbf{Y}\boldsymbol{\lambda} \leq \mathbf{u} \tag{5.16}$$
$$(\mathbf{Y}\boldsymbol{\lambda})_{k'} = \mu$$
$$\boldsymbol{\lambda} \in \Sigma.$$

The problem (5.16) implicitly describes a path on the Pareto boundary parameterized by $\mu$ (see Figure 5.18).

The linear program (5.16) is solved using a Simplex algorithm. Because the LP has $|\mathcal{K}| + 1$ constraints, any basic feasible solution in the Simplex iterations has at most $|\mathcal{K}| + 1$ non-zero elements. Therefore, only $|\mathcal{K}| + 1$ plans enter the convex combination, making the complexity of executing the convex combination predictable and in particular independent of the number of plans in the database.

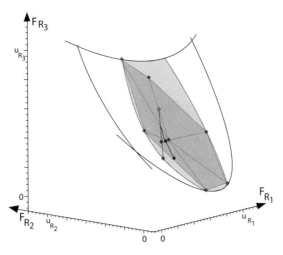

**Fig. 5.18.** The chosen reference points (asterisks) and the corresponding solutions (squares) found in the optimization problem (5.16).

Because $\mathbf{F}(\mathbf{X}\hat{\lambda}) \leq \mathbf{Y}\hat{\lambda}$ for the optimal $\hat{\lambda}$ due to convexity, the resulting $k'^{th}$ criterion value $F_{k'}(\mathbf{X}\hat{\lambda}) < \mu$ is possibly smaller than expected (see also Figure 5.11). If the deviation from the equality constraint is too large, one can use a column generation process to improve the approximation: The matrix $\mathbf{Y}$ is augmented by $\hat{y} := \mathbf{F}(\mathbf{X}\hat{\lambda})$ and the dimension of the vector $\lambda$ is enlarged by one. This results in an improved local approximation of the boundary of $\hat{\mathcal{Y}}_{\mathbf{u}}$, thus improving the accuracy of the equality constraint for the solution of the enlarged optimization problem (5.16).

Depending on the time restrictions and the problem complexity, the solution found through navigation could be used as a starting point for a post-optimization that would push it to the Pareto boundary of the original problem. To accomplish this, a combination of the $\varepsilon$-constraint method and weighted sum or weighted metric method could be used. The point gained can then be added to the database, thus improving the local approximation of $\mathcal{Y}_{\mathbf{u}}$'s Pareto boundary.

### 5.5.3 Possible extensions to the navigation

There are some possible extensions to the navigation making it even more versatile. It is, for example, possible to add or remove criteria at any point during the planning process. Of course having added a new criterion, the solutions in the database are not Pareto optimal with respect to the new set of functions over the original domain $\mathcal{X}$, but they can at least be evaluated under the additional criterion.

The new criterion may then of course be considered in the navigation, and the navigation still selects the best possible choices over the restricted domain $\hat{\mathcal{X}}$. Using post-optimization techniques, the approximation of the now higher dimensional Pareto boundary can again be locally improved, yielding a good local picture of the new Pareto boundary, but revealing an incomplete global picture, i.e., potentially bad estimates for the ideal and nadir point.

The navigation is independent of the way the plans in the database were created. Hence, the database could stem from several manually set up optimizations and the navigation then allows one to mix them. This is in particular relevant, if the clinical case is well-known and the computation of the full set of extreme compromises plus additional plans seems needless.

The independence from the creation process enables the addition of plans at any stage. Therefore, single solutions could be added even after the computation of a database. So automatic computations can be combined with manually set up plans in arbitrary sequence.

### 5.5.4 The user interface for the navigation

The described mechanisms allow a workflow that is a distinct improvement compared with something like a Human Iteration Loop (see Section 5.2). But for implementing the improved workflow, an appropriate visualization and manipulation tool is needed.

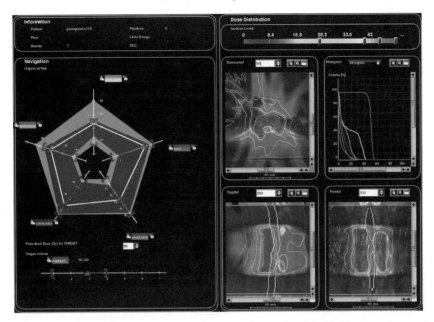

**Fig. 5.19.** The navigation screen. The "star" on the left-hand side shows the active and inactive planning horizon and the current solution. On the right the dose visualization and the dose-volume histogram for the current solution are shown.

Figure 5.19 shows the user interface for the navigation tool. It is divided into two parts. The left-hand side visualizes the database as a whole and embeds the current solution into the database. The right-hand side displays the current plan's dose-volume histogram and the dose distribution on transversal, frontal, and sagittal slices.

The "star" on the left-hand side is composed of axes for the different criteria. The criteria associated with the risk VOIs are combined into a radar plot, whereas the criteria associated with tumor volumes are shown as separate axes. The interval on the axes corresponds with the range of values contained in the database for the respective criterion. The white polygon marks the criterion values for the currently selected plan.

The shaded area represents the planning horizon. It is subdivided into the *active* and the *inactive* planning horizons. The former is bounded on each axis by the maximum and minimum values implied by the currently set restrictions, and the latter is the currently excluded range contained in the database. Note that the line connecting the minimum values of the active planning horizon is the ideal point estimate, and the line connecting the maximum values is the nadir point estimate for the currently set restrictions.

The line representing the currently selected plan has handles called *selector* at each intersection with an axis and triangles for the tumor-related axes. Both can be grabbed with the mouse and moved to carry out the selection mechanism described above. The right-hand side of the screen displays the corresponding plans concurrently. The axes also contain *restrictors* represented by brackets. They can also be grabbed with the mouse and moved to change the upper bound for the corresponding criterion. When the planner moves a restrictor, the active and inactive planning horizon are updated simultaneously.

The visualization is updated several — usually around seven — times a second when a selector or restrictor is moved. This means that around 7 linear problems of type (5.16) are solved and the corresponding convex combinations are carried out every second while the user pulls a selector. For restrictor movements $7|\mathcal{K}|$ linear problems of type (5.11) and the same number of nadir point heuristic problems are solved every second. Hence, instead of waiting for the consequences of a parameter adjustment, the planner is immediately provided with the resulting outcome.

### 5.5.5 Concluding remarks on decision-making

The proposed method offers a level of interactivity that is so far unknown in radiation therapy planning. Neither is there a need to choose weights, classify the criteria with regard to the level of satisfaction, or explicit choice of a reference point. Nor is it necessary to wait for the outcome of the corresponding decision. The systems thus offers the possibility to overcome the Human Iteration Loop, which is standard for current inverse IMRT planning. Furthermore, we believe that working with criterion values only requires less experience in using the planning system than do approaches based on abstract information like weights.

The system offers two complementary mechanisms: one to change the current plan and one to change the feasible region. Combining the two, the planner can successively adapt the current solution and the feasible region to his or her current state of mind. In the end, the feasible region is narrowed down to the *a posteriori* clinically relevant domain, and the current solution is set to the planner's favorite among that set.

The real-time response to any changes regarding the current solution and planning horizon allow the user to get a feeling for the Pareto boundary. Observing the changes in the criterion values implied by a modification of one of the criteria gives the planner a feel for the sensitivity and thus for the local interrelation. Observing the changes in the active planning horizon reveals the global connection between the criteria complementing the planner's mental picture of the Pareto boundary.

The concurrent update of the visualizations of the current dose distribution on the right-hand side of the navigation screen allows the planner to apply quality measures on the solutions that were not modeled into the optimization problem. The system thus acknowledges the existence of further clinical criteria that are relevant for the planner's final decision.

In summary, the decision-making process for the treatment planning problem described in this paper is a distinct improvement over the processes currently in action. Furthermore, its application is not limited to IMRT planning and could be used for other reverse engineering processes as well.

## 5.6 Clinical Examples

This section presents the multicriteria optimization paradigm as it can be realized in daily clinical practice. We will illustrate the new multicriteria treatment planning by two clinical cases representing the most important indications for IMRT, namely prostate and head-and-neck cancer.

### 5.6.1 Prostate cancer

Prostate cancer is the most frequent cancer in men in the Western world. Studies showed that prostate cancer patients with still localized disease but a high risk — which is derived from a histological grading score and the concentration of the prostate specific antigen, PSA — will benefit from a high-dose prostate irradiation. However, this dose is limited by the rectum, which is located directly dorsal to the prostate, implying the risk of rectal bleeding and incontinence [52].

Using IMRT instead of conventional radiotherapy, the dose distribution can be better tailored to the target volume, lowering the rectum toxicity [87]. But even with IMRT, every treatment plan will be a clinical compromise between the dose in the target volume and the dose to the rectum. Other structures involved in prostate treatment planning are the bladder and the femoral heads.

For the sake of simplicity, in the following we will only consider one target volume and the rectum as main structures for this kind of planning problem. After the planner has defined the organ contours and the beam geometry, the multicriteria planning program calculates the plan database. Because there are only two structures to consider, the database consists of the equibalanced solution, two extreme compromises, and, in this example, 17 intermediate plans summing up to 20 solutions, which were computed in approximately 10 minutes.

Because in this case the Pareto front is only two-dimensional, it can also be plotted and shown completely in a graph, see Figure 5.20.

All plans are normalized to the same mean target dose, which greatly facilitates the comparison of different plans, because now only the homogeneity of the target dose distribution, represented by the standard deviation sigma, has to be judged against the rectum dose, which is represented by the equivalent uniform dose, EUD.

The planning horizon can be seen as the range on the axes between the respective coordinates of the extreme compromises in Figure 5.20. In this

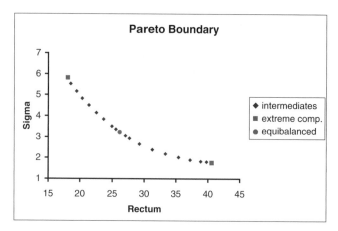

**Fig. 5.20.** Standard deviation in the target against EUD in the rectum.

example, the EUD of the rectum reaches from 18.0 Gy to 40.6 Gy. If the lowest dose to one risk VOI is still too high to be acceptable, then the planner knows immediately without any further calculation that the target dose has to be reduced by re-normalizing the database.

Now the interactive planning process begins. We will defer the description of a planning scenario to the next case, as navigating among the solutions in 2 dimensions is rather straightforward.

### 5.6.2 Head-and-neck cancer

Treatment planning for head-and-neck cancer can be a very challenging task. The primary tumor can be located anywhere in the naso- and oropharyngeal area, and regularly the lymphatic nodal stations have to be irradiated because they are at risk of containing microscopic tumor spread. This results in big, irregular-shaped target volumes with several risk VOIs nearby.

The salivary glands are such risk VOIs that are quite radiosensitive. The tolerance dose of the biggest salivary gland, the parotid gland, is approximately a mean dose of 26 Gy [49]. The goal should be to spare at least one of the parotid glands. Otherwise, the patient might suffer from xerostomia (a completely dry mouth), which can significantly reduce the quality of life. Other normal structures that have to be considered are (depending on the specific case), e.g., the brain stem, the spinal cord, the esophagus and the lungs.

If there is macroscopic tumor left, it can be considered as an additional target volume to be treated with a higher dose. This is known as simultaneously integrated boost concept (SIB) [39, 48] and further increases the complexity of the planning problem.

In Figure 5.21(a), the case of a lymphoepithelioma originating from the left eustachian tube is shown. The database for this case contains 25 solutions

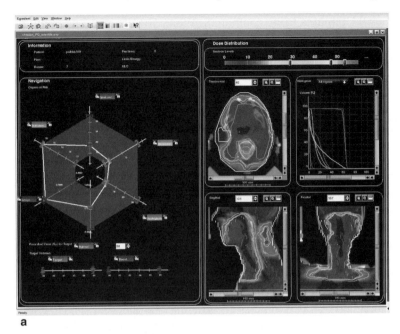

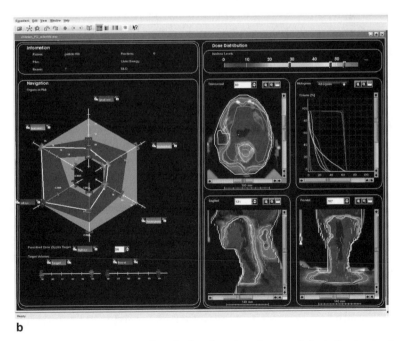

**Fig. 5.21.** Navigation screens for the head-and-neck case: (a) at the beginning of the planning process and (b) after some restrictions have been made — note the significant difference in the remaining planning domain.

and took 11 minutes to be computed. Again, the complete planning horizon can be seen at first sight, and a first solution is presented to the planner.

Now the interactive planning process begins. By dragging either the target homogeneity slider or one of the EUD sliders with the mouse, the treatment planner can quickly explore all compromises between target dose and the doses in critical volumes that are achievable with the given setup geometry.

While dragging one of the navigation sliders, the user wanders along the Pareto boundary, and all information in the navigator window like the iso-dose distribution and the dose-volume histogram is updated in real-time. The program provides the possibility of *locking* or *restricting* an organ to exclude unwanted parts from the navigation. By clicking the *lock* option for a specific structure, all solutions with worse criterion values for the chosen organ than the current one are excluded from further exploration. This is visualized by a reduced planning horizon, see Figure 5.21(b). It allows for narrowing down the solution space to the area that is of highest clinical relevance. Of course, the lock can be reversed at any time, bringing back the broader planning horizon.

Complex planning problems can be interactively explored this way, and the best clinical solution can be found in a short amount of time.

Unfortunately, the dynamics of changing the plan and the effect of direct feedback is impossible to demonstrate in this printed chapter. There is a smooth transition between the curves of the DVH display, and the planner can decide quickly for the best clinical compromise.

### 5.6.3 General remarks

It is important to note that in daily practice, the mathematical details of the implementation as they were described in the previous sections are almost completely hidden from the treatment planner. Instead, we strived for an interface as clean and easy to use as possible so that the planner can focus on the specific case in all its clinical, not mathematical, complexity. This is a crucial aspect for a broad acceptance in the radio-oncological community.

Because many hospitals worldwide already have introduced IMRT into their clinical routine, the new planning scheme as proposed in this chapter also has to be integrated into existing workflows. Treatment planning in radio-therapy departments is usually a close collaboration between physicians and physicists. After the images for treatment planning were acquired, the outlines of the tumor target volume and the risk VOIs are defined. Then the beam geometry and the intensity maps are determined, which is the core part of the treatment planning process. When a certain plan is chosen for treatment, it is dosimetrically verified using hospital-dependent verification procedures, and finally the patient is treated. Today several commercial IMRT treatment plan-ning programs exist for determining the beam setup and the intensity maps, but all of them share the drawbacks of single-objective optimization men-tioned previously. The new multicriteria planning program is able to replace this core part of the workflow while leaving all other parts before and after it

unchanged. The result is an improved plan quality and consequently probably better clinical outcome. At the same time, radiotherapy planning is made
easier to handle with reduced time requirements, facilitating an even broader
introduction of IMRT in radiotherapy departments.

## 5.7 Research Topics

The framework presented here is implemented in a prototype software by the
Fraunhofer ITWM. An academic version of it is available on the web page
http://www.project-mira.net. While the concepts have been tested and
validated at clinical sites like the DKFZ in Heidelberg and the Department
of Radiation Oncology at the Massachusetts General Hospital, there are still
many topics that need to be addressed to improve IMRT planning.

The beam geometry optimization problem was addressed in the introduction. Finding procedures to produce good directions is an ongoing research
effort at the ITWM.

While the optimization of the intensity maps is itself a challenging problem, there are still difficulties concerning the application of a planned treatment. Because the optimized intensity maps are to be delivered with the help
of an MLC, a sequencing algorithm has to determine the configuration of
such hardware. There exist approaches to control the resulting "complexity"
of applying a treatment plan depending on the MLC hardware and method of
application. One approach to this end has been the incorporation of sequencing into the intensity map optimization problem. Romeijn et al. propose a
column generation scheme for the convex planning problem with linear constraints in [58]. The solutions resulted in sequences with a low number of
shapes – one measure of complexity in static sequencing.

An open question is, for example, the impact of the interpolation of plans
in our navigation routines on such approaches to reduce the complexity.

Another direction is the recent movement toward the dynamic plan adap-
tion to the organ geometry known as adaptive or 4D planning [33, 84]. During a treatment, the organ geometry in a patient changes. The impact of an
altered geometry to the quality of a plan may be detrimental if target regions
meant to receive high dose are close to critical structures. These changes are
usually grouped into the *interfraction* [45, 63, 82, 83, 84] and *intrafraction*
[33, 34, 41, 43, 56, 73, 89] changes.

The former are due to the patient losing weight or the tumor becoming
smaller over the time of treatment. They may be reacted upon by a short
re-optimization of the existing plans. The old plans should provide excellent
starting points in the optimization given that the changes are on a relatively
small scale.

The latter are due to breathing or digestion and are harder to tackle. Some
approaches try to anticipate forthcoming changes and incorporate that into

the planning. Doing so, the optimization is very similar to the planning for interfraction changes.

The complications faced by any reactive scheme that monitors the movements of all critical structure during treatment and adjusts the plans online are rather involved. In the future, however, with increasing sophistication of the devices used to deliver treatment, these questions need consideration and practical answers.

## Acknowledgment

This work was supported in part by the National Institutes of Health, grant CA103904-01A1.

## References

[1] M.V. Abramova. Approximation of Pareto set on the basis of inexact information. *Moscow University Computational Mathematics and Cybernetics*, 2:62–69, 1986.

[2] C.B. Barber and H. Huhdanpaa. *Qhull manual*, 2003. Available from http://www.qhull.org, information and data accessed on May 17, 2008.

[3] H.P. Benson. An outer approximation algorithm for generating all efficient extreme points in the outcome set of a multiple objective linear programming problem. *Journal of Global Optimization*, 13(1):1–24, 1998.

[4] H.P. Benson and S. Sayin. Optimization over the efficient set: Four special cases. *Journal of Optimization Theory and Applications*, 80(1):3–18, 1994.

[5] H.P. Benson and E. Sun. Pivoting in an outcome polyhedron. *Journal of Global Optimization*, 16:301–323, 2000.

[6] K.H. Borgwardt. *The Simplex Method - A Probabilistic Analysis*, volume I of *Algorithms and Combinatorics*. Springer, Berlin, 1987.

[7] T.R. Bortfeld. Dosiskonformation in der Tumortherapie mit externer ionisierender Strahlung: Physikalische Möglichkeiten und Grenzen. *Habilitationsschrift, Deutsches Krebsforschungszentrum, Heidelberg*, 1995.

[8] T.R. Bortfeld and W. Schlegel. Optimization of beam orientations radiation therapy: some theoretical considerations. *Physics in Medicine and Biology*, 35:1423–1434, 1993.

[9] T.R. Bortfeld, J. Stein, and K. Preiser. Clinically relevant intensity modulation optimization using physical criteria. In D.D. Leavitt and G. Starkschall, editors, *Proceedings of the XIIth ICCR*, Salt Lake City. Medical Physics Publishing, Madison, Wisconsin, 1997.

[10] A. Brahme. Dosimetric precision requirements in radiation therapy. *Acta Radiol Oncol*, 23:379–391, 1984.

[11] A. Brahme. Treatment optimization using physical and radiobiological objective functions. In A. Smith, editor, *Radiation Therapy Physics*. Springer, Berlin, 1995.

[12] J. Buchanan and L. Gardiner. A comparison of two reference point methods in multiple objective mathematical programming. *European Journal on Operational Research*, 149(1):17–34, 2003.

[13] R.E. Burkard, H. Leitner, R. Rudolf, T. Siegl, and E. Tabbert. Discrete optimization models for treatment planning in radiation therapy. In H. Hutten, editor, *Science and Technology for Medicine, Biomedical Engineering in Graz*, Pabst, Lengerich, 1995.

[14] O.L. Chernykh. Approximation of the Pareto-hull of a convex set by polyhedral sets. *Computational Mathematics and Mathematical Physics*, 35(8):1033–1039, 1995.

[15] C. Cotrutz, M. Lahanas, C. Kappas, and D. Baltas. A multiobjective gradient-based dose optimization algorithm for external beam conformal radiotherapy. *Physics in Medicine and Biology*, 46:2161–2175, 2001.

[16] I. Das. An improved technique for choosing parameters for Pareto surface generation using normal-boundary intersection. In *WCSMO-3 Proceedings, Buffalo, New York*, 1999.

[17] I. Das and J. Dennis. A closer look at the drawbacks of minimizing weighted sums of objectives for Pareto set generation in multicriteria optimization problems. *Structural Optimization*, 14(1):63–69, 1997.

[18] J.P. Dauer. Optimization over the efficient set using an active constraint approach. *Zeitschrift für Operations Research*, 35:185–195, 1991.

[19] M. Ehrgott. *Multicriteria Optimization*. Springer, Berlin, 2000.

[20] M. Ehrgott and D. Tenfelde-Podehl. Computation of ideal and nadir values and implications for their use in MCDM methods. *European Journal on Operational Research*, 151:119–139, 2003.

[21] B. Emami, J. Lyman, A. Brown, L. Coia, M. Goitein, J.E. Munzenrieder, B. Shank, L.J. Solin, and M. Wesson. Tolerance of normal tissue to therapeutic irradiation. *International Journal of Radiation Oncology Biology Physics*, 21:109–122, 1991.

[22] J. Evans, R.D. Plante, D.F. Rogers, and R.T. Wong. Aggregation and disaggregation techniques and methodology in optimization. *Operations Research*, 39(4):553–582, 1991.

[23] J. Fliege and A. Heseler. Constructing approximations to the efficient set of convex quadratic multiobjective problems. Technical Report 211, Fachbereich Mathematik, Universität Dortmund, Dortmund, Germany, 2002.

[24] A.M. Geoffrion. Proper efficiency and the theory of vector maximization. *Journal of Mathematical Analysis and Applications*, 22:618–630, 1968.

[25] A. Gustafsson, B.K. Lind, and A. Brahme. A generalized pencil beam algorithm for optimization of radiation therapy. *Medical Physics*, 21(3):343–356, 1994.

[26] O.C.L. Haas, K.J. Burnham, and J.A. Mills. Adaptive error weighting scheme to solve the inverse problem in radiotherapy. In *12th International Conference on Systems Engineering ICSE97*, Coventry University, 1997.

[27] O.C.L. Haas, J.A. Mills, K.J. Burnham, and C.R. Reeves. Multi objective genetic algorithms for radiotherapy treatment planning. In *IEEE International Conference on Control Applications*, 1998.

[28] H.W. Hamacher and K.-H. Küfer. Inverse radiation therapy planning - a multiple objective optimization approach. *Discrete Applied Mathematics*, 118:145–161, 2002.

164     K.-H. Küfer et al.

[29] A. Holder. Partitioning multiple objective optimal solutions with applications in radiotherapy design. *Technical Report, Trinity University, Mathematics*, 54, 2001.

[30] R. Horst, L.D. Muu, and N.V. Thoai. A decomposition algorithm for optimization over efficient sets. *Forschungsbericht Universität Trier*, 97-04:1–13, 1997.

[31] R. Horst and N.V. Thoai. Utility function programs and optimization over the efficient set in multiple-objective decision making. *Journal of Optimization Theory and Applications*, 92(3):605–631, 1997.

[32] H. Isermann and R.E. Steuer. Computational experience concerning payoff tables and minimum criterion values over the efficient set. *European Journal on Operational Research*, 33:91–97, 1987.

[33] P. Keall. 4-dimensional computed tomography imaging and treatment planning. *Seminars in Radiation Oncology*, 14:81–90, 2004.

[34] P. Keall, V. Kini, S. Vedam, and R. Mohan. Motion adaptive x-ray therapy: a feasibility study. *Physics in Medicine and Biology*, 46:1–10, 2001.

[35] K. Klamroth, J. Tind, and M.M. Wiecek. Unbiased approximation in multicriteria optimization. *Mathematical Methods of Operations Research*, 56(3):413–437, 2002.

[36] P. Korhonen, S. Salo, and R.E. Steuer. A heuristic for estimating nadir criterion values in multiple objective linear programming. *Operations Research*, 45(5):751–757, 1997.

[37] K.-H. Küfer, A. Scherrer, M. Monz, F.V. Alonso, H. Trinkaus, T.R. Bortfeld, and C. Thieke. Intensity-modulated radiotherapy - a large scale multi-criteria programming problem. *OR Spectrum*, 25:223–249, 2003.

[38] M. Lahanas, E. Schreibmann, and D. Baltas. Multiobjective inverse planning for intensity modulated radiotherapy with constraint-free gradient-based optimization algorithms. *Physics in Medicine and Biology*, 48:2843–2871, 2003.

[39] A. Lauve, M. Morris, R. Schmidt-Ullrich, Q. Wu, R. Mohan, O. Abayomi, D. Buck, D. Holdford, K. Dawson, L. Dinardo, and E. Reiter. Simultaneous integrated boost intensity-modulated radiotherapy for locally advanced head-and-neck squamous cell carcinomas: II–clinical results. *International Journal of Radiation Oncology Biology Physics*, 60(2):374–387, 2004.

[40] E.K. Lee, T. Fox, and I. Crocker. Simultaneous beam geometry and intensity map optimization in intensity-modulated radiation therapy. *International Journal of Radiation Oncology Biology Physics*, 64(1):301–320, 2006.

[41] D.A. Low, M. Nystrom, E. Kalinin, P. Parikh, J.F. Dempsey, J.D. Bradley, S. Mutic, S.H. Wahab, T. Islam, G. Christensen, D.G. Politte, and B.R. Whiting. A method for the reconstruction of four-dimensional synchronized CT scans acquired during free breathing. *Medical Physics*, 30:1254–1263, 2003.

[42] L.T. Luc and L.D. Muu. Global optimization approach to optimizing over the efficient set. In P. Gritzmann et al., editors, *Recent Advances in Optimization. Proceedings of the 8th French-German conference on Optimization*, pages 183–195, 1997.

[43] A.E. Lujan, J.M. Balter, and R.K. Ten Haken. A method for incorporating organ motion due to breathing into 3d dose calculations in the liver: sensitivity to variations in motion. *Medical Physics*, 30:2643–2649, 2003.

[44] E. Marchi and J.A. Oviedo. Lexicographic optimality in the multiple objective linear programming: The nucleolar solution. *European Journal on Operational Research*, 57(3):355–359, 1992.

[45] A.A. Martinez, D. Yan, D. Lockman, D. Brabbins, K. Kota, M. Sharpe, D.A. Jaffray, F. Vicini, and J. Wong. Improvement in dose escalation using the process of adaptive radiotherapy combined with three-dimensional conformal or intensity-modulated beams for prostate cancer. *International Journal of Radiation Oncology Biology Physics*, 50(5):1226–1234, 2001.

[46] G. Meedt, M. Alber, and F. Nüsslin. Non-coplanar beam direction optimization for intensity-modulated radiotherapy. *Physics in Medicine and Biology*, 48:2999–3019, 2003.

[47] K. Miettinen. *Nonlinear Multiobjective Optimization*. Kluwer, Boston, 1999.

[48] R. Mohan, W. Wu, M. Manning, and R. Schmidt-Ullrich. Radiobiological considerations in the design of fractionation strategies for intensity-modulated radiation therapy of head and neck cancers. *International Journal of Radiation Oncology Biology Physics*, 46(3):619–630, 2000.

[49] M.W. Münter, C.P. Karger, S.G. Hoffner, H. Hof, C. Thilmann, V. Rudat, S. Nill, M. Wannenmacher, and J. Debus. Evaluation of salivary gland function after treatment of head-and-neck tumors with intensity-modulated radiotherapy by quantitative pertechnetate scintigraphy. *International Journal of Radiation Oncology Biology Physics*, 58(1):175–184, 2004.

[50] V.N. Nefëdov. On the approximation of a Pareto set. *USSR Computational Mathematics and Mathematical Physics*, 24(4):19–28, 1984.

[51] A. Niemierko. Reporting and analyzing dose distributions: a concept of equivalent uniform dose. *Medical Physics*, 24:103–110, 1997.

[52] A. Pollack, G.K. Zagars, J.A. Antolak, D.A. Kuban, and I.I. Rosen. Prostate biopsy status and PSA nadir level as early surrogates for treatment failure: analysis of a prostate cancer randomized radiation dose escalation trial. *International Journal of Radiation Oncology Biology Physics*, 54(3):677–685, 2002.

[53] A. Pugachev, A.L. Boyer, and L. Xing. Beam orientation optimization in intensity-modulated radiation treatment planning. *Medical Physics*, 27:1238–1245, 2000.

[54] A. Pugachev, J.G. Li, A.L. Boyer, S.L. Hancock, Q.T. Le, S.S. Donaldson, and L. Xing. Role of beam orientation in intensity-modulated radiation therapy. *International Journal of Radiation Oncology Biology Physics*, 50(2):551–560, 2001.

[55] H. Reuter. An approximation method for the efficiency set of multiobjective programming problems. *Optimization*, 21(6):905–911, 1990.

[56] E. Rietzel, G.T.Y. Chen, N.C. Choi, and C.G. Willett. Four-dimensional image-based treatment planning: target volume segmentation and dose calculation in the presence of respiratory motion. *International Journal of Radiation Oncology Biology Physics*, 61:1535–1550, 2005.

[57] H.E. Romeijn, R.K. Ahuja, J.F. Dempsey, and A. Kumar. A new linear programming approach to radiation therapy treatment planning problems. *Operations Research*, 54(2):201–216, 2006.

[58] H.E. Romeijn, R.K. Ahuja, J.F. Dempsey, and A. Kumar. A column generation approach to radiation therapy planning using aperture modulation. *SIAM Journal on Optimization*, 15(3):838–862, 2005.

[59] H.E. Romeijn, J.F. Dempsey, and J.G. Li. A unifying framework for multicriteria fluence map optimization models. *Physics in Medicine and Biology*, 49:1991–2013, 2004.

[60] S. Ruzika and M.M. Wiecek. Approximation methods in multiobjective programming. *Journal of Optimization Theory and Applications*, 126(3):473–501, 2005.

[61] J.K. Sankaran. On a variant of lexicographic multi-objective programming. *European Journal on Operational Research*, 107(3):669–674, 1998.

[62] S. Sayin. A procedure to find discrete representations of the efficient set with specified coverage errors. *Operations Research*, 51:427–436, 2003.

[63] B. Schaly, J.A. Kempe, G.S. Bauman, J.J. Battista, and J. Van Dyk. Tracking the dose distribution in radiation therapy by accounting for variable anatomy. *Physics in Medicine and Biology*, 49(5):791–805, 2004.

[64] B. Schandl, K. Klamroth, and M.M. Wiecek. Norm-based approximation in multicriteria programming. *Applied Mathematics and Computation*, 44(7): 925–942, 2002.

[65] A. Scherrer and K.-H. Küfer. Accelerated IMRT plan optimization using the adaptive clustering method. *Linear Algebra and its Applications*, 2008, to appear.

[66] W. Schlegel and A. Mahr. 3D Conformal Radiation Therapy - Multimedia Introduction to Methods and Techniques. Multimedia CD-ROM, Springer, Berlin, 2001.

[67] D.M. Shepard, M.C. Ferris, G.H. Olivera, and T.R. Mackie. Optimizing the delivery of radiation therapy to cancer patients. *SIAM Review*, 41:721–744, 1999.

[68] R.S. Solanki, P.A. Appino, and J.L. Cohon. Approximating the noninferior set in multiobjective linear programming problems. *European Journal on Operational Research*, 68(3):356–373, 1993.

[69] J. Stein, R. Mohan, X.H. Wang, T.R. Bortfeld, Q. Wu, K. Preiser, C.C. Ling, and W. Schlegel. Number and orientations of beams in intensity-modulated radiation treatments. *Medical Physics*, 24(2):149–160, 1997.

[70] R.E. Steuer and F.W. Harris. Intra-set point generation and filtering in decision and criterion space. *Computers & Operations Research*, 7:41–53, 1980.

[71] C. Thieke, T.R. Bortfeld, and K.-H. Küfer. Characterization of dose distributions through the max and mean dose concept. *Acta Oncologica*, 41:158–161, 2002.

[72] H. Trinkaus and K.-H. Küfer. *Vorbereiten der Auswahl von Steuergrößen für eine zeitlich und räumlich einzustellende Dosisverteilung eines Strahlengerätes.* Fraunhofer Institut für Techno- und Wirtschaftsmathematik. Patent erteilt am 8 Mai 2003 unter DE 101 51 987 A.

[73] A. Trofimov, E. Rietzel, H.M. Lu, B. Martin, S. Jiang, G.T.Y. Chen, and T.R. Bortfeld. Temporo-spatial IMRT optimization: concepts, implementation and initial results. *Physics in Medicine and Biology*, 50:2779–2798, 2005.

[74] V.M. Voinalovich. External approximation to the Pareto set in criterion space for multicriterion linear programming tasks. *Kibernetika i vycislitel'naja technika*, 62:89–94, 1984.

[75] A.R. Warburton. Quasiconcave vector maximization: Connectedness of the sets of Pareto-optimal alternatives. *Journal of Optimization Theory and Applications*, 40:537–557, 1983.

[76] S. Webb. *The physics of conformal radiotherapy.* Institute of Physics Publishing Ltd, Bristol, U.K., 1997.

[77] S. Webb. *Intensity-Modulated Radiation Therapy.* Institute of Physics Publishing Ltd, Bristol, U.K., 2001.

[78] A.P. Wierzbicki. A mathematical basis for satisficing decision making. In *Organizations: multiple agents with multiple criteria*, volume 190, pages 465–485. Springer, Berlin, 1981.

[79] A.P. Wierzbicki. A mathematical basis for satisficing decision making. *Mathematical Modelling*, 3:391–405, 1982.

[80] A.P. Wierzbicki. On the completeness and constructiveness of parametric characterizations to vector optimization problems. *OR Spektrum*, 8:73–87, 1986.

[81] Y. Yamamoto. Optimization over the efficient set: overview. *Journal of Global Optimization*, 22:285–317, 2002.

[82] D. Yan, D.A. Jaffray, and J.W. Wong. A model to accumulate fractionated dose in a deforming organ. *International Journal of Radiation Oncology Biology Physics*, 44:665–675, 1999.

[83] D. Yan and D. Lockman. Organ/patient geometric variation in external beam radiotherapy and its effects. *Medical Physics*, 28:593–602, 2001.

[84] D. Yan, F. Vicini, J. Wong, and A. Martinez. Adaptive radiation therapy. *Physics in Medicine and Biology*, 42:123–132, 1997.

[85] P.L. Yu. A class of solutions for group decision problems. *Management Science*, 19(8):936–946, 1973.

[86] C. Zakarian and J.O. Deasy. Beamlet dose distribution compression and reconstruction using wavelets for intensity modulated treatment planning. *Medical Physics*, 31(2):368–375, 2004.

[87] M.J. Zelefsky, Z. Fuks, M. Hunt, Y. Yamada, C. Marion, C.C. Ling, H. Amols, E.S. Venkatraman, and S.A. Leibel. High-dose intensity modulated radiation therapy for prostate cancer: early toxicity and biochemical outcome in 772 patients. *International Journal of Radiation Oncology Biology Physics*, 53(5):1111–1116, 2002.

[88] M. Zeleny. Compromise programming. In JL Cochrane and M Zeleny, editors, *Multiple Criteria Decision Making*, pages 262–301. University of South Carolina Press, Columbia, South Carolina, 1973.

[89] T. Zhang, R. Jeraj, H. Keller, W. Lu, G.H. Olivera, T.R. McNutt, T.R. Mackie, and B. Paliwal. Treatment plan optimization incorporating respiratory motion. *Medical Physics*, 31:1576–1586, 2004.

# 6

# Algorithms for Sequencing Multileaf Collimators

Srijit Kamath[1], Sartaj Sahni[1], Jatinder Palta[2], Sanjay Ranka[1], and Jonathan Li[2]

[1] Department of Computer and Information Science and Engineering, University of Florida, Gainesville, Florida 32611-6120
`srijitk@ufl.edu`, {`sahni,ranka`}`@cise.ufl.edu`

[2] Department of Radiation Oncology, University of Florida, Gainesville, Florida 32610-0385
{`paltajr,lijg`}`@ufl.edu`

**Abstract.** In delivering radiation therapy for cancer treatment, it is desirable to deliver high doses of radiation to a target, while permitting only a low dosage to the surrounding healthy tissues. In recent years, the development of intensity modulated radiation therapy (IMRT) has made this possible. IMRT may be delivered by several techniques. The delivery of IMRT with a multileaf collimator (MLC) requires the delivery of radiation from several beam orientations. The intensity profile for each beam direction is described as a MLC leaf sequence, which is developed using a leaf sequencing algorithm. Important considerations in developing a leaf sequence for a desired intensity profile include maximizing the monitor unit (MU) efficiency (equivalently minimizing the beam-on time) and minimizing the total treatment time subject to the leaf movement constraints of the MLC model. Common leaf movement constraints include minimum and maximum leaf separation and leaf interdigitation. The problem of generating leaf sequences free of tongue-and-groove underdosage also imposes constraints on permissible leaf configurations. In this chapter, we present an overview of recent advances in leaf sequencing algorithms.

## 6.1 Introduction

### 6.1.1 Problem description

The objective of radiation therapy for cancer treatment is to deliver high doses of radiation to the target volume while limiting radiation dose on the surrounding healthy tissues. For example, for head and neck tumors, it is necessary for radiation to be delivered so that the exposure of the spinal cord, optic nerve, salivary glands, or other important structures is minimized. In recent years, this has been made possible due to the development of conformal radio therapy. In conformal therapy, treatment is delivered using a set of radiation beams that are positioned such that the shape of the dose distribution

P.M. Pardalos, H.E. Romeijn (eds.), *Handbook of Optimization in Medicine*, 169
Springer Optimization and Its Applications 26, DOI: 10.1007/978-0-387-09770-1_6,
© Springer Science+Business Media LLC 2009

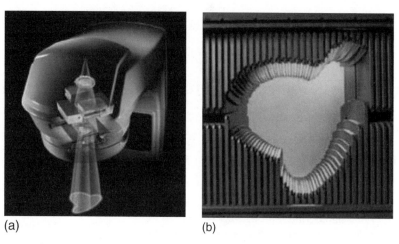

(a)                              (b)

**Fig. 6.1.** (a) A linear accelerator and (b) a multileaf collimator (the figures are from http://www.lexmed.com/medical_services/IMRT.htm).

"conforms" in three dimensions to the shape of the tumor. This is typically achieved by positioning beams of varying shapes from different directions so that each beam is shaped to conform to the projection of the target volume from the beam's-eye view and and to avoid the organs at risk in the vicinity of the target.

Intensity modulated radiation therapy (IMRT) is the state of the art in conformal radiation therapy. IMRT permits the intensity of a radiation beam to be varied across a treatment area, thereby improving the dose conformity. Radiation is delivered using a medical linear accelerator (Figure 6.1(a)). A rotating gantry containing the accelerator structure can rotate around the patient who is positioned on an adjustable treatment couch. Modulation of the beam fluence can be achieved by several techniques. In compensator-based IMRT, the beam is modulated with a preshaped piece of material called a compensator (modulator). The degree of modulation of the beam varies depending on the thickness of the material through which the beam is attenuated. The computer determines the shape of each modulator in order to deliver the desired beam. This type of modulation requires the modulator to be fabricated and then manually inserted into the tray mount of a linear accelerator. In tomotherapy-based IMRT, the linear accelerator travels in multiple circles all the way around the gantry ring to deliver the radiation treatment. The beam is collimated to a narrow slit, and the intensity of the beam is modulated during the gantry movement around the patient. Care must be taken to ensure that adjacent circular arcs do not overlap and thereby do not overdose tissues. This type of delivery is referred to as serial tomotherapy. A modification of serial tomotherapy is helical tomotherapy. In helical tomotherapy, the treatment couch moves linearly (continuously) through the rotating accelerator gantry. Thus each time the accelerator comes around, it directs the beam on a slightly different plane on the patient. In MLC-based IMRT, the accelerator structure is equipped with a computer-controlled mechanical device called a multileaf collimator (MLC, Figure 6.1(b)) that shapes the radiation beam, so as to deliver the radiation as prescribed by the treatment plan. The MLC may have up to 120 movable leaves that can move along an axis perpendicular to the beam and can be arranged so as to shield or expose parts of the anatomy during treatment. The leaves are arranged in pairs so that each leaf pair forms one row of the arrangement. The set of allowable MLC leaf configurations may be restricted by leaf movement constraints that are manufacturer and/or model dependent.

The first stage in the treatment planning process in IMRT is to obtain accurate three-dimensional anatomical information about the patient. This is achieved using computed tomography (CT) and/or magnetic resonance (MR) imaging. An ideal dose distribution would ensure perfect conformity to the target volume while completely sparing all other tissues. However, such a distribution is impossible to realize in practice. Therefore, doses to targets and tolerable doses for critical structures are prescribed, and an objective function that measures the quality of a plan is developed subject to these dose-based constraints. Next, a set of beam parameters (beam angles, profiles, weights) that optimize this objective are determined using a computer program. This method is called "inverse planning" as resultant dose distribution is first described and the best beam parameters that deliver the distribution (approximately) are then solved for. It is to be noted that inverse planning is a general concept and its implementation details vary vastly among various systems. After the inverse planning in MLC-based IMRT, the delivery of radiation intensity profile for each beam direction is described as a MLC leaf sequence, which is developed using a leaf sequencing algorithm. Important considerations in developing a leaf sequence for a desired intensity profile include maximizing the monitor unit (MU) efficiency (equivalently minimizing the beam-on time) and minimizing the total treatment time subject to the leaf movement constraints of the MLC model. Finally, when the leaf sequences for all beam directions are determined, the treatment is performed from various beam angles sequentially using computer control. In this chapter, we present an overview of recent advances in leaf sequencing algorithms.

### 6.1.2 MLC models and constraints

The purpose of the leaf sequencing algorithm is to generate a sequence of leaf positions and/or movements that faithfully reproduce the desired intensity map once the beam is delivered, taking into consideration any hardware and dosimetric characteristics of the delivery system. The two most common methods of IMRT delivery with computer-controlled MLCs are the segmental multileaf collimator (SMLC) and dynamic multileaf collimator (DMLC). In SMLC, the beam is switched off while the leaves are in motion. In other words, the delivery is done using multiple static segments or leaf settings.

This method is also frequently referred to as the "step and shoot" or "stop and shoot" method. In DMLC, the beam is on while the leaves are in motion. The beam is switched on at the start of treatment and is switched off only at the end of treatment. The fundamental difference between the leaf sequences of these two delivery methods is that the leaf sequence defines a finite set of beam shapes for SMLC and trajectories of opposing pairs of leaves for DMLC.

In practical situations, there are some constraints on the movement of the leaves. The minimum separation constraint requires that opposing pairs of leaves be separated by at least some distance ($S_{\min}$) at all times during beam delivery. In MLCs, this constraint is applied not only to opposing pairs of leaves (intra-pair minimum separation constraint), but also to opposing leaves of neighboring pairs (inter-pair minimum separation constraint). For example, in Figure 6.2, $L1$ and $R1$, $L2$ and $R2$, $L3$ and $R3$, $L1$ and $R2$, $L2$ and $R1$, $L2$ and $R3$, $L3$ and $R2$ are pairwise subject to the constraint. The case with $S_{\min} = 0$ is called interdigitation constraint and is applicable to some MLC models. Wherever this constraint applies, opposite adjacent leaves are not permitted to overlap.

In most commercially available MLCs, there is a tongue-and-groove arrangement at the interface between adjacent leaves. A cross section of two adjacent leaves is depicted in Figure 6.3. The width of the tongue-and-groove region is $l$. The area under this region gets underdosed due to the mechanical arrangement, as it remains shielded if either the tongue or the groove portion of a leaf shields it.

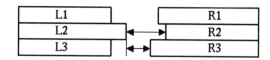

Fig. 6.2. Inter-pair minimum separation constraint.

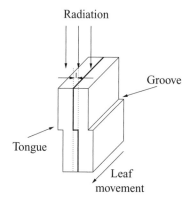

Fig. 6.3. Cross section of leaves.

Maximum leaf spread for leaves on the same leaf bank is one more MLC limitation, according to which no two leaf positions on the same bank can be more than a fixed distance apart throughout the whole leaf sequence. This necessitates a large field (intensity profile) to be split into two or more adjacent abutting sub-fields. This is true for the Varian MLC (Varian Medical Systems, Palo Alto, CA), which has a field size limitation of about 15 cm. The abutting sub-fields are then delivered as separate treatment fields. This often results in longer delivery times, poor MU efficiency, and field matching problems.

This chapter is organized as follows. In Section 6.2, we present leaf sequencing algorithms for the SMLC model. Leaf movement constraints studied include minimum separation constraint (which includes interdigitation as a special case) and the tongue-and-groove constraint (to eliminate the tongue-and-groove effect). In Section 6.3, algorithms for DMLC with or without the interdigitation constraint are developed. In Section 6.4, we study the problem of splitting large intensity modulated fields for models where a maximum leaf spread constraint applies. Finally, in Section 6.5, we provide a summary of recent work on optimizing the number of sements for SMLC delivery.

## 6.2 Algorithms for SMLC

In this section we study the leaf sequencing problem for SMLC. We first introduce the notation that will be used in the remainder of this chapter. We present the leaf sequencing algorithm for a single leaf pair and subsequently extend it for multiple leaf pairs.

### 6.2.1 Single leaf pair

The geometry and coordinate system used are shown in Figure 6.4(a). Consider the delivery of an intensity map produced by the optimizer in the inverse planning stage. It is important to note that the intensity map from the optimizer is always a discrete matrix. The spatial resolution of this matrix is

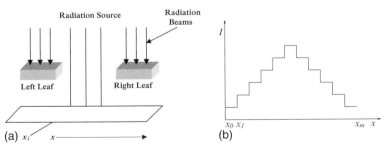

**Fig. 6.4.** (a) Geometry and coordinate system and (b) profile generated by the optimizer.

similar to the smallest beamlet size. The beamlet size typically ranges from 5–10 mm. Let $I(x)$ be the desired intensity profile along the $x$ axis. The discretized profile from the optimizer gives the intensity values at sample points $x_0, x_1, \ldots, x_m$. We assume that the sample points are uniformly spaced and that $\Delta x = x_{i+1} - x_i, 0 \leq i < m$. $I(x)$ is assigned the value $I(x_i)$ for $x_i \leq x < x_{i+1}$, for each $i$. Now, $I(x_i)$ is our desired intensity profile, i.e., $I(x_i)$ is a measure of the number of MUs for which $x_i$, $0 \leq i < m$, needs to be exposed. Figure 6.4(b) shows a profile, which is the output from the optimizer at discrete sample points $x_0, x_1, \ldots, x_m$.

### Movement of leaves

In our analysis, we assume that the leaves are initially at the left most position $x_0$ and that the leaves move unidirectionally from left to right. Figure 6.5 illustrates the leaf trajectory during SMLC delivery. Let $I_l(x_i)$ and $I_r(x_i)$ respectively denote the amount of monitor units (MUs) delivered when the left and right leaves leave position $x_i$. Consider the motion of the left leaf. The left leaf begins at $x_0$ and remains here until $I_l(x_0)$ MUs have been delivered. At this time the left leaf is moved to $x_1$, where it remains until $I_l(x_1)$ MUs have been delivered. The left leaf then moves to $x_3$ where it remains until $I_l(x_3)$ MUs have been delivered. At this time, the left leaf is moved to $x_6$, where it remains until $I_l(x_6)$ MUs have been delivered. The final movement of the left leaf is to $x_7$, where it remains until $I_l(x_7) = I_{\max}$ MUs have been delivered. At this time the machine is turned off. The total beam-on time (which we refer to as *therapy time*), $TT(I_l, I_r)$, is the time needed to deliver $I_{\max}$ MUs. The right leaf moves to $x_2$ when 0 MUs have been delivered; moves to $x_4$ when $I_r(x_2)$ MUs have been delivered; moves to $x_5$ when $I_r(x_4)$ MUs have been delivered; and so on. Note that the machine is off when a leaf is in motion. We make the following observations:

1. All MUs that are delivered along a radiation beam along $x_i$ before the left leaf passes $x_i$ fall on it. The greater the $x$ value, the later the left leaf passes that position. Therefore $I_l(x_i)$ is a non-decreasing function.

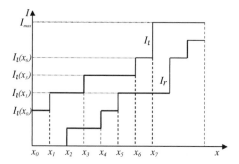

**Fig. 6.5.** Leaf trajectory during SMLC delivery.

2. All MUs that are delivered along a radiation beam along $x_i$ before the right leaf passes $x_i$ are blocked by the leaf. The greater the $x$ value, the later the right leaf passes that position. Therefore $I_r(x_i)$ is also a non-decreasing function.

From these observations, we notice that the net amount of MUs delivered at a point is given by $I_l(x_i) - I_r(x_i)$, which must be the same as the desired profile $I(x_i)$.

## Optimal unidirectional algorithm for one pair of leaves

When the movement of leaves is restricted to only one direction, both the left and right leaves move along the positive $x$ direction, from left to right (Figure 6.4(a)). Once the desired intensity profile, $I(x_i)$ is known, our problem becomes that of determining the individual *intensity profiles* to be delivered by the left and right leaves, $I_l$ and $I_r$, such that:

$$I(x_i) = I_l(x_i) - I_r(x_i), 0 \leq i \leq m. \tag{6.1}$$

We refer to $(I_l, I_r)$ as the *treatment plan* (or simply *plan*) for $I$. Once we obtain the plan, we will be able to determine the movement of both left and right leaves during the therapy. For each $i$, the left leaf can be allowed to pass $x_i$ when the source has delivered $I_l(x_i)$ MUs. Also, we can allow the right leaf to pass $x_i$ when the source has delivered $I_r(x_i)$ MUs. In this manner, we obtain *unidirectional leaf movement profiles* for a plan.

From equation (6.1), we see that one way to determine $I_l$ and $I_r$ from the given target profile $I$ is to begin with $I_l(x_0) = I(x_0)$ and $I_r(x_0) = 0$; examine the remaining $x_i$s from left to right; increase $I_l$ whenever $I$ increases; and increase $I_r$ whenever $I$ decreases. Once $I_l$ and $I_r$ are determined, the leaf movement profiles are obtained as explained in the previous section. The resulting algorithm is shown in Figure 6.6. Figure 6.7 shows a profile and the corresponding plan obtained using the algorithm. Clearly, the complexity of the algorithm is $O(m)$.

Ma et al. [14] show that Algorithm SINGLEPAIR obtains plans that are optimal in therapy time. Their proof relies on the results of Boyer and Strait [4], Spirou and Chui [16], and Stein et al. [17]. Kamath et al. [8] provide a much simpler proof.

**Theorem 1 (Kamath et al. [8]).** *Algorithm SINGLEPAIR obtains plans that are optimal in therapy time. Let $inc1, inc2, \ldots, inck$ be the indices of the points at which the desired profile $I(x_i)$ increases, i.e., $I(x_{inci}) > I(x_{inci-1})$. The therapy time for the plan $(I_l, I_r)$ generated by Algorithm SINGLEPAIR is $\sum_{i=1}^{k}[I(x_{inci}) - I(x_{inci-1})]$, where $I(x_{inc1-1}) = 0$.*

*Proof.* Let $\Delta i = I(x_{inci}) - I(x_{inci-1})$. Suppose that $(I_L, I_R)$ is a plan for $I(x_i)$ (not necessarily that generated by Algorithm SINGLEPAIR). From the

176     S. Kamath et al.

Algorithm SINGLEPAIR
$I_l(x_0) = I(x_0)$
$I_r(x_0) = 0$

For $j = 1$ to $m$ do
  If $(I(x_j) \geq I(x_{j-1})$
    $I_l(x_j) = I_l(x_{j-1}) + I(x_j) - I(x_{j-1})$
    $I_r(x_j) = I_r(x_{j-1})$
  Else
    $I_r(x_j) = I_r(x_{j-1}) + I(x_{j-1}) - I(x_j)$
    $I_l(x_j) = I_l(x_{j-1})$
  End If
End for

**Fig. 6.6.** Obtaining a unidirectional plan.

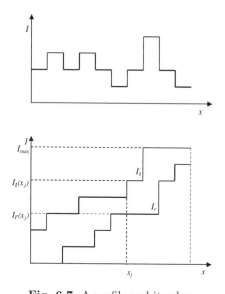

**Fig. 6.7.** A profile and its plan.

unidirectional constraint, it follows that $I_L(x_i)$ and $I_R(x_i)$ are non-decreasing functions of $x$. Because $I(x_i) = I_L(x_i) - I_R(x_i)$ for all $i$, we get

$$\Delta i = (I_L(x_{inci}) - I_R(x_{inci})) - (I_L(x_{inci-1}) - I_R(x_{inci-1}))$$
$$= (I_L(x_{inci}) - I_L(x_{inci-1})) - (I_R(x_{inci}) - I_R(x_{inci-1}))$$
$$\leq I_L(x_{inci}) - I_L(x_{inci-1}).$$

Summing up $\Delta i$, we get

$$\sum_{i=1}^{k}[I(x_{inci}) - I(x_{inci-1})] \le \sum_{i=1}^{k}[I_L(x_{inci}) - I_L(x_{inci-1})]$$
$$= TT(I_L, I_R).$$

Because the therapy time for the plan $(I_l, I_r)$ generated by Algorithm SIN-GLEPAIR is $\sum_{i=1}^{k}[I(x_{inci}) - I(x_{inci-1})]$, it follows that $TT(I_l, I_r)$ is minimum.
$\square$

**Theorem 2 (Kamath et al. [8]).** *If the optimal plan $(I_l, I_r)$ violates the minimum separation constraint, then there is no plan for $I$ that does not violate the minimum separation constraint.*

### 6.2.2 Multiple leaf pairs

We use a single pair of leaves to deliver intensity profiles defined along the axis of the pair of leaves. However, in a real application, we need to deliver intensity profiles defined over a 2D region. Each pair of leaves is controlled independently. If there are no constraints on the leaf movements, we divide the desired profile into a set of parallel profiles defined along the axes of the leaf pairs. Each leaf pair $i$ then delivers the plan for the corresponding intensity profile $I_i(x)$. The set of plans of all leaf pairs forms the solution set. We refer to this set as the *treatment schedule* (or simply *schedule*). In this section, we present leaf sequencing algorithms for SMLC with and without constraints. The constraints considered are (i) minimum separation constraint and (ii) tongue-and-groove constraint and (optionally) interdigitation constraint. These algorithms are from Kamath et al. [8] and Kamath et al. [9].

#### Optimal schedule without the minimum separation constraint

Assume we have $n$ pairs of leaves. For each pair, we have $m$ sample points. The input is represented as a matrix with $n$ rows and $m$ columns, where the $i^{th}$ row represents the desired intensity profile to be delivered by the $i^{th}$ pair of leaves. We apply Algorithm SINGLEPAIR to determine the optimal plan for each of the $n$ leaf pairs. This method of generating schedules is described in Algorithm MULTIPAIR (Figure 6.8). Because the complexity of Algorithm SINGLEPAIR is $O(m)$, it follows that the complexity of Algorithm MULTI-PAIR is $O(mn)$.

**Theorem 3 (Kamath et al. [8]).** *Algorithm MULTIPAIR generates schedules that are optimal in therapy time.*

Boland et al. [3] and Ahuja and Hamacher [1] have developed network flow algorithms that generate schedules that are optimal in therapy time. Baatar et al. [2] also present optimal therapy time algorithms.

Algorithm MULTIPAIR

For$(i = 1; i \leq n; i++)$

Apply Algorithm SINGLEPAIR to the $i^{\text{th}}$ pair of leaves to obtain plan $(I_{il}, I_{ir})$ that delivers the intensity profile $I_i(x)$.

End For

**Fig. 6.8.** Obtaining a schedule.

## Optimal algorithm with inter-pair minimum separation constraint

The schedule generated by Algorithm MULTIPAIR may violate both the intra- and inter-pair minimum separation constraints. If the schedule has no violations of these constraints, it is the desired optimal schedule. If there is a violation of the intra-pair constraint, then it follows from Theorem 2 that there is no schedule that is free of constraint violation. So, assume that only the inter-pair constraint is violated. We eliminate all violations of the inter-pair constraint starting from the left end, i.e., from $x_0$. To eliminate the violations, we modify those plans of the schedule that cause the violations. We scan the schedule from $x_0$ along the positive $x$ direction looking for the least $x_v$ at which is positioned a right leaf (say $R_u$) that violates the inter-pair separation constraint. After rectifying the violation at $x_v$ with respect to $R_u$, we look for other violations. Because the process of eliminating a violation at $x_v$ may, at times, lead to new violations at $x_j, x_j < x_v$, we need to retract a certain distance (we will show that this distance is $S_{min}$, the minimum leaf separation) to the left, every time a modification is made to the schedule. We now restart the scanning and modification process from the new position. The process continues until no inter-pair violations exist. Algorithm MINSEPARATION (Figure 6.9) outlines the procedure.

Let $M = ((I_{1l}, I_{1r}), (I_{2l}, I_{2r}), \ldots, (I_{nl}, I_{nr}))$ be the schedule generated by Algorithm MULTIPAIR for the desired intensity profile. Let $N(p) = ((I_{1lp}, I_{1rp}), (I_{2lp}, I_{2rp}), \ldots, (I_{nlp}, I_{nrp}))$ be the schedule obtained after Step 2 of Algorithm MINSEPARATION is applied $p$ times to the input schedule $M$. Note that $M = N(0)$.

To illustrate the modification process, we use an example (see Figure 6.10). To make things easier, we only show two neighboring pairs of leaves. Suppose that the $(p+1)^{\text{st}}$ violation occurs when the right leaf of pair $u$ is positioned at $x_v$ and the left leaf of pair $t, t \in \{u-1, u+1\}$, arrives at $x_u, x_v - x_u < S_{min}$. Let $x'_u = x_v - S_{min}$. To remove this inter-pair separation violation, we modify $(I_{tlp}, I_{trp})$. The other profiles of $N(p)$ are not modified. The new $I_{tlp}$ (i.e., $I_{tl(p+1)}$) is as defined below

$$I_{tl(p+1)}(x) = \begin{cases} I_{tlp}(x) & x_0 \leq x < x'_u \\ \max\{I_{tlp}(x), I_{tl}(x) + \Delta I\} & x'_u \leq x \leq x_m \end{cases}$$

Algorithm MINSEPARATION
//assume no intra-pair violations exist

$x = x_0$
While (there is an inter-pair violation) do

1. Find the least $x_v$, $x_v \geq x$, such that a right leaf is positioned at $x_v$ and this right leaf has an inter-pair separation violation with one or both of its neighboring left leaves. Let $u$ be the least integer such that the right leaf $R_u$ is positioned at $x_v$ and $R_u$ has an inter-pair separation violation. Let $L_t$ denote the left leaf (or one of the left leaves) with which $R_u$ has an inter-pair violation. Note that $t \in \{u-1, u+1\}$.
2. Modify the schedule to eliminate the violation between $R_u$ and $L_t$.
3. If there is now an intra-pair separation violation between $Rt$ and $L_t$, no feasible schedule exists, terminate.
4. $x = x_v - S_{\min}$

End While

**Fig. 6.9.** Obtaining a schedule under the constraint.

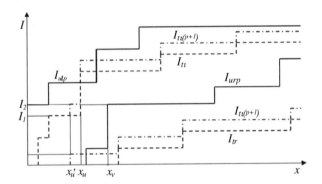

**Fig. 6.10.** Eliminating a violation.

where $\Delta I = I_{urp}(x_v) - I_{tl}(x'_u) = I_2 - I_1$. $I_{tr(p+1)}(x) = I_{tl(p+1)}(x) - I_t(x)$, where $I_t(x)$ is the target profile to be delivered by the leaf pair $t$. Because $I_{tr(p+1)}$ differs from $I_{trp}$ for $x \geq x'_u = x_v - S_{\min}$, there is a possibility that $N(p+1)$ has inter-pair separation violations for right leaf positions $x \geq x'_u = x_v - S_{\min}$. Because none of the other right leaf profiles are changed from those of $N(p)$ and because the change in $I_{tl}$ only delays the rightward movement of the left leaf of pair $t$, no inter-pair violations are possible in $N(p+1)$ for $x < x'_u = x_v - S_{\min}$. One may also verify that as $I_{tl0}$ and $I_{tr0}$ are non-decreasing functions of $x$, so also are $I_{tlp}$ and $I_{trp}$, $p > 0$.

For $N(p), p \geq 0$ and every leaf pair $j, 1 \leq j \leq n$, define $I_{jlp}(x_{-1}) = I_{jrp}(x_{-1}) = 0, \Delta_{jlp}(x_i) = I_{lp}(x_i) - I_{lp}(x_{i-1}), 0 \leq i \leq m$ and $\Delta_{jrp}(x_i) =$

$I_{rp}(x_i) - I_{rp}(x_{i-1}), 0 \leq i \leq m$. Notice that $\Delta_{jlp}(x_i)$ gives the time (in monitor units) for which the left leaf of pair $j$ stops at position $x_i$. Let $\Delta_{jlp}(x_i)$ and $\Delta_{jrp}(x_i)$ be zero for all $x_i$ when $j = 0$ as well as when $j = n+1$.

**Lemma 1 (Kamath et al. [8]).** *For every* $j, 1 \leq j \leq n$ *and every* $i, 1 \leq i \leq m$,

$$\Delta_{jlp}(x_i) \leq \max\{\Delta_{jl0}(x_i), \Delta_{(j-1)rp}(x_i + S_{\min}), \Delta_{(j+1)rp}(x_i + S_{\min})\}. \quad (6.2)$$

*Proof.* The proof is by induction on $p$. For the induction base, $p = 0$. Putting $p = 0$ into the right side of equation (6.2), we get

$$\max\{\Delta_{jl0}(x_i), \Delta_{(j-1)r0}(x_i + S_{\min}), \Delta_{(j+1)r0}(x_i + S_{\min})\} \geq \Delta_{jl0}(x_i).$$

For the induction hypothesis, let $q \geq 0$ be any integer and assume that equation (6.2) holds when $p = q$. In the induction step, we prove that the equation holds when $p = q + 1$. Let $t, u$, and $x_v$ be as in iteration $p - 1$ of the **while** loop of algorithm MINSEPARATION. After this iteration, only $\Delta_{tlp}$ and $\Delta_{trp}$ are different from $\Delta_{tl(p-1)}$ and $\Delta_{tr(p-1)}$, respectively. Furthermore, only $\Delta_{tlp}(x_w)$ and $\Delta_{trp}(x_w)$, where $x_w = x_v - S_{\min}$ may be larger than the corresponding values after iteration $p - 1$. At all but at most one other $x$ value (where $\Delta$ may have decreased), $\Delta_{tlp}$ and $\Delta_{trp}$ are the same as the corresponding values after iteration $p - 1$.

Because $x_v$ is the right leaf position for the leftmost violation, the left leaf of pair $t$ arrives at $x_w = x_v - S_{\min}$ after the right leaf of pair $u$ arrives at $x_v = x_w + S_{\min}$. After the modification made to $I_{tl(p-1)}$, the left leaf of pair $t$ leaves $x_w$ at the same time as the right leaf of pair $u$ leaves $x_w + S_{\min}$. Therefore, $\Delta_{tlp}(x_w) \leq \Delta_{ur(p-1)}(x_w + S_{\min}) = \Delta_{urp}(x_w + S_{\min})$.

The induction step now follows from the induction hypothesis and the observation that $u \in \{t - 1, t + 1\}$. □

**Lemma 2 (Kamath et al. [8]).** *For every* $j, 1 \leq j \leq n$ *and every* $i, 1 \leq i \leq m$,

$$\Delta_{jrp}(x_i) = \Delta_{jlp}(x_i) - (I_j(x_i) - I_j(x_{i-1})) \quad (6.3)$$

*where* $I_j(x_{-1}) = 0$.

*Proof.* We examine $N(p)$. The monitor units delivered by leaf pair $j$ at $x_i$ are $I_{jlp}(x_i) - I_{jrp}(x_i)$ and the units delivered at $x_{i-1}$ are $I_{jlp}(x_{i-1}) - I_{jrp}(x_{i-1})$. Therefore,

$$I_j(x_i) = I_{jlp}(x_i) - I_{jrp}(x_i) \quad (6.4)$$
$$I_j(x_{i-1}) = I_{jlp}(x_{i-1}) - I_{jrp}(x_{i-1}). \quad (6.5)$$

Subtracting equation (6.5) from equation (6.4), we get

$$I_j(x_i) - I_j(x_{i-1}) = (I_{jlp}(x_i) - I_{jlp}(x_{i-1})) - (I_{jrp}(x_i) - I_{jrp}(x_{i-1}))$$
$$= \Delta_{jlp}(x_i) - \Delta_{jrp}(x_i).$$

The lemma follows from this equality. □

Notice that once a right leaf $u$ moves past $x_m$, no separation violation with respect to this leaf is possible. Therefore, $x_v$ (see algorithm MINSEPA-RATION) $\leq x_m$. Hence, $\Delta_{jlp}(x_i) \leq \Delta_{jl0}(x_i)$, and $\Delta_{jrp}(x_i) \leq \Delta_{jr0}(x_i), x_m - S_{\min} \leq x_i \leq x_m, 1 \leq j \leq n$. Starting with these upper bounds, which are independent of $p$, on $\Delta_{jrp}(x_i)$, $x_m - S_{\min} \leq x_i \leq x_m$ and using equations (6.2) and (6.3), we can compute an upper bound on the remaining $\Delta_{jlp}(x_i)$s and $\Delta_{jrp}(x_i)$s (from right to left). The remaining upper bounds are also independent of $p$. Let the computed upper bound on $\Delta_{jlp}(x_i)$ be $U_{jl}(x_i)$. It follows that the therapy time for $(I_{jlp}, I_{jrp})$ is at most $T_{\max}(j) = \sum_{0 \leq i \leq m} U_{jl}(x_i)$. Therefore, the therapy time for $N(p)$ is at most $T_{\max} = \max_{1 \leq j \leq n}\{T_{\max}(j)\}$.

**Theorem 4 (Kamath et al. [8]).** *Algorithm MINSEPARATION always terminates.*

*Proof.* As noted above, Lemmas 1 and 2 provide an upper bound, $T_{\max}$ on the therapy time of any schedule produced by algorithm MINSEPARATION. It is easy to verify that

$$I_{il(p+1)}(x) \geq I_{ilp}(x), 0 \leq i \leq n, x_0 \leq x \leq x_m$$
$$I_{ir(p+1)}(x) \geq I_{irp}(x), 0 \leq i \leq n, x_0 \leq x \leq x_m$$

and that

$$I_{tl(p+1)}(x'_u) > I_{tlp}(x'_u)$$
$$I_{tr(p+1)}(x'_u) > I_{trp}(x'_u).$$

Notice that even though a $\Delta$ value (proof of Lemma 1) may decrease at an $x_i$, the $I_{ilp}$ and $I_{irp}$ values never decrease at any $x_i$ as we go from one iteration of the while loop of MINSEPARATION to the next. Because $I_{tl}$ increases by at least one unit at at least one $x_i$ on each iteration, it follows that the while loop can be iterated at most $mnT_{\max}$ times. □

**Theorem 5 (Kamath et al. [8]).**

(a) When Algorithm MINSEPARATION terminates in step 3, there is no feasible schedule.
(b) Otherwise, the schedule generated is feasible and is optimal in therapy time for unidirectional schedules.

## Elimination of tongue-and-groove effect with or without interdigitation constraint

Figure 6.11 shows a beam's-eye view of the region to be treated by two adjacent leaf pairs, $t$ and $t + 1$. Consider the shaded rectangular areas $A_t(x_i)$ and $A_{t+1}(x_i)$ that require exactly $I_t(x_i)$ and $I_{t+1}(x_i)$ MUs to be delivered, respectively. The tongue-and-groove overlap area between the two leaf pairs

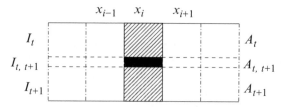

**Fig. 6.11.** Tongue-and-groove effect.

over the sample point $x_i$, $A_{t,t+1}(x_i)$, is colored black. Let the amount of MUs delivered in $A_{t,t+1}(x_i)$ be $I_{t,t+1}(x_i)$. Ignoring leaf transmission, the following lemma is a consequence of the fact that $A_{t,t+1}(x_i)$ is exposed only when both $A_t(x_i)$ and $A_{t+1}(x_i)$ are exposed.

**Lemma 3 (Kamath et al. [9]).** $I_{t,t+1}(x_i) \leq \min\{I_t(x_i), I_{t+1}(x_i)\}$, $0 \leq i \leq m$, $1 \leq t < n$, where $m$ is the number of sample points along each row and $n$ is the number of leaf pairs.

Schedules in which $I_{t,t+1}(x_i) = \min\{I_t(x_i), I_{t+1}(x_i)\}$ are said to be free of tongue-and-groove underdosage effects.

The following lemma provides a necessary and sufficient condition for a unidirectional schedule to be free of tongue-and-groove underdosage effects.

**Lemma 4 (Kamath et al. [9]).** *A unidirectional schedule is free of tongue-and-groove underdosage effects if and only if,*

*(a)* $I_t(x_i) = 0$ *or* $I_{t+1}(x_i) = 0$, *or*
*(b)* $I_{tr}(x_i) \leq I_{(t+1)r}(x_i) \leq I_{(t+1)l}(x_i) \leq I_{tl}(x_i)$, *or*
*(c)* $I_{(t+1)r}(x_i) \leq I_{tr}(x_i) \leq I_{tl}(x_i) \leq I_{(t+1)l}(x_i)$,

*for* $0 \leq i \leq m$, $1 \leq t < n$.

*Proof.* It is easy to see that any schedule that satisfies the above conditions is free of tongue-and-groove underdosage effects. So what remains is for us to show that every schedule that is free of tongue-and-groove underdosage effects satisfies the above conditions. Consider any such schedule. If condition (a) is satisfied at every $i$ and $t$, the proof is complete. So assume $i$ and $t$ such that $I_t(x_i) \neq 0$ and $I_{t+1}(x_i) \neq 0$ exist. We need to show that either (b) or (c) is true for this value of $i$ and $t$. Because the schedule is free of tongue-and-groove effects,

$$I_{t,t+1}(x_i) = \min\{I_t(x_i), I_{t+1}(x_i)\} > 0. \tag{6.6}$$

From the unidirectional constraint, it follows that $A_{t,t+1}(x_i)$ first gets exposed when both right leaves pass $x_i$, and it remains exposed until the first of the left leaves passes $x_i$. Further, if a left leaf passes $x_i$ before a neighboring right leaf passes $x_i$, $A_{t,t+1}(x_i)$ is not exposed at all. So,

$$I_{t,t+1}(x_i) = \max\{0, I_{(t,t+1)l}(x_i) - I_{(t,t+1)r}(x_i)\} \tag{6.7}$$

where

$$I_{(t,t+1)r}(x_i) = \max\{I_{tr}(x_i), I_{(t+1)r}(x_i)\}$$

and

$$I_{(t,t+1)l}(x_i) = \min\{I_{tl}(x_i), I_{(t+1)l}(x_i)\}.$$

From equations (6.6) and (6.7), it follows that

$$I_{t,t+1}(x_i) = I_{(t,t+1)l}(x_i) - I_{(t,t+1)r}(x_i). \tag{6.8}$$

Consider the case $I_t(x_i) \geq I_{t+1}(x_i)$. Suppose that $I_{tr}(x_i) > I_{(t+1)r}(x_i)$. It follows that $I_{(t,t+1)r}(x_i) = I_{tr}(x_i)$ and $I_{(t,t+1)l}(x_i) = I_{(t+1)l}(x_i)$. Now from equation (6.8), we get

$$\begin{aligned}
I_{t,t+1}(x_i) &= I_{(t+1)l}(x_i) - I_{tr}(x_i) \\
&< I_{(t+1)l}(x_i) - I_{(t+1)r}(x_i) \\
&= I_{t+1}(x_i) \\
&\leq I_t(x_i).
\end{aligned}$$

Thus $I_{t,t+1}(x_i) < \min\{I_t(x_i), I_{t+1}(x_i)\}$, which contradicts equation (6.6). So

$$I_{tr}(x_i) \leq I_{(t+1)r}(x_i). \tag{6.9}$$

Now, suppose that $I_{tl}(x_i) < I_{(t+1)l}(x_i)$. From $I_t(x_i) \geq I_{t+1}(x_i)$, it follows that $I_{(t,t+1)l}(x_i) = I_{tl}(x_i)$ and $I_{(t,t+1)r}(x_i) = I_{(t+1)r}(x_i)$. Hence, from equation (6.8), we get

$$\begin{aligned}
I_{t,t+1}(x_i) &= I_{tl}(x_i) - I_{(t+1)r}(x_i) \\
&< I_{(t+1)l}(x_i) - I_{(t+1)r}(x_i) \\
&= I_{t+1}(x_i) \\
&\leq I_t(x_i).
\end{aligned}$$

Thus $I_{t,t+1}(x_i) < \min\{I_t(x_i), I_{t+1}(x_i)\}$, which contradicts equation (6.6). So

$$I_{tl}(x_i) \geq I_{(t+1)l}(x_i). \tag{6.10}$$

From equations (6.9) and (6.10), we can conclude that when $I_t(x_i) \geq I_{t+1}(x_i)$, (b) is true. Similarly one can show that when $I_{t+1}(x_i) \geq I_t(x_i)$, (c) is true. $\square$

Lemma 4 is equivalent to saying that the time period for which a pair of leaves (say pair $t$) exposes the region $A_{t,t+1}(x_i)$ is completely contained by the time period for which pair $t + 1$ exposes region $A_{t,t+1}(x_i)$, or vice versa, whenever $I_t(x_i) \neq 0$ and $I_{t+1}(x_i) \neq 0$. Note that if either $I_t(x_i)$ or $I_{t+1}(x_i)$ is zero, the containment is not necessary. We will refer to the necessary and sufficient condition of Lemma 4 as the *tongue-and-groove constraint condition*. Schedules that satisfy this condition will be said to satisfy the tongue-and-groove constraint. van Santvoort and Heijmen [18] present an algorithm that generates schedules that satisfy the tongue-and-groove constraint for DMLC.

The schedule generated by Algorithm MULTIPAIR (Kamath et al. [8]) may violate the tongue-and-groove constraint. If the schedule has no tongue-and-groove constraint violations, it is the desired optimal schedule. If there are violations in the schedule, we eliminate all violations of the tongue-and-groove constraint starting from the left end, i.e., from $x_0$. To eliminate the violations, we modify those plans of the schedule that cause the violations. We scan the schedule from $x_0$ along the positive $x$ direction looking for the least $x_w$ at which there exist leaf pairs $u$, $t$, $t \in \{u-1, u+1\}$ that violate the constraint at $x_w$. After rectifying the violation at $x_w$, we look for other violations. Because the process of eliminating a violation at $x_w$ may at times lead to new violations at $x_w$, we need to search afresh from $x_w$ every time a modification is made to the schedule. However, a bound of $O(n)$ can be proved on the number of violations that can occur at $x_w$. After eliminating all violations at a particular sample point, $x_w$, we move to the next point, i.e., we increment $w$ and look for possible violations at the new point. We continue the scanning and modification process until no tongue-and-groove constraint violations exist. Algorithm TONGUEANDGROOVE (Figure 6.12) outlines the procedure.

Let $M = ((I_{1l}, I_{1r}), (I_{2l}, I_{2r}), \ldots, (I_{nl}, I_{nr}))$ be the schedule generated by Algorithm MULTIPAIR for the desired intensity profile. Let $N(p) = ((I_{1lp}, I_{1rp}), (I_{2lp}, I_{2rp}), \ldots, (I_{nlp}, I_{nrp}))$ be the schedule obtained after step 2 of Algorithm TONGUEANDGROOVE is applied $p$ times to the input schedule $M$. Note that $M = N(0)$.

To illustrate the modification process, we use examples. To make things easier, we only show two neighboring pairs of leaves. Suppose that the $(p+1)^{\text{th}}$ violation occurs between the leaves of pair $u$ and pair $t = u+1$ at $x_w$. Note that $I_{tlp}(x_w) \neq I_{ulp}(x_w)$, as otherwise, either (b) or (c) of Lemma 4 is true. In case $I_{tlp}(x_w) > I_{ulp}(x_w)$, swap $u$ and $t$. Now, we have $I_{tlp}(x_w) < I_{ulp}(x_w)$. *In the sequel, we refer to these $u$ and $t$ values as the $u$ and $t$ of Algorithm TONGUE-ANDGROOVE.* From Lemma 4 and the fact that a violation has occurred, it follows that $I_{trp}(x_w) < I_{urp}(x_w)$. To remove this tongue-and-groove constraint violation, we modify $(I_{tlp}, I_{trp})$. The other profiles of $N(p)$ are not modified.

Algorithm TONGUEANDGROOVE

$x = x_0$

While (there is a tongue-and-groove violation) do

1. Find the least $x_w$, $x_w \geq x$, such that there exist leaf pairs $u$, $u+1$, that violate the tongue-and-groove constraint at $x_w$.
2. Modify the schedule to eliminate the violation between leaf pairs $u$ and $u+1$.
3. $x = x_w$

End While

**Fig. 6.12.** Obtaining a schedule under the tongue-and-groove constraint.

The new plan for pair $t$, $(I_{tl(p+1)}, I_{tr(p+1)})$ is as defined below. If $I_{ulp}(x_w) - I_{tlp}(x_w) \le I_{urp}(x_w) - I_{trp}(x_w)$, then

$$I_{tl(p+1)}(x) = \begin{cases} I_{tlp}(x) & x_0 \le x < x_w \\ I_{tlp}(x) + \Delta I & x_w \le x \le x_m \end{cases} \tag{6.11}$$

where $\Delta I = I_{ulp}(x_w) - I_{tlp}(x_w)$. $I_{tr(p+1)}(x) = I_{tl(p+1)}(x) - I_t(x)$, where $I_t(x)$ is the target profile to be delivered by the leaf pair $t$. Otherwise,

$$I_{tr(p+1)}(x) = \begin{cases} I_{trp}(x) & x_0 \le x < x_w \\ I_{trp}(x) + \Delta I' & x_w \le x \le x_m \end{cases} \tag{6.12}$$

where $\Delta I' = I_{urp}(x_w) - I_{trp}(x_w)$. $I_{tl(p+1)}(x) = I_{tr(p+1)}(x) + I_t(x)$, where $I_t(x)$ is the target profile to be delivered by the leaf pair $t$. The former case is illustrated in Figure 6.13 and the latter is illustrated in Figure 6.14. Note that our strategy for plan modification is similar to that used by van Santvoort and Heijmen [18] to eliminate a tongue-and-groove violation for dynamic multileaf collimator plans.

Because $(I_{tl(p+1)}, I_{tr(p+1)})$ differs from $(I_{tlp}, I_{trp})$ for $x \ge x_w$, there is a possibility that $N(p+1)$ is involved in tongue-and-groove violations for $x \ge x_w$. Because none of the other leaf profiles are changed from those of $N(p)$, no tongue-and-groove constraint violations are possible in $N(p+1)$ for $x < x_w$. One may also verify that as $I_{tl0}$ and $I_{tr0}$ are non-decreasing functions of $x$, so also are $I_{tlp}$ and $I_{trp}$, $p > 0$.

**Theorem 6 (Kamath et al. [9]).** *Algorithm TONGUEANDGROOVE generates schedules free of tongue-and-groove violations that are optimal in therapy time for unidirectional schedules.*

The elimination of tongue-and-groove constraint violations does not guarantee elimination of interdigitation constraint violations. Therefore the sched-

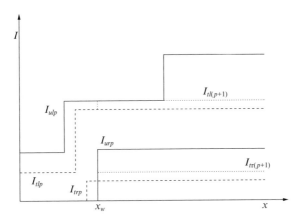

**Fig. 6.13.** Tongue-and-groove constraint violation: case **1**.

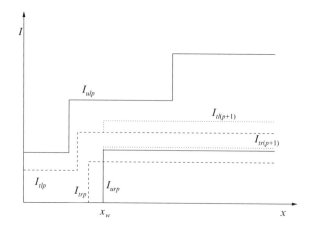

**Fig. 6.14.** Tongue-and-groove constraint violation: case 2 (close parallel, dotted, and solid line segments overlap, they have been drawn with a small separation to enhance readability).

ule generated by Algorithm TONGUEANDGROOVE may not be free of inter-digitation violations. The algorithm we propose for obtaining schedules that simultaneously satisfy both constraints, Algorithm TONGUEANDGROOVE-ID, is similar to Algorithm TONGUEANDGROOVE. The only difference between the two algorithms lies in the definition of the constraint condition. To be precise, we make the following definition.

**Definition 1 (Kamath et al. [9]).** *A unidirectional schedule is said to satisfy the tongue-and-groove-id constraint if*

*(a)* $I_{tr}(x_i) \leq I_{(t+1)r}(x_i) \leq I_{(t+1)l}(x_i) \leq I_{tl}(x_i)$, *or*
*(b)* $I_{(t+1)r}(x_i) \leq I_{tr}(x_i) \leq I_{tl}(x_i) \leq I_{(t+1)l}(x_i)$,

*for* $0 \leq i \leq m$, $1 \leq t < n$.

The only difference between this constraint and the tongue-and-groove constraint is that this constraint enforces condition (a) or (b) above to be true at all sample points $x_i$ including those at which $I_t(x_i) = 0$ and/or $I_{t+1}(x_i) = 0$.

**Lemma 5 (Kamath et al. [9]).** *A schedule satisfies the tongue-and-groove-id constraint iff it satisfies the tongue-and-groove constraint and the interdigitation constraint.*

*Proof.* It is obvious that the tongue-and-groove-id constraint subsumes the tongue-and-groove constraint. If a schedule has a violation of the interdigitation constraint, $\exists$ $i, t$, $I_{(t+1)l}(x_i) < I_{tr}(x_i)$ or $I_{tl}(x_i) < I_{(t+1)r}(x_i)$. From Definition 1, it follows that schedules that satisfy the tongue-and-groove-id constraint do not violate the interdigitation constraint. Therefore a schedule that satisfies the tongue-and-groove-id constraint satisfies the tongue-and-groove constraint and the interdigitation constraint.

For the other direction of the proof, consider a schedule $O$ that satisfies the tongue-and-groove constraint and the interdigitation constraint. From the fact that $O$ satisfies the tongue-and-groove constraint and from Lemma 4 and Definition 1, it only remains to be proved that for schedule $O$,

(a) $I_{tr}(x_i) \leq I_{(t+1)r}(x_i) \leq I_{(t+1)l}(x_i) \leq I_{tl}(x_i)$, or
(b) $I_{(t+1)r}(x_i) \leq I_{tr}(x_i) \leq I_{tl}(x_i) \leq I_{(t+1)l}(x_i)$,

whenever $I_t(x_i) = 0$ or $I_{t+1}(x_i) = 0$, $0 \leq i \leq m$, $1 \leq t < n$. When $I_t(x_i) = 0$,

$$I_{tl}(x_i) = I_{tr}(x_i). \tag{6.13}$$

Since $O$ satisfies the interdigitation constraint,

$$I_{tr}(x_i) \leq I_{(t+1)l}(x_i) \tag{6.14}$$

and

$$I_{(t+1)r}(x_i) \leq I_{tl}(x_i). \tag{6.15}$$

From equations (6.13), (6.14), and (6.15), we get $I_{(t+1)r}(x_i) \leq I_{tr}(x_i) = I_{tl}(x_i) \leq I_{(t+1)l}(x_i)$. Thus (b) is true whenever $I_t(x_i) = 0$. Similarly, (a) is true whenever $I_{t+1}(x_i) = 0$. Therefore, $O$ satisfies the tongue-and-groove-id constraint. $\qquad\square$

**Theorem 7 (Kamath et al. [9]).** *Algorithm TONGUEANDGROOVE-ID generates schedules free of tongue-and-groove-id violations that are optimal in therapy time for unidirectional schedules.*

*In the remainder of this section we will use "algorithm" to mean Algorithm TONGUEANDGROOVE or Algorithm TONGUEANDGROOVE-ID and "violation" to mean tongue-and-groove constraint violation or tongue-and-groove-id constraint violation (depending on which algorithm is considered) unless explicitly mentioned.*

The execution of the algorithm starts with schedule $M$ at $x = x_0$ and sweeps to the right, eliminating violations from the schedule along the way. The modifications applied to eliminate a violation at $x_w$, prescribed by equations (6.11) and (6.12), modify one of the violating profiles for $x \geq x_w$. From the unidirectional nature of the sweep of the algorithm, it is clear that the modification of the profile for $x > x_w$ can have no consequence on violations that may occur at the point $x_w$. Therefore it suffices to modify the profile only at $x_w$ at the time the violation at $x_w$ is detected. The modification can be propagated to the right as the algorithm sweeps. This can be done by using an $(n \times m)$ matrix $A$ that keeps track of the amount by which the profiles have been raised. $A(j, k)$ denotes the cumulative amount by which the $j^{\text{th}}$ leaf pair profiles have been raised at sample point $x_k$ from the schedule $M$ generated using Algorithm MULTIPAIR. When the algorithm has eliminated all violations at each $x_w$, it moves to $x_{w+1}$ to look for possible violations. It first sets the $(w+1)^{\text{st}}$ column of the modification matrix equal to the $w^{\text{th}}$ column to reflect rightward propagation of the modifications. It then looks for and eliminates violations at $x_{w+1}$ and so on.

The process of detecting the violations at $x_w$ merits further investigation. We show that if one carefully selects the order in which violations are detected and eliminated, the number of violations at each $x_w$, $0 \le w \le m$ will be $O(n)$.

**Lemma 6 (Kamath et al. [9]).** *The algorithm can be implemented such that $O(n)$ violations occur at each $x_w$, $0 \le w \le m$.*

*Proof.* The bound is achieved using a two-pass scheme at $x_w$. In pass one, we check adjacent leaf pairs $(\mathbf{1},2), (2,3), \ldots, (n-1,n)$, in that order, for possible violations at $x_w$. In pass two, we check for violations in the reverse order, i.e., $(n-1,n), (n-2,n-1), \ldots, (1,2)$. So each set of adjacent pairs $(i, i+1)$, $\mathbf{1} \le i < n$ is checked exactly twice for possible violations. It is easy to see that if a violation is detected in pass one, either the profile of leaf pair $i$ or that of leaf pair $i+1$ may be modified (raised) to eliminate the violation. However, in pass two only the profile of pair $i$ may be modified. This is because the profile of pair $i$ is not modified between the two times it is checked for violations with pair $i + 1$. The profile of pair $i + \mathbf{1}$, on the other hand, could have been modified between these times as a result of violations with pair $i+2$. Therefore in pass two, only $i$ can be a candidate for $t$ (where $t$ is as explained in the algorithm) when pairs $(i, i + \mathbf{1})$ are examined. From this it also follows that when pairs $(i-\mathbf{1}, i)$ are subsequently examined in pass two, the profile of pair $i$ will not be modified. Because there is no violation between adjacent pairs $(1, 2), (2, 3), \ldots, (i, i+\mathbf{1})$ at that time and none of these pairs is ever examined again, it follows that at the end of pass two there can be no violations between pairs $(i, i + \mathbf{1})$, $1 \le i < n$.     □

**Lemma 7 (Kamath et al. [9]).** *For the execution of the algorithm, the time complexity is $O(nm)$.*

*Proof.* Follows from Lemma 6 and the fact that there are $m$ sample points.□

## 6.3 Algorithms for DMLC

### 6.3.1 Single leaf pair

**Movement of leaves**

We assume that $I(x_0) > 0$ and $I(x_m) > 0$ and that when the beam delivery begins, the leaves can be positioned anywhere. We also assume that the leaves can move with any velocity $v$, $-v_{\max} \le v \le v_{\max}$, where $v_{\max}$ is the maximum allowable velocity of the leaves. Figure 6.15 illustrates the leaf trajectory during DMLC delivery. $I_l(x_i)$ and $I_r(x_i)$, respectively, denote the amount of

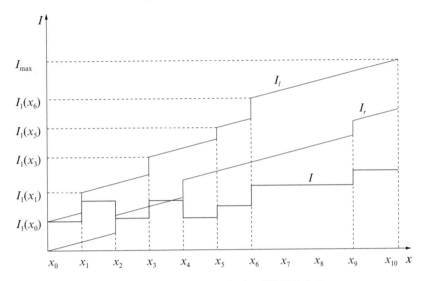

**Fig. 6.15.** Leaf trajectory during DMLC delivery.

MUs delivered when the left and right leaves leave position $x_i$. The total therapy time, $TT(I_l, I_r)$, is the time needed to deliver $I_{max}$ MUs. Note that the machine is on throughout the treatment. All MUs that are delivered along a radiation beam along $x_i$ before the left leaf passes $x_i$ fall on it, and all MUs that are delivered along a radiation beam along $x_i$ before the right leaf passes $x_i$ are blocked by the leaf. Thus the amount of MUs delivered at a point is given by $I_l(x_i) - I_r(x_i)$, which must be the same as $I(x_i)$.

## Maximum velocity constraint

As noted earlier, the velocity of leaves cannot exceed some maximum limit (say $v_{max}$) in practice. This implies that the leaf profile cannot be horizontal at any point. From Figure 6.15, observe that the time needed for a leaf to move from $x_i$ to $x_{i+1}$ is $\geq (x_{i+1} - x_i)/v_{max}$. If $\Phi$ is the flux density of MUs from the source, the number of MUs delivered in this time along a beam is $\geq \Phi \cdot (x_{i+1} - x_i)/v_{max}$. Thus $I_l(x_{i+1}) - I_l(x_i) \geq \Phi * (x_{i+1} - x_i)/v_{max} = \Phi \cdot \Delta x/v_{max}$. The same is true for the right leaf profile $I_r$.

## Optimal unidirectional algorithm for one pair of leaves

As in the case of SMLC, the problem is to find plan $(I_l, I_r)$ such that:

$$I(x_i) = I_l(x_i) - I_r(x_i), 0 \leq i \leq m. \tag{6.16}$$

Of course, $I_l$ and $I_r$ are subject to the maximum velocity constraint. For each $i$, the left leaf can be allowed to pass $x_i$ when the source has delivered

$I_l(x_i)$ MUs, and the right leaf can be allowed to pass $x_i$ when the source has delivered $I_r(x_i)$ MUs. In this manner we obtain *unidirectional leaf movement profiles* for a plan.

Similar to the case of SMLC, one way to determine $I_l$ and $I_r$ from the given target profile $I$ is to begin from $x_0$; set $I_l(x_0) = I(x_0)$ and $I_r(x_0) = 0$; examine the remaining $x_i$s to the right; increase $I_l$ at $x_i$ whenever $I$ increases and by the same amount (in addition to the minimum increase imposed by the maximum velocity constraint); and similarly increase $I_r$ whenever $I$ decreases. This can be done until we reach $x_m$. This yields Algorithm DMLC-SINGLEPAIR. The time complexity of Algorithm DMLC-SINGLEPAIR is $O(m)$. Note that we move the leaves at the maximum velocity $v_{\max}$ whenever they are to be moved. The resulting algorithm is shown in Figure 6.16. Figure 6.15 shows a profile $I$ and the corresponding plan $(I_l, I_r)$ obtained using Algorithm DMLC-SINGLEPAIR. Ma et al. [14] show that Algorithm DMLC-SINGLEPAIR obtains plans that are optimal in therapy time. Their proof relies on the results of Boyer and Strait [4], Spirou and Chui [16], and Stein et al. [17]. Kamath et al. [10] provide a much simpler proof.

**Theorem 8 (Kamath et al. [10]).** *Algorithm DMLC-SINGLEPAIR obtains plans that are optimal in therapy time.*

*Proof.* Let $I(x_i)$ be the desired profile. Let $0 = inc0 < inc1 < \ldots < inck$ be the indices of the points at which $I(x_i)$ increases. Thus $x_{inc0}, x_{inc1}, \ldots, x_{inck}$ are the points at which $I(x)$ increases (i.e., $I(x_{inci}) > I(x_{inci-1})$, assume that $I(x_{-1} = 0)$). Let $\Delta i = I(x_{inci}) - I(x_{inci-1})$, $i \geq 0$. Suppose that $(I_L, I_R)$ is a plan for $I(x_i)$ (not necessarily the plan generated by Algorithm DMLC-SINGLEPAIR). Because $I(x_i) = I_L(x_i) - I_R(x_i)$ for all $i$, we get

$$
\begin{aligned}
\Delta i &= (I_L(x_{inci}) - I_R(x_{inci})) - (I_L(x_{inci-1}) - I_R(x_{inci-1})) \\
&= (I_L(x_{inci}) - I_L(x_{inci-1})) - (I_R(x_{inci}) - I_R(x_{inci-1})) \\
&= (I_L(x_{inci}) - I_L(x_{inci-1}) - \Phi \cdot \Delta x / v_{\max}) - \\
&\quad (I_R(x_{inci}) - I_R(x_{inci-1}) - \Phi \cdot \Delta x / v_{\max}).
\end{aligned}
$$

Algorithm DMLC-SINGLEPAIR

$I_l(x_0) = I(x_0)$
$I_r(x_0) = 0$

For $j = 1$ to $m$ do
  If $(I(x_j) \geq I(x_{j-1}))$
    $I_l(x_j) = I_l(x_{j-1}) + I(x_j) - I(x_{j-1}) + \Phi \cdot \Delta x / v_{\max}$
    $I_r(x_j) = I_r(x_{j-1}) + \Phi \cdot \Delta x / v_{\max}$
  Else
    $I_r(x_j) = I_r(x_{j-1}) + I(x_{j-1}) - I(x_j) + \Phi \cdot \Delta x / v_{\max}$
    $I_l(x_j) = I_l(x_{j-1}) + \Phi \cdot \Delta x / v_{\max}$
End for

Fig. 6.16. Obtaining a unidirectional plan.

Note that from the maximum velocity constraint $I_R(x_{inci}) - I_R(x_{inci-1}) \geq \Phi \cdot \Delta x / v_{max}$, $i \geq 1$. Thus $I_R(x_{inci}) - I_R(x_{inci-1}) - \Phi \cdot \Delta x / v_{max} \geq 0$, $i \geq 1$, and $\Delta i \leq I_L(x_{inci}) - I_L(x_{inci-1}) - \Phi \cdot \Delta x / v_{max}$. Also, $\Delta 0 = I(x_0) - I(x_{-1}) = I(x_0) \leq I_L(x_0) - I_L(x_{-1})$, where $I_L(x_{-1}) = 0$. Summing up $\Delta i$, we get $\sum_{i=0}^{k}[I(x_{inci}) - I(x_{inci-1})] \leq \sum_{i=0}^{k}[I_L(x_{inci}) - I_L(x_{inci-1})] - k \cdot \Phi \cdot \Delta x / v_{max}$. Let $S_1 = \sum_{i=0}^{k}[I_L(x_{inci}) - I_L(x_{inci-1})]$. Then, $S_1 \geq \sum_{i=0}^{k}[I(x_{inci}) - I(x_{inci-1})] + k \cdot \Phi \cdot \Delta x / v_{max}$. Let $S_2 = \sum[I_L(x_j) - I_L(x_{j-1})]$, where the summation is carried out over indices $j$ ($0 \leq j \leq m$) such that $I(x_j) \leq I(x_{j-1})$. There are a total of $m + 1$ indices of which $k + 1$ do not satisfy this condition. Thus there are $m - k$ indices $j$ at which $I(x_j) \leq I(x_{j-1})$. At each of these $j$, $I_L(x_j) \geq I_L(x_{j-1}) + \Phi \cdot \Delta x / v_{max}$. Hence, $S_2 \geq (m - k) \cdot \Phi \cdot \Delta x / v_{max}$. Now, we get $S_1 + S_2 = \sum_{i=0}^{m}[I_L(x_i) - I_L(x_{i-1})] \geq \sum_{i=0}^{k}[I(x_{inci}) - I(x_{inci-1})] + m \cdot \Phi \cdot \Delta x / v_{max}$. Finally, $TT(I_L, I_R) = I_L(x_m) = I_L(x_m) - I_L(x_{-1}) = \sum_{i=0}^{m}[I_L(x_i) - I_L(x_{i-1})] \geq \sum_{i=0}^{k}[I(x_{inci}) - I(x_{inci-1})] + m \cdot \Phi \cdot \Delta x / v_{max} = TT(I_l, I_r)$. Hence, the treatment plan $(I_l, I_r)$ generated by DMLC-SINGLEPAIR is optimal in therapy time. $\square$

### 6.3.2 Multiple leaf pairs

We present multiple leaf pair sequencing algorithms for DMLC without constraints and with the interdigitation constraint. These algorithms are from Kamath et al. [10].

#### Optimal schedule without constraints

For sequencing of multiple leaf pairs, we apply Algorithm DMLC-SINGLE-PAIR to determine the optimal plan for each of the $n$ leaf pairs. This method of generating schedules is described in Algorithm DMLC-MULTIPAIR (Figure 6.17). The complexity of Algorithm DMLC-MULTIPAIR is $O(mn)$. Note that as $x_0$, $x_m$ are not necessarily non-zero for any row, we replace $x_0$ by $x_l$ and $x_m$ by $x_g$ in Algorithm DMLC-SINGLEPAIR for each row, where $x_l$ and $x_g$, respectively, denote the first and last non-zero sample points of that row. Also, for rows that contain only zeroes, the plan simply places the corresponding leaves at the rightmost point in the field (call it $x_{m+1}$).

**Theorem 9 (Kamath et al. [10]).** *Algorithm DMLC-MULTIPAIR generates schedules that are optimal in therapy time.*

Algorithm DMLC-MULTIPAIR

$\text{For}(i = 1; i \leq n; i + +)$

    Apply Algorithm DMLC-SINGLEPAIR to the $i^{\text{th}}$ pair of leaves to obtain plan $(I_{il}, I_{ir})$ that delivers the intensity profile $I_i(x)$.

End For

**Fig. 6.17.** Obtaining a schedule.

**Optimal algorithm with interdigitation constraint**

The schedule generated by Algorithm DMLC-MULTIPAIR may violate the interdigitation constraint. Note that no intra-pair constraint violations can occur for $S_{\min} = 0$. Thus the interdigitation constraint is essentially an inter-pair constraint. If the schedule has no interdigitation constraint violations, it is the desired optimal schedule. If there are violations in the schedule, we eliminate all violations of the interdigitation constraint starting from the left end, i.e., from $x_0$. To eliminate the violations, we modify those plans of the schedule that cause the violations. We scan the schedule from $x_0$ along the positive $x$ direction looking for the least $x_v$ at which is positioned a right leaf (say $R_u$) that violates the inter-pair separation constraint. After rectifying the violation at $x_v$ with respect to $R_u$, we look for other violations. Because the process of eliminating a violation at $x_v$ may at times lead to new violations involving right leaves positioned at $x_v$, we need to search afresh from $x_v$ every time a modification is made to the schedule. We now continue the scanning and modification process until no interdigitation violations exist. Algorithm DMLC-INTERDIGITATION (Figure 6.18) outlines the procedure.

Let $M = ((I_{1l}, I_{1r}), (I_{2l}, I_{2r}), \ldots, (I_{nl}, I_{nr}))$ be the schedule generated by Algorithm DMLC-MULTIPAIR for the desired intensity profile. Let $N(p) = ((I_{1lp}, I_{1rp}), (I_{2lp}, I_{2rp}), \ldots, (I_{nlp}, I_{nrp}))$ be the schedule obtained after Step 2 of Algorithm DMLC-INTERDIGITATION is applied $p$ times to the input schedule $M$. Note that $M = N(0)$.

To illustrate the modification process, we use examples. There are two types of violations that may occur. Call them Type 1 and Type 2 violations and call the corresponding modifications Type 1 and Type 2 modifications. To make things easier, we only show two neighboring pairs of leaves. Suppose that the $(p+1)^{\text{st}}$ violation occurs between the right leaf of pair $u$, which is positioned at $x_v$, and the left leaf of pair $t, t \in \{u-1, u+1\}$.

Algorithm DMLC-INTERDIGITATION

$x = x_0$

While (there is an interdigitation violation) do

1. Find the least $x_v$, $x_v \geq x$, such that a right leaf is positioned at $x_v$ and this right leaf has an interdigitation violation with one or both of its neighboring left leaves. Let $u$ be the least integer such that the right leaf $R_u$ is positioned at $x_v$ and $R_u$ has an interdigitation violation. Let $L_t$ denote the left leaf with which $R_u$ has an interdigitation violation. Note that $t \in \{u-1, u+1\}$. In case $R_u$ has violations with two adjacent left leaves, we let $t = u - 1$.
2. Modify the schedule to eliminate the violation between $R_u$ and $L_t$.
3. $x = x_v$

End While

**Fig. 6.18.** Obtaining a schedule under the constraint.

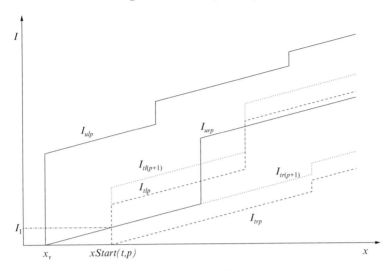

**Fig. 6.19.** Eliminating a Type 1 violation.

In a Type 1 violation, the left leaf of pair $t$ starts its sweep at a point $xStart(t,p) > x_v$ (see Figure 6.19). To remove this interdigitation violation, modify $(I_{tlp}, I_{trp})$ to $(I_{tl(p+1)}, I_{tr(p+1)})$ as follows. We let the leaves of pair $t$ start at $x_v$ and move them at the maximum velocity $v_{max}$ toward the right, until they reach $xStart(t,p)$. Let the number of MUs delivered when they reach $xStart(t,p)$ be $I_1$. Raise the profiles $I_{tlp}(x)$ and $I_{trp}(x)$, $x \geq xStart(t,p)$, by an amount $I_1 = \Phi \cdot (xStart(t,p) - x_v)/v_{max}$. We get

$$I_{tl(p+1)}(x) = \begin{cases} \Phi \cdot (x - x_v)/v_{max} & x_v \leq x < xStart(t,p) \\ I_{tlp}(x) + I_1 & x \geq xStart(t,p) \end{cases}$$

$$I_{tr(p+1)}(x) = I_{tl(p+1)}(x) - I_t(x)$$

where $I_t(x)$ is the target profile to be delivered by the leaf pair $t$.

A Type 2 violation occurs when the left leaf of pair $t$, which starts its sweep from $x \leq x_v$, passes $x_v$ before the right leaf of pair $u$ passes $x_v$ (Figure 6.20). In this case, $I_{tl(p+1)}$ is as defined below

$$I_{tl(p+1)}(x) = \begin{cases} I_{tlp}(x) & x < x_v \\ I_{tlp}(x) + \Delta I & x \geq x_v \end{cases}$$

where $\Delta I = I_{urp}(x_v) - I_{tlp}(x_v) = I_3 - I_2$. Once again, $I_{tr(p+1)}(x) = I_{tl(p+1)}(x) - I_t(x)$, where $I_t(x)$ is the target profile to be delivered by the leaf pair $t$.

In both Type 1 and Type 2 modifications, the other profiles of $N(p)$ are not modified. Because $I_{tr(p+1)}$ differs from $I_{trp}$ for $x \geq x_v$, there is a possibility that $N(p+1)$ has inter-pair separation violations for right leaf positions $x \geq x_v$. Because none of the other right leaf profiles are changed from those of

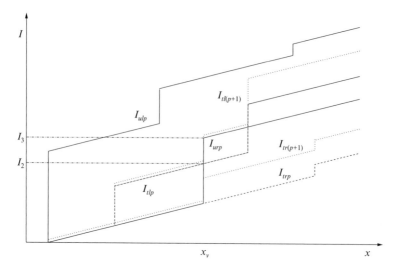

**Fig. 6.20.** Eliminating a Type 2 violation (close parallel dotted and solid line segments overlap; they have been drawn with a small separation to enhance readability).

$N(p)$ and because the change in $I_{tl}$ only delays the rightward movement of the left leaf of pair $t$, no interdigitation violations are possible in $N(p+1)$ for $x < x_v$. One may also verify that as $I_{tl0}$ and $I_{tr0}$ are feasible plans that satisfy the maximum velocity constraints, so also are $I_{tlp}$ and $I_{trp}$, $p > 0$.

**Lemma 8 (Kamath et al. [10]).** *In case of a Type 1 violation, $(I_{tlp}, I_{trp})$ is the same as $(I_{tl0}, I_{tr0})$.*

*Proof.* Let $p$ be such that there is a Type 1 violation. Let $t$, $u$, and $v$ be as in Algorithm DMLC-INTERDIGITATION. If $(I_{tlp}, I_{trp})$ is different from $(I_{tl0}, I_{tr0})$, leaf pair $t$ was modified in an earlier iteration (say iteration $q < p$) of the while loop of Algorithm DMLC-INTERDIGITATION. Let $v(q)$ be the $v$ value in iteration $q$. If iteration $q$ was a Type 1 violation, then $xStart(t, p) \le xStart(t, q+1) = x_{v(q)} \le x_v$. So, iteration $p$ cannot be a Type 1 violation. If iteration $q$ was a Type 2 violation, $xStart(t, p) \le xStart(t, q) \le x_{v(q)} \le x_v$. Again, iteration $p$ cannot be a Type 1 violation. Hence, there is no prior iteration $q$, $q < p$, when the profiles $(I_{tl}, I_{tr})$ were modified.    □

**Lemma 9 (Kamath et al. [10]).** *For the execution of Algorithm DMLC-INTERDIGITATION*

*(a) $O(n)$ Type 1 violations can occur.*
*(b) $O(n^2 m)$ Type 2 violations can occur.*
*(c) Let $T_{\max}$ be the optimal therapy time for the input matrix. The time complexity is $O(mn + n \min\{nm, T_{\max}\})$.*

*Proof.*

(a) It follows from Lemma 8 that each leaf pair can be involved in at most one Type 1 violation as pair $t$, i.e., the pair whose profile is modified. Hence, the number of Type 1 violations is $\leq n$.

(b) We first obtain a bound on the number of Type 2 violations at a fixed $x_v$. Let $u$, $t$ be as in Algorithm DMLC-INTERDIGITATION. Note that $u$ is chosen to be the least possible index. Let $u_i$ be the value of $u$ in the $i^{\text{th}}$ iteration of Algorithm DMLC-INTERDIGITATION at $x_v$. $t_i$ is defined similarly. Let $u_i^{\max} = \max_{j \leq i}\{u_j\}$. If $t_i = u_i - 1$, it is possible that $u_{i+1} = t_i = u_i - 1$ and $t_{i+1} = u_i - 2$. Note that in this case, $t_{i+1} \neq u_i = u_{i+1} + 1$. Next, it is possible that $u_{i+2} = u_i - 2$ and $t_{i+2} = u_{i-3}$ (again $t_{i+2} \neq u_i - 1 = u_{i+2} + 1$). In general, one may verify that $t_i = u_i + 1$ is possible only if $u_i^{\max} = u_i$. If $t_i = u_i + 1$, then $u_{i+1} \geq t_i = u_i + 1$, since the violation between $u_i$ and $t_i$ has been eliminated and no profiles with an index less than $t_i$ have been changed during iteration $i$ at $x_v$. It is also easy to verify that $t_i = 1, u_i = 2 \Rightarrow u_{i+1} \geq u_i^{\max}, u_{i+2}^{\max} > u_i^{\max}$. From this and $t_i \in \{u_i + 1, u_i - 1\}$ it follows that $u_{i+u_i^{\max}}^{\max} > u_i^{\max}$. We know that $u_1^{\max} \geq 1$. It follows that $u_2^{\max} \geq 2$, $u_4^{\max} \geq 3$, $u_7^{\max} \geq 4$ and in general, $u_{(i(i+1)/2)+1}^{\max} \geq i + 1$. Clearly, for the last violation (say $j^{\text{th}}$) at $x_v$, $u_j^{\max} \leq n$ and for this to be true, $j = O(n^2)$. So the number of Type 2 violations at $x_v$ is $O(n^2)$. Because $x_v$ has to be a sample point, there are $m$ possible choices for it. Hence, the total number of Type 2 violations is $O(n^2 m)$.

(c) Because the input matrix contains only integer intensity values, each violation modification raises the profile for one pair of leaves by at least one unit. Hence, if $T_{\max}$ is the optimal therapy time, no profile can be raised more than $T_{\max}$ times. Therefore, the total number of violations that Algorithm DMLC-INTERDIGITATION needs to repair is at most $nT_{\max}$. Combining this bound with those of (a) and (b), we get $O(\min\{n^2 m, nT_{\max}\})$ as a bound on the total number of violations repaired by Algorithm DMLC-INTERDIGITATION. By proper choice of data structures and programming methods it is possible to implement Algorithm DMLC-INTERDIGITATION so as to run in $O(mn + n\min\{nm, T_{\max}\})$ time. $\square$

Note that Lemma 9 provides two upper bounds on the complexity of Algorithm DMLC-INTERDIGITATION: $O(n^2 m)$ and $O(n\max\{m, T_{\max}\})$. In most practical situations, $T_{\max} < nm$ and so $O(n\max\{m, T_{\max}\})$ can be considered a tighter bound.

**Theorem 10 (Kamath et al. [10]).** *Algorithm DMLC-INTERDIGITA-TION generates DMLC schedules free of interdigitation violations that are optimal in therapy time for unidirectional schedules.*

## 6.4 Field Splitting Without Feathering

In this section, we deviate slightly from our earlier notation and assume that the sample points are $x_1, x_2, \ldots, x_m$ rather than $x_0, x_1, \ldots, x_m$. All other notation remains unchanged. The notation and algorithms are from Kamath et al. [11]. Recently, Wu [20] has also developed efficient algorithms for field splitting problems.

### 6.4.1 Optimal field splitting for one leaf pair

#### Delivering a profile using one field

An intensity profile $I$ can be delivered in optimal therapy time using the plan generated by Algorithm SINGLEPAIR. Algorithm SINGLEPAIR can be directly used to obtain plans when $I$ is deliverable using a single field. Let $l$ be the least index such that $I(x_l) > 0$ and let $g$ be the greatest index such that $I(x_g) > 0$. We will assume without loss of generality that $l = 1$. Thus the width of the profile is $g$ sample points, where $g$ can vary for different profiles. Assuming that the maximum allowable field width is $w$ sample points, $I$ is deliverable using one field if $g \leq w$; $I$ requires at least two fields for $g > w$; $I$ requires at least three fields for $g > 2w$. The case where $g > 3w$ is not studied as it never arises in clinical cases. The objective of field splitting is to split a profile so that each of the resulting profiles is deliverable using a single field. Further, it is desirable that the total therapy time is minimized, i.e., the sum of optimal therapy times of the resulting profiles is minimized. We will call the problem of splitting the profile $I$ of a single leaf pair into 2 profiles each of which is deliverable using one field such that the sum of their optimal therapy times is minimized as the $S2$ (single pair 2 field split) problem. The sum of the optimal therapy times of the two resulting profiles is denoted by $S2(I)$. $S3$ and $S3(I)$ are defined similarly for splits into 3 profiles. The problem $S1$ is trivial, as the input profile need not be split and is to be delivered using a single field. Note that $S1(I)$ is the optimal therapy time for delivering the profile $I$ in a single field. From Theorem 1, $S1(I) = \sum_{i=1}^{q} [I(x_{inci}) - I(x_{inci-1})]$, where $inc1, inc2, \ldots, incq$ are the indices of the points at which $I(x_i)$ increases.

#### Splitting a profile into two

Suppose that a profile $I$ is split into two profiles. Let $j$ be the index at which the profile is split. As a result, we get two profiles, $P_j$ and $S_j$. $P_j(x_i) = I(x_i)$, $1 \leq i < j$, and $P_j(x_i) = 0$, elsewhere. $S_j(x_i) = I(x_i)$, $j \leq i \leq g$, and $S_j(x_i) = 0$, elsewhere. $P_j$ is a *left profile* and $S_j$ is a *right profile* of $I$.

**Lemma 10 (Kamath et al. [11]).** *Let $S1(P_j)$ and $S1(S_j)$ be the optimal therapy times, respectively, for $P_j$ and $S_j$. Then $S1(P_j) + S1(S_j) = S1(I) + \hat{I}(x_j)$, where $\hat{I}(x_j) = \min\{I(x_{j-1}), I(x_j)\}$.*

We illustrate Lemma 10 using the example of Figure 6.21. The optimal therapy time for the profile $I$ is the sum of increments in intensity values of successive sample points. However, if $I$ is split at $x_3$ into $P_3$ and $S_3$, an additional therapy time of $\hat{I}(x_3) = \min\{I(x_2), I(x_3)\} = I(x_3)$ is required for treatment. Similarly, if $I$ is split at $x_4$ into $P_4$ and $S_4$, an additional therapy time of $\hat{I}(x_4) = \min\{I(x_3), I(x_4)\} = I(x_3)$ is required. Lemma 10 leads to an $O(g)$ algorithm (Algorithm $S2$, Figure 6.22) for $S2$. It is evident from Lemma 10 that if the width of the profile is less than the maximum allowable field width ($g \leq w$), the profile is best delivered using a single field. If $g > 2w$ two fields are insufficient. Thus it is useful to apply Algorithm $S2$ only for $w < g \leq 2w$. Once the profile $I$ is split into two as determined by Algorithm $S2$, the left and right profiles are delivered using separate fields. The total therapy time is $S2(I) = S1(P_j) + S1(S_j)$, where $j$ is the split point.

### Splitting a profile into three

Suppose that a profile $I$ is split into three profiles. Let $j$ and $k$, $j < k$, be the indices at which the profile is split. As a result we get three profiles $P_j$, $M_{(j,k)}$ and $S_k$, where $P_j(x_i) = I(x_i)$, $1 \leq i < j$, $M_{(j,k)}(x_i) = I(x_i)$, $j \leq i < k$, and $S_k(x_i) = I(x_i)$, $k \leq i \leq g$. $P_j$, $M_{(j,k)}$ and $S_j$ are zero at all other points. $P_j$ is a *left profile*, $M_{(j,k)}$ is a *middle profile* of $I$, and $S_k$ is a *right profile*.

**Lemma 11 (Kamath et al. [11]).** *Let $S1(P_j)$, $S1(M_{(j,k)})$ and $S1(S_k)$ be the optimal therapy times, respectively, for $P_j$, $M_{(j,k)}$ and $S_k$. Then*

$$S1(P_j) + S1(M_{(j,k)}) + S1(S_k)$$
$$= S1(I) + \min\{I(x_{j-1}), I(x_j)\} + \min\{I(x_{k-1}), I(x_k)\}$$
$$= S1(I) + \hat{I}(x_j) + \hat{I}(x_k).$$

Lemma 11 motivates Algorithm $S3$ (Figure 6.23) for $S3$. Note that for Algorithm $S3$ to split $I$ into three profiles that are each deliverable in one field, it must be the case that $g \leq 3w$. Once the profile $I$ is split into three as determined by Algorithm $S3$, the resulting profiles are delivered using separate fields. The minimum total therapy time is $S3(I) = S1(P_j) + S1(M_{(j,k)}) + S1(S_k)$. Algorithm $S3$ examines at most $g^2$ candidates for $(j, k)$. Thus the complexity of the algorithm is $O(g^2)$.

### Bounds on optimal therapy time ratios

The following bounds have been proved on ratios of optimal therapy times.

**Lemma 12 (Kamath et al. [11]).**

(a) $1 \leq S2(I)/S1(I) \leq 2$
(b) $1 \leq S3(I)/S1(I) \leq 3$
(c) $0.5 < S3(I)/S2(I) < 2$.

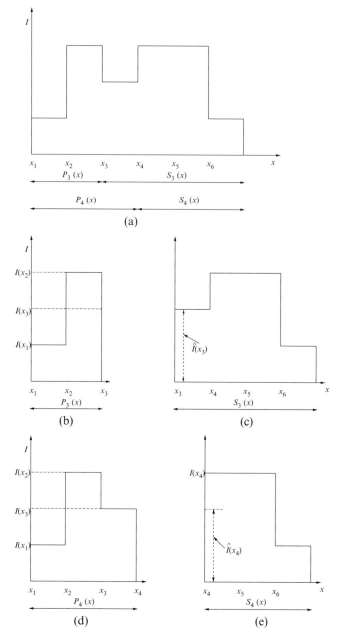

**Fig. 6.21.** Splitting a profile (a) into two; (b) and (c) show the left and right profiles resulting from a split at $x_3$; (d) and (e) show the left and right profiles resulting from a split at $x_4$

Algorithm $S2$

Compute $\hat{I}(x_i) = \min\{I(x_{i-1}), I(x_i)\}$, for $g - w < i \leq w + 1$.
Split the field at a point $x_j$ where $\hat{I}(x_j)$ is minimized for $g - w < j \leq w + 1$.

**Fig. 6.22.** Splitting a single row profile into two.

Algorithm $S3$

Compute $\hat{I}(x_i) = \min\{I(x_{i-1}), I(x_i)\}$, for $1 < i \leq w + 1$, $g - w < i \leq g$.
Split the field at two points $x_j$, $x_k$ such that $1 \leq j \leq w + 1$, $g - w < k \leq g$, $0 < k - j \leq w$, and $\hat{I}(x_j) + \hat{I}(x_k)$ is minimized.

**Fig. 6.23.** Splitting a single row profile into three.

Lemma 12 tells us that the optimal therapy times can at most increase by factors of 2 and 3, respectively, as a result of a splitting a single leaf pair profile into 2 and 3. Also, the optimal therapy time for a split into 2 can be at most twice that for a split into 3 and vice versa.

### 6.4.2 Optimal field splitting for multiple leaf pairs

The input intensity matrix (say $I$) for the leaf sequencing problem is obtained using the inverse planning technique. The matrix $I$ consists of $n$ rows and $m$ columns. Each row of the matrix specifies the number of monitor units (MUs) that need to be delivered using one leaf pair. Denote the rows of $I$ by $I_1, I_2, \ldots, I_n$. For the case where $I$ is deliverable using one field, the leaf sequencing problem has been well studied in the past. The algorithm that generates optimal therapy time schedules for multiple leaf pairs (Algorithm MULTIPAIR) applies algorithm SINGLEPAIR independently to each row $I_i$ of $I$. Without loss of generality, assume that the least column index containing a non-zero element in $I$ is **1** and the largest column index containing a non-zero element in $I$ is $g$. If $g > w$, the profile will need to be split. We define problems $M1$, $M2$, and $M3$ for multiple leaf pairs as being analogous to $S1$, $S2$, and $S3$ for single leaf pair. The optimal therapy times $M1(I)$, $M2(I)$, and $M3(I)$ are also defined similarly.

### Splitting a profile into two

Suppose that a profile $I$ is split into two profiles. Let $x_j$ be the column at which the profile is split. This is equivalent to splitting each row profile $I_i$, $1 \leq i \leq n$, at $j$ as defined for single leaf pair split. As a result, we get two profiles, $P_j$ (left) and $S_j$ (right). $P_j$ has rows $P_j^1, P_j^2, \ldots, P_j^n$ and $S_j$ has rows $S_j^1, S_j^2, \ldots, S_j^n$.

**Lemma 13 (Kamath et al. [11]).** *Suppose $I$ is split into two profiles at $x_j$. The optimal therapy time for delivering $P_j$ and $S_j$ using separate fields is $\max_i\{S1(P_j^i)\} + \max_i\{S1(S_j^i)\}$.*

Algorithm $M2$

Compute $\max_i\{S1(P_j^i)\} + \max_i\{S1(S_j^i)\}$ for $g - w < j \le w + 1$.
Split the field at a point $x_j$ where $\max_i\{S1(P_j^i)\} + \max_i\{S1(S_j^i)\}$ is minimized for $g - w < j \le w + 1$.

**Fig. 6.24.** Splitting a multiple row profile into two.

*Proof.* The optimal therapy time schedule for $P_j$ and $S_j$ are obtained using Algorithm MULTIPAIR. The therapy times are equal to $\max_i\{S1(P_j^i)\}$ and $\max_i\{S1(S_j^i)\}$, respectively. Thus the total therapy time is $\max_i\{S1(P_j^i)\} + \max_i\{S1(S_j^i)\}$. $\qquad\Box$

From Lemma 13, it follows that the $M2$ problem can be solved by finding the index $j$, $1 < j \le g$ such that $\max_i\{S1(P_j^i)\} + \max_i\{S1(S_j^i)\}$ is minimized (Algorithm $M2$, Figure 6.24).

From Theorem 1, $S1(P_j^i) = \sum_{inci \le j}[I(x_{inci}) - I(x_{inci-1})]$. For each $i$, $S1(P_1^i), S1(P_2^i), \ldots, S1(P_g^i)$ can all be computed in a total of $O(g)$ time progressively from left to right. Thus the computation of $S1$s (optimal therapy times) of all left profiles of all $n$ rows of $I$ can be done in $O(ng)$ time. The same is true of right profiles. Once these values are computed, step (1) of Algorithm $M2$ is applied. $\max_i\{S1(P_j^i)\} + \max_i\{S1(S_j^i)\}$ can be found in $O(n)$ time for each $j$ and hence in $O(ng)$ time for all $j$ in the permissible range. Thus the time complexity of Algorithm $M2$ is $O(ng)$.

## Splitting a profile into three

Suppose that a profile $I$ is split into three profiles. Let $j$, $k$, $j < k$, be the indices at which the profile is split. Once again, this is equivalent to splitting each row profile $I_i$, $1 \le i \le n$ at $j$ and $k$ as defined for single leaf pair split. As a result, we get three profiles $P_j$, $M_{(j,k)}$, and $S_k$. $P_j$ has rows $P_j^1, P_j^2, \ldots, P_j^n$, $M_{(j,k)}$ has rows $M_{(j,k)}^1, M_{(j,k)}^2, \ldots, M_{(j,k)}^n$, and $S_k$ has rows $S_k^1, S_k^2, \ldots, S_k^n$.

**Lemma 14 (Kamath et al. [11]).** *Suppose $I$ is split into three profiles by splitting at $x_j$ and $x_k$, $j < k$. The optimal therapy time for delivering $P_j$, $M_{(j,k)}$, and $S_k$ using separate fields is $\max_i\{S1(P_j^i)\} + \max_i\{S1(M_{(j,k)}^i)\} + \max_i\{S1(S_k^i)\}$.*

*Proof.* Similar to that of Lemma 13. $\qquad\Box$

Algorithm $M3$ (Figure 6.25) solves the $M3$ problem. The complexity analysis is similar to that of Algorithm $M2$. In this case though, $O(g^2)$ pairs of split points have to be examined. It is easy to see that the time complexity of Algorithm $M3$ is $O(ng^2)$.

Algorithm $M3$

Compute $\max_i\{S1(P_j^i)\} + \max_i\{S1(M_{(j,k)}^i)\} + \max_i\{S1(S_k^i)\}$ for $1 < j \leq w + 1$, $g - w < k \leq g$, $0 < k - j \leq w$.
Split the field at two points $x_j$, $x_k$, such that $1 < j \leq w + 1$, $g - w < k \leq g$, $0 < k-j \leq w$, and $\max_i\{S1(P_j^i)\}+\max_i\{S1(M_{(j,k)}^i)\}+\max_i\{S1(S_k^i)\}$ is minimized.

**Fig. 6.25.** Splitting a multiple row profile into three.

## Bounds on optimal therapy time ratios

The following bounds have been proved on ratios of optimal therapy times.

### Lemma 15 (Kamath et al. [11]).

$(a)\, 1 \leq M2(I)/M1(I) \leq 2$
$(b)\, 1 \leq M3(I)/M1(I) < 3$
$(c)\, 0.5 < M3(I)/M2(I) < 2$

Lemma 15 tells us that the optimal therapy times can at most increase by factors of 2 and 3, respectively, as a result of splitting a field into 2 and 3. Also, the optimal therapy time for a split into 2 can be at most twice that for a split into 3 and vice versa. These bounds give us the potential benefits of designing MLCs with larger maximal aperture so that large fields do not need to be split.

## Tongue-and-groove effect and interdigitation

Algorithms $M2$ and $M3$ may be extended to generate optimal therapy time fields with elimination of tongue-and-groove underdosage and (optionally) the interdigitation constraint on the leaf sequences. Consider the algorithms for delivering an intensity matrix $I$ using a single field with optimal therapy time while eliminating the tongue-and-groove underdosage (Algorithm TONGUEANDGROOVE) and also while simultaneously eliminating the tongue-and-groove underdosage and interdigitation constraint violations (Algorithm TONGUEANDGROOVE-ID). Denote these problems by $M1'$ and $M1''$, respectively ($M2'$, $M2''$, $M3'$, and $M3''$ are defined similarly for splits into two and three fields). Let $M1'(I)$ and $M1''(I)$, respectively, denote the optimal therapy times required to deliver $I$ using the leaf sequences generated by these algorithms. To solve problem $M2'$, we need to determine $x_j$ where $M1'(P_j) + M1'(S_j)$ is minimized for $g - w < j \leq w + 1$. Note that this is similar to Algorithm $M2$. Using the fact that $M1'$ can be solved in $O(nm)$ time for an intensity profile with $n$ rows and $m$ columns (Lemma 7, Kamath et al. [8]), and by computing $M1'(P_j)$ and $M1'(S_j)$ progressively from left to right, it is possible to solve $M2'$ in $O(ng)$ time. In case of $M3'$, we need to find $x_j$, $x_k$, such that $1 < j \leq w + 1$, $g - w < k \leq g$, $0 < k - j \leq w$, and $M1'(P_j) + M1'(M_{(j,k)}) + M1'(S_k)$ is minimized. $M3'$ can be solved in $O(ng^2)$ time. The solutions for $M2''$ and $M3''$ are now obvious.

### 6.4.3 Field splitting with feathering

One of the problems associated with field splitting is the field matching problem that occurs in the field junction region due to uncertainties in setup and organ motion. To illustrate the problem, we use an example. Consider the single leaf pair intensity profile of Figure 6.26(a). Due to width limitations, the profile needs to be split. Suppose that it is split at $x_j$. Further suppose

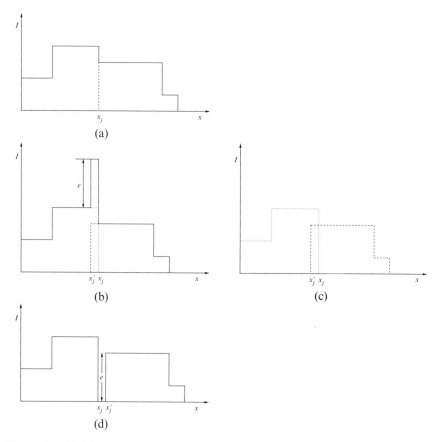

(a)

(b)

(c)

(d)

**Fig. 6.26.** Field matching problem: The profile in (a) is the desired profile. It is split into two fields at $x_j$. Due to incorrect field matching, the left end of right field is positioned at point $x_j'$ instead of $x_j$ and the fields may overlap as in (c) or may be separated as in (d). In (c), the dotted line shows the left profile, and the dashed line shows the right profile. (b) shows these profiles as well as the delivered profile in this case in bold. In (d), the left and right fields are separated, and their two profiles together constitute the delivered profile, which is shown in bold. The delivered profiles in these cases vary significantly from the desired profile in the junction region. $e$ is the maximum intensity error in the junction region, i.e., the maximum deviation of delivered intensity from the desired intensity.

that the left field is delivered accurately and that the right field is misaligned so that its left end is positioned at $x'_j$ rather than $x_j$. Due to incorrect field matching the actual profile delivered may be, for example, either of the profiles shown in Figure 6.26(b) or Figure 6.26(d), depending on the direction of error. In Figure 6.26(b), the region between $x'_j$ and $x_j$ gets overdosed and is a *hotspot*. In Figure 6.26(d), the region between $x_j$ and $x'_j$ gets underdosed and is a *coldspot*.

One way to partially eliminate the field matching problem is to use the "feathering" technique. In this technique, the large field is not split at one sample point into two non-overlapping fields. Instead, the profiles to be delivered by the two fields resulting from the split overlap over a central *feathering region*. The beam splitting algorithm proposed by Wu et al. [19] splits a large field with feathering, such that in the feathering region the sum of the split fields equals the desired intensity profile. Figure 6.27(a) shows a split of the profile of Figure 6.26 with feathering. Figures 6.27(c) and 6.27(d) show the effect of field matching problem on the split with feathering. The extent of field mismatches is the same as those in Figures 6.26(b) and 6.26(d), respectively. Note that while the profile delivered in the case with feathering is not the exact profile either, the delivered profile is *less sensitive* to mismatch compared with the case when it is split without feathering as in Figure 6.26. In other words, the purpose of feathering is to lower the magnitude of *maximum intensity error* $e$ in the delivered profile from the desired profile over all sample points in the junction region.

In this section, we extend our field splitting algorithms to incorporate feathering. In order to do so, we define a feathering scheme similar to that of Wu et al. [19]. However, there are two differences between the splitting algorithm we propose and the algorithm of Wu et al. [19]. First, our feathering scheme is defined for profiles discretized in space and in MUs as is the profile generated by the optimizer. Second, the feathering scheme we propose defines the profile values in the feathering region, which is centered at some sample point called the *split point* for that split. Thus given a split point, our scheme will specify how to split the large field with a feathering region that is centered at that point. The split point to be used in the actual split will be determined by a splitting algorithm that takes into account the feathering scheme. In contrast, Wu et al. [19] always choose the center of the intensity profile as the split point, as they do not optimize the split with respect to any objective.

We study how to split a single leaf pair profile into two (three) fields using our feathering scheme such that the sum of the optimal therapy times of the individual fields is minimized. We will denote this minimization problem by *S2F* (*S3F*). The extension of the methods develped for the multiple leaf pairs problems (*M2F* and *M3F*) is straightforward and is therefore not discussed separately.

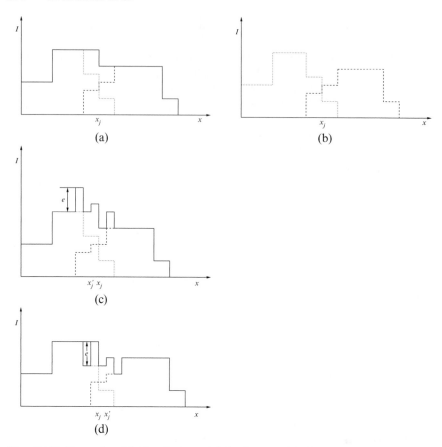

**Fig. 6.27.** Example of field splitting with feathering: (a) shows a split of the profile of Figure 6.26 with feathering. The dotted line shows the right part of the left profile, and the dashed line shows the left part of the right profile. The left and right profiles are shown separately in (b). (c) and (d) show the effect of field matching problem on the split with feathering. The extent of field mismatches in (c) and (d) are the same as those in Figures 6.26(b) and 6.26(d), respectively, i.e., the distances between $x_j$ and $x'_j$ are the same as in Figure 6.26. Note that the maximum intensity error $e$ reduces in both cases with feathering.

## Splitting a profile into two

Let $I$ be a single leaf pair profile. Let $x_j$ be the split point and let $P_j$ and $S_j$ be the profiles resulting from the split. $P_j$ is a *left profile* and $S_j$ is a *right profile* of $I$. The feathering region spans $x_j$ and $d-1$ sample points on either side of $x_j$, i.e., the feathering region stretches from $x_{j-d+1}$ to $x_{j+d-1}$. $P_j$ and $S_j$ are defined as follows

Algorithm $S2F$

Find $P_i$ and $S_i$ using equations (6.17) and (6.18), for $g - w + d \leq i \leq w - d + 1$.
Split the field at a point $x_j$ where $S1(P_j) + S1(S_j)$ is minimized for $g - w + d \leq j \leq w - d + 1$.

**Fig. 6.28.** Splitting a single row profile into two with feathering.

$$P_j(x_i) = \begin{cases} I_j(x_i) & 1 \leq i \leq j - d \\ \lceil I_j(x_i) \cdot (j + d - i)/2d \rceil & j - d < i < j + d \\ 0 & j + d \leq i \leq g \end{cases} \quad (6.17)$$

$$S_j(x_i) = \begin{cases} 0 & 1 \leq i \leq j - d \\ I_j(x_i) - P_j(x_i) & j - d < i < j + d \\ I_j(x_i) & j + d \leq i \leq g. \end{cases} \quad (6.18)$$

Note that the profiles overlap over the $2d - 1$ points $j - d + 1, j - d + 2, \ldots, j + d - 2, j + d - 1$. Therefore, for the profile $I$ of width $g$ to be deliverable using two fields, it must be the case that $g \leq 2w - 2d + 1$. Because $P_j$ needs to be delivered using one field, the split point $x_j$ and at least $d - 1$ points to the right of it should be contained in the first field, i.e., $j + d - 1 \leq w \Rightarrow j \leq w - d + 1$. Similarly, as $S_j$ has to be delivered using one field $j - (d - 1) > g - w \Rightarrow j \geq g - w + d$. These range restrictions on $j$ lead to an algorithm for the $S2F$ problem. Algorithm $S2F$, which solves problem $S2F$, is described in Figure 6.28. Note that the $P_i$s and $S_i$s can all be computed in a single left to right sweep in $O(d)$ time at each $i$. Thus the time complexity of Algorithm $S2F$ is $O(dg)$.

## Splitting a profile into three

Suppose that a profile $I$ is split into three profiles with feathering. Let $j$ and $k$, $j < k$, be the two split points. As a result, we get three profiles $P_j$, $M_{(j,k)}$, and $S_k$, where $P_j$ is a *left profile*, $M_{(j,k)}$ is a *middle profile* of $I$, and $S_k$ is a *right profile*. In this case, there are two feathering regions, each of which spans across $2d - 1$ sample points centered at the corresponding split point. One feathering region stretches from $x_{j-d+1}$ to $x_{j+d-1}$ and the other from $x_{k-d+1}$ to $x_{k+d-1}$. $P_j$, $M_{(j,k)}$, and $S_j$ are defined as follows

$$P_j(x_i) = \begin{cases} I_j(x_i) & 1 \leq i \leq j - d \\ \lceil I_j(x_i) \cdot (j + d - i)/2d \rceil & j - d < i < j + d \\ 0 & j + d \leq i \leq g \end{cases} \quad (6.19)$$

$$M_{(j,k)}(x_i) = \begin{cases} 0 & 1 \leq i \leq j - d \\ I_j(x_i) - P_j(x_i) & j - d < i < j + d \\ I_j(x_i) & j + d \leq i \leq k - d \\ \lceil I_k(x_i) \cdot (k + d - i)/2d \rceil & k - d < i < k + d \\ 0 & k + d \leq i \leq g \end{cases} \quad (6.20)$$

Algorithm $S3F$

Find $P_j$, $M_{(j,k)}$ and $S_k$ using equations (6.19), (6.20) and (6.21), for $g-2w+3d-1 \leq j \leq w - d + 1$, $g - w + d \leq k \leq 2w - 3d + 2$ and $k - j \leq w - 2d + 1$.
Split the field at two points $x_j$, $x_k$, where $S1(P_j)+S1(M_{(j,k)})+S1(S_j)$ is minimized, subject to $g - 2w + 3d - 1 \leq j \leq w - d + 1$, $g - w + d \leq k \leq 2w - 3d + 2$ and $k - j \leq w - 2d + 1$.

**Fig. 6.29.** Splitting a single row profile into three with feathering.

$$S_j(x_i) = \begin{cases} 0 & 1 \leq i \leq k - d \\ I_j(x_i) - M_{(j,k)}(x_i) & k - d < i < k + d \\ I_j(x_i) & k + d \leq i \leq g. \end{cases} \qquad (6.21)$$

The profiles $P_j$ and $M_{(j,k)}$ overlap over $2d - 1$ points, as do $M_{(j,k)}$ and $S_k$. For the profile $I$ to be deliverable using three fields, it must be the case that $g \leq 3w - 2(2d - 1) = 3w - 4d + 2$. Also, it is undesirable for the two feathering regions to overlap. Thus $g \geq 4d - 2$. For the feathering regions to be well defined and for the split to be useful, it can be shown that $g - 2w + 3d - 1 \leq j \leq w - d + 1$ and that $g - w + d \leq k \leq 2w - 3d + 2$. Also, $k - j + 1 + 2(d - 1) \leq w \Rightarrow k - j \leq w - 2d + 1$. Using these ranges for $j$ and $k$, we arrive at Algorithm $S3F$ (Figure 6.29), which can be implemented to solve problem $S3F$ in $O(dg^2)$ time.

## Tongue-and-groove effect and interdigitation

The algorithms for $M2F$ and $M3F$ may be further extended to generate optimal therapy time fields with elimination of tongue-and-groove underdosage and (optionally) the interdigitation constraint on the leaf sequences as is done for field splits without feathering in Section 6.4.2. The definitions of problems $M2F'$ ($M3F'$) and $M2F''$ ($M3F''$), respectively, for splits into two (three) fields are similar to those made in Section 6.4.2 for splits without feathering.

# 6.5 Minimizing the Number of Segments

Several algorithms have been proposed for minimizing the total number of segments required for treatment using SMLC. Some of these algorithms are designed to minimize the number of segments without explicitly considering the number of MUs in the optimization. Xia and Verhey [21] propose two classes of such algorithms: sliding window algorithms and reducing level algorithms. In sliding window algorithms, the left-most columns of the intensity matrix are initially exposed so as to reduce the residual intensities in these columns to zero. The columns are progressively exposed from left to right and the residual intensities become zero from left to right during treatment. In the reducing level algorithms, the intensity level of each segment is calculated as a function of the maximum intensity level in the residual matrix. Once the intensity level is calculated, a mask pattern that determines which area of the matrix is to be exposed using that intensity level is found. The process of calculating the intensity level and mask pattern is iteratively performed until the residual intensity matrix becomes zero. Xia and Verhey [21] propose multiple schemes for determining the intensity level and mask pattern. Que [15] proposes variations of the schemes of Xia and Verhey [21] and also compares the performance of some of the published leaf sequencing algorithms. Chen et al. [5] and Luan et al. [13] have proposed geometric and graph theoretic algorithms to minimize number of segments. Recently, algorithms have also been developed that minimize the number of segments while also using the optimum (minimum) number of MUs. Langer et al. [12] develop an integer programming formulation to minimize the number of segments subject to optimal MUs. Engel [6] and Kalinowski [7] have developed algorithms that heuristically minimize the number of segments while optimizing the number of MUs. Below we describe the approaches of Langer et al. [12] and Engel [6].

### 6.5.1 Algorithm of Langer et al. [12]

Let $T$ be the minimum number of MUs needed to deliver the profile $I$. $T$ may be computed using the expression in the proof of Theorem 3. Let $I_{i,j}$ be the desired number of MUs for the sample point $(i, j)$ on the $i^{\text{th}}$ row, $j^{\text{th}}$ column. During each unit of time $t$, $1 \leq t \leq T$, either one MU is delivered to a sample point or the sample point is shielded so that it receives no MUs. Let $l^t_{i,j}$, $r^t_{i,j}$ and $d^t_{i,j}$ be binary variables. The variable $l^t_{i,j}$ takes the value 1 if the $i^{\text{th}}$ left leaf shields position $(i, j)$ during the $t$th unit of time. Similarly, $r^t_{i,j}$ takes the value 1 if the $i^{\text{th}}$ right leaf shields position $(i, j)$ during the $t^{\text{th}}$ unit of time. If neither leaf shields this position, then the variable $d^t_{i,j}$ takes the value 1 and the sample point receives one MU. We have the following relationship

$$r^t_{i,j} + l^t_{i,j} = 1 - d^t_{i,j} \tag{6.22}$$

where $r^t_{i,j}, l^t_{i,j}, d^t_{i,j} \in \{0, 1\}$. From the geometry of the leaves we have,

$$r^t_{i,j} \leq r^t_{i,j+1} \tag{6.23}$$

and

$$l^t_{i,j+1} \leq l^t_{i,j}. \tag{6.24}$$

Because the number of MUs delivered at each sample point must match the desired number of MUs,

$$\sum_{i=1}^{T} d^t_{i,j} = I_{i,j}. \tag{6.25}$$

Variables $c^t_{i,j}$ and $u^t_{i,j}$ are used to keep track of changes in the state of sample points. $c^t_{i,j}$ takes the value 1 if position $(i, j)$ is not shielded during time unit $t$ and is shielded during time unit $t + 1$. Similarly, $u_{i,j}^t$ takes the value 1 if $(i, j)$ is shielded during time unit $t$ and is not shielded during time unit $t + 1$,

$$-c_{i,j}^t \leq d_{i,j}^{t+1} - d_{i,j}^t \leq u_{i,j}^t \qquad (6.26)$$

where $u_{i,j}^t, c_{i,j}^t \in \{0, 1\}$. If there is a change in the state of $I_{i,j}$ from time unit $t$ to $t + 1$, then the variable $s_{i,j}^t$ is 1,

$$u_{i,j}^t + c_{i,j}^t = s_{i,j}^t \qquad (6.27)$$

where $s_{i,j}^t \in \{0, 1\}$. If at least one sample point changes state between successive time units, we have a new segment. This is indicated by the variable $g^t$ being set to 1,

$$\sum_{i=1}^{n} \sum_{j=1}^{m} s_{i,j}^t \leq mng^t. \qquad (6.28)$$

The number of segments is minimized by minimizing the sum of the $g^t$s

$$\min \sum_{t=1}^{T-1} g^t. \qquad (6.29)$$

The minimum number of segments subject to minimum number of MUs can be found by minimizing the objective of equation (6.29) subject to equations (6.22)–(6.28).

Unidirectional leaf movement can be enforced using the following constraints

$$r_{i,j}^t - r_{i,j}^{t+1} \geq 0 \qquad (6.30)$$
$$l_{i,j}^{t+1} - l_{i,j}^t \geq 0. \qquad (6.31)$$

The interdigitation constraint is described as follows

$$l_{i+1,j}^t + r_{i,j}^t \leq 1 \qquad (6.32)$$
$$l_{i-1,j}^t + r_{i,j}^t \leq 1. \qquad (6.33)$$

Finally, the tongue-and-groove constraint is enforced by the following inequalities

$$-1 \leq d_{i+1,j}^t + d_{i,j}^{t'} - d_{i,j}^t - d_{i+1,j}^{t'} \leq 1, (t \neq t'). \qquad (6.34)$$

Note that when the unidirectional leaf movement constraint is applied with no other additional constraint, the minimum number of MUs is no more than without the constraint. This follows from the fact that Algorithm MULTI-PAIR generates a schedule that is optimal in MUs. For the problem with the unidirectional and interdigitation constraints, the value of $T$ is equal to the number of MUs required for delivering the profile $I$ using the schedule

generated by Algorithm MINSEPARATION with $S_{min} = 0$. Similarly, for
the problem with the unidirectional and tongue-and-groove constraints, the
value of $T$ is equal to the number of MUs required for delivering the profile $I$
using the schedule generated by Algorithm TONGUEANDGROOVE. It is to
be noted that these integer programming solutions use a very large number
of variables. Therefore, these solutions currently are applicable only for very
small matrices and are not practical for most clinical matrices.

### 6.5.2 Algorithm of Engel [6]

Engel [6] has proposed an algorithm that generates schedules that are optimal
in MUs and also heuristically minimizes the number of segments simultane-
ously. Let $d_{i,j} = I_{i,j} - I_{i,j-1}$, where $I_{i,0} = I_{i,m+1} = 0$. For each row $I_i$, the
TNMU (total number of monitor units)-row complexity, $C_i(I)$ is defined as
follows:

$$C_i(I) = \sum_{j=1}^{m+1} \max\{0, d_{i,j}\}. \tag{6.35}$$

From Theorem 1, it follows that $C_i(I)$, which equals the number of MUs for
the plan generated by Algorithm SINGLEPAIR, is the minimum number of
MUs required to deliver the intensity profile of row $I$. The TNMU complexity
of $I$, $C(I)$ is defined as

$$C(I) = \max_{1 \le i \le n} \{C_i(I)\}. \tag{6.36}$$

This is the minimum number of MUs required to deliver profile $I$ and is also
the number of MUs for the schedule generated by Algorithm MULTIPAIR.
From this fact, it follows that the class of algorithms of Figure 6.30 (Algorithm
MIN-TNMU) always yields a schedule with optimal MUs.

In Algorithm MIN-TNMU, $S$ is a binary matrix. It is represented as an
$n$-tuple $(S_1, S_2, \ldots, S_n)$, where $S_i = [l_i, r_i]$ is an interval, and

$$S_{ij} = \begin{cases} 1 \ j \in S_i \\ 0 \text{ otherwise.} \end{cases}$$

Algorithm MIN-TNMU

While $I \ne 0$

>Find $u > 0$ and a segment $S$ such that $I' = I - uS$ is nonnegative and $C(I') = C(I) - u$.
>Output $(u, S)$.
>$I = I - uS$.

End While

**Fig. 6.30.** Obtaining a schedule that heuristically minimizes number of segments
subject to optimal number of MUs.

The pair $(u, S)$ where $u$ and $S$ are as in step 2 of Algorithm MIN-TNMU is called an *admissible segmentation pair*. Note that for $C(I') = C(I) - u$ to be true, we require that $C_i(I - uS) \le C(I) - u$, $1 \le i \le n$. The objective now is to find a schedule that uses mimimum MUs and also minimizes the number of segments. The first strategy used is as follows. Take the largest possible $u$ in each iteration of Algorithm MIN-TNMU, i.e., the greatest number $u_{max}$ for which there exists a segment $S$ such that $(u_{max}, S)$ is an admissible segmentation pair. Call an interval $S_i = [l_i, r_i]$ an *essential interval* if $S_i = \Phi$ or $(d_{i,l_i} > 0$ and $d_{i,r_i+1} < 0)$. To determine $u_{max}$, as described below, it can be shown that it suffices to consider segments $S$ for which the $S_i$s are all essential intervals. Let $v(S_i)$ be defined as follows.

$$v(S_i) = \begin{cases} g_i(I) & S_i = \Phi \\ g_i(I) + \min\{d_{i,l_i}, -d_{i,r_i+1}\} & l_i \le r_i \text{ and } g_i(I) \le |d_{i,l_i} + d_{i,r_i+1}| \\ (d_{i,l_i} - d_{i,r_i+1} + g_i(I))/2 & l_i \le r_i \text{ and } g_i(I) > |d_{i,l_i} + d_{i,r_i+1}| \end{cases}$$

where $g_i(I) = C(I) - C_i(I)$. It can be shown that $u \le v(S_i)$ follows from $C_i(I - uS) \le C(I) - u$. Also, note that $I - uS \ge 0$ from which it follows that $u \le w(S_i)$, $1 \le i \le n$, where

$$w(S_i) = \begin{cases} \infty & S_i = \Phi \\ \min_{l_i \le j \le r_i} I_{i,j} & l_i \le r_i. \end{cases}$$

Let $u(S_i) = \min\{v(S_i), w(S_i)\}$, $1 \le i \le n$. Let $u_i = \max\{u(S_i)\}$ where the max is taken over all essential intervals $S_i$ for row $i$ of $I$. It can be shown that $u_{max} = \min_{1 \le i \le n} u_i$. Using the aforementioned results, it is possible to compute $u_{max}$ and a segment $S$ such that $(u_{max}, S)$ is an admissible segmentation pair. Engel [6] also briefly discusses other choices for admissible segmentation pairs, one of which results in a fewer number of segments than the process above.

Kalinowski [7] has extended the work of Engel [6] to account for the interdigitation constraint.

## 6.6 Conclusion

In this chapter, we have reviewed some of the recent work on leaf sequencing algorithms for multileaf collimation. The algorithms minimize the number of MUs and/or the number of segments. Most of the algorithms have also been adapted to account for machine dependent leaf movement constraints that include the interdigitation constraint, the tongue-and-groove constraint, and the maximum field width constraint.

## Acknowledgment

This work was supported in part by the National Library of Medicine under grant LM06659-03.

# References

[1] R. Ahuja and H. Hamacher. A network flow algorithm to minimize beam-on time for unconstrained multileaf collimator problems in cancer radiation therapy. *Networks*, 45:36–41, 2005.

[2] D. Baatar, H. Hamacher, M. Ehrgott, and G. Woeginger. Decomposition of integer matrices and multileaf collimator sequencing. *Discrete Applied Mathematics*, 152:6–34, 2004.

[3] N. Boland, H. Hamacher, and F. Lenzen. Minimizing beam-on time in cancer radiation treatment using multileaf collimators. *Networks*, 43:226–240, 2004.

[4] A. Boyer and J. Strait. Delivery of intensity modulated treatments with dynamic multileaf collimators. In D. Leavitt and G. Starkschall, editors, *Proceedings of the XIIth International Conference on the use of Computers in Radiation Therapy, Salt Lake City, Utah*, pages 13–15. Medical Physics Publising, Madison, Wisconsin, 1997.

[5] D. Chen, X. Hu, S. Luan, C. Wang, and X. Wu. Geometric algorithms for static leaf sequencing problems in radiation therapy. *International Journal of Computational Geometry and Applications*, 14:311–339, 2004.

[6] E. Engel. A new algorithm for optimal multileaf collimator field segmentation. Preprint, Fachbereich Mathematik, Universitaet Rostock, Rostock, Germany, 2003.

[7] T. Kalinowski. An algorithm for optimal multileaf collimator field segmentation with interleaf collision constraint. Preprint, Fachbereich Mathematik, Universitaet Rostock, Rostock, Germany, 2003.

[8] S. Kamath, S. Sahni, J. Li, J. Palta, and S. Ranka. Leaf sequencing algorithms for segmented multileaf collimation. *Physics in Medicine and Biology*, 48:307–324, 2003.

[9] S. Kamath, S. Sahni, J. Li, J. Palta, and S. Ranka. Optimal leaf sequencing with elimination of tongue-and-groove underdosage. *Physics in Medicine and Biology*, 49:N7–N19, 2004.

[10] S. Kamath, S. Sahni, J. Palta, and S. Ranka. Algorithms for optimal sequencing of dynamic multileaf collimators. *Physics in Medicine and Biology*, 49:33–54, 2004.

[11] S. Kamath, S. Sahni, S. Ranka, J. Li, and J. Palta. Optimal field splitting for large intensity-modulated fields. *Medical Physics*, 31:3314–3323, 2004.

[12] M. Langer, V. Thai, and L. Papiez. Improved leaf sequencing reduces segments or monitor units needed to deliver imrt using multileaf collimators. *Medical Physics*, 28:2450–2458, 2001.

[13] S. Luan, C. Wang, D. Chen, X. Hu, S. Naqvi, C. Yu, and C. Lee. A new mlc segmentation algorithm/software for step-and-shoot imrt delivery. *Medical Physics*, 31:695–707, 2004.

[14] L. Ma, A. Boyer, L. Xing, and C. Ma. An optimized leaf-setting algorithm for beam intensity modulation using dynamic multileaf collimators. *Physics in Medicine and Biology*, 26:2390–2396, 1998.

[15] W. Que. Comparison of algorithms for multileaf collimator field segmentation. *Medical Physics*, 26:2390–2396, 1999.

[16] S. Spirou and C. Chui. Generation of arbitrary intensity profiles by dynamic jaws or multileaf collimators. *Medical Physics*, 21:1031–1041, 1994.

212    S. Kamath et al.

[17] J. Stein, T. Bortfeld, B. Doerschel, and W. Schegel. Dynamic x-ray compensation for conformal radiotherapy by means of multileaf collimation. *Radiotherapy and Oncology*, 32:163–167, 1994.

[18] J. van Santvoort and B. Heijmen. Dynamic multileaf collimation without "tongue-and-groove" underdosage effects. *Physics in Medicine and Biology*, 41:2091–2105, 1996.

[19] Q. Wu, M. Arnfield, S. Tong, Y. Wu, and R. Mohan. Dynamic splitting of large intensity-modulated fields. *Physics in Medicine and Biology*, 45:1731–1740, 2000.

[20] X. Wu. Efficient algorithms for intensity map splitting problems in radiation therapy. In *Proceedings of the 11th international computing and combinatorics conference, Kunming, China*, pages 504–513, 2005.

[21] P. Xia and L. Verhey. Multileaf collimator leaf sequencing algorithm for intensity modulated beams with multiple static segments. *Medical Physics*, 25:1424–1434, 1998.

# 7

# Image Registration and Segmentation Based on Energy Minimization

Michael Hintermüller[1] and Stephen L. Keeling[2]

[1] Department of Mathematics, University of Sussex, Mantell Building, Falmer,
Brighton BN1 9RF, United Kingdom
`m.hintermueller@sussex.ac.uk`
[2] Department of Mathematics and Scientific Computing, University of Graz,
Heinrichstraße 36, A-8010 Graz, Austria
`stephen.keeling@uni-graz.at`

**Abstract.** Variational methods for image registration and image segmentation based on energy minimization are presented. In image registration, approaches that aim at minimizing a similarity measure + an appropriate regularization of the displacement field are investigated. Also, image interpolation problems based on optical flow techniques are considered. Several possible similarity measures as well as regularization terms are discussed. Corresponding optimality conditions (Euler–Lagrange equations) are derived, and numerical methods and graphical illustrations of various computational outcomes are presented. In a second part, approaches to image segmentation are introduced. General concepts such as region and edge growing are characterized and formulated as minimization problems. Then, two main paradigms are discussed: geodesic active contours (snakes) and the Mumford–Shah approach. Both techniques contain the edge set, which is a geometrical object, as the unknown quantity such that the minimization problem can be cast as a shape optimization problem. In order to cope with this aspect, techniques from shape sensitivity analysis are introduced. Finally, their numerical realization within a level set framework is highlighted.

## 7.1 Image Registration

Separate images of related objects are compared or aligned by at least implicitly conceiving a correspondence between like points. For example, two given images may be of a single patient at different times, such as during a mammography examination involving repeated imaging after the injection of a contrast agent [55]. On the other hand, the images may be of a single patient viewed by different imaging modalities, such as by magnetic resonance and computed tomography to provide complementary information for image-guided surgery [21]. In fact, images of two separate patients may even be compared to evaluate the extent of pathology of one in relation to the other [61]. Similarly,

P.M. Pardalos, H.E. Romeijn (eds.), *Handbook of Optimization in Medicine*,
Springer Optimization and Its Applications 26, DOI: 10.1007/978-0-387-09770-1_7,

an image of a patient may be compared to an idealized atlas in order to identify or segment tissue classes based upon a detailed segmentation of the atlas [61]. When an explicit coordinate transformation connecting like points is constructed, images are said to be *registered*. When a parameterized transformation permits images to be morphed one to the other, images are said to be *interpolated*. Because many applications involve the processing of sets as opposed to pairs of images, it is also of interest to consider methods for registering and interpolating image sequences.

Because the term registration is often used rather loosely in the context of its applications, it may be useful to elaborate on the above description of what registration is by stating what it is not. Note that by manipulating intensities alone, it is possible to warp or morph one image into another without having an explicit coordinate transformation identifying like image points. Thus, image registration is not image morphing but can be used for such an application. Similarly, a *continuous* warping of one image to another can be achieved without registration, but a parameterized coordinate transformation can be used to interpolate between images. Also, when complementary information in separate imaging modalities is superimposed, images are said to be *fused*. Because fusion too can be achieved by manipulating intensities alone, fused images need not be registered but rather *can* be fused by registration.

In order to compute a transformation that matches given images, two main ingredients are combined. First, there must be a measure of image similarity to quantify the extent to which a prospective transformation has achieved the matching goal [21]. Secondly, owing to the ill-posed nature of the registration problem, very pathological transformations are possible but not desired, and therefore a measure of transformation regularity is required [47]. Typically, one determines the desired transformation by minimizing an energy functional consisting of a weighted sum of these two measures.

The simplest image similarity measure is the sum of squared intensity differences, which is natural when images are related by a simple misalignment. Statistical measures have also been employed, and the correlation coefficient has been recognized as ideal when the intensities of the two images are related by a linear rescaling [62]. Also, the adaptation of thermodynamic entropy for information theory has suggested mutual information as an image similarity measure [43, 64], and a heuristically based normalized mutual information has been found to work very well in practice [60]. In [65], it is found in practice that highly accurate registrations of multimodal brain images can be achieved with information-theoretic measures. Nevertheless, as recognized in [54], mutual information contains no local spatial information, and random pixel perturbations leave underlying entropies unchanged. Higher order entropies including probabilities of neighboring pixel pairs can be employed to achieve superior results for non-rigid registration [54]; however, the message is that local spatial information in an image similarity measure is advantageous. In [17], Gauss maps are used to perform morphological, i.e., contrast invariant, image matching. Image level sets are also matched in [18] by using a Mumford–Shah formulation for registration. Higher order derivatives of the optical flow equation residual are penalized for an image similarity measure in [63] to obtain optical flows that do not require image structures to maintain a temporally constant brightness. In [11], the optical flow equation residual is replaced by a contrast invariant similarity measure that aligns level sets. In [33], the constant brightness assumption is circumvented without differential formulations by simply composing intensities with scaling functions.

The simplest approach to achieving regularity in a registration transformation is to use a low-dimensional parameterization. Before computing a very general type of registration transformation, many practitioners often consider first how well one of two natural classes of parameterized transformations manage to match given images: rigid and affine transformations. A rigid transformation is a sum of a translation and a rotation. An affine transformation is a sum of a translation and a matrix multiplication that is no longer constrained to be conformal or isometric. A registration or interpolation method may be called *generalized rigid* or *generalized affine* if it selects a rigid or an affine transformation, respectively, when one fits the given images [34]. The motivation for considering rigid or affine transformations, and generalizations thereof, lies in their applicability in two important categories of biomedical imaging. First, generalized rigid registration and interpolation are of particular interest, for instance, to facilitate medical examination of dynamic imaging data because of the ubiquity of rigid objects in the human body. Second, generalized affine registration and interpolation are of particular interest, for instance, for object reconstruction from histological data as histological sections may be affinely deformed in the process of slicing. A leading application and demand for non-rigid registration is for mammographic image sequences in which tissue deformations are less rigid and more elastic [55]. This observation has motivated the development of registration methods based on linear elasticity [19], [53]. Some authors relax rigidity by constraining transformations to be conformal or isometric [24]. Others employ a local rigidity constraint [40] or allow identified objects to move as rigid bodies [42].

### 7.1.1 Variational framework

Image registration and interpolation can be visualized using the illustration in Figure 7.1 for 2D images, in which two given images $I_0$ and $I_1$ are situated respectively on the front and back faces of a box $Q = \Omega \times (0,1)$ where a generic cross section of $Q$ is denoted by $\Omega = (0,1)^N$. In particular, the front and back faces of $Q$ are denoted by $\Omega_0$ and $\Omega_1$, on which $I_0$ and $I_1$ are situated, respectively. The rectangular spatial coordinates in $\Omega$ are denoted by $\boldsymbol{x} = (x_1, \ldots, x_N)$ and the depth or temporal coordinate by $z$.

The surfaces shown in Figure 7.1 are surfaces in which all but one of the curvilinear coordinates $\boldsymbol{\xi} = (\xi_1, \ldots, \xi_N)$ are constant, and the intersection of these surfaces represents a trajectory through $Q$ connecting like points in $I_0$ and $I_1$. The coordinates $\boldsymbol{\xi}(\boldsymbol{x}, z)$ are initialized in $\Omega_0$ so that $\boldsymbol{\xi}(\boldsymbol{x}, 0) = \boldsymbol{x}$ holds,

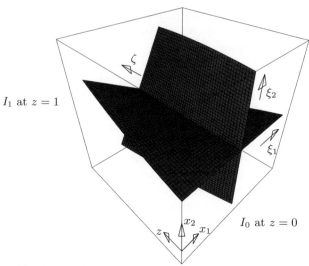

**Fig. 7.1.** The domain $Q$ with 2D images $I_0$ and $I_1$ on the front and back faces $\Omega_0$ and $\Omega_1$, respectively. Curvilinear coordinates are defined to be constant on trajectories connecting like points in $I_0$ and $I_1$.

and therefore the displacement vector within $Q$ is $\boldsymbol{d}(\boldsymbol{x}, z) = \boldsymbol{x} - \boldsymbol{\xi}(\boldsymbol{x}, z)$. The curvilinear coordinate system is completed by parameterizing a trajectory in the depth direction according to $\zeta = z$. Thus, a trajectory emanating from the point $\boldsymbol{\xi} \in \Omega_0$ is denoted by $\boldsymbol{x}(\boldsymbol{\xi}, \zeta)$. The coordinates in $\Omega_1$ of the finite displacement from coordinates $\boldsymbol{\xi}$ in $\Omega_0$ are written as $\boldsymbol{x}(\boldsymbol{\xi}) = \boldsymbol{x}(\boldsymbol{\xi}, 1)$. For those points in $Q$ situated on a trajectory joined to $\Omega_1$ but not necessarily to $\Omega_0$, let $\boldsymbol{y} = (y_1, \ldots, y_N)$ and $\boldsymbol{\eta} = (\eta_1, \ldots, \eta_N)$ be the counterparts to $\boldsymbol{x}$ and $\boldsymbol{\xi}$ defined so that $\boldsymbol{\eta}(\boldsymbol{y}, 1) = \boldsymbol{y}$ holds in $\Omega_1$; thus, a trajectory emanating from the point $\boldsymbol{\eta} \in \Omega_1$ is denoted by $\boldsymbol{y}(\boldsymbol{\eta}, \zeta)$, and the finite displacement from $\Omega_1$ to $\Omega_0$ is written as $\boldsymbol{y}(\boldsymbol{\eta}) = \boldsymbol{y}(\boldsymbol{\eta}, 0)$. A trajectory tangent is given by $(u_1, \ldots, u_N, 1)$ in terms of the optical flow defined as $\boldsymbol{u} = (u_1, \ldots, u_N) = \boldsymbol{x}_\zeta$. Since it is not assumed that every point in $\Omega_0$ finds a like point in $\Omega_1$, let the subsets of $\Omega_0$ and $\Omega_1$ with respect to which trajectories extend completely through the full depth of $Q$ be denoted respectively by $\Omega_0^c = \{\boldsymbol{\xi} \in \Omega_0 : \boldsymbol{x}(\boldsymbol{\xi}, \zeta) \in Q, 0 < \zeta < 1\}$ and $\Omega_1^c = \{\boldsymbol{\eta} \in \Omega_1 : \boldsymbol{y}(\boldsymbol{\eta}, \zeta) \in Q, 0 < \zeta < 1\}$. For those trajectories extending incompletely through $Q$, define $\Omega_0^i = \Omega_0 \backslash \Omega_0^c$ and $\Omega_1^i = \Omega_0 \backslash \Omega_1^c$.

To perform image registration using a finite displacement field $\boldsymbol{x}$, a functional of the following form can be minimized:

$$J(\boldsymbol{x}) = \mathcal{S}(\boldsymbol{x}) + \mathcal{R}(\boldsymbol{x}) \tag{7.1}$$

where $\mathcal{S}(\boldsymbol{x})$ is an image similarity measure depending upon the given images $I_0$ and $I_1$, and $\mathcal{R}(\boldsymbol{x})$ is a regularity measure of the transformation $\boldsymbol{x}$. To perform image registration and interpolation using an optical flow field $\boldsymbol{u}$ and an interpolated intensity $I$, a functional of the following form can be minimized:

$$J(\boldsymbol{u}, I) = \mathcal{S}(\boldsymbol{u}, I) + \mathcal{R}(\boldsymbol{u}) \tag{7.2}$$

where the intensity field $I$ is constrained by the boundary conditions:

$$I(\boldsymbol{x}, 0) = I_0(\boldsymbol{x}), \quad I(\boldsymbol{x}, 1) = I_1(\boldsymbol{x}) \tag{7.3}$$

and $\mathcal{S}(\boldsymbol{u}, I)$ quantifies the variation of intensity $I$ in the flow direction $(\boldsymbol{u}, 1)$ while $\mathcal{R}(\boldsymbol{x})$ is a regularity measure of the optical flow $\boldsymbol{u}$. Trajectories through the domain $Q$ are defined by integrating the optical flow under boundary conditions, i.e., by solving:

$$\boldsymbol{x}(\boldsymbol{\xi}, \zeta) = \boldsymbol{\xi} + \int_0^\zeta \boldsymbol{u}(\boldsymbol{x}(\boldsymbol{\xi}, \rho), \rho) d\rho, \qquad \boldsymbol{\xi} \in \Omega_0, \quad \zeta \in [0, 1] \tag{7.4}$$

and a similar equation for $\boldsymbol{y}(\boldsymbol{\eta}, \zeta)$ with $\boldsymbol{\eta} \in \Omega_1$ and $\zeta \in [0, 1]$. A registration is given by the coordinate transformation $\boldsymbol{x}(\boldsymbol{\xi}, 1)$ and by the inverse transformation $\boldsymbol{y}(\boldsymbol{\eta}, 0)$. The given images $I_0$ and $I_1$ are interpolated by the intensity $I$.

### 7.1.2 Similarity measures

The simplest similarity measure involves the squared differences $[I_0(\boldsymbol{\xi}) - I_1(\boldsymbol{x}(\boldsymbol{\xi}))]^2$ over $\Omega_0^c$. However, as discussed in detail in [34], $\Omega_0^c$ depends upon $\boldsymbol{x}(\boldsymbol{\xi})$. To avoid having to differentiate the domain with respect to the displacement for optimization, it is assumed that the images $I_0$ and $I_1$ can be continued in $\boldsymbol{R}^N$ by their respective *background* intensities, $I_0^\infty$ and $I_1^\infty$, which are understood as those intensities for which no active signal is measured. For simplicity, it is assumed here that the background intensities are zero. With such continuations, a similarity measure can be defined in terms of the sum of squared differences as follows:

$$\mathcal{S}_1(\boldsymbol{x}) = \int_{\Omega_0} [I_0(\boldsymbol{\xi}) - I_1(\boldsymbol{x}(\boldsymbol{\xi}))]^2 d\boldsymbol{\xi} \tag{7.5}$$

where here and below $I_1(\boldsymbol{x}(\boldsymbol{\xi}))$, $\boldsymbol{\xi} \in \Omega_0^i$ is understood as zero. So that $\mathcal{S}_1$ is independent of the order in which images are given, a similar integral over $\Omega_1$ may be added in (7.5) in which $I_0(\boldsymbol{y}(\boldsymbol{\eta}))$, $\boldsymbol{\eta} \in \Omega_1^i$ is understood as zero.

As illustrated in Figure 7.2, the finite displacements discussed above in connection with (7.5) can be written equivalently in terms of trajectories passing at least partly through $Q$ and some impinging upon the side of the box:

$$\Gamma = \partial Q \backslash \{\Omega_0 \cup \Omega_1\}. \tag{7.6}$$

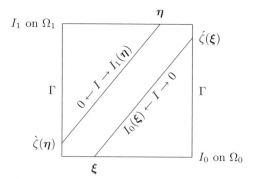

**Fig. 7.2.** $\acute{\zeta}(\boldsymbol{\xi})$ and $\grave{\zeta}(\boldsymbol{\eta})$ denote the $\zeta$ coordinates at which trajectories emanating respectively from $\boldsymbol{\xi} \in \Omega_0^i$ and $\boldsymbol{\eta} \in \Omega_1^i$ meet $\Gamma$.

The corresponding intensity differences can be written equivalently in terms of integrals of $[dI/d\zeta]^2$ for an intensity $I$ satisfying the boundary conditions (7.3) as well as those illustrated in Figure 7.2:

$$I = 0 \text{ on } \Gamma. \tag{7.7}$$

Once such integrals of $[dI/d\zeta]^2$ are transformed from the *Lagrangian* (trajectory following) form to the *Eulerian* (local) counterpart, $dI/d\zeta = \nabla I \cdot \boldsymbol{u} + I_z$, and transformation Jacobians such as $1/\det[\nabla_{\boldsymbol{\xi}}\boldsymbol{x}]$ are neglected, the following penalty on the optical flow equation residual [31] is obtained:

$$S_2(\boldsymbol{u}, I) = \int_Q [\nabla I \cdot \boldsymbol{u} + I_z]^2 \, d\boldsymbol{x} dz \tag{7.8}$$

subject to the boundary conditions (7.3) and (7.7).

To circumvent a constant brightness condition along trajectories, which in the present context involves minimizing the variation of the intensity $I$ along a trajectory, the similarity may be defined in terms of intensity derivatives as follows [63]:

$$S_3(\boldsymbol{u}, I) = \int_Q [\nabla|\nabla I| \cdot \boldsymbol{u} + |\nabla I|_z]^2 \, d\boldsymbol{x} dz. \tag{7.9}$$

To avoid the use of derivatives, the given data may instead be composed with scaling functions so that the intensity $I$ in (7.8) is constrained by the following modification of (7.3):

$$I = \sigma_0(I_0) \text{ on } \Omega_0, \quad I = I_1 \text{ on } \Omega_1 \tag{7.10}$$

in which only $I_0$ is scaled, and $I_1$ may be scaled similarly [33]. Furthermore, both of the given images may be scaled reciprocally in order that the registration be independent of image order [33].

The simplest statistical image similarity measure is the correlation coefficient:

$$S_4(\boldsymbol{x}) = \int_{\Omega_0} \left[ \frac{I_0(\boldsymbol{\xi}) - \mu(I_0)}{\sigma(I_0)} \right] \left[ \frac{I_1(\boldsymbol{x}(\boldsymbol{\xi})) - \mu(I_1 \circ \boldsymbol{x})}{\sigma(I_1 \circ \boldsymbol{x})} \right] d\boldsymbol{\xi} \qquad (7.11)$$

where $\mu(I_0) = \int_{\Omega_0} I_0 dx / \text{meas}(\Omega_0)$, with $\text{meas}(\Omega_0)$ representing the Lebesgue measure of $\Omega_0$, and $\sigma(I_0) = \mu([I_0 - \mu(I_0)]^2)$ denote the mean value and variance of $I_0$ respectively. Also, $I_1 \circ \boldsymbol{x}$ denotes the composition of $I_1$ and $\boldsymbol{x}$. So that $S_4$ is independent of the order in which images are given, a similar integral over $\Omega_1$ may be added in (7.11) as with (7.5). The similarity measures (7.5) and (7.11) coincide when they are restricted to pure translation [47]. A more complex statistical image similarity measure is the mutual information:

$$S_5(\boldsymbol{x}) = H(I_0) + H(I_1 \circ \boldsymbol{x}) - H(I_0, I_1 \circ \boldsymbol{x}) \qquad (7.12)$$

where, for images taking values in the interval $[0, 1]$, the entropy $H(A)$ of image $A$ and the joint entropy $H(A, B)$ of images $A$ and $B$ are given by:

$$H(A) = - \int_0^1 p(A = a) \log[p(A = a)] da$$

$$\qquad (7.13)$$

$$H(A, B) = - \int_0^1 \int_0^1 p(A = a, B = b) \log[p(A = a, B = b)] da db.$$

Here, $p(A = a)$ denotes the probability that the image $A$ assumes the intensity $a$, and $p(A = a, B = b)$ denotes the probability that the images $A$ and $B$ assume the intensities $a$ and $b$ simultaneously. So that $S_5$ is independent of the order in which images are given, the sum $H(I_0 \circ \boldsymbol{y}) + H(I_1) - H(I_0 \circ \boldsymbol{y}, I_1)$ may be added in (7.12) as with (7.5) and (7.11). For a simple example of mutual information, let $A$ and $B$ be the following $2 \times 2$ images:

$$A = \begin{array}{|c|c|} \hline 0 & 0 \\ \hline 1 & 1 \\ \hline \end{array} \quad B = \begin{array}{|c|c|} \hline 0 & 1 \\ \hline 0 & 1 \\ \hline \end{array}. \qquad (7.14)$$

So the intensity values are $\{a_i\} = \{0, 0, 1, 1\}$ and $\{b_j\} = \{0, 1, 0, 1\}$ and their probabilities are $p(A = a_i) = \frac{1}{2} = p(B = b_j)$. Also, there are precisely four intensity pairs $\{(0,0), (0,1), (1,0), (1,1)\}$, each with probability $p(A = a_i, B = b_j) = \frac{1}{4}$. The entropies of the two images are the same, $H(A) = H(B) = \log(2)$. The joint entropy is $H(A, B) = \log(4)$, which is larger than the joint entropy of $A$ with itself, $H(A, A) = \log(2)$. Thus, a transformation which rotates image $B$ to be aligned with the image $A$ would minimize the mutual information. Note that this similarity measure operates purely on intensity values and on pairs of intensity values, and it involves no local spatial information. Such spatial information can be incorporated by defining higher order entropies involving probabilities of neighboring pixel pairs [54]. On the other hand, the variational treatment of (7.12) and variations of it are more complicated than that of similarity measures such as (7.5) with an explicit spatial orientation [47].

When finitely many clearly matching points are identified manually, or else from particular features found in the images $I_0$ and $I_1$, these landmarks:

$$\mathcal{E}_\ell(x) = x(\xi_\ell) - x_\ell = 0, \quad \ell = 1, \ldots, L \tag{7.15}$$

may be used as constraints in the optimization process for determining the registration or interpolation. On the other hand, these landmarks may be used exclusively to determine a parametric registration by minimizing the sum of squared differences of the landmark residuals [22]:

$$S_6(x) = \sum_{\ell=1}^{L} |x_\ell - x(\xi_\ell)|^2. \tag{7.16}$$

### 7.1.3 Regularity measures

The most easily determined registrations are those that are parametric and low-dimensional. For instance, a transformation $x$ could be computed as a combination of thin plate spline functions:

$$x(\xi) = \sum_{m=1}^{N+1} \alpha_m P_m(\xi) + \sum_{\ell=1}^{L} \beta_\ell U(\xi - \xi_\ell) \tag{7.17}$$

where $U(\xi) = |\xi|^{4-N} \log|\xi|$ for $N$ even (or $U(\xi) = |\xi|^{4-N}$ for $N$ odd) and $\{P_m\}$ is a basis for linear functions. The transformation in (7.17) that minimizes the following regularity measure:

$$\mathcal{R}_1(x) = \sum_{|\alpha|=2} \frac{2!}{\alpha!} \int_{\Omega_0} |\partial_\xi^\alpha x|^2 d\xi \tag{7.18}$$

under the constraints in (7.15) is given by solving systems of the form [47]:

$$\begin{pmatrix} K & B^T \\ B & 0 \end{pmatrix} \begin{pmatrix} \bar{\alpha} \\ \bar{\beta} \end{pmatrix} = \begin{pmatrix} \bar{x} \\ 0 \end{pmatrix} \tag{7.19}$$

for the $i$th component $(x)_i$ of $x$ according to:

$$K_{ij} = U(\xi_i - \xi_j), \quad B_{im} = P_m(\xi_i), \quad \bar{\alpha} = \{(\alpha_m)_i\}_{m=1}^{N+1},$$
$$\bar{\beta} = \{(\beta_\ell)_i\}_{\ell=1}^{L}, \quad \bar{x} = \{(x_\ell)_i\}_{\ell=1}^{L}. \tag{7.20}$$

In (7.18), $2!/\alpha!$ is the multinomial coefficient for a multi-index $\alpha$. Although such registrations are easily computed and are often used, the transformation can be pathological enough as to fail to be diffeomorphic [47].

On the other hand, the transformation $x$ has been expressed in terms of *piecewise* polynomial splines and determined by minimizing a weighted sum of the regularity measure (7.18) and the similarity measure (7.12) as seen in [55]. Particularly because of the non-uniqueness of minimizers, the iterative solution of such minimization problems is typically started with the rigid or affine transformation

$$\text{rigid: } \boldsymbol{x}(\boldsymbol{\xi}) = \boldsymbol{\tau} + e^{W}\boldsymbol{\xi}, \quad W = -W^{\mathrm{T}}, \qquad \text{affine: } \boldsymbol{x}(\boldsymbol{\xi}) = \boldsymbol{\tau} + A\boldsymbol{\xi} \qquad (7.21)$$

which minimizes the similarity measure.

The kernel of the regularity measure (7.18) selects affine transformations and it thus provides *generalized affine* registration in the sense that an affine transformation is selected when one fits the data. On the other hand, the kernel of (7.18) does not necessarily select a rigid transformation. In order to select rigid transformations, it is necessary to consider the full non-linearized elastic potential energy in a regularity measure of the following form [48]:

$$\mathcal{R}_2(\boldsymbol{x}) = \int_{\Omega_0} |\nabla_{\boldsymbol{\xi}} \boldsymbol{x}^{\mathrm{T}} \nabla_{\boldsymbol{\xi}} \boldsymbol{x} - I|^2 d\boldsymbol{\xi}. \qquad (7.22)$$

However, the corresponding optimality system is quite complex, and *generalized rigid* registration is achieved more easily below with optical flow [37]. A convenient alternative to (7.22) is given by linearized elastic potential energy [19, 53]:

$$\mathcal{R}_3(\boldsymbol{x}) = \mathcal{R}_3(\boldsymbol{d} + I) = \int_{\Omega_0} \left[ \lambda[\nabla \cdot \boldsymbol{d}]^2 + \tfrac{1}{2}\mu|\nabla \boldsymbol{d}^T + \nabla \boldsymbol{d}|^2 \right] d\boldsymbol{\xi} \qquad (7.23)$$

although it does not seleted rigid transformations [37]. A visco-elastic fluid model is adopted with the regularity measure [15, 47]:

$$\mathcal{R}_4(\boldsymbol{x}) = \mathcal{R}_4(\boldsymbol{d} + I) = \int_{\Omega_0} \left[ \lambda[\nabla \cdot \boldsymbol{d}_t]^2 + \tfrac{1}{2}\mu|\nabla \boldsymbol{d}_t^T + \nabla \boldsymbol{d}_t|^2 \right] d\boldsymbol{\xi} \qquad (7.24)$$

where the transformation $\boldsymbol{x}$ is considered to depend on time $t$. In this case, the optimality system for the functional, say, $J(\boldsymbol{x}) = \mathcal{S}_1(\boldsymbol{x}) + \nu\mathcal{R}_4(\boldsymbol{x})$ leads to an evolution equation which may be solved to steady state allowing the regularizing effect of (7.24) to diminish with time.

By using optical flow, generalized affine registration and interpolation is achieved with the regularity measure [34]:

$$\mathcal{R}_5(\boldsymbol{u}) = \int_Q \left[ \sum_{|\alpha|=2} \frac{2!}{\alpha!} |\partial_{\boldsymbol{x}}^\alpha \boldsymbol{u}|^2 + \gamma|\boldsymbol{u}_z|^2 \right] d\boldsymbol{x} dz \qquad (7.25)$$

and generalized rigid registration and interpolation is achieved using [37]:

$$\mathcal{R}_6(\boldsymbol{u}) = \int_Q \left[ |\nabla_{\boldsymbol{x}} \boldsymbol{u}^{\mathrm{T}} + \nabla_{\boldsymbol{x}} \boldsymbol{u}|^2 + \gamma|\boldsymbol{u}_z|^2 \right] d\boldsymbol{x} dz. \qquad (7.26)$$

Although it is shown in [34] that non-autonomous flows are theoretically possible with these regularity measures, a $z$-dependence is not found in practice. Thus, these integrals over $Q$ can be replaced with integrals over $\Omega$ after setting $\gamma = \infty$.

### 7.1.4 Optimality conditions

As an example of image registration by finite displacements, consider the minimization of the following functional:

$$\mathcal{J}_{11}(\boldsymbol{x}) = \mathcal{S}_1(\boldsymbol{x}) + \nu\mathcal{R}_1(\boldsymbol{x}). \tag{7.27}$$

This functional is stationary when $\boldsymbol{x}$ satisfies:

$$0 = \frac{1}{2}\frac{\delta\mathcal{J}_{11}}{\delta\boldsymbol{x}}(\boldsymbol{x},\bar{\boldsymbol{x}}) = \mathcal{B}_{11}(\boldsymbol{x},\bar{\boldsymbol{x}}) - \mathcal{F}_{11}(\boldsymbol{x},\bar{\boldsymbol{x}}), \quad \forall\bar{\boldsymbol{x}} \in H^2(\Omega,\boldsymbol{R}^N) \tag{7.28}$$

where $H^m(\Omega,\boldsymbol{R}^N)$ is the Sobolev space of functions mapping $\Omega$ into $\boldsymbol{R}^N$ with Lebesgue square integrable derivatives up to order $m$, and $\mathcal{B}_{11}$ and $\mathcal{F}_{11}$ are defined by [34]:

$$\mathcal{B}_{11}(\boldsymbol{x},\bar{\boldsymbol{x}}) = \nu\sum_{|\alpha|=2}\frac{2!}{\alpha!}\int_{\Omega_0}[\partial_\xi^\alpha\boldsymbol{x}]\cdot[\partial_\xi^\alpha\bar{\boldsymbol{x}}]d\boldsymbol{\xi} \tag{7.29}$$

$$\mathcal{F}_{11}(\boldsymbol{x},\bar{\boldsymbol{x}}) = \int_{\Omega_0}[I_0(\boldsymbol{\xi}) - I_1(\boldsymbol{x}(\boldsymbol{\xi}))]\nabla_x I_1(\boldsymbol{x}(\boldsymbol{\xi}))^{\mathrm{T}}\bar{\boldsymbol{x}}(\boldsymbol{\xi})d\boldsymbol{\xi}. \tag{7.30}$$

The form $\mathcal{F}_{11}$ contains a similar term over $\Omega_1$ when $\mathcal{S}_1$ contains the corresponding term mentioned in relation to (7.5). The transformation $\boldsymbol{x}$ satisfying (7.28) can be computed by the following quasi-Newton iteration [34]:

$$\begin{cases} \mathcal{N}_{11}(d\boldsymbol{x}_k,\boldsymbol{x}_k,\bar{\boldsymbol{x}}) = -[\mathcal{B}_{11}(\boldsymbol{x}_k,\bar{\boldsymbol{x}}) - \mathcal{F}_{11}(\boldsymbol{x}_k,\bar{\boldsymbol{x}})], \quad \forall\bar{\boldsymbol{x}} \in H^2(\Omega_0,\boldsymbol{R}^N) \\ \boldsymbol{x}_{k+1} = \boldsymbol{x}_k + \theta d\boldsymbol{x}_k \end{cases} \tag{7.31}$$

for $k = 0, 1, 2, \ldots$, where:

$$\mathcal{N}_{11}(d\boldsymbol{x}_k,\boldsymbol{x}_k,\bar{\boldsymbol{x}}) = \mathcal{B}_{11}(d\boldsymbol{x}_k,\bar{\boldsymbol{x}})$$
$$+ \int_{\Omega_0}[\nabla_x I_1(\boldsymbol{x}_k(\boldsymbol{\xi}))\cdot d\boldsymbol{x}_k(\boldsymbol{\xi})][\nabla_x I_1(\boldsymbol{x}_k(\boldsymbol{\xi}))\cdot\bar{\boldsymbol{x}}(\boldsymbol{\xi})]d\boldsymbol{\xi} \tag{7.32}$$

and $\theta$ is chosen by a line search to minimize $\mathcal{S}_1$ [26]. Note that no additional boundary conditions are imposed by restricting the domain of the forms defined above, and thus natural boundary conditions hold.

As an example of image registration and interpolation by optical flow, consider the minimization of the following functional:

$$\mathcal{J}_{26}(\boldsymbol{u}, I) = \mathcal{S}_2(\boldsymbol{u}, I) + \nu\mathcal{R}_6(\boldsymbol{u}). \tag{7.33}$$

This functional is stationary in the optical flow $\boldsymbol{u}$ for fixed $I$ when $\boldsymbol{u}$ satisfies:

$$0 = \frac{1}{2}\frac{\delta\mathcal{J}_{26}}{\delta\boldsymbol{u}}(\boldsymbol{u},\bar{\boldsymbol{u}}) = \mathcal{B}_{26}(\boldsymbol{u},\bar{\boldsymbol{u}}) - \mathcal{F}_{26}(\bar{\boldsymbol{u}}), \quad \forall\bar{\boldsymbol{u}} \in H^1(Q,\boldsymbol{R}^N), \tag{7.34}$$

where $\mathcal{B}_{26}$ and $\mathcal{F}_{26}$ are defined by [37]:

$$\mathcal{B}_{26}(\boldsymbol{u}, \bar{\boldsymbol{u}}) = \int_Q \left[ (\nabla I \cdot \boldsymbol{u})(\nabla I \cdot \bar{\boldsymbol{u}}) + \gamma \left( \boldsymbol{u}_z \cdot \bar{\boldsymbol{u}}_z \right) \right] d\boldsymbol{x} dz$$

$$+ \int_Q \tfrac{1}{2} \left( \nabla \boldsymbol{u}^{\mathrm{T}} + \nabla \boldsymbol{u} \right) : \left( \nabla \bar{\boldsymbol{u}}^{\mathrm{T}} + \nabla \bar{\boldsymbol{u}} \right) d\boldsymbol{x} dz \qquad (7.35)$$

$$\mathcal{F}_{26}(\bar{\boldsymbol{u}}) = -\int_Q I_z \nabla I \cdot \bar{\boldsymbol{u}} d\boldsymbol{x} dz. \qquad (7.36)$$

Note that no additional boundary conditions are imposed by restricting the domain of these forms, and thus natural boundary conditions hold.

The optimality condition for $\mathcal{J}_{26}$ with respect to the intensity $I$ involves solving the equation $d^2 I/d\zeta^2 + (\nabla \cdot \boldsymbol{u}) dI/d\zeta = 0$ with boundary conditions as seen in Figure 7.2. When this condition is formulated and solved in a Eulerian fashion, the resulting interpolated images lose clarity between $\Omega_0$ and $\Omega_1$ [37]. Thus, the optimality condition on the intensity should be formulated in a Lagrangian fashion. Specifically, the functional $\mathcal{J}_{26}$ is stationary in the intensity $I$ for fixed $\boldsymbol{u}$ when $I$ satisfies the following in terms of quantities defined below [37]:

$$I(\boldsymbol{x}(\boldsymbol{\xi}, \zeta), \zeta) = \begin{cases} I_0(\boldsymbol{\xi})[1 - U(\boldsymbol{\xi}, \zeta, 1)] + I_1(\boldsymbol{x}(\boldsymbol{\xi}, 1)) U(\boldsymbol{\xi}, \zeta, 1), & \boldsymbol{\xi} \in \Omega_0^c \\ I_0(\boldsymbol{\xi})[1 - U(\boldsymbol{\xi}, \zeta, \acute{\zeta})], & \boldsymbol{x}(\boldsymbol{\xi}, \acute{\zeta}) \in \Gamma, \boldsymbol{\xi} \in \Omega_0^i \end{cases}$$

$$(7.37)$$

$$I(\boldsymbol{y}(\boldsymbol{\eta}, \zeta), \zeta) = \begin{cases} I_1(\boldsymbol{\eta})[1 - V(\boldsymbol{\eta}, 0, \zeta)] + I_0(\boldsymbol{y}(\boldsymbol{\eta}, 0)) V(\boldsymbol{\eta}, 0, \zeta), & \boldsymbol{\eta} \in \Omega_1^c \\ I_1(\boldsymbol{\eta})[1 - V(\boldsymbol{\eta}, \grave{\zeta}, \zeta)], & \boldsymbol{y}(\boldsymbol{\eta}, \grave{\zeta}) \in \Gamma, \boldsymbol{\eta} \in \Omega_1^i. \end{cases}$$

$$(7.38)$$

As illustrated in Figure 7.2, the parameters $\acute{\zeta}$ and $\grave{\zeta}$ denote the $\zeta$ coordinates at which trajectories emanating respectively from $\Omega_0^i$ and $\Omega_1^i$ meet $\Gamma$. Then, $U$ and $V$ are defined by:

$$U(\boldsymbol{\xi}, \zeta, \acute{\zeta}) = \frac{\tilde{U}(\boldsymbol{\xi}, \zeta) - \tilde{U}(\boldsymbol{\xi}, 0)}{\tilde{U}(\boldsymbol{\xi}, \acute{\zeta}) - \tilde{U}(\boldsymbol{\xi}, 0)},$$

$$\tilde{U}(\boldsymbol{\xi}, \zeta) = \int_{\zeta_0}^{\zeta} \exp\left[ -\int_{\zeta_0}^{\varrho} \nabla \cdot \boldsymbol{u}(\boldsymbol{x}(\boldsymbol{\xi}, \rho), \rho) d\rho \right] d\varrho, \qquad (7.39)$$

for $\boldsymbol{\xi} \in \Omega_0$, $\zeta \in [0, \acute{\zeta}]$, and arbitrary $\zeta_0 \in [0, \acute{\zeta}]$, and:

$$V(\boldsymbol{\eta}, \grave{\zeta}, \zeta) = \frac{\tilde{V}(\boldsymbol{\eta}, 1) - \tilde{V}(\boldsymbol{\eta}, \zeta)}{\tilde{V}(\boldsymbol{\eta}, 1) - \tilde{V}(\boldsymbol{\eta}, \grave{\zeta})},$$

$$\tilde{V}(\boldsymbol{\eta}, \zeta) = \int_{\zeta_0}^{\zeta} \exp\left[ -\int_{\zeta_0}^{\varrho} \nabla \cdot \boldsymbol{u}(\boldsymbol{y}(\boldsymbol{\eta}, \rho), \rho) d\rho \right] d\varrho, \qquad (7.40)$$

for $\boldsymbol{\eta} \in \Omega_1$, $\zeta \in [\grave{\zeta}, 1]$, and arbitrary $\zeta_0 \in [\grave{\zeta}, 1]$. These formulas can be easily interpreted by considering the case that the transformation is rigid and thus

$\nabla \cdot \boldsymbol{u} = 0$ holds. In this case, $I$ must satisfy $d^2I/d\zeta^2 = 0$ and $U(\boldsymbol{\xi}, \zeta, 1) = \zeta$ and $V(\boldsymbol{\eta}, 0, \zeta) = (1 - \zeta)$ hold.

If the similarity measure $\mathcal{S}_2$ of $\mathcal{J}_{26}$ is modified to incorporate intensity scaling as seen in (7.10), then under the simplifying assumption that the given images are piecewise constant the functional $\mathcal{J}_{26}$ is stationary in the scaling function $\sigma_0$ for fixed optical flow $\boldsymbol{u}$ and intensity $I$ when $\sigma_0$ satisfies [33]:

$$\sigma_0(\iota) = \int_{I_0(\boldsymbol{\xi})=\iota} \mathcal{I}_1(\boldsymbol{\xi})\mathcal{U}(\boldsymbol{\xi})d\boldsymbol{\xi} \bigg/ \int_{I_0(\boldsymbol{\xi})=\iota} \mathcal{U}(\boldsymbol{\xi})d\boldsymbol{\xi} \tag{7.41}$$

where the morphing of $I_1$ into $\Omega_0$ is given by:

$$\mathcal{I}_1(\boldsymbol{\xi}) = \begin{cases} I_1(\boldsymbol{x}(\boldsymbol{\xi}, 1)), & \boldsymbol{\xi} \in \Omega_0^c \\ 0, & \boldsymbol{\xi} \in \Omega_0^i \end{cases} \tag{7.42}$$

and $\mathcal{U}$ is defined by:

$$\mathcal{U}(\boldsymbol{\xi}) = \begin{cases} \displaystyle\int_0^1 U_\zeta^2(\boldsymbol{\xi}, \zeta, 1) \det(\nabla_{\boldsymbol{\xi}}\boldsymbol{x}) \, d\zeta, & \boldsymbol{\xi} \in \Omega_0^c \\[2mm] \displaystyle\int_0^{\acute{\zeta}(\boldsymbol{\xi})} U_\zeta^2(\boldsymbol{\xi}, \zeta, \acute{\zeta}(\boldsymbol{\xi})) \det(\nabla_{\boldsymbol{\xi}}\boldsymbol{x}) \, d\zeta, & \boldsymbol{\xi} \in \Omega_0^i. \end{cases} \tag{7.43}$$

These formulas can be easily interpreted by considering the case that the transformation is rigid. It follows from $\nabla \cdot \boldsymbol{u} = 0$ that $U(\boldsymbol{\xi}, \zeta, 1) = \zeta$, $U_\zeta(\boldsymbol{\xi}, \zeta, 1) = 1$ and $\det(\nabla_{\boldsymbol{\xi}}\boldsymbol{x}) = 1$ hold. Thus, $\mathcal{U}(\boldsymbol{\xi}) = 1$ holds. The formula (7.41) determines the value of $\sigma_0(\iota)$ as the average value of morphed image over the level set $I_0(\boldsymbol{\xi}) = \iota$. When the transformation is not rigid, the value of $\sigma_0(\iota)$ is a weighted average of the morphed image over the level set.

To incorporate the landmark constraints (7.15) into the determination of finite displacements, $\mathcal{J}_{11}$ for instance may be augmented to form the following Lagrangian functional [53, 20]:

$$\mathcal{L}_{11}(\boldsymbol{x}, \boldsymbol{\lambda}) = \tfrac{1}{2}\mathcal{J}_{11}(\boldsymbol{x}) + \sum_{\ell=1}^L \boldsymbol{\lambda}_\ell^T \mathcal{E}_\ell(\boldsymbol{x}). \tag{7.44}$$

This Lagrangian functional is stationary in $(\boldsymbol{x}, \boldsymbol{\lambda}_1, \ldots, \boldsymbol{\lambda}_L)$ when the following hold:

$$\begin{cases} \mathcal{B}_{11}(\boldsymbol{x}, \bar{\boldsymbol{x}}) - \mathcal{F}_{11}(\boldsymbol{x}, \bar{\boldsymbol{x}}) + \displaystyle\sum_{\ell=1}^L \boldsymbol{\lambda}^T \bar{\boldsymbol{x}}(\boldsymbol{\xi}_\ell) = 0, \, \forall \bar{\boldsymbol{x}} \in H^2(\Omega, \boldsymbol{R}^N), \\[3mm] \boldsymbol{x}(\boldsymbol{\xi}_\ell) - \boldsymbol{x}_\ell = 0, \, \ell = 1, \ldots, L. \end{cases} \tag{7.45}$$

For this, let $U_\ell(\boldsymbol{\xi}) = U(\boldsymbol{\xi} - \boldsymbol{\xi}_\ell)$ be the solution to $\mathcal{B}_{11}(U_\ell, \bar{U}) + \bar{U}(\boldsymbol{\xi}_\ell) = 0, \, \forall \bar{U} \in H^2(\Omega, \boldsymbol{R})$; cf. (7.17). Also, let $\tilde{\boldsymbol{x}}$ be the solution to $\mathcal{B}_{11}(\tilde{\boldsymbol{x}}, \bar{\boldsymbol{x}}) -$

$\mathcal{F}_{11}(x, \bar{x}) = 0$, $\forall \bar{x} \in H^2(\Omega, \mathbf{R}^N)$. Then determine the Lagrange multipliers $\{\lambda_\ell\}$ algebraically from the condition that $x = \tilde{x} + \sum_{\ell=1}^{L} \lambda_\ell U_\ell(\xi)$ satisfy the landmark constraints (7.15). Thus, $(x, \lambda_1, \ldots, \lambda_L)$ satisfy (7.45). Of course, $\tilde{x}$ and $x$ depend upon each other, and these may be computed iteratively. Note that the formulation of landmark constraints for optical flow is more complicated [36].

### 7.1.5 Processing image sequences

An image sequence may be registered or interpolated of course by processing the images only pairwise and concatenating the results. On the other hand, a coupling among images may be introduced as follows; see also [47].

The images of a sequence $\{I_k\}_{k=0}^{K}$ can be registered simultaneously using finite displacements $\{x_k\}_{k=1}^{K}$ by minimizing:

$$\mathcal{J}_{11}^{(K)}(x_1, \ldots, x_K) = \sum_{k=1}^{K} \mathcal{S}_1^{(k)}(x_1, \ldots, x_K) + \mathcal{R}_1(x_k) \tag{7.46}$$

where:

$$\mathcal{S}_1^{(k)}(x_1, \ldots, x_K) = \sum_{|k-j|=1} \int_{\Omega_k} [I_j(x_j(\xi)) - I_k(x_k(\xi))]^2 \, d\xi \tag{7.47}$$

where all images are extended by their background intensities (here as before assumed to be zero) outside their domains, $\Omega_l$, which are additional counterparts to $\Omega_0$ and $\Omega_1$ depicted in Figure 7.1. The end indices $k = 0$ and $k = K$ in (7.46) correspond with pairwise registration with the single near neighbor. When (7.46) has been minimized, the point $x_i(\xi) \in \Omega_i$ has been matched to the point $x_j(\xi) \in \Omega_j$. To minimize $\mathcal{J}_{11}^{(K)}$ with respect to $x_k$ while all other transformations are held fixed, replace $\mathcal{F}_{11}$ in (7.28) and (7.31) with $\mathcal{F}_{11}^{(k)} = -\frac{1}{2}\delta \mathcal{S}_1^{(k)}/\delta x$:

$$\mathcal{F}_{11}^{(k)}(x_k, \bar{x}) = \int_{\Omega_k} \sum_{|k-j|=1} [I_j(x_j(\xi)) - I_k(x_k(\xi))] \nabla_x I_k(x_k(\xi))^{\mathrm{T}} \bar{x}(\xi) d\xi. \tag{7.48}$$

The functional of (7.46) can be minimized by freezing all current transformations except for one, minimizing the functional with respect to the selected transformation, updating that transformation immediately (Gauss–Seidel strategy) or else updating all transformations simultaneously (Jacobi strategy), and then repeating the process until the updates have converged. Known transformations can remain frozen as fixed boundary conditions, e.g., at one or both of the end indices $k = 0$ and $k = K$ in (7.46) when the position of one or both of the end images $I_0$ and $I_K$ is known.

The calculation (7.48) shows that $\mathcal{J}_{11}^{(L)}$ is just as well minimized with respect to $x_k$ by registering the image $I_k$ with the image

$$I_{k_n}(\boldsymbol{\xi}) = \sum_{|j-k|=1} I_j(\boldsymbol{x}_j(\boldsymbol{\xi})) / \sum_{|j-k|=1} 1.$$

Analogously, the images $\{I_k\}_{k=0}^{K}$ can be registered simultaneously by computing autonomous optical flows $\{\boldsymbol{u}_k\}_{k=0}^{K}$ for the image pairs $\{[I_k, I_{k_n}]\}_{k=0}^{K}$ according to pairwise procedures, where the transformations $\{\boldsymbol{x}_k\}_{k=0}^{K}$ are computed by using their respective flows in (7.4). Then the flows and their corresponding transformations can be updated repeatedly until convergence, where known transformations can remain frozen as fixed boundary conditions as discussed above.

The images $\{I_k\}_{k=0}^{K}$ can be interpolated from autonomous optical flows $\{\boldsymbol{u}_k\}_{k=0}^{K-1+M}$ using the semi-discretization defined on $Q^{(K)} = \Omega \times (0, K)$:

$$\boldsymbol{u}(\boldsymbol{x}, z) = \sum_{k=0}^{K-1+M} \boldsymbol{u}_k(\boldsymbol{x}) \chi_k^M(z) \tag{7.49}$$

where $\{\chi_k^M\}_{k=0}^{K-1+M}$ is a basis for the canonical B-splines of degree $M$ defined on the grid $\{[k, k+1]\}_{k=0}^{K-1}$ of $[0, K]$ [30]. Then the transformations are given by natural modifications of (7.4) replacing $\Omega_0$, $\Omega_1$ and $\zeta \in [0, 1]$ with $\Omega_k$, $\Omega_{k+1}$ and $\zeta \in [k, k+1]$. Also, the intensity $I$ is given by natural modifications of (7.37) and (7.38) replacing $\Omega_0$, $\Omega_1$ and $\Gamma$ with $\Omega_k$, $\Omega_{k+1}$ and $\Gamma^{(K)} = \partial Q^{(K)} \setminus \{\Omega_0 \cup \Omega_K\}$. For instance, for $M = 0$, $\chi_k^0$ is the characteristic function for the interval $[k, k+1]$, and the above procedure corresponds with pairwise interpolation of the given images. When smoother trajectories and greater coupling among images are desired, higher order splines can be used in (7.49), and (7.34) can be solved for $\{\boldsymbol{u}_k\}_{k=0}^{K-1+M}$ with $\gamma = 0$ and $Q$ replaced by $Q^{(K)}$.

### 7.1.6 Numerical methods

The most costly computations required to solve the optimality systems of the previous subsection are those involved in solving (7.28) and (7.34). It is shown in [37] that the trajectory integrations must be performed from every point in $Q$ where the intensity $I$ is needed, and trajectories must be extended in both directions toward $\Omega_0$ and $\Omega_1$ in order to connect values of $I_0$ and $I_1$; nevertheless, such integrations can be vectorized and obtained remarkably quickly. All other computations are even less expensive than those required for (7.28) and (7.34).

For the numerical solution of the finite displacement problem (7.28) or of the (autonomous) optical flow problem (7.34), it is useful to consider the common structure among such problems, which is found often in image processing. Specifically, the boundary value problems have the form:

Find $\varphi \in H^{\kappa}(\Omega, \boldsymbol{R}^N)$ such that:
$$\nu(\mathcal{D}^{(\kappa)}\varphi, \mathcal{D}^{(\kappa)}\psi)_{L^2(\Omega, \boldsymbol{R}^N)} + (\boldsymbol{g} \cdot \varphi, \boldsymbol{g} \cdot \psi)_{L^2(\Omega, \boldsymbol{R}^N)} = (\boldsymbol{f}, \psi)_{L^2(\Omega, \boldsymbol{R}^N)}, \tag{7.50}$$
$$\text{for all } \psi \in H^{\kappa}(\Omega, \boldsymbol{R}^N)$$

where $g \in L^\infty(\Omega, \boldsymbol{R}^N)$ and $\boldsymbol{f} \in L^2(\Omega, \boldsymbol{R}^N)$ have the same compact support in $\Omega$, and $\mathcal{D}^{(\kappa)}$ is a differential operator of order $\kappa$. The $\mathcal{D}^{(\kappa)}$-regularization term as well as the $g$-data term are both indefinite on $H^\kappa(\Omega, \boldsymbol{R}^N)$, but the sum is bounded and coercive. With additional homogeneous boundary conditions, the $\mathcal{D}^{(\kappa)}$-regularization term can be made definite and therefore numerically better conditioned, but such artificial boundary conditions would corrupt a generalized rigid or generalized affine approach. Whereas Fourier methods have been used for similar systems [19, 47], multigrid methods [25, 26, 34] can be used with comparable speed and greater generality, for instance, to accommodate the natural boundary conditions associated with (7.50).

A geometric multigrid formulation is developed in [34] for (7.28) and (7.34) and is based upon [23]. The usual multigrid strategy is generally to enhance a convergent relaxation scheme by using its initial and rapid smoothing of small scales on finer grids and then to transfer the problem progressively to coarser grids before relaxation is decelerated. The principal ingredients of the strategy include the definition of a smoothing relaxation scheme and the definition of a coarse grid representation of the problem, which can be used to provide an improvement or correction on a finer grid.

For the representation of the boundary value problem on progressively coarser grids, (7.50) can be formulated on a nested sequence of finite element subspaces $S^\kappa_{2h}(\Omega, \boldsymbol{R}^N) \subset S^\kappa_h(\Omega, \boldsymbol{R}^N)$ such as tensor products of the B-splines illustrated in Figure 7.3. Then the finite element approximation to the solution $\varphi$ of (7.50) is $\varphi_h \in S^\kappa_h(\Omega, \boldsymbol{R}^N)$ defined by replacing $H^\kappa(\Omega, \boldsymbol{R}^N)$ in (7.50) with $S^\kappa_h(\Omega, \boldsymbol{R}^N)$. This finite-dimensional formulation is expressed as $A_h \Phi_h = F_h$ where $A_h$ is the matrix representation of the differential operator in the finite element basis, and $\Phi_h$ and $F_h$ are vectors of finite element basis coefficients for $\varphi_h$ and $\boldsymbol{f}$, respectively. Let $K_h$ denote the mapping from coefficients to functions so that $\Phi_h = K_h \varphi_h$ holds. Also, let $I_{2h}$ denote the injection operator

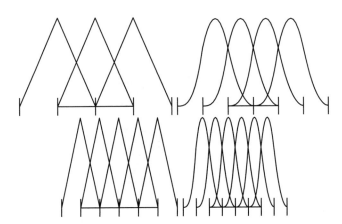

**Fig. 7.3.** Examples of nested finite elements spaces $S^\kappa_{2h}(\Omega, \boldsymbol{R}^1) \subset S^\kappa_h(\Omega, \boldsymbol{R}^1)$ of degree 1 (left column) and 2 (right column).

from $S_{2h}^{\kappa}(\Omega, \boldsymbol{R}^N)$ into $S_h^{\kappa}(\Omega, \boldsymbol{R}^N)$. Then the coarse grid matrix $A_{2h}$ is computed from the fine grid matrix $A_h$ according to the Galerkin approximation $A_{2h} = R_h^{2h} A_h E_{2h}^h$ where $E_{2h}^h$ and $R_h^{2h}$ are the canonical expansion and restriction operators satisfying $I_{2h} K_{2h} = K_h E_{2h}^h$ and $(E_{2h}^h)^* = R_h^{2h}$. In words, $E_{2h}^h$ produces *coefficients* $\boldsymbol{\Phi}_h = E_{2h}^h \boldsymbol{\Phi}_{2h}$ from coefficients $\boldsymbol{\Phi}_{2h}$ so that the *function* $K_h \boldsymbol{\Phi}_h$ is identical to the function $K_{2h} \boldsymbol{\Phi}_{2h}$. With the coarse grid problem and the intergrid transfer operators defined, it remains to identify a suitable relaxation scheme and to define the multigrid iteration.

Because the bilinear form in (7.50) is symmetric and coercive, the matrices, $A_h = \mathcal{D}_h + \mathcal{L}_h + \mathcal{L}_h^{\mathrm{T}}$, are symmetric and positive definite, where $\mathcal{D}_h$ is strictly diagonal and $\mathcal{L}_h$ is strictly lower triangular. Thus, it is natural to use a symmetric relaxation scheme such as the symmetric successive over-relaxation,

$$
\begin{aligned}
\boldsymbol{\Phi}_h^{k+1} &= S_h \boldsymbol{\Phi}_h^k + \omega W_h^{-1} F_h, \qquad S_h = I - \omega W_h^{-1} A_h, \\
W_h &= (\mathcal{D}_h + \mathcal{L}_h) \mathcal{D}_h^{-1} (\mathcal{D}_h + \mathcal{L}_h^{\mathrm{T}}), \quad \omega \in (0, 2).
\end{aligned}
\tag{7.51}
$$

As discussed in detail in [35], this relaxation scheme can be vectorized for implementation in systems such as IDL or MATLAB by using a multicolored ordering of cells as illustrated in Figure 7.4 for a stencil diameter of 3 cells. In general, for a stencil diameter of $(2\kappa + 1)$, define a set of *same-color* cells as those that are separated from one another in any of $N$ coordinate directions by exactly $\kappa$ cells. These cells have stencils that do not weight any other cells in the set; thus, the strategy is to update such sets of cells simultaneously in the relaxation. Such same-color cells are ordered along coordinate directions within that color, and then ordered sequentially among the colors. With such a multicolored ordering, the relaxation scheme can be implemented by performing a Jacobi iteration on same-colored cells while looping in one direction and then the other over the colors.

$$
\begin{aligned}
&\text{for } c = 1, \ldots, (\kappa + 1)^N \text{ and then } c = (\kappa + 1)^N, \ldots, 1 \text{ do:} \\
&\boldsymbol{\Phi}_h^c \leftarrow \boldsymbol{\Phi}_h^c - [\mathcal{D}_h^{-1}(A_h \boldsymbol{\Phi}_h - F_h)]^c.
\end{aligned}
\tag{7.52}
$$

**Fig. 7.4.** A multicolor ordering of cells for a stencil diameter of $2\kappa + 1(=3)$ in which same-color cells are separated from one another in any of $N(=2)$ coordinate directions by exactly $\kappa(=1)$ cells.

In this way, same-colored cells are updated simultaneously. Similarly, the known stencil diameter can also be used to advantage to vectorize the computation of elements of the coarse grid matrix [35].

With the above ingredients, a symmetric two-grid cycle $\text{TGC}(h, \sigma)$ is obtained by:

(1) performing $\sigma$ relaxation steps to update $\Phi_h$,
(2) computing the coarse-grid residual $D_{2h} = R_{2h}^h(F - A_h\Phi_h)$,
(3) solving on the coarse grid $A_{2h}\Psi_{2h} = D_{2h}$,
(4) correcting on the fine grid $\Phi_h \leftarrow \Phi_h + E_{2h}^h\Psi_{2h}$, and finally
(5) performing another $\sigma$ relaxation steps to update $\Phi_h$.

Then a symmetric multigrid cycle $\text{MGC}(h, \sigma, \tau)$ is defined as with the two-grid cycle except that step 3 in $\text{TGC}(h, \sigma)$ is recursively replaced with $\tau$ iterations of $\text{MGC}(2h, \sigma, \tau)$ unless $2h$ is large enough that the the coarse grid problem may easily be solved exactly.

### 7.1.7 Computational examples

Here examples of generalized affine and generalized rigid image registration and interpolation are shown together with examples of intensity scaling. Shown in Figure 7.5 are two given images on the far left and on the far right, which may be related by either an affine or by a rigid transformation. The results of minimizing $\mathcal{S}_2 + \nu\mathcal{R}_5$ and $\mathcal{S}_2 + \nu\mathcal{R}_6$ to register and to interpolate between the given images are shown respectively in the top and bottom rows. The figure shows that $\mathcal{R}_5$ and $\mathcal{R}_6$ produce affine and rigid transformations respectively when such transformations fit the data.

In the case when the given data are not related by such a simple transformation, e.g., by a rigid transformation, Figure 7.6 shows that the departure from rigidity may be controlled by the regularization parameter $\nu$. Specifically, the result for larger $\nu$ is strongly rigid while the result for smaller $\nu$ is called weakly rigid [37]. Also, strong or weak rigidity may be controlled locally by incorporating $\nu$ into the regularization penalty $\mathcal{R}_6$ as a distributed parameter; see [48].

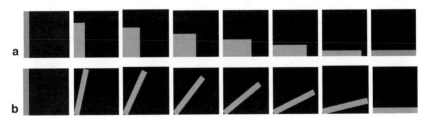

**Fig. 7.5.** Images on the far left and right, which may be related by either an affine or by a rigid transformation, are registered and interpolated by minimizing (a) $\mathcal{S}_2 + \nu\mathcal{R}_5$ and (b) $\mathcal{S}_2 + \nu\mathcal{R}_6$.

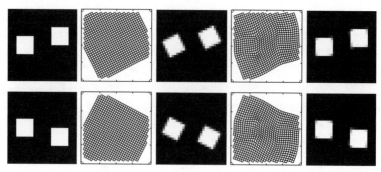

**Fig. 7.6.** The given images are shown in the far left column. These images are registered by minimizing $S_2 + \nu \mathcal{R}_6$, and the results for large $\nu$ are shown in the second and third columns, and the results for smaller $\nu$ are shown in the fourth and fifth columns. In each case, registration results are illustrated by applying the transformation, as well as its inverse, first to a uniform grid and then to the given image situated on the front or on the back face of $Q$ in Figure 7.1.

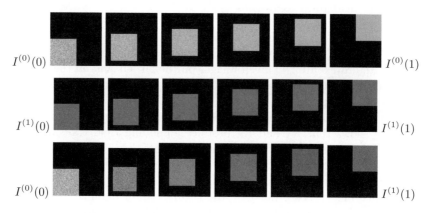

**Fig. 7.7.** The image sequences, $I^{(0)}(t)$ and $I^{(1)}(t)$, $t = 0, .2, .4, .6, .8, 1$, are shown in the top two rows. The given raw images are at the upper left $I^{(0)}(0)$ and at the middle right $I^{(1)}(1)$. The intensity scaling of (7.10) and (7.41) transforms the upper left $I^{(0)}(0)$ into the middle left image $I^{(1)}(0)$ and the middle right $I^{(1)}(1)$ into the upper right image $I^{(0)}(1)$. Registration and interpolation are then performed independently in the top two rows by minimizing $S_2 + \nu \mathcal{R}_6$. The convex convex combination $(t-1)I^{(0)}(t) + tI^{(1)}(t)$ of these sequences gives the interpolation shown in the third row.

Finally, the intensity scaling approach of (7.10) and (7.41) is illustrated in Figure 7.7. Let the upper image sequence here be denoted by $I^{(0)}(t)$, $0 \leq t \leq 1$, and the middle image sequence by $I^{(1)}(t)$, $0 \leq t \leq 1$, where the given raw images are at the upper left $I^{(0)}(0)$ and at the middle right $I^{(1)}(1)$. These two images have different histograms and different noise levels, but the intensity scaling of (7.10) and (7.41) transforms the upper left $I^{(0)}(0)$

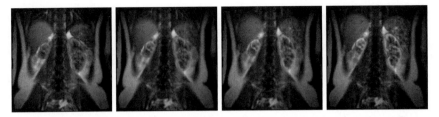

**Fig. 7.8.** Shown on the left and on the right are two given raw magnetic resonance images measured in the course of respiration and the introduction of contrast agent. The images in-between have been interpolated by minimizing $\mathcal{S}_2 + \nu\mathcal{R}_6$ using the scaling approach of (7.10) and (7.41).

into the middle left image $I^{(1)}(0)$ and the middle right $I^{(1)}(1)$ into the upper right image $I^{(0)}(1)$. Once the images are rescaled in this way, registration and interpolation may be performed independently in the top two rows by minimizing $\mathcal{S}_2 + \nu\mathcal{R}_6$. The interpolation between the given images is then given by the convex combination $(t-1)I^{(0)}(t) + tI^{(1)}(t)$ as shown in the third row of Figure 7.7. This is precisely the procedure used to interpolate between the raw magnetic resonance images shown on the left and on the right in Figure 7.8. These raw images have been measured in the course of respiration and they have different histograms because of the appearance of contrast agent [33, 34]. The two raw images shown in Figure 7.8 are part of the larger sequence `http://math.uni-graz.at/keeling/respfilm1.mpg` which is interpolated as seen in `http://math.uni-graz.at/keeling/respfilm2.mpg`.

## 7.2 Edge Detection and Image Segmentation

Variational methods in image segmentation have a natural relation to energy minimization and partial differential equations. Some of the advantages of using variational methods are that

- they allow common formulations by assembling "energy" and/or data fidelity terms in a real-valued objective (or energy) functional $J$ over the set of edges (or segmentations);
- on the other hand, many PDE-based segmentation and edge-detection techniques can be interpreted as (approximate) minimization of certain energy functionals;
- having an energy functional to be minimized is related to the fact that, at the same time, it can serve as a measure for comparing different segmentations (we would say that segmentation $\Gamma_1$ is better than $\Gamma_2$ if $J(\Gamma_1) < J(\Gamma_2)$);
- they allow one to introduce a scale $\alpha$, which typically determines the amount of image detail that is kept during the segmentation (multiscale analysis, scale space).

The multiscale principle works as follows: Assume that $I_0(x)$ is some given gray-scale image defined on a square or rectangle $\Omega \subset \mathbb{R}^2$. Let $\{I_\alpha\}$ denote a sequence of images approximating the original $I_0$. An element $I_\alpha$ will contain edges with scale exceeding $\alpha$. By $S_\alpha : I_0 \mapsto I_\alpha$ we denote the solution operator of some variational method and by $\Gamma_\alpha$ the pertinent edge set. The multiscale principle is based on

- Consistency (or fidelity), i.e., $I_\alpha \to I_0$ as $\alpha \to 0$.
- Strong monotonicity (or strong causality), i.e., $\Gamma_\alpha \subset \Gamma_{\underline{\alpha}}$ if $\alpha > \underline{\alpha}$.
- Euclidean invariance, i.e., isometric mappings do not influence the result.

In general, one may also consider a weaker monotonicity principle involving the solution operator $S_\alpha$; see [49].

### 7.2.1 Region and edge growing

Region growing methods create a partitioning of the image in homogeneous regions by starting with small "seed regions" that are then grown by some homogeneity criterion. Edge growing methods start with an initial fine scale edge set that is then connected iteratively depending on orientation and proximity criteria. Hybrid growing methods combine these two aspects.

One of the simplest energy functionals in the region growing context capturing the amount of information contained in a segmentation $\Gamma$ measures the amount of boundaries in $\Gamma$ and their smoothness as well as the smoothness of $I_\alpha$ on each region:

$$J_{\mathrm{MS}}(I, \Gamma) = \frac{1}{2} \int_{\Omega \backslash \Gamma} (I - I_0)^2 dx + \frac{\alpha}{2} \int_{\Omega \backslash \Gamma} |\nabla I|^2 dx + \gamma \int_\Gamma 1 \, d\mathcal{H}_1, \quad (7.53)$$

where $\mathcal{H}_1$ denotes the one-dimensional Hausdorff measure restricted to $\Gamma$. The scales $\alpha$ and $\gamma$ are both assumed to be positive. The term attached to $\alpha$ penalizes variations of $\nabla I$ from zero, i.e., violation of homogeneity, and the $\gamma$-term weighs the length of the edge set $\Gamma$. In that sense these two energies regularize non-smooth images. The first term reflects data fidelity and represents a distance measure to the given (possibly smoothed) image data $I_0$. If $I$ is imposed to be constant on each region, then the $\alpha$-term vanishes and the energy reduces to the energy in the piecewise constant case:

$$J_{\mathrm{pcMS}}(I, \Gamma) = \frac{1}{2} \int_{\Omega \backslash \Gamma} (I - I_0)^2 dx + \gamma \int_\Gamma 1 \, d\mathcal{H}_1. \quad (7.54)$$

The functional in (7.53) is known as the Mumford–Shah functional. The difficulty when minimizing $J_{\mathrm{MS}}$ is due to the different nature of the unknowns. In fact, $I$ is a function defined on $\Omega$ whereas $\Gamma$ is a one-dimensional set. As a result, both variables have to be treated differently in theory as well as in numerical approaches.

Appropriate energies in the context of edge growing appear to be one-dimensional equivalents of the Mumford–Shah energy. Hence, in the smoothing part, these energies contain an integral depending on the length and the curvature as well as the length of the boundary. The one-dimensional analogue of the latter energy is the cardinality of the tips of curves:

$$J_{\mathrm{EG}}^s(\Gamma_\alpha) = \int_{\Gamma_\alpha} (1 + \kappa(s)^2)ds + \gamma\,\mathrm{card}(\partial\Gamma_\alpha), \tag{7.55}$$

where $\partial\Gamma_\alpha$ represents the tips of curves and $\kappa$ denotes the curvature. Notice that $J_{\mathrm{EG}}^s$ models the smoothing part of the energy. The fidelity term captures the quality of the approximation of $\Gamma_0$ by $\Gamma_\alpha$. The possibly simplest way to measure this proximity is given by $\mathrm{length}(\Gamma_\alpha \setminus \Gamma_0)$. However, as $I_0$ might have been pre-smoothed by some filtering method, this term should be augmented by the fidelity of $\Gamma_\alpha$ to edges. This can be done by

$$- \int_{\Gamma_\alpha} |\nabla I(s)|^2 ds.$$

In addition, one can measure the strength of an edge by considering the derivative across the edge, i.e.,

$$- \int_{\Gamma_\alpha} \left| \frac{\partial I}{\partial \boldsymbol{n}}(s) \right| ds,$$

where $\boldsymbol{n}$ denotes the normal to $\Gamma_\alpha$. The overall energy now reads

$$J_{\mathrm{EG}}(\Gamma_\alpha) = \alpha J_{\mathrm{EG}}^s(\Gamma_\alpha) - \int_{\Gamma_\alpha} |\nabla I(s)|^2 ds - \int_{\Gamma_\alpha} \left| \frac{\partial I}{\partial \boldsymbol{n}}(s) \right| ds + \mathrm{length}(\Gamma_\alpha \setminus \Gamma_0). \tag{7.56}$$

The functional may be simplified, but one needs to keep at least one term of each of the constituents of $J_{\mathrm{EG}}$. For instance, minimizing

$$J_{\mathrm{EG}}^0(\Gamma_\alpha) = \alpha\,\mathrm{card}(\partial\Gamma_\alpha) + \mathrm{length}(\Gamma_\alpha \setminus \Gamma_0) \tag{7.57}$$

is related to a continuous version of the traveling salesman problem. The classic snake model by Kass, Witkin and Terzopolous [32] is given by

$$\tilde{J}_{\mathrm{EG}}(\Gamma_\alpha) = \alpha \int_{\Gamma_\alpha} (1 + \kappa(s)^2)ds - \int_{\Gamma_\alpha} |\nabla I(s)|^2 ds. \tag{7.58}$$

### 7.2.2 Snakes, geodesic active contours, and level set methods

In the previous section, we introduced the classic snake or deformable active contour model in (7.58). Assume that $\Gamma$ is the union of a finite or countable number of closed piecewise $\mathcal{C}^1$-curves $C_j$ in $\mathbb{R}^2$.

**Snakes**

The snake model uses parameterized curves in $P_{\boldsymbol{c}} = \{\boldsymbol{c} : [t_0, T] \to \Omega :$ $\boldsymbol{c}$ piecewise $\mathbb{C}^1, \boldsymbol{c}(t_0) = \boldsymbol{c}(T)\}$ and considers the energy minimization

$$\text{minimize } \tilde{J}(\boldsymbol{c}) = \alpha \tilde{J}_{\text{int}}(\boldsymbol{c}) + \tilde{J}_{\text{ext}}(\boldsymbol{c}) \text{ over } \boldsymbol{c} \in P_{\boldsymbol{c}}, \tag{7.59}$$

where $\tilde{J}_{\text{int}}$ models the internal or smoothing energy whereas $\tilde{J}_{\text{ext}}$ represents the external energy or fidelity. In fact,

$$\tilde{J}_{\text{int}}(\boldsymbol{c}) = \alpha \left( \int_{t_0}^T |\boldsymbol{c}'(t)|^2 dt + \beta \int_{t_0}^T |\boldsymbol{c}''(t)|^2 dt \right), \tag{7.60}$$

$$\tilde{J}_{\text{ext}}(\boldsymbol{c}) = \int_{t_0}^T g^2(|\nabla I(\boldsymbol{c}(t))|)dt. \tag{7.61}$$

Here, $g$ is a mapping from $[0, +\infty)$ to $[0, +\infty)$ that is monotonically decreasing and satisfies $g(0) = 1$ and $\lim_{z \to \infty} g(z) = 0$. Typical choices are

$$g(z) = \frac{1}{1 + z^k} \quad k = 1, 2.$$

Hence, $g(|\nabla I|)$ will be zero at ideal edges (infinite gradient of the intensity map $I$) and positive elsewhere. The mapping

$$\boldsymbol{x} \mapsto g(|\nabla|)(\boldsymbol{x})$$

is called an *edge detector*. The fidelity term therefore attracts the curve $\boldsymbol{c}$ to the edge set.

Because $\Omega$ is bounded, it can be shown that there exists a minimizer $\boldsymbol{c}^* = (c_1^*, c_2^*) \in P_{\boldsymbol{c}}$ of (7.59). It can be characterized by the corresponding necessary first order optimality condition (also called the *Euler–Lagrange equations*):

$$\alpha(-\boldsymbol{c}'' + \beta \boldsymbol{c}^{(\text{iv})}) + g(|\nabla I(\boldsymbol{c})|)g'(|\nabla I(\boldsymbol{c})|)\boldsymbol{p}(\boldsymbol{c})\frac{d}{d\boldsymbol{c}}(\nabla I(\boldsymbol{c})) = 0,$$

$$\boldsymbol{p}(\boldsymbol{c}) \in \partial|\nabla I(\boldsymbol{c})|,$$

$$\boldsymbol{c}(t_0) = \boldsymbol{c}(T).$$

Above $\partial$ denotes the subdifferential from convex analysis, i.e.,

$$\partial|\nabla I(\boldsymbol{c})| \in \begin{cases} \{\nabla I(\boldsymbol{c})/|\nabla I(\boldsymbol{c})|\} & \text{if } |\nabla I(\boldsymbol{c})| > 0, \\ \bar{B}(0; 1) & \text{if } |\nabla I(\boldsymbol{c})| = 0, \end{cases}$$

where $\bar{B}(0; 1)$ denotes the closed unit ball in $\mathbb{R}^2$. In general, one cannot expect a unique minimizer due to the non-convexity of $\tilde{J}$. Hence, by solving the Euler–Lagrange equations, one will typically find a local minimizer only.

Drawbacks of the snake model are due to the fact that it is a non-intrinsic model, i.e., the solution depends on the chosen parametrization and that it cannot handle topological changes, i.e., only one object can be detected.

**Geodesic active contours**

In (7.60), the term involving $c''$ aims at minimizing the squared curvature. It turns out that the model with $\beta = 0$ also aims at reducing the curvature. But this simpler model still remains non-intrinsic as it depends on the chosen parametrization. In order to overcome this drawback, at about the same time, in [12, 13] and [38, 39] the following functional was introduced:

$$J_{\text{gAC}}(c(t)) = \int_{t_0}^{T} g(|\nabla I(c(t))|)\,|c'(t)|\,dt. \tag{7.62}$$

It can readily be seen that $J_{\text{gAC}}$ is intrinsic, i.e., the energy does not change under parameter changes. This is due to the fact that the Euclidean length is weighted by the term $g(|\nabla I(c(t))|)$, which induces a Riemannian metric. In [5], it was shown that minimizing $\tilde{J}$ with $\beta = 0$ is equivalent to minimizing $J_{\text{gAC}}$.

Embedding $C = c(t) \in P_c$ into the family of curves defined by $c(\omega, t)$ with $\omega \geq 0$ and $c(0, t) = c(t)$ as well as $c(\omega, t_0) = c(\omega, T)$ and $\frac{\partial c(\omega, t_0)}{\partial t} = \frac{\partial c(\omega, T)}{\partial t}$, using calculus of variations one finds that $J_{\text{gAC}}(c(\omega, t))$ decreases most in the direction

$$\frac{\partial c(\omega, t)}{\partial t} = (\kappa\, g - \nabla g \cdot N)N, \tag{7.63}$$

where $N$ denotes the unit normal to the curve, $\kappa$ is the curvature, and $\nabla g$ represents the gradient of the mapping $x \mapsto g(|\nabla I(x)|)$. Note that if $c$ matches with the edges in $I$ (where $g = 0$), then $\frac{\partial c(\omega, t)}{\partial t} = 0$, i.e., a stationary point for $J_{\text{gAC}}$ is found.

A modified version of (7.63) that aims at an increase of the convergence speed as well as improved detection of non-convex objects is given by

$$\frac{\partial c(\omega, t)}{\partial t} = ((\kappa + \mu)\, g - \nabla g \cdot N)\, N, \tag{7.64}$$

with a constant $\mu$ such that $\kappa + \mu$ has a constant sign; see [12, 13]. The term $\mu\, g$ represents a "driving force" that, depending on the sign, either helps to expand or deflate the propagating curve (or *contour*). To some extent this may be considered to act as a regularization to overcome noise in the image.

**Level-set method**

Compared with parametrization-based approaches, in numerical practice it turned out that a realization of (7.63) or (7.64) within a level set framework has several advantages: It allows flexibility in representing the curve (or iterative approximations thereof) numerically, and it is numerically robust as it operates on a fixed (Eulerian) grid. In fact, techniques based on parameterizations of $c$ may suffer from the need of (expensive) re-parametrizations in case of topological changes and from numerical instabilities as discretization

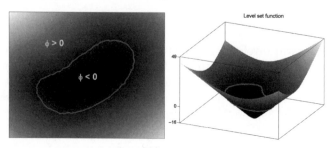

**Fig. 7.9.** The closed curve in the left plot is represented as the zero-level set of a level set function–here a signed distance function–in the right plot.

points on approximations of $c$ may cluster or have gaps during an iterative procedure.

In their seminal work [52], Osher and Sethian propose the following approach; see also [51, 58]. A closed curve $c$ in $\mathbb{R}^2$ can be represented as the zero-level set of a *level set function* (or, according to [51], a *geometrical implicit function*) $\phi : \mathbb{R}^2 \times \mathbb{R}^+ \to \mathbb{R}$ in the following way:

$$\phi(\boldsymbol{x}(t), t) = 0 \quad \text{for all } \boldsymbol{x}(t) \text{ and } t \geq 0, \tag{7.65}$$

where $\boldsymbol{x}(t)$ represents a point on $c(t) \subset \mathbb{R}^2$. Further it is assumed that the sign convention shown in Figure 7.9 holds true. Assuming that $\phi$ is sufficiently regular, one differentiates (7.65) with respect to $t$. This yields

$$\phi_t(\boldsymbol{x}(t), t) + \nabla\phi(\boldsymbol{x}(t), t) \cdot \boldsymbol{x}'(t) = 0 \quad \text{for all } t \geq 0. \tag{7.66}$$

Next it is supposed that $c$ (or, equivalently, $\boldsymbol{x}$) travels with velocity $F$ in the unit outward normal direction to $c$, i.e.,

$$\boldsymbol{x}'(t) = \frac{\partial c(t)}{\partial t} = F(t, \boldsymbol{x}(t), \boldsymbol{x}'(t), \ldots) \boldsymbol{N}(\boldsymbol{x}(t), t).$$

Inserting this form of $\boldsymbol{x}'$ into (7.66) gives

$$\phi_t(\boldsymbol{x}(t), t) + \nabla\phi(\boldsymbol{x}(t), t) \cdot (F(t, \boldsymbol{x}(t), \boldsymbol{x}'(t), \ldots) \boldsymbol{N}(\boldsymbol{x}(t), t)) = 0 \quad \text{for all } t \geq 0. \tag{7.67}$$

In view of the sign convention according to Figure 7.9, the unit outward normal is given by

$$\boldsymbol{N}(\boldsymbol{x}(t), t) = -\frac{\nabla\phi(\boldsymbol{x}(t), t)}{|\nabla\phi(\boldsymbol{x}(t), t)|}.$$

This yields the equation

$$\phi_t(\boldsymbol{x}(t), t) - F(t, \boldsymbol{x}(t), \boldsymbol{x}'(t), \ldots)|\nabla\phi(\boldsymbol{x}(t), t)| = 0 \quad \text{for all } t \geq 0 \tag{7.68}$$

which can readily be extended to a domain $\Omega \subset \mathbb{R}^2$ containing $c$:

$$\phi_t(\boldsymbol{x}, t) - F|\nabla\phi(\boldsymbol{x}, t)| = 0 \quad \text{for all } t \geq 0 \text{ and } \boldsymbol{x} \in \Omega. \tag{7.69}$$

This partial differential equation (PDE) is typically combined with a homogeneous Neumann boundary condition for $\phi$ on $]0, +\infty[ \times \partial\Omega$ and the initial condition $\phi(\boldsymbol{x}, 0) = \phi_0(\boldsymbol{x})$. A popular choice of $\phi_0$ is the signed distance function

$$\phi_0(\boldsymbol{x}) := \begin{cases} d(\boldsymbol{x}, \boldsymbol{c}_0) & \text{if } \boldsymbol{x} \text{ is outside } \boldsymbol{c}_0, \\ -d(\boldsymbol{x}, \boldsymbol{c}_0) & \text{if } \boldsymbol{x} \text{ is inside } \boldsymbol{c}_0, \end{cases}$$

where $\boldsymbol{c}_0$ denotes the initial curve and $d(\boldsymbol{x}, \boldsymbol{c}_0)$ is the Euclidean distance of $\boldsymbol{x}$ to $\boldsymbol{c}_0$. The PDE (7.69) together with its boundary and initial condition is of Hamilton–Jacobi type.

In the context of image segmentation the velocity $F$ is given by

$$F(t, \boldsymbol{x}(t), \boldsymbol{x}'(t), \boldsymbol{x}''(t)) := \kappa\, g - \nabla g \cdot \frac{\nabla\phi}{|\nabla\phi|}, \tag{7.70}$$

in case of minimizing $J_{\mathrm{gAC}}$, or, if $F$ is based on (7.64), by

$$F(t, \boldsymbol{x}(t), \boldsymbol{x}'(t), \boldsymbol{x}''(t)) := (\kappa + \mu)\, g - \nabla g \cdot \frac{\nabla\phi}{|\nabla\phi|}, \tag{7.71}$$

The equation (7.69) is then solved until steady state. Note that this procedure corresponds to a steepest descent method for minimizing $J_{\mathrm{gAC}}$ or a modification of thereof in case of (7.71), as we shall see later.

The curvature $\kappa$ can be written as

$$\kappa = \mathrm{div}\left(\frac{\nabla\phi}{|\nabla\phi|}\right);$$

see, e.g., [41]. Hence, (7.69) with the velocity in (7.71) becomes

$$\phi_t = g(|\nabla I|)\left(\mathrm{div}\left(\frac{\nabla\phi}{|\nabla\phi|}\right) + \mu\right)|\nabla u| + \nabla g \cdot \nabla\phi. \tag{7.72}$$

Note that the natural extension of $F(t, \cdot)$ to $\Omega$ is used above. Clearly, in (7.72) the first term of the sum in the right-hand side above is zero, when the zero-level of $\phi$ is equal to an (ideal) object boundary or contour due to $g = 0$. The second term attracts the contour toward object boundaries; see [6] for more details also on a solution theory of the Hamilton–Jacobi equation based on viscosity solutions.

## Shape optimization and edge detector–based segmentation

The first papers that utilized the level-set concept along the lines indicated in the previous section are [10, 13, 39, 44, 45].

The velocity function $F$ proposed in [10] is given by

$$F = g\left(\operatorname{div}\left(\frac{\nabla\phi}{|\nabla\phi|}\right) + \mu\right), \tag{7.73}$$

In contrast with (7.73), as we have shown previously the velocity function

$$F = \operatorname{div}\left(g\frac{\nabla\phi}{|\nabla\phi|}\right) = g\operatorname{div}\left(\frac{\nabla\phi}{|\nabla\phi|}\right) + \frac{1}{|\nabla\phi|}\langle\nabla g, \nabla\phi\rangle, \tag{7.74}$$

can be interpreted as the gradient direction for the cost functional

$$J_{\mathrm{gAC}}(\boldsymbol{c}(t)) = \int_{t_0}^{T} g(|\nabla I(\boldsymbol{c}(t))|)\,|\boldsymbol{c}'(t)|\,dt. \tag{7.75}$$

In [28] it is argued that $J_{\mathrm{gAC}}$ is equivalent to

$$J_{\mathrm{gAC}}(\Gamma) = \int_{\Gamma} g\,dS, \tag{7.76}$$

where $S$ denotes the arc-length measure on $\Gamma$. In (7.76) $\boldsymbol{c}$ is replaced by $\Gamma \subset \Omega$, which represents now a genuine geometrical variable. It is assumed that $\Gamma = \partial\Omega$ is the boundary of an open set $\Omega \subset \mathbb{R}^2$. We call $\Gamma$ a *contour* in $\mathbb{R}^2$. The "driving force" $\mu$ can be modeled by adding the term

$$\mu \int_{\Omega} g\,d\boldsymbol{x}$$

to $J_{\mathrm{gAC}}(\Gamma)$. This gives

$$J_{\mathrm{gAC}}^{\mu}(\Gamma) = \int_{\Gamma} g\,dS + \mu \int_{\Omega} g\,d\boldsymbol{x}, \tag{7.77}$$

where $\Omega$ denotes the subset of $\mathbb{R}^2$ with boundary $\partial\Omega = \Gamma$.

This view opens a new perspective on energy-minimization–based image segmentation. The functional $J_{\mathrm{gAC}}^{\mu}$ represents a so-called *shape functional* as the unknown quantity is a geometric object (shape). The sensitivity analysis of $J_{\mathrm{gAC}}^{\mu}$ (such as computing first and second order derivatives) can be performed by means of shape sensitivity techniques; see [16, 59] and the many references therein.

Next we summarize some of the basic principles of shape sensitivity analysis that will then be applied to $J_{\mathrm{gAC}}^{\mu}$. Suppose that $\boldsymbol{V} : \mathbb{R}^2 \to \mathbb{R}^2$ is a given smooth vector field with compact support in $\mathbb{R}^2$. We study perturbations of $\Gamma$ by means of an initial value problem with right-hand side given by the perturbation vector field $\boldsymbol{V}$. In fact, we consider

$$\begin{cases} \boldsymbol{x}'(t) = \boldsymbol{V}\left(\boldsymbol{x}(t)\right) \\ \boldsymbol{x}(0) = \mathbf{x}, \end{cases} \tag{7.78}$$

with $\mathbf{x} \in \mathbb{R}^2$ given. The flow (or time-$t$ map) with respect to $V$ is defined as the mapping $T_t : \mathbb{R}^2 \to \mathbb{R}^2$, with

$$T_t(\mathbf{x}) = \boldsymbol{x}(t), \tag{7.79}$$

where $\boldsymbol{x}(t)$ is the solution to (7.78) at time $t$. For a contour $\Gamma$, we define

$$\Gamma_t = \{T_t(\mathbf{x}) : \mathbf{x} \in \Gamma\} = T_t(\Gamma) \tag{7.80}$$

and similarly $\Omega_t = T_t(\Omega)$ for an arbitrary open set $\Omega$. If $V \in \mathcal{C}_0^k(\mathbb{R}^2, \mathbb{R}^2)$, then $T_t \in \mathcal{C}^k(\mathbb{R}^2, \mathbb{R}^2)$. Thus, smoothness properties of $\Gamma$ are inherited by $\Gamma_t$ provided that the vector field $V$ is smooth enough.

Suppose we are given a functional $\mathcal{J} : C \to \mathbb{R}$, where $C$ is an appropriate set of contours. We define the *Eulerian (semi)derivative* of $\mathcal{J}$ at a contour $\Gamma$ in direction of a perturbation vector field $V$ by

$$d\mathcal{J}(\Gamma; V) = \lim_{t \downarrow 0} \frac{1}{t}\Big(\mathcal{J}(\Gamma_t) - \mathcal{J}(\Gamma)\Big). \tag{7.81}$$

Let $B$ be a Banach space of perturbation vector fields. The functional $\mathcal{J}$ is *shape differentiable at* $\Gamma$ in $B$ if $d\mathcal{J}(\Gamma; V)$ exists for all $V \in B$ and the mapping $V \mapsto d\mathcal{J}(\Gamma; V)$ is linear and continuous on $B$. We use the analogous definition for functionals $\mathcal{J}(\Omega)$, which depend on an open set $\Omega$ as independent variable instead of a contour $\Gamma$.

Next we present some results from shape calculus that will become useful later on. For a domain integral with a domain-independent integrand $\varphi \in W_{\text{loc}}^{1,1}(\mathbb{R}^2)$, with $\Omega \subset \mathbb{R}^2$ open and bounded, one has that

$$\mathcal{J}(\Omega) = \int_\Omega \varphi \, d\boldsymbol{x}$$

is shape differentiable for perturbation vector fields $V \in \mathcal{C}_0^1(\mathbb{R}^2; \mathbb{R}^2)$. The Eulerian semiderivative of $\mathcal{J}$ is given by

$$d\mathcal{J}(\Omega; V) = \int_\Omega \text{div}\,(\varphi\,V)\,d\boldsymbol{x}. \tag{7.82a}$$

If $\Gamma = \partial\Omega$ is of class $\mathcal{C}^1$, then

$$d\mathcal{J}(\Omega; V) = \int_\Gamma \varphi\,V \cdot N\,dS, \tag{7.82b}$$

where $N$ denotes the exterior unit normal vector to $\Omega$; see Propositions 2.45 and 2.46 in [59].

For a vector field $V \in \mathcal{C}_0^1(\mathbb{R}^2; \mathbb{R}^2)$ and an open set of class $\mathcal{C}^2$ with boundary $\Gamma$, the tangential divergence of $V$ is defined by

$$\text{div}_{\,\Gamma} V = (\text{div}\,V - \langle DV \cdot N, N \rangle)\Big|_{\Gamma}, \tag{7.83}$$

where $DV$ denotes the Jacobian matrix of $V$. With this definition we are able to study the Eulerian semiderivative of the boundary functional

$$J(\Gamma) = \int_\Gamma \varphi \, dS \tag{7.84}$$

where $\varphi \in W^{2,1}_{\mathrm{loc}}(\mathbb{R}^2)$ and $\Gamma$ is a contour of class $C^1$. In fact this functional is shape differentiable for perturbation vector fields $V \in C^1_0(\mathbb{R}^2; \mathbb{R}^2)$ with

$$dJ(\Gamma; V) = \int_\Gamma \left( \nabla\varphi \cdot V + \varphi \operatorname{div}_\Gamma V \right) dS; \tag{7.85}$$

see Sections 2.18 and 2.19 in [59].

Using tangential calculus (see Sections 2.19 and 2.20 in [16, 59] or Section 2 in [28]), one can show that the Eulerian derivative of the cost functional (7.84) is equivalent to

$$dJ(\Gamma; V) = \int_\Gamma \left( \frac{\partial\varphi}{\partial n} + \varphi\kappa \right) V \cdot N \, dS. \tag{7.86}$$

It is also useful to be able to calculate sensitivities for more general functionals of the form

$$J(\Omega) = \int_\Omega \varphi(\Omega, x) \, dx \tag{7.87}$$

or

$$J(\Gamma) = \int_\Gamma \psi(\Gamma, x) \, dS(x), \tag{7.88}$$

where the functions $\varphi(\Omega) : \Omega \to \mathbb{R}$ and $\psi(\Gamma) : \Gamma \to \mathbb{R}$ themselves depend on the geometric variables $\Omega$ and $\Gamma$, respectively. In this case, formulas (7.82) and (7.86) have to be corrected by terms that take care of the derivatives of $\varphi$ and $\psi$ with respect to $\Omega$ or $\Gamma$. For this purpose, the following two variants of derivatives of a geometry dependent function with respect to the geometry are considered: Suppose $\psi(\Gamma) \in B(\Gamma)$ for all $\Gamma \in C$, where $B(\Gamma)$ is some appropriate Banach space of functions on $\Gamma$ and let $V \in C^1_0(\mathbb{R}^2, \mathbb{R}^2)$. We set $\psi^t := \psi(\Gamma_t) \circ T_t(V)$ and $\psi^0 := \psi(\Gamma)$, and we assume that $\psi^t \in B(\Gamma_t)$ for all $0 < t < T$ with some $T > 0$. If the limit

$$\dot\psi(\Gamma; V) = \lim_{t\downarrow 0} \frac{1}{t} \left( \psi^t - \psi^0 \right) \tag{7.89}$$

exists in the strong (weak) topology on $B(\Gamma)$, then $\dot\psi(\Gamma; V)$ is called the strong (weak) *material derivative* of $\psi$ at $\Gamma$ in direction $V$.

The analogous definition holds for functions $\varphi(\Omega)$ that are defined on open sets and not on contours.

The material derivative is the derivative of $\varphi$ (or $\psi$) with respect to the geometry for a moving (Lagrangian) coordinate system. Let us first consider the case of a domain function $\varphi : \Omega \to \mathbb{R}$. It is easily seen that, for the special case where $\varphi$ is independent of $\Omega$, we find

$$\dot{\varphi}(\Omega; \boldsymbol{V}) = \dot{\varphi}(\boldsymbol{V}) = \nabla \varphi \cdot \boldsymbol{V}.$$

For a function, that does not depend on $\Omega$, any reasonable derivative with respect to $\Omega$ in a fixed (Eulerian) coordinate system must be 0. It is therefore natural to subtract the term $\nabla \varphi \cdot \boldsymbol{V}$ from $\dot{\varphi}$ to define a derivative of $\varphi$ with respect to $\Omega$ in a stationary coordinate system. This is the idea of the following definition: Suppose, the weak material derivative $\dot{\varphi}(\Omega; \boldsymbol{V})$ and the expression $\nabla \varphi(\Omega) \cdot \boldsymbol{V}$ exist in $B(\Omega)$. Then, we set

$$\varphi'(\Omega; \boldsymbol{V}) = \dot{\varphi}(\Omega; \boldsymbol{V}) - \nabla \varphi \cdot \boldsymbol{V} \qquad (7.90)$$

and we call $\varphi'(\Omega; \boldsymbol{V})$ the *shape derivative* of $\varphi$ at $\Omega$ in direction $\boldsymbol{V}$.

Note that

$$\varphi'(\Omega; \boldsymbol{V}) = \varphi'(\boldsymbol{V}) = 0$$

for any function $\varphi$ which does not depend on $\Omega$.

For boundary functions $\psi(\Gamma) : \Gamma \to \mathbb{R}$, the expression $\nabla \psi \cdot \boldsymbol{V}$ does not make sense. In this case, we define the shape derivative as

$$\psi'(\Gamma; \boldsymbol{V}) = \dot{\psi}(\Gamma; \boldsymbol{V}) - \nabla_\Gamma \psi \cdot \boldsymbol{V}\big|_\Gamma, \qquad (7.91)$$

where the tangential gradient $\nabla_\Gamma \psi$ is defined by

$$\nabla_\Gamma \psi = \nabla \tilde{\psi}\big|_\Gamma - \frac{\partial \tilde{\psi}}{\partial \boldsymbol{n}} \boldsymbol{N} \qquad (7.92)$$

on $\Gamma$, where $\tilde{\psi}$ denotes an arbitrary smooth extension of $\psi$. It can be shown that the definition (7.92) does not depend on the specific choice of the extension.

With these definitions, the Eulerian derivatives for the shape functionals (7.87) and (7.88) are computed as follows:

- Suppose $\varphi = \varphi(\Omega)$ is given such that the weak $L^1$-material derivative $\dot{\varphi}(\Omega; \boldsymbol{V})$ and the shape derivative $\varphi'(\Omega; \boldsymbol{V}) \in L^1(\Omega)$ exist. Then, the cost functional (7.87) is shape differentiable and

$$d\mathcal{J}(\Omega; \boldsymbol{V}) = \int_\Omega \varphi'(\Omega; \boldsymbol{V}) \, d\boldsymbol{x} + \int_\Gamma \varphi \boldsymbol{V} \cdot \boldsymbol{N} \, dS. \qquad (7.93)$$

- For boundary functions $\psi(\Gamma)$ we get under the same technical assumptions for the cost functional (7.88):

$$d\mathcal{J}(\Gamma; \boldsymbol{V}) = \int_\Gamma \psi'(\Gamma; \boldsymbol{V}) \, dS + \int_\Gamma \kappa \psi \boldsymbol{V} \cdot \boldsymbol{N} \, dS. \qquad (7.94)$$

If $\psi(\Gamma) = \varphi(\Omega)\big|_\Gamma$, then we have

$$d\mathcal{J}(\Gamma; \boldsymbol{V}) = \int_\Gamma \varphi'(\Omega; \boldsymbol{V})\big|_\Gamma \, dS + \int_\Gamma \left( \frac{\partial \varphi}{\partial \boldsymbol{n}} + \kappa \varphi \right) \boldsymbol{V} \cdot \boldsymbol{N} \, dS. \qquad (7.95)$$

Finally, the Hadamard–Zolesio structure theorem [16, Theorem 3.6 and Corollary 1, p. 348f] states that the Eulerian semiderivative of a domain or boundary functional has always a representation of the form

$$dJ(\Omega; \boldsymbol{V}) = \left\langle G, \boldsymbol{V} \cdot \boldsymbol{N} \right\rangle_{\mathcal{C}^{-k}(\Gamma),\mathcal{C}^{k}(\Gamma)} = \left\langle G\boldsymbol{N}, \boldsymbol{V} \right\rangle_{\mathcal{C}_2^{-k}(\Gamma),\mathcal{C}_2^{k}(\Gamma)}, \qquad (7.96)$$

that is, the Eulerian derivative is concentrated on $\Gamma$ and can be identified with the *normal vector field* $G\boldsymbol{N}$ on $\Gamma$. The expression

$$D_\Gamma J(\Omega) = G\boldsymbol{N} \qquad (7.97)$$

is called the *shape gradient* of $J$ at $\Omega$.

Using shape sensitivity one finds that the Eulerian semiderivative of

$$J^{\mu}_{\text{gAC}}(\Gamma) = \int_\Gamma g\, dS + \mu \int_\Omega g\, d\boldsymbol{x}, \qquad (7.98)$$

is given by

$$dJ^{\mu}_{\text{gAC}}(\Gamma; \boldsymbol{V}) = \left\langle D_\Gamma J, \boldsymbol{V} \right\rangle = \int_\Gamma \left\langle \left( \frac{\partial g}{\partial \boldsymbol{n}} + g\,(\kappa + \mu) \right) \boldsymbol{N}, \boldsymbol{V} \right\rangle dS. \qquad (7.99)$$

Observe that (7.99) coincides with (7.72).

In order to establish a Newton-type method, the second Eulerian semiderivative needs to be studied. For this purpose, let

$$d^2 J(\Gamma; \boldsymbol{V}; \boldsymbol{W}) = d\big(dJ(\Omega; \boldsymbol{V})\big)(\Omega; \boldsymbol{W}),$$

be the second Eulerian semiderivative of the cost functional $J : C \to \mathbb{R}$. In general, the second Eulerian semiderivative is not symmetric in the two arguments $\boldsymbol{V}$ and $\boldsymbol{W}$ and it depends not only on $\boldsymbol{V}|_\Gamma$ and $\boldsymbol{W}|_\Gamma$. From the subsequent computation we shall see, however, that for perturbation vector fields $(\boldsymbol{V}_F, \boldsymbol{V}_G)$ of the form (7.109), the second Eulerian semiderivative is symmetric in $(\boldsymbol{V}_F, \boldsymbol{V}_G)$ and depends only on $F$ and $G$.

In fact, let $F : \Gamma \to \mathbb{R}$ and $G : \Gamma \to \mathbb{R}$ be given functions. A one-to-one correspondence between scalar velocity functions and a certain class of perturbation vector fields is as follows. Let $\tilde{F}$ and $\tilde{G}$ denote extensions of $F$ and $G$, respectively, which are constructed as solutions to the transport equations

$$\langle \nabla\tilde{F}, \nabla b_\Gamma \rangle = 0 \text{ on } \mathbb{R}^2; \quad \tilde{F}\big|_\Gamma = F \qquad (7.100)$$

and

$$\langle \nabla\tilde{G}, \nabla b_\Gamma \rangle = 0 \text{ on } \mathbb{R}^2; \quad \tilde{G}\big|_\Gamma = G. \qquad (7.101)$$

Here the *signed distance function* $b_\Omega$ of a bounded open set $\Omega \subset \mathbb{R}^2$ is defined as

$$b_\Omega(\boldsymbol{x}) = d_\Omega(\boldsymbol{x}) - d_{\mathbb{R}^2 \setminus \Omega}(\boldsymbol{x}) \qquad (7.102)$$

with the *distance function* $d_A$ of a subset $A \subset \mathbb{R}^2$ defined as

$$d_A(\boldsymbol{x}) = \inf_{\boldsymbol{y} \in A} |\boldsymbol{y} - \boldsymbol{x}|. \qquad (7.103)$$

Whenever $\Gamma = \partial\Omega$, $d_\Omega$ can be expressed in terms of $\Gamma$:

$$b_\Omega(\boldsymbol{x}) = \begin{cases} d_\Gamma(\boldsymbol{x}) & \text{for } \boldsymbol{x} \in \text{int}(\mathbb{R}^2 \setminus \Omega). \\ 0 & \text{for } \boldsymbol{x} \in \Gamma, \\ -d_\Gamma(\boldsymbol{x}) & \text{for } \boldsymbol{x} \in \Omega. \end{cases} \tag{7.104}$$

We shall use the notation $b_\Gamma = b_\Omega$. Note in particular that

$$b_\Gamma\big|_\Gamma = 0. \tag{7.105}$$

By Rademacher's theorem, $b_\Gamma$ is differentiable almost everywhere on $\mathbb{R}^2$ with $|\nabla b_\Gamma| = 1$ a.e. on $\mathbb{R}^2 \setminus \Gamma$. If $\text{meas}(\Gamma) = 0$, then

$$|\nabla b_\Gamma|^2 = 1 \text{ a.e. on } \mathbb{R}^2. \tag{7.106}$$

Further, $\nabla b_\Gamma$ can be considered as an extension of the unit normal vector field $\boldsymbol{N}$ to a neighborhood of $\Gamma$. One has

$$\nabla b_\Gamma\big|_\Gamma = \boldsymbol{N}. \tag{7.107}$$

Moreover, the second fundamental form of $\Gamma$ can be expressed in terms of $b_\Gamma$. For a $C^2$-submanifold $\Gamma \subset \mathbb{R}^2$, there holds

$$\Delta b_\Gamma\big|_\Gamma = \kappa. \tag{7.108}$$

Summarizing, important geometrical information such as normals and curvature of $\Gamma$ can be expressed by means of $b_\Gamma$.

Coming back to (7.100) and (7.101), note that $\Gamma$ is non-characteristic with respect to the transport equation. Thus, (7.100) and (7.101) have unique solutions, at least locally in some neighborhood of $\Gamma$ that is small enough such that the characteristics of (7.100) (which are straight lines) do not intersect. Based on these (unique) solutions, the perturbation vector fields

$$\boldsymbol{V}_F = \tilde{F} \nabla b_\Gamma, \text{ and } \boldsymbol{V}_G = \tilde{G} \nabla b_\Gamma \tag{7.109}$$

are defined on some neighborhood of $\Gamma$ on which $\tilde{F}$, $\tilde{G}$, and $\nabla b_\Gamma$ are smooth. Outside this neighborhood we assume that $\boldsymbol{V}_F$ and $\boldsymbol{V}_G$ are extended in some smooth way. Note that the construction of $\boldsymbol{V}_F$ and $\boldsymbol{V}_G$ is such that

$$\boldsymbol{V}_F \cdot \boldsymbol{N} = F \text{ and } \boldsymbol{V}_G \cdot \boldsymbol{N} = G \text{ on } \Gamma. \tag{7.110}$$

With these considerations, the second Eulerian semiderivative of $J_{\text{gAC}}^\mu$ is given by

$$d^2 J_{\text{gAC}}^\mu(\Gamma; F, G) = \int_\Gamma \left[ \left( \frac{\partial^2 g}{\partial n^2} + (2\kappa + \mu) \frac{\partial g}{\partial n} + \mu\,\kappa\,g \right) F\,G + g\nabla_\Gamma F \cdot \nabla_\Gamma G \right] dS. \tag{7.111}$$

**Level-set–based descent framework in shape optimization**

Based on the results collected in the previous section, a level-set–based descent method for minimizing $J^{\mu}_{\text{gAC}}$ is as follows.

**Level-set–based descent method.**

1. **Initialization.** Choose an initial (closed) contour $\Gamma_0$. Initialize the level set function $\phi^0$ such that $\Gamma_0$ is the zero level set of $\phi^0$; set $k = 0$. Choose a bandwidth $w \in \mathbb{N}$.
2. **Descent direction.** Find the zero level set $\Gamma_k$ of the actual level set function $\phi^k$. Solve

$$B(\Gamma_k; F^k, G) = -dJ^{\mu}_{\text{gAC}}(\Gamma^k; V_G) \quad \text{for all } G$$

   to obtain the descent direction $F^k$.
3. **Extension.** Extend $F^k$ to a band around the actual zero level set $\Gamma_k$ with bandwidth $w$ yielding $F^k_{ext}$.
4. **Update.** Perform a time step in the level set equation with speed function $F^k_{ext}$ to update $\phi^k$ on the band. Let $\hat{\phi}^{k+1}$ denote this update.
5. **Reinitialization.** Reinitialize $\hat{\phi}^{k+1}$ in order to obtain a signed distance function $\phi^{k+1}$ with zero level set given by the zero level set of $\hat{\phi}^{k+1}$. Set $k = k + 1$ and go to (2).

Subsequently the steps of the above algorithm are explained in some detail.

(i) In order to reduce the computational burden, usually $\phi^k$ is not defined on all of $\Omega$. Rather it is defined only in a band around $\Gamma_k$; see Figure 7.10. Given $\Gamma_0$, the initialization is done by utilizing the fast marching technique [56, 57] on the band around $\Gamma_0$ for solving the Eikonal equation

$$|\nabla \phi^0| = 1 \quad \text{with} \quad \phi^0 = 0 \text{ on } \Gamma_0.$$

Hence, the level set function is given by the signed distance function, i.e., $\phi^0 = b_{\Gamma_0}$. The same procedure is used for reinitialization. The latter step is necessary due to the fact that after several time steps in the level set equation and in particular after large time steps, the signed distance nature of the level set function is lost.

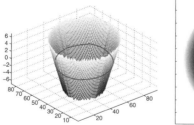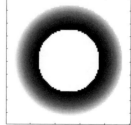

**Fig. 7.10.** A signed distance function defined only on a band around the zero-level set (red).

(ii) In step 2, a positive definite bilinear form $B(\Gamma; \cdot, \cdot)$ is used. If $B = \mathrm{id}$, then $B(\Gamma_k; F^k, G) = \langle F^k, G \rangle$ and $F^k$ corresponds with the direction of steepest descent (negative shape gradient direction). If

$$B(\Gamma_k; F^k, G) = d^2 J_{\mathrm{gAC}}^\mu(\Gamma_k; F^k, G)$$

is chosen (this is only possible if $d^2 J_{\mathrm{gAC}}^\mu(\Gamma_k; \cdot, \cdot)$ is positive definite), then $F^k$ corresponds with a shape Newton direction. Shape Newton–like directions are obtained by modifying $d^2 J_{\mathrm{gAC}}^\mu$ such that the resulting bilinear form $B$ is positive definite. In [28], it is demonstrated that at the expense of solving a (one-dimensional) elliptic partial differential equation on the manifold $\Gamma_k$, Newton-type methods usually require a significantly smaller number of iterations until successful termination and they are less parameter dependent. The latter aspect is related to the "variable-metric"-like aspect by choosing $B$ in dependence on $\Gamma_k$.

(iii) By the Hadamard–Zolesio structure theorem, the shape gradient and the shape Hessian are concentrated on $\Gamma_k$. Hence $F^k$ is also concentrated on $\Gamma_k$. The level set equation at $t$, however, is defined in $\Omega$ (or at least on a band around $\Gamma_k$). Hence, $F^k$ has to be extended onto $\Omega$ or the given band. Let $F_{\mathrm{ext}}^k$ denote the corresponding extension velocity. This extension is computed by solving the transport equation

$$\nabla F_{\mathrm{ext}}^k \cdot \nabla \phi^k = 0, \quad F_{\mathrm{ext}}^k|_{\Gamma_k} = F^k. \qquad (7.112)$$

(iv) Finally, the algorithm is stopped as soon as $\|F^k\|_{\Gamma_k}$ is smaller than some user-specified stopping tolerance.

(v) Details on a possible discretization as well as on an Armijo-type line search procedure for performing the time step in the level set equation can be found in [27, 28].

### 7.2.3 Approaches based on the Mumford–Shah functional

A widely used model in image segmentation and simultaneous denoising was introduced by Mumford and Shah in [50]:

$$J_{\mathrm{MS}}(I, \Gamma) = \frac{1}{2} \int_{\Omega \setminus \Gamma} (I - I_0)^2 d\boldsymbol{x} + \frac{\alpha}{2} \int_{\Omega \setminus \Gamma} |\nabla I|^2 d\boldsymbol{x} + \gamma \int_\Gamma 1 \, d\mathcal{H}_1, \qquad (7.113)$$

where $I$ denotes a reconstruction of the image, and $\Gamma$ represents the edge or discontinuity set in $I$. The minimization of the Mumford–Shah functional is delicate as it involves the two unknowns $I$ and $\Gamma$, which are of entirely different nature. Whereas $I$ is a function defined on a subset of $\mathbb{R}^n$, $\Gamma$ is a geometrical variable, a one-dimensional set.

Compared with the edge detector–based approach highlighted in Section 7.2.2, the Mumford–Shah approach has the ability to successfully segment images even without clear edges; see Figure 7.11, which shows the

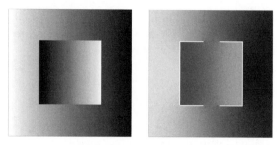

**Fig. 7.11.** Left: Original image. Right: Edge detector–based segmentation; see Section 7.2.2.

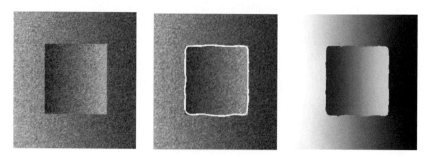

**Fig. 7.12.** Left: Noisy image. Middle: Mumford–Shah based segmentation. Right: Denoising result, i.e., concatenation of the reconstructions $I_k$, $k = 1, 2$.

result (right plot) for the edge detector–based approach. The Mumford–Shah result will be discussed later; see Figure 7.12.

With respect to existence of a minimizing pair $(I, \Gamma)$ in (7.113) the application of classic arguments in the calculus of variations utilizing minimizing sequences, compactness and lower semicontinuity properties of the objective functional fail because the map $\Omega \mapsto \int_{\Gamma} 1 \, d\mathcal{H}_1$ is not lower semicontinuous with respect to any compact topology. Combining the results in [2, 3], existence of a solution of (7.113) is shown with $I \in W^{1,2}(\Omega \setminus \Gamma) \cap L^{\infty}(\Omega)$ and $\Gamma \subset \Omega$, $\Gamma$ closed, and $\int_{\Gamma} 1 \, d\mathcal{H}_1 < +\infty$. Here $W^{1,2}(\Omega \setminus \Gamma)$ denotes the usual Sobolev space of square integrable functions over $\Omega \setminus \Gamma$, which admit a square integrable first derivative in the generalized sense; see [1]. For further regularity considerations concerning $\Gamma$, see [50] and, e.g., [7].

Several approaches for the numerical solution of (7.113) are available. In this context, the discretization of the edge set $\Gamma$ represents a significant challenge. Many approaches, therefore, try to avoid this difficulty by approximating the minimization of the Mumford–Shah functional by problems with functions as the only unknowns. However, there are very recent techniques which keep $\Gamma$ as a geometric variable, which utilize techniques from shape sensitivity analysis for computing sensitivities of $J_{\mathrm{MS}}$ with respect to $\Gamma$ and which use the level-set method as a numerical tool; see Section 7.2.2.

An approach avoiding the explicit use of $\Gamma$ is due to Ambrosio and Tortorelli [4]. Their technique replaces $\Gamma$ by an auxiliary function $\omega$ approximating the characteristic function $(1 - \chi_\Gamma)$, where $\chi_\Gamma(\boldsymbol{x}) = 1$ if $\boldsymbol{x} \in \Gamma$ and $\chi_\Gamma(\boldsymbol{x}) = 0$ else. The corresponding functional is

$$J^\epsilon_{\mathrm{MS}}(I,\omega) = \frac{1}{2}\int_\Omega (I - I_0)^2 d\boldsymbol{x} + \int_\Omega \omega^2 |\nabla I|^2 d\boldsymbol{x} + \int_\Omega \left( \epsilon |\nabla \omega|^2 + \frac{1}{4\epsilon}(v-1)^2 \right) d\boldsymbol{x}. \tag{7.114}$$

Note the dependence on $\epsilon$. In [46], formal arguments for $J^\epsilon_{\mathrm{MS}}(I,\omega)$ approaching $J_{\mathrm{MS}}(I,\Gamma)$ as $\epsilon \to 0$ are provided. A rigorous treatment can be found in [8].

Further approaches avoiding the explicit use of $\Gamma$ are based on second-order singular perturbations [8], the introduction of nonlocal terms [9], or use approximations by finite differences [14]. See [6] for an excellent overview, further details and references on these approximation techniques for the Mumford–Shah functional.

In [29], the edge set $\Gamma$ is kept as an explicit variable, and a shape sensitivity–based minimization of (7.113) is proposed. It is assumed that $\Gamma$ is the boundary of an open set $\Omega_1 \subset \Omega$ and that the minimization problem for $J_{\mathrm{MS}}(I,\Gamma)$ can be written as

$$\inf_{(I,\Gamma)\in W^{1,2}(\Omega\setminus\Gamma)\times\mathcal{E}} J_{\mathrm{MS}}(I,\Gamma) = \inf_{\Gamma\in\mathcal{E}} \min_{I\in W^{1,2}(\Omega\setminus\Gamma)} J_{\mathrm{MS}}(I,\Gamma), \tag{7.115}$$

where $\mathcal{E}$ represents the set of admissible edges. Let $\Omega_2$ denote the complement of $\Omega_1$ in $\Omega$. The splitting of the minimization process in (7.115) allows one to consider, for fixed $\Gamma$, the minimization problem

$$\min_{I\in W^{1,2}(\Omega\setminus\Gamma)} J_{\mathrm{MS}}(I,\Gamma). \tag{7.116}$$

Its Euler–Lagrange equations are

$$\begin{cases} -\alpha \Delta I_k(\Gamma) + I_k(\Gamma) = I_0 \text{ on } \Omega_k, \\ \frac{\partial I_k(\Gamma)}{\partial \boldsymbol{n}_k} = 0 \text{ on } \partial\Omega_k \end{cases} \tag{7.117}$$

for $k = 1, 2$. Here $\frac{\partial}{\partial \boldsymbol{n}_k}$ denotes the derivative with respect to the exterior normal direction $\boldsymbol{N}_k$ to $\partial\Omega_k$. On $\Gamma$ we have $\boldsymbol{N}_1 = -\boldsymbol{N}_2$. With this, (7.115) can be formulated as the shape optimization problem of minimizing the functional

$$\hat{J}_{\mathrm{MS}}(\Gamma) = \sum_{k=1}^2 \int_{\Omega_k} \left( \frac{1}{2}|I_k(\Gamma) - I_0|^2 + \frac{\alpha}{2}|\nabla I_k(\Gamma)|^2 \right) d\boldsymbol{x} + \gamma \int_\Gamma 1\, d\mathcal{H}_1 \tag{7.118}$$

over $\Gamma \in \mathcal{E}$.

The Eulerian semiderivative of $\hat{J}_{\mathrm{MS}}$ is given by

$$d\hat{J}_{\mathrm{MS}}(\Gamma; \boldsymbol{V}_F) = \int_\Gamma \left( \frac{1}{2}\left[|I - I_0|^2\right] + \frac{\alpha}{2}\left[|\nabla_\Gamma I(\Gamma)|^2\right] + \gamma\kappa \right) F\, d\mathcal{H}_1 \tag{7.119}$$

with $\boldsymbol{V}_F$ according to (7.109) and $[\![|I - I_0|^2]\!] = |I_1 - I_0|^2 - |I_2 - I_0|^2$ and $[\![|\nabla_\Gamma I(\Gamma)|^2]\!] = |\nabla_\Gamma I_1(\Gamma)|^2 - |\nabla_\Gamma I_2(\Gamma)|^2$ the jumps of $|I - I_0|^2$ and $|\nabla_\Gamma I|^2$, respectively, across $\Gamma$.

With this information and the choice $B = $ id, a level-set–based shape gradient method for the minimization of $J_{\mathrm{MS}}$ utilizing the descent algorithm in Section 7.2.2 can be employed. As before, the level set method is used to represent and update the geometry within an iterative scheme.

In [29], a shape Hessian–based choice for $B$ is proposed and numerically realized. In contrast with Newton-type methods in the context of edge detectors (see Section 7.2.2), the Hessian in the case of the Mumford–Shah functional admits no explicit discrete representation. Rather its application to a perturbation velocity (this corresponds with a "matrix-times-vector"-product in the discrete setting) is available at reasonable computational cost. In Figure 7.12 the result obtained by the level-set–based descent algorithm in Section 7.2.2 using a shape Newton–based minimization of the $J_{\mathrm{MS}}$ is shown. From this figure, the simultaneous segmentation and denoising ability of the Mumford–Shah approach becomes apparent; see [29] for details.

## Acknowledgment

The first author gratefully acknowledges financial support from the Austrian Science Fund FWF under START-program Y305 "Interfaces and Free Boundaries."

## References

[1] R.A. Adams. *Sobolev spaces*. Academic Press, New York-London, 1975. Pure and Applied Mathematics, Vol. 65.

[2] L. Ambrosio. A compactness theorem for a new class of functions of bounded variation. *Bollettino dell'Unione Matematica Italiana*, VII(4):857–881, 1989.

[3] L. Ambrosio. Existence theory for a new class of variational problems. *Archive for Rational Mechanics and Analysis*, 111:291–322, 1990.

[4] L. Ambrosio and V.M. Tortorelli. Approximation of functionals depending on jumps by elliptic functionals via Γ-convergence. *Communications on Pure & Applied Mathematics*, 43(8):999–1036, 1990.

[5] G. Aubert and L. Blanc-Féraud. Some remarks on the equivalence between 2d and 3d classical snakes and geodesic active contours. *International Journal of Computer Vision*, 34(1):19–28, 1999.

[6] G. Aubert and P. Kornprobst. *Mathematical Problems in Image Processing*. Springer-Verlag, New York, 2002. Partial differential equations and the calculus of variations.

[7] A. Bonnet. On the regularity of the edge set of Mumford–Shah minimizers. *Progress in Nonlinear Differential Equations*, 25:93–103, 1996.

[8]  A. Braides. *Approximation of Free-Discontinuity Problems*, volume 1694 of *Lecture Notes in Mathematics*. Springer-Verlag, Berlin, 1998.

[9]  A. Braides and G. Dal Maso. Non-local approximation of the Mumford-Shah functional. *Calculus of Variations and Partial Differential Equations*, 5(4):293–322, 1997.

[10]  V. Caselles, F. Catté, T. Coll, and F. Dibos. A geometric model for active contours in image processing. *Numerische Mathematik*, 66(1):1–31, 1993.

[11]  V. Caselles and L. Garrido. A contrast invariant approach to motion estimation. Scale Space 2005 7-9 April, Hofgeismar, Germany, 2005.

[12]  V. Caselles, R. Kimmel, and G. Sapiro. Geodesic active contours. In *Proceedings of the 5th International Conference on Computer Vision*, pages 694–699. IEEE Computer Society Press, 1995.

[13]  V. Caselles, R. Kimmel, and G. Sapiro. Geodesic active contours. *International Journal of Computer Vision*, 22(1):61–79, 1997.

[14]  A. Chambolle. Image segmentation by variational methods: Mumford and Shah functional and the discrete approximations. *SIAM Journal on Applied Mathematics*, 55(3):827–863, 1995.

[15]  G.E. Christensen. *Deformable Shape Models for Anatomy*. PhD Thesis. Sever Institute of Technology, Washington University, 1994.

[16]  M. Delfour and J.-P. Zolesio. *Shapes and Geometries. Analysis, Differential Claculus and Optimization*. SIAM Advances in Design and Control. SIAM, Philadelphia, 2001.

[17]  M. Droske and M. Rumpf. A variational approach to non-rigid morphological registration. *SIAM Journal on Applied Mathematics*, 64:668–687, 2004.

[18]  M. Droske and W. Ring. A mumford-shah level-set approach for geometric image registration. Preprint 99, DFP-SPP 1114, April 2005.

[19]  B. Fischer and J. Modersitzki. Fast inversion of matrices arising in image processing. *Numerical Algorithms*, 22:1–11, 1999.

[20]  B. Fischer and J. Modersitzki. Combination of automatic non-rigid and landmark based registration: the best of both worlds. In M. Sonka and J.M. Fitzpatrick, editors, *Medical Imaging 2003: Image Processing. Proceedings of the SPIE 5032*, pages 1037–1048, 2003.

[21]  M. Fitzpatrick, D.L.G. Hill, and C.R. Maurer Jr. Image registration. In M. Sonka and J.M. Fitzpatrick, editors, *Handbook of Medical Imaging*, volume II, Chapter 8. SPIE Press, 2000.

[22]  M. Fitzpatrick and J.B. West. Predicting error in rigid-body point-based registration. *IEEE Transactions on Medical Imaging*, 17:694–702, 1998.

[23]  W. Hackbusch. *Iterative Solution of Large Sparse Systems of Equations*. Springer, Berlin, 1993.

[24]  S. Haker. Mass preserving mappings and image registration. *MICCAI 2001*, pages 120–127, 2001.

[25]  S. Henn. A multigrid method for a fourth-order diffusion equation with application to image processing. *SIAM Journal on Scientific Computing*, 27(3):831–849, 2005.

[26]  S. Henn and K. Witsch. Iterative multigrid regularization techniques for image matching. *SIAM Journal on Scientific Computing*, 23(4):1077–1093, 2001.

[27]  M. Hintermüller and W. Ring. Numerical aspects of a level set based algorithm for state constrained optimal control problems. *Computer Assisted Mechanics and Engineering Sciences Journal*, 3, 2003.

[28] M. Hintermüller and W. Ring. A second order shape optimization approach for image segmentation. *SIAM Journal on Applied Mathematics*, 64(2):442–467, 2003.

[29] M. Hintermüller and W. Ring. An inexact Newton-CG-type active contour approach for the minimization of the Mumford-Shah functional. *Journal of Mathematical Imaging and Vision*, 20:19–42, 2004.

[30] K. Höllig. *Finite Element Methods with B-Splines*, volume 26 of *Frontiers in Applied Mathematics*. SIAM, Philadelphia, 2003.

[31] B.K.P. Horn and B.G. Schunck. Determining optical flow. *Artificial Intelligence*, 23:185–203, 1981.

[32] M. Kass, A. Witkin, and D. Terzopoulos. Snakes; active contour models. *International Journal of Computer Vision*, 1:321–331, 1987.

[33] S.L. Keeling. Image similarity based on intensity scaling. *Journal on Mathematical Imaging and Vision*, 29:21–34, 2007.

[34] S.L. Keeling. Generalized rigid and generalized affine image registration and interpolation by geometric multigrid. *Journal of Mathematical Imaging and Vision*, 29:163–183, 2007.

[35] S.L. Keeling and G. Haase. Geometric multigrid for high-order regularizations of early vision problems. *Applied Mathematics and Computation*, 184:536–556, 2007.

[36] S.L. Keeling. Medical image registration and interpolation by optical flow with maximal rigidity. In O. Scherzer, editor, *Mathematical Methods in Registration for Applications in Industry and Medicine*, Mathematics in Industry. Springer, Berlin, 1995.

[37] S.L. Keeling and W. Ring. Medical image registration and interpolation by optical flow with maximal rigidity. *Journal of Mathematical Imaging and Vision*, 23:47–65, 2005.

[38] S. Kichenassamy, A. Kumar, P. Olver, A. Tannenbaum, and A. Yezzi. Gradient flows and geometric active contour models. In *Proceedings of the 5th International Conference on Computer Vision*. IEEE Computer Society Press, 1995.

[39] S. Kichenassamy, A. Kumar, P. Olver, A. Tannenbaum, and A. Yezzi. Conformal curvature flows: from phase transitions to active vision. *Archive for Rational Mechanics and Analysis*, 134:275–301, 1996.

[40] M. Lefébure and L.D. Cohen. Image registration, optical flow and local rigidity. *Journal of Mathematical Imaging and Vision*, 14(2):131–147, 2001.

[41] M.M. Lipschutz. *Differential Geometry, Theory and Problems*. Schaum's Outline Series. McGraw-Hill, New York, 1969.

[42] J.A. Little and D.L.G. Hill. Deformations incorporating rigid structures. *Computer Vision and Image Understanding*, 66(2):223–232, 1997.

[43] F. Maes and A. Collignon. Multimodality image registration by maximization of mutual information. *IEEE Transactions on Medical Imaging*, 16:187—209, 1997.

[44] R. Malladi, J. Sethian, and B.C. Vemuri. Evolutionary fronts for topology independent shape modeling and recovery. In *Proceedings of the 3rd ECCV, Stockholm, Sweden*, pages 3–13. 1994.

[45] R. Malladi, J. Sethian, and B.C. Vemuri. Shape modeling with front propagation: A level set approach. *IEEE Transactions on Pattern Analysis and Machine Intelligence*, 13:158–175, 1995.

[46] R. March. Visual reconstructions with discontinuities using variational methods. *Image and Vision Computing*, 10:30–38, 1992.

[47] J. Modersitzki. *Numerical Methods for Image Registration*. Oxford University Press, Oxford, 2004.

[48] J. Modersitzki. Flirt with rigidity – image registration with a local non-rigidity penalty. *International Journal of Computer Vision*, pages 1–18, 2007.

[49] J.-M. Morel and S. Solimini. *Variational Methods in Image Segmentation*, volume 14 of *Progress in Nonlinear Differential Equations and their Applications*. Birkhäuser Boston Inc., Boston, Massachusetts, 1995. With seven image processing experiments.

[50] D. Mumford and J. Shah. Optimal approximations by piecewise smooth functions and associated variational problems. *Communications on Pure & Applied Mathematics*, 42(5):577–685, 1989.

[51] S.J. Osher and R.P. Fedkiw. *Level Set Methods and Dynamic Implicit Surfaces*. Springer Verlag, New York, 2002.

[52] S.J. Osher and J.A. Sethian. Fronts propagating with curvature-dependent speed: algorithms based on Hamilton-Jacobi formulations. *Journal on Computational Physics*, 79(1):12–49, 1988.

[53] W. Peckar, C. Schnörr, K. Rohr, and H.S. Stiehl. Parameter-free elastic deformation approach for 2d and 3d registration using prescribed displacements. *Journal of Mathematical Imaging and Vision*, 10:143–162, 1999.

[54] D. Rueckert, B. Clarkson, D.L.G. Hill, and D.J. Hawkes. Non-rigid registration using higher-order mutual information. In K.M. Hanson, editor, *Medical Imaging 2000: Image Processing, Proceedings of SPIE*, volume 3979, pages 438–447, 2000.

[55] D. Rueckert, L.I. Sonoda, C. Hayes, D.L.G. Hill, M.O. Leach, and D.J. Hawkes. Non-rigid registration using free-form deformations: Application to breast mr images. *IEEE Transactions on Medical Imaging*, 18(8):712–721, 1999.

[56] J.A. Sethian. A fast marching level set method for monotonically advancing fronts. *Proceedings of the National Academy of Sciences*, 93(4):1591–1595, 1996.

[57] J.A. Sethian. Fast marching methods. *SIAM Review*, 41(2):199–235 (electronic), 1999.

[58] J.A. Sethian. *Level Set Methods and Fast Marching Methods*. Cambridge University Press, Cambridge, second edition, 1999. Evolving interfaces in computational geometry, fluid mechanics, computer vision, and materials science.

[59] J. Sokołowski and J.-P. Zolésio. *Introduction to Shape Optimization*. Springer-Verlag, Berlin, 1992. Shape sensitivity analysis.

[60] C. Studholme, D.L.G. Hill, and D.J. Hawkes. An overlap invariant entropy measure of 3d medical image alignment. *Pattern Recognition*, 32:71–86, 1999.

[61] P.M. Thompson, M.S. Mega, K.L. Narr, E.R. Sowell, R.E. Blanton, and A.W. Toga. Brain image analysis and atlas construction. In M. Sonka and J.M. Fitzpatrick, editors, *Handbook of Medical Imaging*, volume II. SPIE Press, 2000.

[62] P.A. Viola. *Alignment by Maximization of Mutual Information: Foundations and extensions*. PhD thesis, Massachusetts Institute of Technology, Cambridge, Massachusetts, 1995.

[63] J. Weickert, A. Bruhn, N. Papenberg, and T. Brox. Variational optical flow computation: From continuous models to algorithms. In O. Scherzer, editor, *Mathematical Method for Registration and Applications to Medical Imaging*, volume 10 of *Mathematics in Industry*. Springer, Berlin, 2006.

[64] W.M. Wells III, P. Viola, H. Atsumi, S. Nakajima, and R. Kikinis. Multi-modal volume registration by maximization of mutual information. *Medical Image Analysis*, 1:35–51, 1996.

[65] J. West, J.M. Fitzpatrick, M.Y. Wang, B.M. Dawant, C.R. Maurer Jr., R.M. Kessler, R.J. Maciunas, C. Barillot, D. Lemoine, A. Collignon, F. Maes, P. Suetens, D. Vandermeulen, P.A. van den Elsen, S. Napel, T.S. Sumanaweera, B. Harkness, P.F. Hemler, D.L.G. Hill, D.J. Hawkes, C. Studholme, J.B.A. Maintz, M.A. Viergever, G. Malandain, X. Pennec, M.E. Noz, G.Q. Maguire Jr., M. Pollack, C.A. Pelizzari, R.A. Robb, D. Hanson, and R.P. Woods. Comparison and evaluation of retrospective intermodality brain image registration techniques. *Journal of Computer Assisted Tomography*, 21:554–566, 1997.

# 8

## Optimization Techniques for Data Representations with Biomedical Applications

Pando G. Georgiev[1] and Fabian J. Theis[2]

[1] University of Cincinnati, Cincinnati, Ohio 45221
   `pgeorgie@ececs.uc.edu`
[2] University of Regensburg, D-93040 Regensburg, Germany
   `fabian@theis.name`

**Abstract.** We present two methods for data representations based on matrix factorization: Independent Component Analysis (ICA) and Sparse Component Analysis (SCA). Our presentation focuses on the mathematical foundation of ICA and SCA based on optimization theory, which appears to be enough for rigorous justification of the methods, although the ICA methods usually are justified by principles from physics, such as entropy maximization, minimization of mutual information, and so forth. We illustrate the methods with examples from biomedicine, especially from functional Magneto Resonance Imaging.

## 8.1 Introduction

A fundamental question in data analysis, signal processing, data mining, neuroscience, biomedicine, and so forth, is how to represent a large data set $\mathbf{X}$ (given in form of a $(m \times N)$-matrix) in appropriate ways suitable for efficient processing and analysis. A useful approach is a linear matrix factorization:

$$\mathbf{X} = \mathbf{AS}, \quad \mathbf{A} \in \mathbb{R}^{m \times n}, \mathbf{S} \in \mathbb{R}^{n \times N}, \tag{8.1}$$

where the unknown matrices $\mathbf{A}$ (dictionary) and $\mathbf{S}$ (source signals) have some specific properties, for instance:

1. the rows of $\mathbf{S}$ are (discrete) random variables, which are statistically independent as much as possible – this is the *Independent Component Analysis* (ICA) problem;
2. $\mathbf{S}$ contains as many zeros as possible – this is the sparse representation or *Sparse Component Analysis* (SCA) problem.

There is a large amount of papers devoted to ICA problems (see for instance [13, 29] and references therein) but mostly for the case $m \geq n$. We refer to [5, 34, 45, 48, 52] and references therein for some recent papers on SCA and underdetermined ICA ($m < n$).

P.M. Pardalos, H.E. Romeijn (eds.), *Handbook of Optimization in Medicine*,
Springer Optimization and Its Applications 26, DOI: 10.1007/978-0-387-09770-1_8,
© Springer Science+Business Media LLC 2009

A related problem is the so-called *Blind Source Separation* (BSS) problem, in which we know *a priori* that a representation such as in equation (8.1) exists and the task is to recover the sources (and the mixing matrix) as accurately as possible. A fundamental property of the complete BSS problem is that such a recovery (under assumptions 1 above and non-Gaussianity of the sources) is possible up to permutation and scaling of the sources, which makes the BSS problem so attractive. A similar property holds under some sparsity assumptions, which we will describe later.

## 8.2 Independent Component Analysis

The ICA problem and the induced BSS problem has received wide attention because of their potential applications in various fields such as biomedical signal analysis and processing (EEG, MEG, fMRI), speech enhancement, geophysical data processing, data mining, wireless communications, image processing, and so forth.

Since the introduction of ICA by Hérault and Jutten [27], various methods have been proposed to solve the BSS problem (see [2, 3, 10, 9, 15, 16, 30, 43] for an (incomplete) list of some of the most popular methods). Good textbook-level introductions to ICA are given in [29, 13]. A comprehensive description of the mathematics in ICA can be found in [42].

An alternative formulation of the problem is as follows: we can observe sensor signals (random variables) $\mathbf{x}(k) = [x_1(k), \ldots, x_m(k)]^T$, which are described as

$$\mathbf{x}(k) = \mathbf{A}\mathbf{s}(k) \quad k = 1, 2, \ldots \tag{8.2}$$

where $\mathbf{s}(k) = [s_1(k), \ldots, s_n(k)]^T$ is a vector of unknown source signals and $\mathbf{A}$ is $n \times n$ non-singular unknown mixing matrix.

Our objective is to estimate the source signals sequentially one-by-one or simultaneously assuming that they are statistically independent.

The uniqueness of such estimation (up to permutation and scaling), or identifiability of the linear ICA model, is justified in the literature by the Skitovitch–Darmois theorem [41, 17]. Whereas this theorem is probabilistic in nature, an elementary lemma from optimization theory (although with a non-elementary proof) can serve the same purpose — rigorous justification of the identifiability of ICA model, when maximization of the cumulants is used. We will present an elementary proof of identifiability of the linear ICA model, based on the properties of the cumulants.

### 8.2.1 Extraction via maximization of the absolute value of the cumulants

Maximization of non-Gaussianity is one of the basic ICA estimation principles (see [13, 29]). This principle is explained by the central limit theorem, according to which, sums of non-Gaussian random variables are closer to Gaussian than the original ones. Therefore, a linear combination $y = \mathbf{w}^\top \mathbf{x} = \sum_{i=1}^{n} w_i x_i$ of the observed mixture variables (which is a linear combination of the independent components as well, because of the linear mixing model) will be maximally non-Gaussian if it equals one of the independent components. Below we give rigorous mathematical proof of this statement. The task how to find such a vector $\mathbf{w}$, which gives one independent component, and therefore should be one (scaled) row of the inverse of the mixing matrix $\mathbf{A}$, is the main task of (sequential) ICA. We will describe an optimization problem for this task.

Recall the following formula for the cross-cumulants of the random variables $x_1, \ldots, x_n$ in terms of moments (see for instance [40, p. 292]):

$$\mathbf{cum}(x_1, \ldots, x_n)$$
$$= \sum_{(p_1, \ldots, p_m)} (-1)^{m-1} (m-1)! E\left[\prod_{i \in p_1} x_i\right] \cdots E\left[\prod_{i \in p_m} x_i\right], \qquad (8.3)$$

where the summation is taken over all possible partitions $\{p_1, \ldots, p_m\}$, $m = 1, \ldots, n$ of the set of the natural numbers $\{1, \ldots, n\}$; $\{p_i\}_{i=1}^{m}$ are disjoint subsets, which union is $\{1, \ldots, n\}$, $E$ is the expectation operator (see [40] for properties of the cumulants). For $n = 4$, the above formula gives

$$\mathbf{cum}\{x_i, x_j, x_k, x_l\} = \quad E(x_i x_j x_k x_l) - E(x_i x_j)E(x_k x_l)$$
$$- E(x_i x_k)E(x_j x_l) - E(x_i x_l)E(x_k x_j).$$

The following property of the cumulants is used essentially in derivation of fixed point algorithm [30] and its generalization below: if $s_i$, $i = 1, \ldots, n$ are statistically independent and $c_i$, $i = 1, \ldots, n$ are arbitrary real numbers, then

$$\mathbf{cum}_p\left(\sum_{i=1}^{n} c_i s_i\right) = \sum_{i=1}^{n} c_i^p \mathbf{cum}_p(s_i). \qquad (8.4)$$

Define the function $\varphi : \mathbb{R}^n \to \mathbb{R}$ by

$$\varphi_p(\mathbf{w}) = \mathbf{cum}_p(\mathbf{w}^\top \mathbf{x})$$

where $\mathbf{cum}_p$ means the self-cumulant of order $p$:

$$\mathbf{cum}_p(s) = \mathbf{cum}(\underbrace{s, \ldots, s}_{p}).$$

Then consider the maximization problems $\mathrm{OP}(p)$

$$\text{maximize } |\varphi_p(\mathbf{w})| \text{ subject to } \|\mathbf{w}\| = 1$$

and $\mathrm{DP}(p)$

$$\text{maximize } |\psi_p(\mathbf{c})| \text{ subject to } \|\mathbf{c}\| = 1,$$

where $\psi_p(\mathbf{c}) = \mathbf{cum}_p\left(\sum_{i=1}^{n} c_i s_i\right)$ and $c_i$, $i = 1, \ldots, n$ are the components of the vector $\mathbf{c}$. Denoting $y = \mathbf{w}^\top \mathbf{x}$ and $\mathbf{c} = \mathbf{A}^\top \mathbf{w}$, we have

$$y = \mathbf{c}^\top \mathbf{s} = \sum_{i=1}^{n} c_i s_i$$

and

$$\varphi_p(\mathbf{w}) = \psi_p(\mathbf{A}^\top \mathbf{w}). \tag{8.5}$$

Without loss of generality, we may assume that the matrix $\mathbf{A}$ is orthogonal (assuming that we have performed the well-known preprocessing procedure called "prewhitening," see [13, 29]).

It is easy to see (using (8.5) and the orthogonality of $\mathbf{A}$) that the problems DP$(p)$ and OP$(p)$ are equivalent in the sense that $\mathbf{w}_0$ is a solution of OP$(p)$if and only if $\mathbf{c}_0 = \mathbf{A}^\top \mathbf{w}_0$ is a solution of DP$(p)$.

A very useful observation is the following: if a vector $\mathbf{c}$ contains only one nonzero component, say $c_{i_0} = \pm 1$, then the vector $\mathbf{w} = \mathbf{A}\mathbf{c}$ gives an extraction (say $y(k)$) of the source with index $i_0$, as

$$\begin{aligned} y(k) &:= \mathbf{w}^\top \mathbf{x}(k) \\ &= \mathbf{c}^\top \mathbf{A}^\top \mathbf{x}(k) \\ &= \mathbf{c}^\top \mathbf{s}(k) = s_{i_0}(k) \quad \forall k = 1, 2, \dots. \end{aligned} \tag{8.6}$$

The following lemma shows that the solutions $\mathbf{c}$ of DP$(p)$ have exactly one nonzero element. Thus, we can obtain the vectors $\mathbf{w} = \mathbf{A}\mathbf{c}$ as solutions of the original problem OP$(p)$, and by (8.6) we achieve extraction of one source.

One interesting property of the optimization problem OP$(p)$ is that it has exactly $n$ solutions (up to sign) that are orthonormal and any of them gives extraction of one source signal. The fixed point algorithm [30] finds its solutions one by one.

We note that the idea of maximizing of $\mathbf{cum}_4(\mathbf{w}^\top \mathbf{x})$ in order to extract one source from a linear mixture is already considered in [18], but the proof presented there is quite complicated, whereas our proof here (see Lemma 1) is transparent and contains the case of cumulants of an arbitrary even order. For more general result we refer to [21].

**Lemma 1.** *Consider the optimization problem*

$$minimize \ (maximize) \ \sum_{i=1}^{n} k_i v_i^p \ subject \ to \ |\mathbf{v}| = c > 0,$$

*where $p > 2$ is even and $\mathbf{v} = (v_1, \dots, v_n)$. Denote*

$$\begin{aligned} I^+ &= \{i \in \{1, \dots, n\} : k_i > 0\} \\ I^- &= \{i \in \{1, \dots, n\} : k_i < 0\} \end{aligned}$$

*and $e_i = (0, \dots 0, 1, 0, \dots, 0)$, (1 is the $i^{\text{th}}$ place). Assume that $I^+ \neq \emptyset$ and $I^- \neq \emptyset$.*

*Then the points of local minimum are exactly the vectors $\mathbf{m}_i^\pm = \pm c e_i$, $i \in I^-$ and the points of local maximum are exactly the vectors $\mathbf{M}_j^\pm = \pm c e_j$, $j \in I^+$.*

**Proof.** Applying the Lagrange multipliers theorem for a point of a local optimum $\bar{\mathbf{v}} = (\bar{v}_1, \ldots, \bar{v}_m)$, we write:

$$k_i p \bar{v}_i^{p-1} - 2\lambda \bar{v}_i = 0, i = 1, \ldots, m, \tag{8.7}$$

where $\lambda$ is a Lagrange multiplier.

Multiplying (8.7) by $\bar{v}_i$ and summing, we obtain:

$$p f_{\text{opt.}} = 2\lambda c^2,$$

where $f_{\text{opt.}}$ means the value of $f$ at the local optimum. Hence

$$\lambda = \frac{p}{2c^2} f_{\text{opt.}}. \tag{8.8}$$

From (8.7) we obtain

$$\bar{v}_i \left( k_i p \bar{v}_i^{p-2} - \frac{p}{c^2} f_{\text{opt.}} \right) = 0$$

whence

$$\bar{v}_i \text{ is either } 0, \text{ or } \pm \left( \frac{f_{\text{opt.}}}{k_i c^2} \right)^{\frac{1}{p-2}}. \tag{8.9}$$

*Case 1.*
Assume that $k_{i_0} < 0$ for some index $i_0$ and $\bar{\mathbf{v}}$ is a local minimum. Then obviously $f_{\text{loc.min.}} < 0$. According to the second-order optimality condition [1], a point $\mathbf{x}^0$ is a local minimum if

$$\mathbf{h}^\top L''(\mathbf{x}^0)\mathbf{h} > 0 \quad \forall \mathbf{h} \in K(\mathbf{x}^0) = \{\mathbf{h} : \mathbf{h}^\top \mathbf{x}^0 = 0\}, \mathbf{h} \neq 0,$$

where

$$L(\mathbf{x}) = \sum_{i=1}^{n} k_i x_i^p - \lambda(|\mathbf{x}|^2 - c^2)$$

is the Lagrange function.

In our case, by (8.8) and (8.9) we obtain

$$\mathbf{h}^\top L''(\bar{\mathbf{v}})\mathbf{h} = \sum_{i=1}^{n} (p(p-1)k_i \bar{v}_i^{p-2} - 2\lambda)h_i^2 \tag{8.10}$$

$$= \frac{p}{c^2} f_{\text{loc.min.}} \left[ (p-2) \sum_{i \in I} h_i^2 - \sum_{i \notin I} h_i^2 \right],$$

where $I$ is the set of those indexes $i$, for which $\bar{v}_i$ is different from 0.

We shall check the second order sufficient condition for a local minimum for the points $\mathbf{m}_{i_0}^\pm$. We have

$$K(\mathbf{m}_{i_0}^\pm) = \{\mathbf{h} : h_{i_0} = 0\}.$$

Therefore, for $\mathbf{h} \in K(\mathbf{m}_{i_0}^{\pm})$, $\mathbf{h} \neq 0$ we have

$$\mathbf{h}^{\top} L''(\mathbf{m}_{i_0}^{\pm})\mathbf{h} > 0,$$

as $h_{i_0} = 0$ and $f_{\text{loc.min.}} < 0$, i.e., the second-order sufficient condition is satisfied and $\mathbf{m}_{i_0}^{\pm}$ is a local minimum.

By (8.10), it follows that for any vector $v$ with at least two nonzero elements, the quadratic form (8.10) can take positive and negative values for different values of $h$, i.e., the necessary condition for a local minimum is not satisfied for such a vector.

*Case 2.*
Assume that $k_j > 0$ and $\overline{\mathbf{v}}$ is a local maximum. We apply Case 1 to the function $-f$ and finish the proof.     □

### 8.2.2 A generalization of the fixed point algorithm

Consider the following algorithm:

$$\mathbf{w}(l) = \frac{\varphi_p'(\mathbf{w}(l-1))}{\|\varphi_p'(\mathbf{w}(l-1))\|}, \quad l = 1, 2, \ldots, \tag{8.11}$$

which is a generalization of the fixed point algorithm of Hyvärinen and Oja. The name is derived by the Lagrange equation for the optimization problem $OP(p)$, as (8.11) tries to find a solution of it iteratively, and this solution is a fixed point of the operator defined by the right-hand side of (8.11).

The next theorem gives precise conditions for convergence of the fixed point algorithm of Hyvärinen and Oja and its generalization (8.11) (for a proof, see [21]).

**Theorem 1.** *Assume that $s_i$ are statistically independent, zero mean signals and the mixing matrix $\mathbf{A}$ is orthogonal. Let $p \geq 3$ be a given even integer number, $\mathbf{cum}_p(s_i) \neq 0$, $i = 1, \ldots, n$ and let*

$$I(\mathbf{c}) = \arg \max_{1 \leq i \leq n} c_i \left| \mathbf{cum}_p(s_i) \right|^{\frac{1}{p-2}}.$$

*Denote by $W_0$ the set of all elements $\mathbf{w} \in \mathbb{R}^n$ such that $\|\mathbf{w}\| = 1$. The set $I(\mathbf{A}^{\top}\mathbf{w})$ contains only one element, say $i(\mathbf{w})$, and $c_{i(\mathbf{w})} \neq 0$. Then*

*(a) The complement of $W_0$ has measure zero.*
*(b) If $\mathbf{w}(0) \in W_0$ then*

$$\lim_{l \to \infty} y_l(k) = \pm s_{i_0}(k) \quad \forall k = 1, 2, \ldots,$$

*where $y_l(k) = \mathbf{w}(l)^{\top}\mathbf{x}(k)$ and $i_0 = i(\mathbf{w}(0))$.*
*(c) The rate of convergence in (b) is of order $p - 1$.*

When $p = 4$, we obtain:

$$\varphi_4(\mathbf{w}) = \mathbf{cum}_4(\mathbf{w}^\top \mathbf{x}) = E\{(\mathbf{w}^\top \mathbf{x})^4\} - 3(E\{(\mathbf{w}^\top \mathbf{x})^2\})^2$$

and

$$\varphi_4'(\mathbf{w}) = 4E\{(\mathbf{w}^\top \mathbf{x})^3 \mathbf{x}\} - 12E\{(\mathbf{w}^\top \mathbf{x})^2\}E\{(\mathbf{x}\mathbf{x}^\top)\}\mathbf{w}.$$

We note that if the standard prewhitening procedure is performed (i.e., $E\{\mathbf{x}\mathbf{x}^\top\} = \mathbf{I}_n$, $\mathbf{A}$ is orthogonal), the algorithm (8.11) recovers the fixed-point algorithm of Hyvarinen and Oja, i.e.,

$$\mathbf{w}(l+1) = \frac{E\{(\mathbf{w}(l)^\top \mathbf{x})^3 \mathbf{x}\} - 3\mathbf{w}(l)}{\|E\{(\mathbf{w}(l)^\top \mathbf{x})^3 \mathbf{x}\} - 3\mathbf{w}(l)\|}.$$

Different schemes for deflation are considered, for instance, in [29]. Different fixed point algorithms are described in [28, 31] based on nonlinearity which gives maximization of a nonlinear function different from kurtosis. The maximization problem is

$$\text{maximize } [E\{G(\mathbf{w}^\top x)\} - E\{G(\nu)\}]^2 \text{ subject to } E\{(\mathbf{w}^\top \mathbf{x})^2\} = 1,$$

where $\nu$ is a standard Gaussian variable.

A local solution $\mathbf{w}_0$ of this problem (under some assumptions on the non-linearity of $G$) is such that $\mathbf{w}_0^\top \mathbf{x} = \pm s_i$, i.e., when the linear combination gives one of the independent components. The convergence, however, of any algorithm with $G(u)$ different from $u^3$ (which gives kurtosis maximization) is not so fast as in that case (in which case it is cubic). From other point of view, using nonlinearity gives robustness to outliers in some cases.

### 8.2.3 Separability of linear BSS

Consider the noiseless linear instantaneous BSS model with as many sources as sensors:

$$\mathbf{X} = \mathbf{AS} \tag{8.12}$$

with an independent $n$-dimensional random vector $\mathbf{S}$ and $\mathbf{A} \in \text{Gl}(n)$. Here $\text{Gl}(n)$ denotes the general linear group of $\mathbb{R}^n$, i.e., the group of all invertible $(n \times n)$-matrices.

The task of linear BSS is to find $\mathbf{A}$ and $\mathbf{S}$ given only $\mathbf{X}$. An obvious indeterminacy of this problem is that $\mathbf{A}$ can be found only up to scaling and permutation because for scaling $\mathbf{L}$ and permutation matrix $\mathbf{P}$

$$\mathbf{X} = \mathbf{ALPP}^{-1}\mathbf{L}^{-1}\mathbf{S}$$

and $\mathbf{P}^{-1}\mathbf{L}^{-1}\mathbf{S}$ is also independent. Here, an invertible matrix $\mathbf{L} \in \text{Gl}(n)$ is said to be a *scaling matrix* if it is diagonal. We say two matrices $\mathbf{B}, \mathbf{C}$ are *equivalent*, $\mathbf{B} \sim \mathbf{C}$, if $\mathbf{C}$ can be written as $\mathbf{C} = \mathbf{BPL}$ with a scaling matrix

$\mathbf{L} \in \mathrm{Gl}(n)$ and an invertible matrix with unit vectors in each row (*permutation matrix*) $\mathbf{P} \in \mathrm{Gl}(n)$. Note that $\mathbf{PL} = \mathbf{L'P}$ for some scaling matrix $\mathbf{L'} \in \mathrm{Gl}(n)$, so the order of the permutation and the scaling matrix does not play a role for equivalence. Furthermore, if $\mathbf{B} \in \mathrm{Gl}(n)$ with $\mathbf{B} \sim \mathbf{I}$, then also $\mathbf{B}^{-1} \sim \mathbf{I}$, and more general if $\mathbf{BC} \sim \mathbf{A}$, then $\mathbf{C} \sim \mathbf{B}^{-1}\mathbf{A}$. According to the above, solutions of linear BSS are equivalent. We will show that under mild assumptions on $\mathbf{S}$, there are no further indeterminacies of linear BSS.

$\mathbf{S}$ is said to have a *Gaussian component* if one of the $S_i$ is a one-dimensional *Gaussian*, that is, $p_{S_i}(x) = d \exp(-ax^2 + bx + c)$ with $a, b, c, d \in \mathbb{R}$, $a > 0$.

**Theorem 2 (Separability of linear BSS).** *Let* $\mathbf{A} \in \mathrm{Gl}(n)$ *and* $\mathbf{S}$ *be an independent random vector. Assume one of the following:*

*(i)* $\mathbf{S}$ *has at most one Gaussian or deterministic component and the covariance of* $\mathbf{S}$ *exists.*

*(ii)* $\mathbf{S}$ *has no Gaussian component and its density* $p_{\mathbf{S}}$ *exists and is twice continuously differentiable.*

*Then if* $\mathbf{X} = \mathbf{AS}$ *is again independent,* $\mathbf{A}$ *is equivalent to the identity.*

Thus $\mathbf{A}$ is the product of a scaling and a permutation matrix. The important part of this theorem is assumption (i), which has been used to show separability by Comon [16] and extended by Erikkson and Koivunen [19] based on the Darmois–Skitovitch theorem [17, 41]. Using this theorem, the second part can be easily shown without $\mathbf{C}^2$-densities.

Theorem 2 indeed proves separability of the linear BSS model, because if $\mathbf{X} = \mathbf{AS}$ and $\mathbf{W}$ is a demixing matrix such that $\mathbf{WX}$ is independent, then $\mathbf{WA} \sim \mathbf{I}$, so $\mathbf{W}^{-1} \sim \mathbf{A}$ as desired.

For a proof of the above theorem without having to use the Darmois–Skitovitch theorem we refer to [44].

Now we will give a simple proof of Theorem 2 in the case when $E|s_i|^m < \infty$ for any $i = 1, \ldots, n$ and any natural $m$. By these assumptions, it follows that the cumulants of $s_i$ of any order exist (see [41, p. 289]). Suppose that $\mathbf{S}$ has at most one Gaussian or deterministic component.

We will first show using whitening that $\mathbf{A}$ can be assumed to be orthogonal. For this we can assume $\mathbf{S}$ and $\mathbf{X}$ to have no deterministic component at all (because arbitrary choice of the matrix coefficients of the deterministic components does not change the covariance). Hence by assumption $\mathrm{Cov}(\mathbf{X})$ is diagonal and positive definite, so let $\mathbf{D}_1$ be diagonal invertible with $\mathrm{Cov}(\mathbf{X}) = \mathbf{D}_1^2$. Similarly let $\mathbf{D}_2$ be diagonal invertible with $\mathrm{Cov}(\mathbf{S}) = \mathbf{D}_2^2$. Set $\mathbf{Y} := \mathbf{D}_1^{-1}\mathbf{X}$ and $\mathbf{T} := \mathbf{D}_2^{-1}\mathbf{S}$, i.e., normalize $\mathbf{X}$ and $\mathbf{S}$ to covariance $\mathbf{I}$. Then

$$\mathbf{Y} = \mathbf{D}_1^{-1}\mathbf{X} = \mathbf{D}_1^{-1}\mathbf{AS} = \mathbf{D}_1^{-1}\mathbf{AD}_2\mathbf{T}$$

and $\mathbf{T}$, $\mathbf{D}_1^{-1}\mathbf{AD}_2$ and $\mathbf{Y}$ satisfy the assumption and $\mathbf{D}_1^{-1}\mathbf{AD}_2$ is orthogonal because

$$\begin{aligned}
\mathbf{I} &= \mathrm{Cov}(\mathbf{Y}) \\
&= E(\mathbf{Y}\mathbf{Y}) \\
&= E(\mathbf{D}_1^{-1}\mathbf{A}\mathbf{D}_2\mathbf{T}\mathbf{T}\mathbf{D}_2\mathbf{A}\mathbf{D}_1^{-1}) \\
&= (\mathbf{D}_1^{-1}\mathbf{A}\mathbf{D}_2)(\mathbf{D}_1^{-1}\mathbf{A}\mathbf{D}_2).
\end{aligned}$$

Thus, without loss of generality let $\mathbf{A}$ be orthogonal.

Let $x_i$ and $s_i$ be the components of $\mathbf{X}$ and $\mathbf{S}$ respectively. Because $\{s_i\}$ are independent, using property (8.4) we have:

$$\mathbf{cum}_p(x_i) = \mathbf{cum}_p\left(\sum_{j=1}^{n} a_{ij}s_i\right) \qquad (8.13)$$

$$= \sum_{j=1}^{n} a_{ij}^p \mathbf{cum}_p(s_i).$$

Because $\{x_i\}$ are independent and $\mathbf{S} = \mathbf{A}\mathbf{X}$, using again property (8.4) we obtain:

$$\mathbf{cum}_p(s_i) = \mathbf{cum}_p\left(\sum_{j=1}^{n} a_{ji}x_j\right) \qquad (8.14)$$

$$= \sum_{j=1}^{n} a_{ji}^p \mathbf{cum}_p(x_i).$$

If we denote by $\mathbf{A}^{(p)}$ the matrix with elements $a_{ij}^p$ and put $\mathbf{c}_p(\mathbf{x}) = \big(\mathbf{cum}_p(x_1),$ $\dots, \mathbf{cum}_p(x_n)\big)$ and $\mathbf{c}_p(\mathbf{s}) = \big(\mathbf{cum}_p(s_1), \dots, \mathbf{cum}_p(s_n)\big)$, we have

$$\mathbf{c}_p(\mathbf{x}) = \mathbf{A}^{(p)}\mathbf{c}_p(\mathbf{s})$$

and

$$\mathbf{c}_p(\mathbf{s}) = \left(\mathbf{A}^{(p)}\right)\mathbf{c}_p(\mathbf{x}).$$

Hence,

$$\left\|\mathbf{c}_p(\mathbf{s})\right\| \le \left\|\mathbf{A}^{(p)}\right\|\left\|\mathbf{c}_p(\mathbf{x})\right\| \le \left\|\mathbf{A}^{(p)}\right\|^2\left\|\mathbf{c}_p(\mathbf{s})\right\|. \qquad (8.15)$$

Here we have to note that, by Marcinkiewics's theorem (see [41, p. 288]), the Gaussian distribution is the only distribution with the property that all its cumulants are zero from a certain index onward. Because, by assumption, only one variable from $\{s_i\}$ is Gaussian, by the above remark it follows that there exists a sequence of natural numbers $p_m$ such that $\mathbf{c}_{p_m}(\mathbf{s}) \ne 0$ for every natural number $m$. From (8.15) it follows that $\|\mathbf{A}^{(p_m)}\| \ge 1$ for every natural $m$, and as every element of $\mathbf{A}$ is in the interval $[-1, 1]$ ($\mathbf{A}$ is orthogonal), it follows that at least one element of $\mathbf{A}$, say $a_{i_1 j_1}$ should be 1 or $-1$ (otherwise $\|\mathbf{A}^{(p_m)}\| \to 0$ when $m \to \infty$). The elements $a_{i_1,j}$ and $a_{i,j_1}$ for all $i \ne i_1$ and all $j \ne j_2$ should be zero (as $\mathbf{A}$ is orthogonal). Removing row $i$ and column $j$, and repeating the same reasonings for the remaining system (with dimension $n - 1$), we obtain that another element of $\mathbf{A}$, say $a_{i_2,j_2}$ should be 1 or $-1$ and $a_{i_2,j}$ and $a_{i,j_2}$ for all $i \neq i_2$ and all $j \neq j_2$ should be zero. Repeating this reasoning $n-1$ times, we obtain that $\mathbf{A}$ have to be a sign permutation matrix, i.e., in each row and each column only one element is 1 or $-1$ and the rest are zero, as desired.

We next briefly present the concept of separated functions from [44], which can be seen as a general framework for the algorithms proposed in [44, 35, 51].

**Definition 1.** *A function* $f : \mathbb{R}^n \to \mathbb{C}$ *is said to be* separated *respectively* linearly separated *if there exist one-dimensional functions* $g_1, \ldots, g_n : \mathbb{R} \to \mathbb{C}$ *such that* $f(\mathbf{x}) = g_1(x_1) \ldots g_n(x_n)$ *respectively* $f(\mathbf{x}) = g_1(x_1) + \ldots + g_n(x_n)$ *for all* $\mathbf{x} \in \mathbb{R}^n$.

Note that the functions $g_i$ are uniquely determined by $f$ up to a scalar factor respectively an additive constant. If $f$ is linearly separated, then $\exp f$ is separated. The density function of an independent random vector is separated – this fact provides motivation for the presented method.

Let $\mathbf{C}^m(U, V)$ be the ring of all $m$-times continuously differentiable functions from $U \subset \mathbb{R}^n$ to $V \subset \mathbb{C}$, $U$ open. For a $\mathbf{C}^m$-function $f$, we write $\partial_{i_1} \ldots \partial_{i_m} f := \partial^m f / \partial x_{i_1} \ldots \partial x_{i_m}$ for the $m$-fold partial derivatives. If $f \in \mathbf{C}^2(\mathbb{R}^n, \mathbb{C})$, denote with the symmetric $(n \times n)$-matrix $\mathbf{H}_f(\mathbf{x}) := (\partial_i \partial_j f(\mathbf{x}))_{i,j=1}^n$ the *Hessian* of $f$ at $\mathbf{x} \in \mathbb{R}^n$.

It is an easy fact that linearly separated functions can be classified using their Hessian (if it exists):

**Lemma 2.** *A function* $f \in \mathbf{C}^2(\mathbb{R}^n, \mathbb{C})$ *is linearly separated if and only if* $\mathbf{H}_f(\mathbf{x})$ *is diagonal for all* $\mathbf{x} \in \mathbb{R}^n$.

A similar result for "block diagonal" Hessians has been shown by [35].

Note that Lemma 2 obviously also holds for functions defined on any open parallelepiped $(a_1, b_1) \times \cdots \times (a_n, b_n) \subset \mathbb{R}^n$. Hence an arbitrary real-valued $\mathbf{C}^2$-function $f$ is locally separated at $\mathbf{x}$ with $f(\mathbf{x}) \neq 0$ if and only if the Hessian of $\ln |f|$ is locally diagonal.

Thus for a positive function $f$ the Hessian of its logarithm is diagonal everywhere if it is separated, and it is easy to see that for positive $f$ also the converse holds globally (Theorem 3(ii)). In this case we have for $i \neq j$

$$0 \equiv \partial_i \partial_j \ln f \equiv \frac{f \partial_i \partial_j f - (\partial_i f)(\partial_j f)}{f^2},$$

so $f$ is separated if and only if

$$f \partial_i \partial_j f \equiv (\partial_i f)(\partial_j f)$$

for $i \neq j$ or even $i < j$. This motivates the following definition:

**Definition 2.** *For* $i \neq j$, *the operator*

$$R_{ij} : \mathbf{C}^2(\mathbb{R}^n, \mathbb{C}) \to \mathbf{C}^0(\mathbb{R}^n, \mathbb{C})$$
$$f \mapsto R_{ij}[f] := f \partial_i \partial_j f - (\partial_i f)(\partial_j f)$$

*is called the* $ij$-separator.

**Theorem 3.** *Let* $f \in \mathbf{C}^2(\mathbb{R}^n, \mathbb{C})$.

*(i) If $f$ is separated then $R_{ij}[f] \equiv 0$ for $i \neq j$ or equivalently*

$$f \partial_i \partial_j f \equiv (\partial_i f)(\partial_j f). \tag{8.16}$$

*holds for $i \neq j$.*

*(ii) If $f$ is positive and $R_{ij}[f] \equiv 0$ holds for all $i \neq j$ then $f$ is separated.*

If $f$ is assumed to be only nonnegative, then $f$ is locally separated but not necessarily globally separated (if the support of $f$ has more than one component). Some trivial properties of the separator $R_{ij}$ are listed in the next lemma:

**Lemma 3.** *Let* $f, g \in \mathbf{C}^2(\mathbb{R}^n, \mathbb{C})$, $i \neq j$ *and* $\alpha \in \mathbb{C}$. *Then*

$$R_{ij}[\alpha f] = \alpha^2 R_{ij}[f]$$

*and*

$$R_{ij}[f + g] = R_{ij}[f] + R_{ij}[g] + f\partial_i\partial_j g + g\partial_i\partial_j f - (\partial_i f)(\partial_j g) - (\partial_i g)(\partial_j f).$$

### 8.2.4 Global Hessian diagonalization using kernel-based density approximation

We suggest using kernel-based density estimation to get an energy function with minima at the BSS solutions together with a global Hessian diagonalization in the following (see [44]).

The idea is to construct a measure for separatedness of the densities (hence independence) based on Theorem 3. A possible measure could be the norm of the summed up separators $\sum_{i<j} R_{ij}[f]$. In order for this to be calculable, we only choose a set of points $\mathbf{p}^{(i)}$ where we evaluate the difference, and minimize $\sum_k (\sum_{i<j} R_{ij}[f](\mathbf{p}^{(k)}))^2$ at those points. Although in the linear noiseless case calculation of the Hessian at only one point would be enough, using an energy function of this type ensures using global information of the densities while averaging over possible local errors.

First, we need to approximate the density function. For this, let $\mathbf{X} \in \mathbb{R}^n$ be an $n$-dimensional random vector with $\nu$ i.i.d.-samples $\mathbf{x}^{(1)}, \ldots, \mathbf{x}^{(\nu)} \in \mathbb{R}^n$. Let

$$\phi : \mathbb{R}^n \to \mathbb{R}$$

$$\mathbf{x} \mapsto \frac{1}{\sigma^n \sqrt{(2\pi)^n}} \exp\left(-\frac{1}{2\sigma^2} \|\mathbf{x}\|^2\right)$$

be the $n$-dimensional centered independent Gaussian with fixed variance $\sigma^2 > 0$. For ease of notation denote $\kappa := \frac{1}{2\sigma^2}$.

Define the approximated density $\hat{p}_\mathbf{X}$ of $\mathbf{X}$ by

$$\hat{p}_\mathbf{X}(\mathbf{x}) := \frac{1}{\nu} \sum_{i=1}^{\nu} \phi(\mathbf{x} - \mathbf{x}^{(i)}). \tag{8.17}$$

If $\nu \to \infty$, $\hat{p}_\mathbf{X}$ converges to $p_\mathbf{X}$ in the space of all integrable functions if $\sigma$ is chosen appropriately. This can be shown using the central limit theorem.

The partial derivatives of $\phi$ can be calculated as

$$\partial_i \phi(\mathbf{x}) = -2\kappa x_i \phi(\mathbf{x})$$
$$\partial_i \partial_j \phi(\mathbf{x}) = 4\kappa^2 x_i x_j \phi(\mathbf{x}) \tag{8.18}$$

for $i \neq j$. $\phi$ is separated, so $R[\phi] \equiv 0$. Note that $\hat{p}_\mathbf{X} \in \mathbf{C}^\infty(\mathbb{R}^n, \mathbb{R})$ is positive. Thus according to Theorem 3, $\hat{p}_\mathbf{X}$ is separated if and only if $R_{ij}[\hat{p}_\mathbf{X}] \equiv 0$ for $i < j$. And, as $\hat{p}_\mathbf{X}$ is an approximation of $p_\mathbf{X}$, separatedness of $\hat{p}_\mathbf{X}$ also induces approximate independence of $\mathbf{X}$.

$R_{ij}[\hat{p}_\mathbf{X}]$ can be calculated using Lemma 3 — here $R_{ij}[\phi(\mathbf{x} - \mathbf{x}^{(k)})] \equiv 0$ — and equation (8.18):

$$
\begin{aligned}
R_{ij}[\hat{p}_\mathbf{X}](\mathbf{x}) &= \frac{1}{\nu^2} R_{ij} \left[ \sum_{k=1}^{\nu} \phi(\mathbf{x} - \mathbf{x}^{(k)}) \right] \\
&= \frac{1}{\nu^2} \sum_{k \neq l} \phi(\mathbf{x} - \mathbf{x}^{(k)}) \partial_i \partial_j \phi(\mathbf{x} - \mathbf{x}^{(l)}) \\
&\quad - (\partial_i \phi)(\mathbf{x} - \mathbf{x}^{(k)})(\partial_j \phi)(\mathbf{x} - \mathbf{x}^{(l)}) \\
&= \frac{4\kappa^2}{\nu^2} \sum_{k \neq l} \phi(\mathbf{x} - \mathbf{x}^{(k)}) \phi(\mathbf{x} - \mathbf{x}^{(l)})(x_i^{(k)} - x_i^{(l)})(x_j - x_j^{(l)}) \\
&= \frac{4\kappa^2}{\nu^2} \sum_{k < l} \phi(\mathbf{x} - \mathbf{x}^{(k)}) \phi(\mathbf{x} - \mathbf{x}^{(l)})(x_i^{(k)} - x_i^{(l)})(x_j^{(k)} - x_j^{(l)}).
\end{aligned}
$$

This function is zero for $i < j$ if and only if $\hat{p}_\mathbf{X}$ is separated. For linear BSS, it would be enough to check this at one point "in general position" (see [44, Theorem 3]), but for robustness reason we want to require $R_{ij}[\hat{p}_\mathbf{X}]$ to be zero (or as close to zero as possible) at all sample points. Thus the desired independence estimator can be calculated as

$$E(\mathbf{x}^1, \ldots, \mathbf{x}^{(n)}) := E := \sum_{m=1}^{\nu} \sum_{i<j} (R_{ij}[\hat{p}_\mathbf{X}](\mathbf{x}^{(m)}))^2$$

hence

$$E = (\sigma^2 \nu)^{-4} \sum_{m} \sum_{i<j} \left( \sum_{k<l} \phi(\mathbf{x} - \mathbf{x}^{(k)}) \phi(\mathbf{x} - \mathbf{x}^{(l)})(x_i^{(k)} - x_i^{(l)})(x_j^{(k)} - x_j^{(l)}) \right)^2.$$

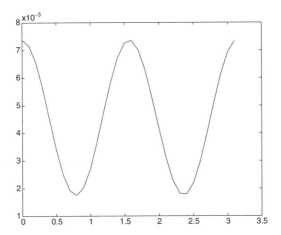

**Fig. 8.1.** Energy function $\mathbf{W} \mapsto E(\mathbf{WX})$ of a mixture $\mathbf{X}$ of two Laplacians using as mixing matrix $\mathbf{A}$ a rotation by 45 degrees. One hundred samples were used, and $E$ is plotted in steps of 0.1. The minima of $E$ clearly lie at $\frac{1}{4}\pi$ and $\frac{3}{4}\pi$ as desired.

Minimizing the function

$$\varepsilon : \mathrm{Gl}(n) \to \mathbb{R}$$
$$\mathbf{W} \mapsto E(\mathbf{Wx}^1, \dots, \mathbf{Wx}^{(n)})$$

then yields the desired demixing matrix with $\mathbf{W}^{-1} \sim \mathbf{A}$ according to Theorem 2, $\varepsilon$ can be minimized using the usual techniques such as for example global search, gradient descent, or fixed-point search. Figure 8.1 shows the energy function of an example mixture of two Laplacians. $E$ is minimal at the points where $\mathbf{WX}$ is independent.

Note that $E$ represents a new approximate measure of independence. Therefore, the linear BSS algorithm can now be readily generalized to nonlinear situations by finding an appropriate parameterization of the possibly nonlinear separating model.

The proposed algorithm in [44] basically performs a global diagonalization of the logarithmic Hessian after prewhitening. Interestingly, this is similar to traditional BSS algorithms based on joint diagonalization such as JADE [10] using cumulant matrices or AMUSE [46] and SOBI [4] employing time-decorrelation. Instead of using a global energy function as proposed above, we could therefore also jointly diagonalize a given set of Hessians (respectively separator matrices as above), see also [51]. Another relation to previously proposed ICA algorithms lies in the kernel approximation technique. Gaussian or generalized Gaussian kernels have already been used in the field of ICA to model the source densities [33, 26] thus giving an estimate of the score function used in Bell–Sejnowski type semiparametric algorithms [3] or enabling direct separation using maximum likelihood parameter estimation. The algorithm in [44] also uses density approximation but employs this for the mixture density, which can be problematic in higher dimensions. A different approach not involving density approximation is direct sample-based Hessian estimation similar to [35].

## 8.3 Other Methods for ICA

In this section, we consider briefly other well-known methods for ICA. We refer to [13] for a comprehensive citation of ICA methods. The choice of a specific method is up to the reader preference. However, for validation purposes it is recommended to use at least two methods.

For fMRI analysis, the following algorithms are tested to give satisfactory results in different applications: the fact ICA algorithm, the natural gradient method, the infomax algorithm, and the JADE algorithm.

### 8.3.1 Likelihood

Denoting the matrix $\mathbf{A}^{-1}$ by $\mathbf{W} = (\mathbf{w}_1, \ldots, \mathbf{w}_m)^\top$, the log-likelihood takes the form [38]:

$$L = \sum_{t=1}^{T} \sum_{i=1}^{m} \log f_i(\mathbf{w}_i^\top \mathbf{x}(t)) + T \ln |\det \mathbf{W}|$$

where the $f_i$ are the density functions of the $s_i$ (here assumed to be known), and the $\mathbf{x}(t), t = 1, \ldots, T$ are the realizations of $\mathbf{x}$. Maximizing $L$ gives an estimation of the demixing matrix $\mathbf{W}$.

### 8.3.2 Network entropy

Another related contrast function was derived from a neural network viewpoint. This was based on maximizing the output entropy (or information flow) of a neural network with non-linear outputs. Assume that $\mathbf{x}$ is the input to the neural network whose outputs are of the form $g_i(\mathbf{w}_i^\top \mathbf{x})$, where the $g_i$ are some non-linear scalar functions, and the $\mathbf{w}_i$ are the weight vectors of the neurons. One then wants to maximize the entropy of the outputs:

$$L_2 = H(g_1(\mathbf{w}_1^\top \mathbf{x}), \ldots, g_m(\mathbf{w}_m^\top \mathbf{x})),$$

where

$$H(\mathbf{y}) = -\int f_\mathbf{y}(\mathbf{x}) \log f_\mathbf{y}(\mathbf{x}) d\mathbf{x}$$

is the differential entropy, $f_\mathbf{y}$ is the pdf of $\mathbf{y}$.

### 8.3.3 Mutual information

Theoretically the most satisfying contrast function in the multi-unit case is, in our view, mutual information. Mutual information $I$ between $m$ (scalar) random variables $y_i, i = 1 \ldots, m$, is

$$I(y_1, y_2, \ldots, y_m) = \sum_i H(y_i) - H(\mathbf{y})$$

where $H$ denotes differential entropy. The mutual information is a natural measure of the dependence between random variables. It is always non-negative, and zero if and only if the variables are statistically independent. Thus the mutual information takes into account the whole dependence structure of the variables.

Important property: for an invertible linear transformation $\mathbf{y} = \mathbf{W}\mathbf{x}$:

$$I(y_1, y_2, \ldots, y_m) = \sum_i H(y_i) - H(\mathbf{x}) - \log|\det \mathbf{W}|.$$

### 8.3.4 Infomax algorithm [3]

$$\mathbf{W}_{l+1} = \mathbf{W}_l + \eta \left[\mathbf{I} - E(\mathbf{g}(\mathbf{y}_l)\mathbf{y}_l^\top)\right] \left(\mathbf{W}_l^{-1}\right)^\top,$$

where $\mathbf{g}$ is a vector function acting componentwise, i.e., $\mathbf{g}(\mathbf{y}) = [g_1(y_1, \ldots, g_n(y_n)]^\top$, $g_i$ are chosen nonlinear functions, $E$ is the expectation operator, and $\mathbf{y}_l = \mathbf{W}_l\mathbf{x}$. The principle of network entropy maximization, or "infomax," is equivalent to maximum likelihood estimation. This equivalence requires $g_i = (\log f_i)' = \frac{f_i'}{f_i}$, where $f_i$ are the probability density functions (pdfs) of the sources.

### 8.3.5 Kullback–Leibler divergence

Defined for two probability densities $p$ and $q$ as

$$\delta(p, q) = \int p(\mathbf{y}) \log \frac{p(\mathbf{y})}{q(\mathbf{y})} d\mathbf{y}.$$

Property: $\delta(p, q) \geq 0$, with equality if and only if $p = q$. We can measure the independence of the $y_i$ as the Kullback–Leibler divergence between the real density $f(\mathbf{y})$ and the factorized density $\tilde{f}(\mathbf{y}) = f_1(y_1)f_2(y_2) \cdots, f_m(y_m)$, where the $f_i(.)$ are the marginal densities of the $y_i$.

### 8.3.6 Natural gradient algorithm [2]

$$\mathbf{W}_{l+1} = \mathbf{W}_l + \eta \left[\mathbf{I} - E(\mathbf{g}(\mathbf{y}_l)\mathbf{y}_l^\top\right] \mathbf{W}_l,$$

where $\mathbf{y}_l = \mathbf{W}_l\mathbf{x}$. Interpretation:

- minimization of the Kullback–Leibler divergence between the real density $f(\mathbf{y})$ and the factorized density using Rimanian (natural) gradient on the manifold of all orthogonal matrices;
- nonlinear decorrelation.

A wide class of algorithms for ICA can be expressed in general form as [13]

$$\frac{d\,\mathbf{W}(t)}{dt} = \mu(t)\,\mathbf{F}(\mathbf{y}(t))\,\mathbf{W}(t), \qquad (8.19)$$

where $\mathbf{y}(t) = \mathbf{W}(t)\mathbf{x}(t)$ and the matrix $\mathbf{F}(\mathbf{y})$ can take different forms, for example $\mathbf{F}(\mathbf{y}) = \mathbf{I}_n - \mathbf{f}(\mathbf{y})\mathbf{g}^\top(\mathbf{y})$ with suitably chosen nonlinearities $\mathbf{f}(\mathbf{y}) = [f(y_1),\ldots,f(y_n)]$ and $\mathbf{g}(\mathbf{y}) = [g(y_1),\ldots,g(y_n)]$ [13].

For some $\mathbf{F}$, the dynamical system (8.19) does not correspond with minimization of any cost function, for example, this is the case of nonholonomic orthogonal learning algorithm [13], where, for specific $\mathbf{F}(\mathbf{y}) = \mathrm{diag}\{\mathbf{f}(\mathbf{y})\mathbf{y}^\top\} - \mathbf{f}(\mathbf{y})\mathbf{y}^\top$, the diagonal elements of $\mathbf{F}(\mathbf{y})$ are zero. The main observation for proving this fact is that for a given diagonal matrix $\mathbf{D}$ (different from the identity matrix), there is no cost function $\mathbf{J}(\mathbf{W})$ such that

$$\frac{\partial \mathbf{J}(\mathbf{W})}{\partial \mathbf{W}} = \mathbf{D}\left(\mathbf{W}^{-1}\right)^\top.$$

This fact follows from the criterion for the existence of potential functions (see [32, Theorem 3.4]).

For such a general case, the algorithm may diverge to infinity, or may converge to zero. The following modification, which stabilizes the Frobenius norm of $\mathbf{W}$, is proposed in [14].

**Theorem 4.** *The learning rule*

$$\frac{d\,\mathbf{W}(t)}{dt} = \mu(t)\,[\mathbf{F}(\mathbf{y}(t)) - \beta\gamma(t)\mathbf{I}_n]\,\mathbf{W}(t), \qquad (8.20)$$

*where $\beta > 0$ is a scaling factor and $\gamma(t) = trace\left(\mathbf{W}^\top(t)\mathbf{F}(\mathbf{y}(t))\mathbf{W}(t)\right) > 0$ stabilizes the Frobenius norm of $\mathbf{W}(t)$ such that $\|\mathbf{W}(t)\|_F^2 = trace(\mathbf{W}^\top(t)\mathbf{W}(t)) \approx \beta^{-1}$.*

### Joint approximate diagonalization of eigen-matrices

Define a fourth-order cumulant matrix $\mathbf{C}_{\mathbf{x},\mathbf{x}_p}^{2,2}(\mathbf{B})$ of the sensor signals as follows [9] (for $p = 0$) [20]:

$$\mathbf{C}_{\mathbf{x},\mathbf{x}_p}^{2,2}(\mathbf{B}) = E\{\mathbf{x}\mathbf{x}^\top \mathbf{x}_p^\top \mathbf{B}\mathbf{x}_p\} - E\{\mathbf{x}\mathbf{x}^\top\}\mathrm{tr}(\mathbf{B}E\{\mathbf{x}_p\mathbf{x}_p^\top\})$$
$$- E\{\mathbf{x}\mathbf{x}_p^\top\}\mathbf{B}E\{\mathbf{x}_p\mathbf{x}^\top\} - E\{\mathbf{x}\mathbf{x}_p^\top\}\mathbf{B}^\top E\{\mathbf{x}_p\mathbf{x}^\top\}.$$

Assume that the additive noise $\mathbf{n}$ has independent Gaussian components (with zero means), which are independent also with $s_i$, $i = 1,\ldots,n$. The $(i,j)^{\text{th}}$ element of $\mathbf{C}_{\mathbf{x},\mathbf{x}_p}^{2,2}(\mathbf{B})$ is

$$C_{\mathbf{x},\mathbf{x}_p}^{2,2}(\mathbf{B})_{i,j} = \sum_{k,l=1}^{n} \mathbf{cum}\{x_i(t), x_j(t), x_k(t-p), x_l(t-p)\}B_{k,l},$$

where $\mathbf{cum}\{x_i(t), x_j(t), x_k(t-p), x_l(t-p)\}$ denotes the fourth-order cumulant. Main property: if $\mathbf{x} = \mathbf{Hs}$ and $\mathbf{s}$ has independent components, then

$$\mathbf{C}_{\mathbf{x},\mathbf{x}_p}^{2,2}(\mathbf{B}) = \mathbf{H}\Delta(\mathbf{B})\mathbf{H}^\top,$$

where $\Delta(\mathbf{B}) = \text{diag}\{\mathbf{cum}_{s_1}(p)\mathbf{H}_{*1}^\top\mathbf{BH}_{*1}, \ldots, \mathbf{cum}_{s_n}(p)\mathbf{H}_{*n}^\top\mathbf{BH}_{*n}\}$, $\mathbf{cum}_{s_i}(p)$ $= \mathbf{cum}\{s_i(k), s_i(k), s_i(k-p), s_i(k-p)\}$ and $\mathbf{H}_{*i}$ denotes the $i^{\text{th}}$ column of $\mathbf{H}$. Therefore, if the mixing matrix $\mathbf{H}$ is orthogonal, we can separate the sources by:

- eigenvalue decomposition of $\mathbf{C}_{\mathbf{x},\mathbf{x}_p}^{2,2}(\mathbf{B})$ (if its eigenvalues are distinct), which estimates $\mathbf{H}$ up to multiplication with permutation and diagonal matrices – this method works if the initial sources are independent enough,
- joint diagonalization of several cumulant matrices: find orthogonal matrix $\mathbf{H}$ such that the matrices

$$\mathbf{H}^\top\mathbf{CH} : \mathbf{C} \in \mathcal{C}$$

   are diagonal as much as possible.

We will consider two classes of $\mathcal{C}$, which give two types of algorithms.

(1) JADE algorithm [10, 11]: $\mathcal{C} = \{\mathbf{C}_{\mathbf{x},\mathbf{x}}^{2,2}(\mathbf{B}), \mathbf{B} \in \mathcal{B}\}$ when $\mathcal{B}$ consists of eigen-matrices $\mathbf{B}$ of the cumulant tensor defined by the linear operator $\mathbf{F}$ on all matrices by $\mathbf{F}(\mathbf{M})_{ij} := \sum_{kl}\mathbf{cum}(x_i, x_j, x_k, x_l)M_{kl}$, i.e., $\mathbf{F}(\mathbf{B}) = \lambda\mathbf{B}$.
(2) JADETD algorithm [20]: when $\mathbf{B} = \mathbf{I}$ (identity matrix) and we jointly duagonalize the class of matrices $\mathcal{C} = \{\mathbf{C}_{\mathbf{x},\mathbf{x}_p}^{2,2}(\mathbf{I}), p = 1, \ldots, L\}$. It can separate colored sources of order 4, which are white of order 2.

## 8.4 Sparse Component Analysis and Blind Source Separation Using Sparseness

In this section, we present SCA and the BSS problem, initiated in [23, 22, 24]. We show that it can be solved if the sources are sufficiently sparse, even if the mixing matrix is singular. More generally, we consider the problem of identifying the source matrix $\mathbf{S} \in \mathbb{R}^{n \times N}$ if a linear mixture $\mathbf{X} = \mathbf{AS}$ is known only, where $\mathbf{A} \in \mathbb{R}^{m \times n}, m \leq n$ and the rank of $\mathbf{A}$ is less than $m$. A sufficient condition for solving this problem is that the level of sparsity of $\mathbf{S}$ is bigger than $m - \text{rank}(\mathbf{A})$, in the sense that the number of zeros in each column of $\mathbf{S}$ is bigger than $m - \text{rank}(\mathbf{A})$.

**Definition 3.** *A vector* $\mathbf{v} \in \mathbb{R}^m$ *is said to be* $k$-*sparse if* $\mathbf{v}$ *has at least* $k$ *zero entries. A matrix* $\mathbf{S} \in \mathbb{R}^{m \times n}$ *is said to be* $k$-*sparse if each column of it is* $k$-*sparse.*

270 P.G. Georgiev and F.J. Theis

The goal of BSS of level $k$ ($k$-BSS) is to decompose a given $m$-dimensional random vector $\mathbf{X}$ into

$$\mathbf{X} = \mathbf{AS} \tag{8.21}$$

with a real $m \times n$-matrix $\mathbf{A}$ and an $n \times N$-dimensional $k$-sparse matrix $\mathbf{S}$. $\mathbf{S}$ is called the *source matrix*, $\mathbf{X}$ the *mixtures*, and $\mathbf{A}$ the *mixing matrix*. We speak of *complete*, *overcomplete*, or *undercomplete* $k$-BSS if $m = n$, $m < n$, or $m > n$ respectively.

Note that in contrast with the ICA model, the above problem is not translation invariant. However it is easy to see that if instead of $\mathbf{A}$ we choose an affine linear transformation, the translation constant can be determined from $\mathbf{X}$ only, as long as the sources are non-determined. Termed differently, this means that instead of assuming $k$-sparseness of the sources, we could also assume that in any column of $\mathbf{S}$ only $n - k$ components are allowed to vary from a previously fixed constant (which can be different for each source).

In the following, we will assume without loss of generality that $m \leq n$: the undercomplete case can be reduced to the complete case by projection of $\mathbf{X}$.

The following theorem is a generalization of a similar one from [23]. Here, for illustrative purposes, we formulate the theorem for the case when the rank of $\mathbf{A}$ is $m - 1$, but its formulation in full generality is straightforward.

**Theorem 5 (Matrix identifiability 1).** *Assume that* $\mathbf{X}$ *satisfies (8.1) and*

1. *every $m - 1$ columns of the matrix $\mathbf{A}$ are linearly independent;*

*the indexes $\{1, \ldots, N\}$ are divided in two groups $\mathcal{N}_1$ and $\mathcal{N}_2$ such that*

2. *vectors from the group $\{\mathbf{S}_1 = \{\mathbf{S}(:, j)\} : j \in \mathcal{N}_1\}$ are sufficiently rich represented in the sense that for any index set of $n - m + 2$ elements $I \subset \{1, \ldots, n\}$ there exist at least $m-1$ vectors $\mathbf{s}_1, \ldots, \mathbf{s}_{m-1}$ from $\mathbf{S}_1$ (depending on $I$) such that each of them has zero elements in places with indexes in $I$ and there exists at least one subgroup of $\{\mathbf{s}_1, \ldots, \mathbf{s}_{m-1}\}$ consisting of $m-2$ linearly independent elements;*
3. *the vectors from the group $\{\mathbf{X}(:, j), j \in \mathcal{N}_2\}$ have the property that no subset of $m - 1$ elements from them lie on a 2-codimensional subspace.*

*Then $\mathbf{A}$ is uniquely determined by $\mathbf{X}$ except for right-multiplication with permutation and scaling matrices, i.e., if $\mathbf{X} = \mathbf{AS} = \hat{\mathbf{A}}\hat{\mathbf{S}}$, then $\mathbf{A} = \hat{\mathbf{A}}\mathbf{PL}$ with a permutation matrix $\mathbf{P}$ and a nonsingular diagonal scaling matrix $\mathbf{L}$.*

**Proof.** It is clear that any column $\mathbf{a}_j$ of the mixing matrix lies in the intersection of all $\binom{n-1}{m-3}$ 2-codimensional subspaces generated by those groups of columns of $\mathbf{A}$, in which $\mathbf{a}_j$ participates.

We will show that these 2-codimensional subspaces can be obtained by the columns $\{\mathbf{X}(:, j), j \in \mathcal{N}_1\}$ under the condition of the theorem. Let $\mathcal{J}$ be the set of all subsets of $\{1, \ldots, n\}$ containing $m - 2$ elements and let $J \in \mathcal{J}$. Note that $\mathcal{J}$ consists of $\binom{n}{m-2}$ elements. We will show that the 2-codimensional subspace (denoted by $H_J$) spanned by the columns of $\mathbf{A}$ with indexes from $J$ can be obtained by some elements from $\{\mathbf{X}(:,j), j \in \mathcal{N}_1\}$. By (2), there exist $m-1$ indexes $\{t_k\}_{k=1}^{m-1} \subset \mathcal{N}_1$ and $m-2$ vectors from the group $\{\mathbf{S}(:,t_k)\}_{k=1}^{m-1}$, which form a basis of the $(m-2)$-dimensional coordinate subspace of $\mathbb{R}^n$ with zero coordinates given by the indexes $\{1, \ldots, n\} \setminus J$. Because of the mixing model, vectors of the form

$$\mathbf{v}_k = \sum_{j \in J} S(j, t_k)\mathbf{a}_j, \quad k = 1, \ldots, m-1,$$

belong to the group $\{\mathbf{X}(:,j) : j \in \mathcal{N}_1\}$. Now, applying condition (1) we obtain that there exists a subgroup of $m-2$ vectors from $\{\mathbf{v}_k\}_{k=1}^{m-1}$ that are linearly independent. This implies that the vectors $\{\mathbf{v}_k\}_{k=1}^{m-1}$ will span the same 2-codimensional subspace $H_J$. By (1) it follows that the 2-codimensional subspaces $H_{J_1}$ and $H_{J_2}$ are different, if the indexes $J_1 \in \mathcal{J}$ and $J_2 \in \mathcal{J}$ are different. By the above reasonings and by (3) it follows that if we cluster the columns of $\mathbf{X}$ in 2-codimensional subspaces containing more than $m-2$ elements from the columns of $\mathbf{X}$, we will obtain $\binom{n}{m-2}$ unique 2-codimensional subspaces, containing all elements of $\{\mathbf{X}(:,j), j \in \mathcal{N}_1\}$ and no elements from $\{\mathbf{X}(:,j), j \in \mathcal{N}_2\}$. Now we cluster the 2-codimensional subspaces obtained in such a way in the smallest number of groups such that the intersection of all 2-codimensional subspaces in one group gives a single one-dimensional subspace. It is clear that such one-dimensional subspace will contain one column of the mixing matrix, the number of these groups is $n$, and each group consists of $\binom{n-1}{m-3}$ 2-codimensional subspaces.

In such a way, we can identify the columns of the mixing matrix up to scaling and permutation. In other words, if $\mathbf{X} = \mathbf{AS} = \hat{\mathbf{A}}\hat{\mathbf{S}}$, then $\mathbf{A} = \hat{\mathbf{A}}\mathbf{PL}$ with a permutation matrix $\mathbf{P}$ and a nonsingular diagonal scaling matrix $\mathbf{L}$. $\square$

In a similar way, we can prove the following generalization of the above theorem.

**Theorem 6 (Matrix identifiability 2).** *Assume that $\mathbf{X}$ satisfies (8.1) and*

*1. every $m-1$ columns of the matrix $\mathbf{A}$ are linearly independent;*

*and the indexes $\{1, \ldots, N\}$ are divided in two groups $\mathcal{N}_1$ and $\mathcal{N}_2$ such that*

*2. vectors from the group $\mathbf{S}_1 = \{\mathbf{S}(:,j)\}, j \in \mathcal{N}_1$ are sufficiently rich represented in the sense that, for any index set of $n - m + 2$ elements $I \subset \{1, \ldots, n\}$, there exist $N_I \geq m$ vectors $\mathbf{s}_1, \ldots, \mathbf{s}_{N_I}$ from $\mathbf{S}_1$ (depending on $I$) such that each of them has zero elements in places with indexes in $I$ and there exists a subset of $\{\mathbf{s}_1, \ldots, \mathbf{s}_{N_I}\}$ containing $m-2$ linearly independent elements;*

*3. the vectors from the group $\{\mathbf{X}(:,j), j \in \mathcal{N}_2\}$ have the property that at most $\min\{N_{I_1}, \ldots, N_{I_p}\} - 1$ of them lie on a common 2-codimensional subspace, where $\{I_1, \ldots, I_p\}$ is the set of all subsets of $\{1, \ldots, n\}$ with $n - m + 2$ elements and $p = \binom{n}{m-2}$.*

*Then* $\mathbf{A}$ *is uniquely determined by* $\mathbf{X}$ *except for right-multiplication with permutation and scaling matrices, i.e., if* $\mathbf{X} = \mathbf{AS} = \hat{\mathbf{A}}\hat{\mathbf{S}}$, *then* $\mathbf{A} = \hat{\mathbf{A}}\mathbf{PL}$ *with a permutation matrix* $\mathbf{P}$ *and a nonsingular diagonal scaling matrix* $\mathbf{L}$.

The proof of Theorem 1 gives the idea for the matrix identification algorithm.

## Algorithm for identification of the mixing matrix (under the assumptions of Theorems 1 or 2)

1. Cluster the columns $\{\mathbf{X}(:,j) : j \in \mathcal{N}_1\}$ in $\binom{n}{m-2}$ groups $\mathcal{H}_k$, $k = 1, \ldots, \binom{n}{m-2}$ such that the span of the elements of each group $\mathcal{H}_k$ produces one 2-codimensional subspace and these 2-codimensional subspaces are different.
2. Calculate any basis of the orthogonal complement of each of these 2-codimensional subspaces.
3. Find all possible groups such that each of them is composed of the elements of $\binom{n-1}{m-3}$ bases in (2), and the vectors in each group span a hyperplane. The number of these hyperplanes gives the number of sources $n$. The normal vectors to these hyperplanes are estimations of the columns of the mixing matrix $\mathbf{A}$ (up to permutation and scaling).

In practical realization, all operations in the above algorithm are performed up to some precision $\varepsilon > 0$.

*Remark 1.* The above algorithm is quite general and allows different realizations. Below we propose another method for matrix identification, based on PCA.

The above theorems shows that we can recover the mixing matrix from the mixtures uniquely, up to permutation and scaling of the columns. The next theorem shows that in this case also the sources $\{\mathbf{S}(:,j) : j \in \mathcal{N}_1\}$ can be recovered uniquely (up to a set of "bad" data points that has measure zero with respect to the "good" data points).

### 8.4.1 Identification of sources

The following theorem is generalization of those in [23] and the proof is the same.

**Theorem 7 (Uniqueness).** *Let* $\mathcal{H}$ *be the set of all* $\mathbf{x} \in \mathbb{R}^m$ *such that the linear system* $\mathbf{As} = \mathbf{x}$ *has a solution with at least* $n - m + k$ *zero components* $(k \geq 1)$. *If any* $m - k$ *columns of* $\mathbf{A}$ *are linearly independent, then there exists a subset* $\mathcal{H}_0 \subset \mathcal{H}$ *with measure zero with respect to* $\mathcal{H}$, *such that for every* $\mathbf{x} \in \mathcal{H} \setminus \mathcal{H}_0$, *this system has no other solution with this property.*

From Theorem 7, it follows that the sources are uniquely identifiable generically, i.e., up to a set with a measure zero, if they compose a matrix that is $(n - m + k)$-sparse, and the mixing matrix is known. Below we present an algorithm based on the observation in Theorem 7.

**Source recovery algorithm:**

1. Identify the the set of $k$-codimensional subspaces $\mathcal{H}$ produced by taking the linear hull of every subsets of the columns of $\mathbf{A}$ with $m - k$ elements;
2. Repeat for $i = 1$ to $N$:
   (a) Identify the space $H \in \mathcal{H}$ containing $\mathbf{x}_i := \mathbf{X}(:, i)$, or, in practical situation with presence of noise, identify the one to which the distance from $\mathbf{x}_i$ is minimal and project $\mathbf{x}_i$ onto $H$ to $\tilde{\mathbf{x}}_i$;
   (b) if $H$ is produced by the linear hull of column vectors $\mathbf{a}_{i_1}, \ldots, \mathbf{a}_{i_{m-k}}$, then find coefficients $\mathbf{L}_{i,j}$ such that

$$\tilde{\mathbf{x}}_i = \sum_{j=1}^{m-k} \mathbf{L}_{i,j}\mathbf{a}_{i_j}.$$

These coefficients are uniquely determined if $\tilde{\mathbf{x}}_i$ doesn't belong to the set $\mathcal{H}_0$ with measure zero with respect to to $\mathcal{H}$ (see Theorem 7);
   (c) Construct the solution $\mathbf{s}_i = \mathbf{S}(:, i)$: it contains $\mathbf{L}_{i,j}$ in the place $i_j$ for $j = 1, \ldots, m - k$, the other its components are zero.

### 8.4.2 A new algorithms for sparse representation based on subspace clustering

Our new sparse representation algorithm makes use of the concept of *skeleton of a finite set of points* defined below.

The solution $\{(\mathbf{n}_i^0, b_i^0)\}_{i=1}^k$ of the minimization problem:

$$\text{minimize} \sum_{j=1}^N \min_{1 \leq i \leq k} |\mathbf{n}_i^\top \mathbf{x}_j - b_i|^l \qquad (8.22)$$

$$\text{subject to} \quad \|\mathbf{n}_i\| = 1, b_i \in \mathbb{R}, i = 1, \ldots, k, \qquad (8.23)$$

defines the $k^{(l)}$-*skeleton* of $\mathbf{X}$, introduced for $l = 1$ in [39] and for $l = 2$ in [6]. It consists of a union of $k$ hyperplanes $H_i = \{x \in \mathbb{R}^m : \mathbf{n}_i^\top x = b_i\}, i = 1, \ldots, k$, such that the sum of minimum distances raised to power $l$ from every point $\mathbf{x}_j$ to them is minimal. We introduce the "affine $r$-subspace skeleton" as the solution of the following minimization problem:

$$\text{minimize} \quad \sum_{j=1}^{N} \min_{1 \leq i \leq k} \sum_{s=1}^{r} |\mathbf{u}_{i,s}^{\top} \mathbf{x}_j - b_i|^l \tag{8.24}$$

$$\text{subject to} \quad \|\mathbf{n}_{i,s}\| = 1, b_i \in \mathbb{R}, i = 1, \ldots, k, \tag{8.25}$$

$$s = 1, \ldots, r, \mathbf{n}_{i,p}^{\top} \mathbf{n}_{i,q} = 0, p \neq q. \tag{8.26}$$

Assuming that the conditions of Theorem 5 are satisfied, it is clear that the representation $\mathbf{X} = \mathbf{AS}$ is $n - m + r$-sparse (each column of $\mathbf{S}$ contains at most $m - r$ non-zero elements), the above two skeletons coincide, $b_i = 0$, and the data points (columns of $\mathbf{X}$) lie on them.

The *Subspace Clustering Algorithm* (see below) finds these skeletons. It can be considered as a generalization of the Bradley–Mangasarian [6] $k$-plane clustering algorithm.

### Subspace clustering algorithm

**Data:** samples $\mathbf{x}(1), \ldots, \mathbf{x}(T)$ (column vectors) of $\mathbf{X}$
**Result:** estimated $k$ groups of orthonormal vectors: $G_i = \{\mathbf{u}_{i,s} \ (s = 1, \ldots, r\}$,
    $i = 1, \ldots, k)$, as $\mathbf{u}_{i,s_1}^{\top} \mathbf{u}_{i,s_2} = 0 \ (s_1 \neq s_2)$
Initialize randomly $k$ groups of orthonormal vectors $G_i = \{\mathbf{u}_{i,s}, \ (s = 1, \ldots, r\}$,
$i = 1, \ldots, k)$, as $\mathbf{u}_{i,s_1}^{\top} \mathbf{u}_{i,s_2} = 0 \ (s_1 \neq s_2)$.
**for** $j \leftarrow 1, \ldots, j_0$ **do**
    *Cluster assignment*
    **for** $t \leftarrow 1, \ldots, T$ **do**
        Add $\mathbf{x}(t)$ to cluster $\mathbf{Y}^{(i)}$, where $i$ is chosen to minimize $\sum_{s=1}^{r} |\mathbf{u}_{i,s}^{\top} \mathbf{x}(t)|^2$
        (distance to the orthogonal complement of $span\{\mathbf{u}_{i,s}\}_{s=1}^{r}$)
    **end**
    Exit if the mean distance to the subspaces is smaller than some preset value.
**end**

*Cluster update*
**for** $i \leftarrow 1, \ldots, k$ **do**
    **for** $s \leftarrow 1, \ldots, r$ **do**
        Define projection matrix $\mathbf{P}$ with rows consisting of an orthonormal basis of the orthogonal complement of $\mathbf{u}_{i,1}, \ldots, \mathbf{u}_{i,s-1}$ (if $s = 1$, then $\mathbf{P} = \mathbf{I}_m$, the identity matrix)
        Calculate projected cluster covariance $\mathbf{C} \leftarrow \mathbf{P}\mathbf{Y}^{(i)}(\mathbf{Y}^{(i)})^{\top}\mathbf{P}^{\top}$
        Choose eigenvector $\mathbf{v}_s$ of $\mathbf{C}$ corresponding with a minimal eigenvalue.
        Set $\mathbf{u}_{i,s} \leftarrow \mathbf{P}^{\top}\mathbf{v}_s$.
    **end**
**end**

The constant $j_0$ is in practice chosen to be sufficiently large. The finite termination of the algorithm is proved in [6, Theorem 3.7].

### 8.4.3 Orthogonal $m$-planes clustering algorithm

In this section, we propose a modification of the $k$-plane clustering algorithm of Bradley and Mangasarian [6]. The idea is to reduce the problem of finding the $m$-skeleton of $\mathbf{X}$ to an orthogonal problem: requiring that the hyperplanes of it are orthogonal, i.e., defined by an orthonormal matrix $\mathbf{W} \in \mathbb{R}^{m \times m}$. This can be done in the following way, if we we assume that the source matrix $\mathbf{S}$ after normalization is semi-orthogonal, i.e., $\tilde{\mathbf{S}}\tilde{\mathbf{S}}^\top = \mathbf{I}$.

Let $\mathbf{XX}^\top = \mathbf{ULU}^\top$ be the eigenvalue decomposition of the matrix $\mathbf{XX}^\top$. Assume that the diagonal elements of $\mathbf{L}$ are positive. Then, denoting $\mathbf{W} = \mathbf{L}^{-1/2}\mathbf{U}^\top \mathbf{A}$ and $\mathbf{Y} = \mathbf{L}^{-1/2}\mathbf{U}^\top \mathbf{X}$, we have

$$\mathbf{Y} = \mathbf{W}\tilde{\mathbf{S}}, \quad \mathbf{WW}^\top = \mathbf{I}, \quad \mathbf{YY}^\top = \mathbf{I}. \tag{8.27}$$

Then the cluster update steps in Bradley–Mangasarian algorithm [6] can be unified in the following optimization problem with orthogonality constraints:

$$\text{minimize} \quad \sum_{i=1}^{m} \mathbf{w}_i^\top \mathbf{Y}^{(i)} (\mathbf{Y}^{(i)})^\top \mathbf{w}_i \tag{8.28}$$

$$\text{under constraints} \quad \mathbf{w}_i \mathbf{w}_j^\top = \delta_{ij}, \tag{8.29}$$

where $\mathbf{Y}^{(i)}$ is the matrix with vector columns, which are elements of the $i^{\text{th}}$ cluster.

### Orthogonal $m$-planes clustering algorithm

**Data:** samples $\mathbf{x}(1), \ldots, \mathbf{x}(T)$ of $\mathbf{X}$
**Result:** estimated orthonormal mixing matrix $\mathbf{W}$ in (8.27)

Initialize randomly $\mathbf{W} = (\mathbf{w}_1, \ldots, \mathbf{w}_n)$ – orthonormal matrix.
**for** $j \leftarrow 1, \ldots, j_0$ **do**
    *Cluster assignment*
    **for** $t \leftarrow 1, \ldots, T$ **do**
        Add $\mathbf{x}(t)$ to cluster $\mathbf{Y}^{(i)}$, where $i$ is chosen to minimize $|\mathbf{w}_i^\top \mathbf{x}(t)|$ (distance to hyperplane given by the $i^{\text{th}}$ column of $\mathbf{W}$).
    **end**
    Exit if the mean distance to the hyperplanes is smaller than some preset value.
**end**
*Matrix update*
**for** $k \leftarrow 1, \ldots, n$ **do**
    **for** $s \leftarrow 1, \ldots, r$ **do**
        Define projection matrix $\mathbf{P}$ with rows consisting of an orthonormal basis of the orthogonal complement of $\mathbf{w}_1, \ldots, \mathbf{w}_{k-1}$
        Calculate projected cluster covariance $\mathbf{C} \leftarrow \mathbf{P}\mathbf{Y}^{(i)}(\mathbf{Y}^{(i)})^\top \mathbf{P}^\top$

Choose eigenvector $\mathbf{v}_k$ of $\mathbf{C}$ corresponding with a minimal eigenvalue.
Set $\mathbf{w}_k \leftarrow \mathbf{P}^\top \mathbf{v}_k$.

**end**

**end**

The constant $j_0$ is in practice chosen to be sufficiently large. The finite termination of the algorithm is proved in [6, Theorem 3.7].

## 8.5 Applications

### 8.5.1 Computer simulation example: underdetermined case

We consider a mixture of 7 artificially created sources (see Figure 8.2 left) – sparsified randomly generated signals with at least 5 zeros in each column – with a randomly generated mixing matrix with dimension $3 \times 7$.

Figure 8.3 gives the mixed signals together with a normalized scatterplot of the mixtures – the data lies in $21 = \binom{7}{2}$ hyperplanes.

Applying the underdetermined matrix recovery algorithm [23] to the mixtures gives the recovered mixing matrix perfectly well, up to permutation and scaling. Applying the source recovery algorithm, we recover the source signals up to permutation and scaling (see Figure 8.3, middle). This figure (right) shows also that the recovery by $l_1$-norm minimization (known as the Basis Pursuit method [12]) does not perform well, even if the mixing matrix is perfectly known.

### 8.5.2 Computer simulation example: subspace clustering algorithm

We created four artificial source signals (nearly Gaussian), sparse of level 2, i.e., each column of the source matrix contains at least 2 zeros

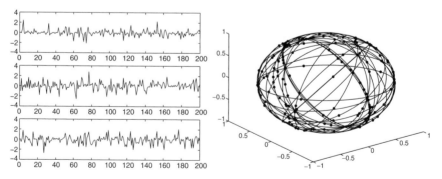

**Fig. 8.2.** Mixed signals (left) and normalized scatter plot (density) of the mixtures (right) together with the 21 data set hyperplanes, visualized by their intersection with the unit sphere in $\mathbb{R}^3$.

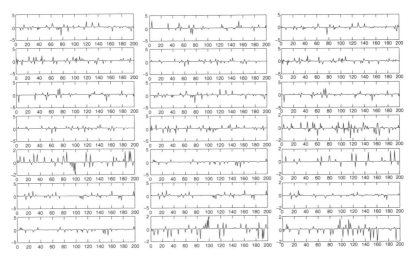

**Fig. 8.3.** The original source signals are shown in the left column. The middle column gives the recovered source signals — the signal-to-noise ratio between the original sources and the recoveries is very high (above 278 dB after permutation and normalization). Note that only 200 samples are enough for excellent separation. The right column shows the recovered source signals using $l_1$-norm minimization and known mixing matrix. Simple comparison confirms that the recovered signals are far from the original ones — the signal-to-noise ratio is only around 4 dB.

(shown in Figure 8.4). They are mixed with a square normalized matrix $\mathbf{H}4$ (each column of it has norm one):

$$\mathbf{H} = \begin{pmatrix} -0.0506 & -0.2818 & 0.5457 & 0.3111 \\ 0.1974 & 0.8497 & -0.4128 & -0.5214 \\ 0.9707 & 0.4291 & 0.6958 & 0.7645 \\ 0.1271 & 0.1200 & 0.2182 & 0.2163 \end{pmatrix}.$$

The mixed signals are shown in Figure 8.5.

We apply our subspace clustering algorithm in order to identify the skeleton of the data points and after that apply our matrix identification algorithm. We obtain an estimation $\mathbf{W}$ of the mixing matrix (after normalization of each column):

$$\mathbf{W} = \begin{pmatrix} 0.5457 & 0.3113 & 0.2819 & 0.0504 \\ -0.4128 & -0.5212 & -0.8498 & -0.1972 \\ 0.6958 & 0.7646 & -0.4289 & -0.9708 \\ 0.2182 & 0.2165 & -0.1199 & -0.1273 \end{pmatrix}.$$

which is very near to $\mathbf{H}$ (up to permutation and sign).

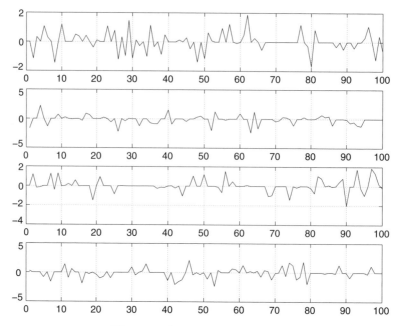

**Fig. 8.4.** Original source signals.

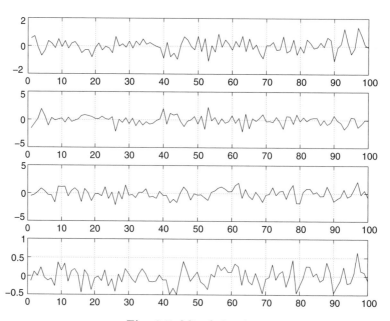

**Fig. 8.5.** Mixed signals.

### 8.5.3 Applications of SCA to fMRI data

**SCA applied to fMRI toy data**

We simulate a low-dimensional example of fMRI data analysis. The typical setup of fMRI experiments is the following: NMR brain imaging techniques are used to record brain activity data over a certain span of time, during which the subject is asked to perform some kind of task (e.g., 5 seconds of activity in the motor cortex followed by 5 seconds of activity in the visual cortex; this iterative procedure is often called *block diagram*). The brain recordings show areas of high and of low brain activity (using the *BOLD effect*). Analysis is performed on the 2D-image slices recorded at the discrete time steps. General linear model (GLM) approaches or ICA-based fMRI analysis then decompose this data set into a certain set of *component maps* (see Figure 8.11), i.e., sets of (hopefully independent) images that are active at certain time steps corresponding to the block diagram.

In the following, we simulate a low-dimensional example of such brain activity recordings. For this we mix three "source component maps" (Figure 8.6) linearly to three mixture images and add some noise.

These mixtures represent our recordings at three different time steps. Only from the recordings, we want to recover the original components or component maps. We want to use an unsupervised approach (not GLM, which requires additional knowledge of the mixing system) but with a different contrast than ICA. We believe that the assumption of *independence* of the component maps does not hold in a lot of situations, so we replace this assumption by *sparseness* of the maps, meaning that at a certain voxel, not all maps are allowed to be active (in the case of as many mixtures as sources).

We consider a mixture of 3 artificially created *non-independent* source images of size $30 \times 30$ — see Figure 8.6 — with the (normalized) mixing matrix

$$\mathbf{A} = \begin{pmatrix} -0.9069 & 0.1577 & 0.4726 \\ -0.2737 & -0.9564 & 0.0225 \\ -0.3204 & -0.2458 & -0.8810 \end{pmatrix}$$

and 4% of additive white noise. The mixtures are shown in Figure 8.7 together with their scatterplot after normalization to unit length.

**Fig. 8.6.** Example: artificial *non-independent* and *non-sparse* source signals.

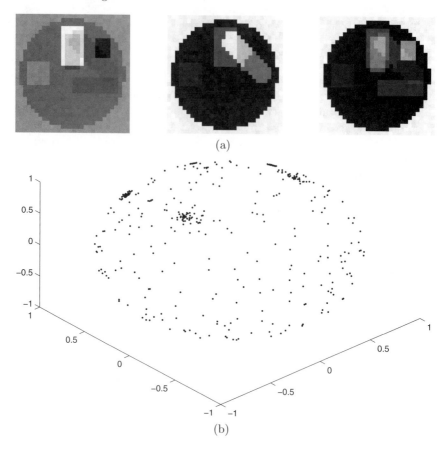

**Fig. 8.7.** Example: mixed signals with 4% additive noise (a), and scatterplot after normalization to unit length (b).

Note that due to the circular "brain region," we have to preprocess the data ("sparsification") by removing the non-brain voxels from the boundary. Then, we apply the matrix identification algorithm. This gives the recovered matrix (after normalization)

$$\hat{\mathbf{A}} = \begin{pmatrix} 0.9110 & 0.1660 & 0.4693 \\ 0.2823 & -0.9541 & 0.0135 \\ 0.3007 & -0.2494 & -0.8829 \end{pmatrix}$$

with low cross-talking error 0.12 and the recovered sources $\hat{\mathbf{S}}$ shown in Figure 8.8, with high signal-to-noise ratio of 28, 27, and 27 dB with respect to the original sources (after permutation and normalization).

This can be enhanced by applying a denoising algorithm to each image. Figure 8.9 shows the application of local PCA denoising with an MDL-

**Fig. 8.8.** Example: recovered source signals. The signal-to-noise ratio between the original sources (Figure 8.6) and the recoveries is high with 28, 27, and 27 dB after permutation and normalization.

**Fig. 8.9.** Example: recovered denoised source signals. Now the SNR is even higher than that in Figure 8.8 (32, 31, and 29 dB after permutation and normalization).

parameter estimation criterion, which gives SNRs of 32, 31, and 29 dB now, so a mean enhancement of around 4 dB has been achieved.

Note that if we apply ICA to the previous example (after sparsification as above — without sparsification ICA performs even worse), the algorithm cannot recover the mixing matrix:

$$\bar{\mathbf{A}} = \begin{pmatrix} 0.6319 & -0.3212 & 0.8094 \\ -0.0080 & -0.8108 & -0.3138 \\ -0.7750 & -0.4893 & 0.4964 \end{pmatrix}$$

has very high cross-talking error of 4.7 with respect to $\mathbf{A}$. Figure 8.10 shows the poorly recovered sources; the SNRs with respect to the sources are only 3.3, 13, and 12 dB respectively. The reason for ICA not being able to recover the sources simply lies in the fact that they were not chosen to be independent.

## SCA applied to real fMRI data

We now analyze the performance of SCA when applied to real fMRI measurements. fMRI data were recorded from six subjects (3 female, 3 male, age 20–37) performing a visual task. In five subjects, five slices with 100 images (TR/TE = 3000/60 ms) were acquired with five periods of rest and five photic

**Fig. 8.10.** Example: poorly recovered source signals using ICA. The signal-to-noise ratio between the original sources (Figure 8.6) and the recoveries is very low with 3.3, 13 and 12 dB after permutation and normalization.

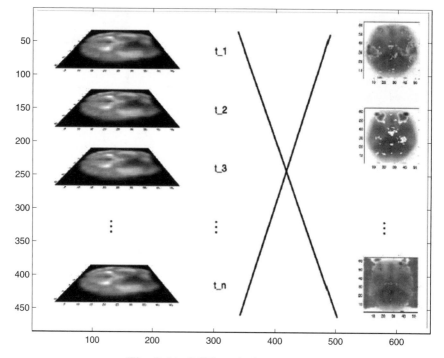

**Fig. 8.11.** fMRI analysis – setting.

simulation periods with rest. Simulation and rest periods comprised 10 repetitions each, i.e., 30 s. Resolution was $3 \times 3 \times 4$ mm. The slices were oriented parallel to the calcarine fissure. Photic stimulation was performed using an 8 Hz alternating checkerboard stimulus with a central fixation point and a dark background with a central fixation point during the control periods [49]. The first scans were discarded for remaining saturation effects. Motion artifacts were compensated by automatic image alignment (AIR, [50]).

Blind Signal Separation, mainly based on ICA, nowadays is a quite common tool in fMRI analysis (see, for example, [37, 36, 8, 47, 7] and references

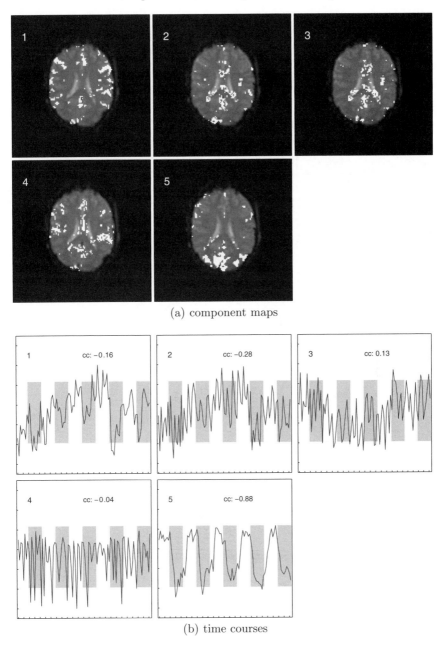

(a) component maps

(b) time courses

**Fig. 8.12.** SCA fMRI analysis. The data was reduced to the first 5 principal components: (a) shows the recovered component maps (white points indicate values stronger than 3 standard deviations) and (b) their time courses. The stimulus component is given in component 5 (indicated by the high cross-correlation $cc = -0.86$ with the stimulus time course, delayed by roughly 2 seconds due to the BOLD effect), which is strongly active in the visual cortex as expected.

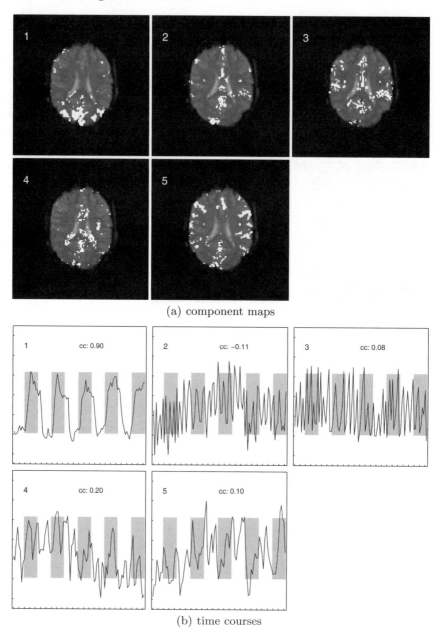

(a) component maps

(b) time courses

**Fig. 8.13.** FastICA result during fMRI analysis of the same data set as in Figure 8.12. The stimulus component is given in component 1 with high stimulus cross-correlation $cc = 0.90$.

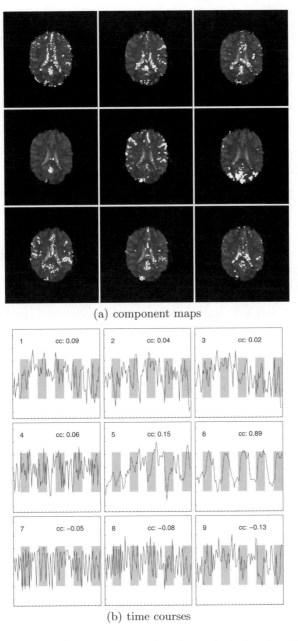

(a) component maps

(b) time courses

**Fig. 8.14.** fMRI analysis by our orthogonal $m$-planes clustering algorithm. The data was reduced to the first 9 principal components. (a) shows the recovered component maps (white points indicate values stronger than 3 standard deviations) and (b) their time courses. The stimulus component is given in component 6 (indicated by the high cross-correlation $cc = 0.89$ with the stimulus time course, delayed by roughly 2 seconds due to the BOLD effect), which is strongly active in the visual cortex as expected.

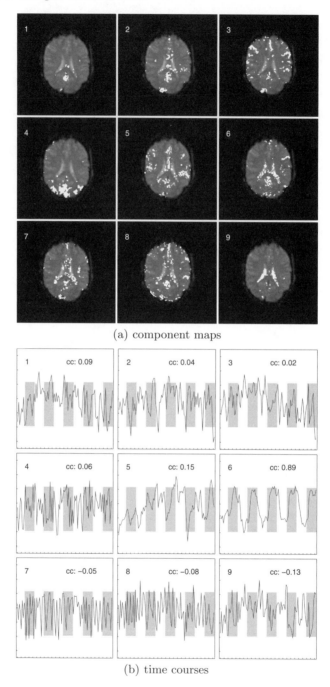

(a) component maps

(b) time courses

**Fig. 8.15.** FastICA result during fMRI analysis of the same data set as in Figure 8.12. The stimulus component is given in component 4 with high stimulus cross-correlation $cc = 0.87$.

therein). We analyze the fMRI data set using as a separation criterion a spatial decomposition of fMRI data images to sparse component maps [25]. Such an approach we consider as very reasonable and advantageous when the stimuli are sparse and dependent, and therefore the ICA methods could not give good results. Due to the availability of fMRI data, it appears that the results of our SCA method and ICA method give similar results, which itself we consider as a surprising fact. Here we again use the matrix identification algorithm and apply the matrix inversion algorithm the estimated matrix for estimation of the sources.

Figure 8.12 shows the performance of SCA method; see the figure legend for interpretation. Using only the first 5 principal components, SCA could recover the stimulus component as well as detect additional components. Figure 8.14 shows the performance of our orthogonal $m$-planes clustering algorithm (see the figure legend for interpretation). Now we use the first 9 principal components and recover the stimulus component and additional components. The performance of SCA algorithms is equally well as fastICA [29] (see Figures 8.13 and 8.15), which is interesting in itself: apparently the two different criteria, sparseness and independence, lead to similar results in this setting. This can be partially explained by noting that all components, mainly the stimulus component, have high kurtoses, i.e., strongly peaked densities.

## 8.6 Conclusion

We presented rigorous mathematical justification based on optimization theory of two basic methods for data representation: Independent Component Analysis and Sparse Component Analysis. Several algorithms and experiments are presented including application to toy and real fMRI data.

## Acknowledgments

The authors would like to thank Dr. Dorothee Auer from the Max Planck Institute of Psychiatry in Munich, Germany, for providing the fMRI data, and Oliver Lange from the Department of Clinical Radiology, Ludwig-Maximilian University, Munich, Germany, for data preprocessing and visualization.

## References

[1] V.M. Alekseev, V.M. Tihomirov, and S.V. Fomin. *Optimal Control*. Nauka, Moscow, 1979 (in Russian).
[2] S. Amari. Natural gradient works efficiently in learning. *Neural Computation*, 10(2):251–276, 1998.

[3] A. Bell and T. Sejnowski. An information-maximisation approach to blind separation and blind deconvolution. *Neural Computation*, 7:1129–1159, 1995.

[4] A. Belouchrani, K. A. Meraim, J.-P. Cardoso, and E. Moulines. A blind source separation technique using second order statistics. *IEEE Transactions on Signal Processing*, 45(2):434–444, 1997.

[5] P. Bofill and M. Zibulevsky. Underdetermined blind source separation using sparse representation. *Signal Processing*, 81(11):2353–2362, 2001.

[6] P.S. Bradley and O.L. Mangasarian. k-Plane clustering. *Journal of Global Optimization*, 16(1):23–32, 2000.

[7] V.D. Calhoun, T. Adali, J.J. Pekar, and G.D. Pearlson. Latency (in)sensitive ICA. group independent component analysis of fMRI data in the temporal frequency domain. *Neuroimage*, 20:1661–1669, 2003.

[8] V.D. Calhoun, L.K. Hansen T. Adali, J. Larsen, and J.J. Pekar. ICA of functional MRI data: an overview. In *4th International Symposium on Independent Component Analysis and Blind Signal Separation (ICA2003)*, pages 281–288, Nara, Japan, April 2003.

[9] J.F. Cardoso. High-order contrasts for independent component analysis. *Neural Computation*, 11(1):157–192, 1999.

[10] J.F. Cardoso and A. Souloumiac. Blind beamforming for non-Gaussian signals. *IEE Proceedings-F*, 140(6):362–370, 1993.

[11] J.F. Cardoso and A. Souloumiac. Jacobi angles for simultanious diagonalization. *SIAM Journal on Matrix Analysis and Applications*, 17(1):161–164, 1996.

[12] S. Chen, D. Donoho, and M. Saunders. Atomic decomposition by basis pursuit. *SIAM Journal on Scientific Computing*, 20(1):33–61, 1998.

[13] A. Cichocki and S. Amari. *Adaptive Blind Signal and Image Processing*. John Wiley, Chichester, U.K., 2002.

[14] A. Cichocki and P. Georgiev. Blind source separation algorithms with matrix constraints. *IEICE Transactions on Fundamentals of Electronics, Communications and Computer Sciences*, E86-A(3):522–531, 2003.

[15] A. Cichocki, R. Unbehauen, and E. Rummert. Robust learning algorithm for blind separation of signals. *Electronics Letters*, 30(17):1386–1387, 1994.

[16] P. Comon. Independent component analysis – a new concept? *Signal Processing*, 36:287–314, 1994.

[17] G. Darmois. Analyse générale des liaisons stochastiques. *Review of the International Statistical Insitute*, 21:2–8, 1953.

[18] N. Delfosse and P. Loubaton. Adaptive blind separation of independent sources: a deflation approach. *Signal Processing*, 45:59–83, 1995.

[19] J. Eriksson and V. Koivunen. Identifiability and separability of linear ICA models revisited. In *4th International Symposium on Independent Component Analysis and Blind Signal Separation (ICA2003)*, pages 23–27, Nara, Japan, April 2003.

[20] P. Georgiev and A. Cichocki. Robust independent component analysis via time-delayed cumulant functions. *IEICE Transactions on Fundamentals of Electronics, Communications and Computer Sciences*, E86-A(3):573–579, 2003.

[21] P. Georgiev, P. Pardalos, and A. Cichocki. Algorithms with high order convergence speed for blind source extraction. *Computational Optimization and Applications*, 38(1):123–131, 2007.

[22] P. Georgiev and F. Theis. Blind source separation of linear mixtures with singular matrices. In *Independent Component Analysis and Blind Signal Separation*, pages 121–128. Springer-Verlag, Heidelberg, Germany, 2004.

[23] P. Georgiev, F. Theis, and A. Cichocki. Sparse component analysis and blind source separation of underdetermined mixtures. *IEEE Transactions on Neural Networks*, 16(4):992–996, 2005.

[24] P.G. Georgiev, F. Theis, and A. Cichocki. Blind source separation and sparse component analysis of overcomplete mixtures. In *Proceedings of the International Conference on Acoustics and Statistical Signal Processing (ICASSP2004)*, Montreal, Canada, 2004.

[25] P.G. Georgiev, F. Theis, A. Cichocki, and H. Bakardjian. Sparse component analysis: a new tool for data mining. In P.M. Denker, V.L. Boginski, and A. Vazacopoulos, editors, *Data Mining in Biomedicine*, Springer, Berlin, 2007.

[26] M. Habl, C. Bauer, C. Puntonet, M. Rodriguez-Alvarez, and E. Lang. Analyzing biomedical signals with probabilistic ICA and kernel-based source density estimation. In M. Sebaaly, editor, *Proceedings of the International Congress on Information Science Innovations (ISI'2001)*, pages 219–225, Canada, 2001. ICSC Academic Press.

[27] J. Hérault and C. Jutten. Space or time adaptive signal processing by neural network models. In J. Denker, editor, *Neural Networks for Computing (Proceedings of the AIP Conference)*, pages 206–211, New York, 1986. American Institute of Physics.

[28] A. Hyvärinen. Fast and robust fixed-point algorithms for independent component analysis. *IEEE Transactions on Neural Networks*, 10(3):626–634, 1999.

[29] A. Hyvärinen, J. Karhunen, and E. Oja. *Independent Component Analysis*. John Wiley & Sons, Hoboken, New Jersey, 2001.

[30] A. Hyvärinen and E. Oja. A fast fixed-point algorithm for independent component analysis. *Neural Computation*, 9(1483–1492), 1997.

[31] A. Hyvärinen and E. Oja. Independent component analysis by general nonlinear Hebbian-like learning rules. *Signal Processing*, 64(3):301–313, 1998.

[32] S. Lang. *Undergraduate Analysis*. Springer, New York, 2nd edition, 1997.

[33] T. Lee and M. Lewicki. The generalized Gaussian mixture model using ICA. In *Proceedings of the International Workshop on Independent Component Analysis (ICA'00)*, pages 239–244, 2000.

[34] T.-W. Lee, M.S. Lewicki, M. Girolami, and T.J. Sejnowski. Blind source separation of more sources than mixtures using overcomplete representations. *IEEE Signal Processing Letters*, 6(4):87–90, 1999.

[35] J. Lin. Factorizing multivariate function classes. *Advances in Neural Information Processing Systems*, 10:563–569, 1998.

[36] M. McKeown, L. Hansen, and T. Sejnowski. Independent component analysis of functional MRI: what is signal and what is noise? *Current Opinion in Neurobiology*, 13:620–629, 2003.

[37] M. McKeown, T. Jung, S. Makeig, G. Brown, S. Kindermann, A. Bell, and T. Sejnowski. Analysis of fmri data by blind separation into independent spatial components. *Human Brain Mapping*, 6:160–188, 1998.

[38] D.-T. Pham, P. Garrat, and C. Jutten. Separation of a mixture of independent sources through a maximum likelihood approach. In *Proceedings of EUSIPCO*, pages 771–774, 1992.

[39] A.M. Rubinov and J. Ugon. Skeletons of finite sets of points. Research working paper 03/06, School of Information Technology & Mathematical Sciences, University of Ballarat, 2003.

[40] A.N. Shiryaev. *Probability*. Graduate Texts in Mathematics. Springer, 1996.

[41] V. Skitovich. On a property of the normal distribution. *Doklady Academii nauk SSSR*, 89(217–219), 1953.

[42] F. Theis. *Mathematics in Independent Component Analysis*. Logos Verlag, Berlin, Germany, 2002.

[43] F. Theis, A. Jung, C. Puntonet, and E. Lang. Linear geometric ICA: Fundamentals and algorithms. *Neural Computation*, 15(1):1–21, 2002.

[44] F.J. Theis. A new concept for separability problems in blind source separation. *Neural Computation*, 16:1827–1850, 2004.

[45] F.J. Theis, E.W. Lang, and C.G. Puntonet. A geometric algorithm for overcomplete linear ICA. *Neurocomputing*, 56:381–398, 2004.

[46] L. Tong, R.-W. Liu, V. Soon, and Y.-F. Huang. Indeterminacy and identifiability of blind identification. *IEEE Transactions on Circuits and Systems*, 38(499–509), 1991.

[47] V.D. Calhoun VD, T. Adali, and J.J. Pekar. A method for comparing group fMRI data using independent component analysis: application to visual, motor and visuomotor tasks. *Journal of Magnetic Resonance Imaging*, 22:1181–1191, 2004.

[48] K. Waheed and F. Salem. Algebraic overcomplete independent component analysis. In *4th International Symposium on Independent Component Analysis and Blind Signal Separation (ICA2003)*, pages 1077–1082, Nara, Japan, April 2003.

[49] A. Wismüller, O. Lange, D. Dersch, G. Leinsinger, K. Hahn, B. Pütz, and D. Auer. Cluster analysis of biomedical image time–series. *International Journal on Computer Vision*, 46(2):102–128, 2002.

[50] R. Woods, S. Cherry, and J. Mazziotta. Rapid automated algorithm for aligning and reslicing PET images. *Journal of Computer Assisted Tomography*, 16(8):620–663, 1992.

[51] A. Yeredor. Blind source separation via the second characteristic function. *Signal Processing*, 80(5):897–902, 2000.

[52] M. Zibulevsky and B.A. Pearlmutter. Blind source separation by sparse decomposition in a signal dictionary. *Neural Computing*, 13(4):863–882, 2001.

# 9

# Algorithms for Genomics Analysis

Eva K. Lee and Kapil Gupta

Center for Operations Research in Medicine and HealthCare, School of Industrial
and Systems Engineering, Georgia Institute of Technology, Atlanta,
Georgia 30332-0205
evakylee@isye.gatech.edu

**Abstract.** The genome of an organism not only serves as its blueprint that holds
the key for diagnosing and curing diseases but also plays a pivotal role in obtaining
a holistic view of its ancestry. Recent years have witnessed a large number of innovations in this field, as exemplified by the Human Genome Project. This chapter
provides an overview of popular algorithms used in genome analysis and in particular
explores two important and deeply interconnected problems: Phylogenetic Analysis and Multiple Sequence Alignment. We also describe our novel graph-theoretical
approach that encompasses a wide variety of genome sequence analysis problems
within a single model.

## 9.1 Introduction

Genomics encompasses the study of genome in human and other organisms.
The rate of innovation in this field has been breathtaking over the past decade,
especially with the completion of the Human Genome Project. The purpose
of this chapter is to review some well-known algorithms that facilitate genome
analysis. The material is presented in a way that is interesting to both the
specialists working in this area and others. Thus, this review includes a brief
sketch of the algorithms to facilitate a deeper understanding of the concepts
involved. The list of problems related to genomics is very extensive; hence
the scope of this chapter is restricted to the following two related important
problems: (1) Phylogenetic Analysis, and (2) Multiple Sequence Alignment.
Readers interested in algorithms used in other fields of computational biology
are recommended to refer to reviews by Abbas and Holmes [1] and Blazewicz
et al. [7].

Genome refers to the complete DNA sequence contained in the cell. DNA
sequence consists of the four nucleotides adenine (A), thymine (T), cytosine
(C), and guanine (G). Associated with each DNA strand (sequence) is a complementary DNA strand of the same length. The strands are complementary
in that each nucleotide in one strand uniquely defines an associated nucleotide

P.M. Pardalos, H.E. Romeijn (eds.), *Handbook of Optimization in Medicine,*
Springer Optimization and Its Applications 26, DOI: 10.1007/978-0-387-09770-1_9,
© Springer Science+Business Media LLC 2009

in the other: A and T are always paired, and C and G are always paired. Each pairing is referred to as a base pair; and bound complementary strands make up a DNA molecule. Typically, the number of base pairs in a DNA molecule is between thousands and billions, depending on the complexity of a given organism. For example, a bacterium contains about 600,000 base pairs, whereas humans and mice have some 3 billion base pairs. Among humans, 99.9% of base pairs are the same between any two unrelated persons. But that leaves millions of single-letter differences, which provide genetic variation between people.

Understanding the DNA sequence is extremely important. It is considered as the blueprint for an organism's structure and function. The sequence order underlies all of life's diversity, even dictating whether an organism is human or another species such as yeast or a fruit fly. It helps in understanding the evolution of mankind, identifying genetic diseases, and creating new approaches for treating and controlling those diseases. In order to achieve these goals, the research in genome analysis has rapidly progressed over the past decade.

The rest of this chapter is organized as follows. Section 9.2 discusses techniques used to infer the evolutionary history of species, and Section 9.3 presents Multiple Sequence Alignment problem and recent advances. In Section 9.4, we describe our research effort for advancing genomics analysis through the design of a novel graph-theoretical approach for representing a wide variety of genomic sequence analysis problems within a single model. We summarize our theoretical findings and present computational models based on two integer programming formulations. Finally, Section 9.5 summarizes the interdependence and the pivotal role played by the above-mentioned two problems in computational biology.

## 9.2 Phylogenetic Analysis

Phylogenetic Analysis is a major aspect of genome research. It refers to the study of evolutionary relationships of a group of organisms. These hierarchical relationships among organisms arising through evolution are usually represented by a phylogenetic tree (Figure 9.1). The idea of using trees to

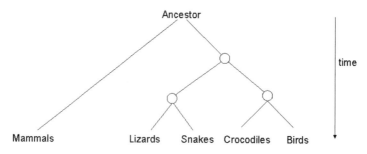

**Fig. 9.1.** An example of evolutionary tree.

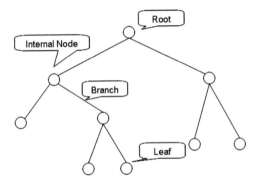

**Fig. 9.2.** Tree terminology.

represent evolution dates back to Darwin. Both rooted and unrooted tree representation have been used in practice [17]. The branches of tree represents the time of divergence and the root represents the ancestral sequence (Figure 9.2).

The study of phylogenies and processes of evolution by the analysis of DNA or amino acid sequence data is called Molecular Phylogenetics. In this study, we will focus on methods that use DNA sequence data. There are two processes involved in inferring both rooted and unrooted trees. First is estimating the branching structure or topology of the tree. Second is estimating the branch lengths for a given tree. Currently, there are wide varieties of methods available to conduct this analysis ([55, 79, 16, 19]). These available approaches can be classified into three broad groups: (i) Distance Methods; (ii) Parsimony Methods; and (iii) Maximum Likelihood Methods. Below, we will discuss each of them in detail.

### 9.2.1 Methods based on pairwise distance

In distance methods, an evolutionary distance $d_{ij}$ is computed between each pair $i$, $j$ of sequences, and a phylogenetic tree is constructed from these pairwise distances. There are many different ways of defining pairwise evolutionary distance used for this purpose. Most of the approaches estimate the number of nucleotides substitutions per site, but other measures have also been used [71, 70]. The most popular one is Jukes–Cantor distance [37], which defines $d_{ij}$ as $-\frac{3}{4}\log(1 - \frac{4f}{3})$, where $f$ is the fraction of sites where nucleotides differ in the pairwise alignment.

There are a large number of distance methods for constructing evolutionary trees [77]. In this chapter, we discuss methods based on *cluster analysis* and *neighbor joining*.

### Cluster analysis: UPGMA

The conceptually simplest and most known distance method is UPGMA (Unweighted Pair Group Method using Arithmetic averages) developed by

Sokal and Michener [66]. Given a matrix of pairwise distances between each pair of sequences, it starts with assigning each sequence to its own cluster. The distances between the clusters are defined as $d_{ij} = \frac{1}{|C_i||C_j|} \sum_{p \in C_i, q \in C_j} d(p, q)$ where $C_i$ and $C_j$ denote sequences in clusters $i$ and $j$, respectively. At each stage in the process, the least distant pair of clusters are merged to create a new cluster. This process continues until only one cluster is left. Given $n$ sequences, the general schema of UPGMA is shown in Algorithm 1.

---

**Algorithm 1** UPGMA

---

1: INPUT: Distance matrix $d_{ij}, 1 \leq i, j \leq n$
2: **for** $i = 1$ to $n$ **do**
3:     Define singleton cluster $C_i$ comprising of sequence $i$
4:     Place cluster $C_i$ as a tree leaf at height zero
5: **end for**
6: **repeat**
7:     Determine two clusters $i$, $j$ such that $d_{ij}$ is minimal.
8:     Merge these two clusters to form a new cluster $k$ having distance from other clusters defined as the weighted average of the comprised two clusters. If $C_k$ is the union of two clusters $C_i$ and $C_j$, and if $C_l$ is any other cluster, then $d_{kl} = \frac{d_{il}|C_i| + d_{jl}|C_j|}{|C_i| + |C_j|}$
9:     Define a node $k$ at height $\frac{d_{ij}}{2}$ with daughter nodes $i$ and $j$
10: **until** just a single cluster remains

---

The time and space complexity of UPGMA is $O(n^2)$, as there are $n - 1$ iterations of complexity $O(n)$. A number of approaches have been developed that are motivated by UPGMA. Li [52] developed a similar approach, which also makes corrections for unequal rates of evolution among lineages. Klotz and Blanken [43] presented a method where a present-day sequence serves as an ancestor in order to determine the tree regardless of the rates of evolution of the sequences involved.

## Neighbor joining

Neighbor Joining (NJ) is another very popular algorithm based on pairwise distances [63]. This approach yields an unrooted tree and overcomes the assumption of UPGMA method that the same rate of evolution applies to each branch.

Given a matrix of pairwise distances between each pair of sequences $d_{ij}$, it first defines modified distance matrix $\bar{d}_{ij}$. This matrix is calculated by subtracting average distances to all other sequences from the $d_{ij}$ and thus compensating for long edges. In each stage, the two nearest nodes (minimal $\bar{d}_{ij}$) of the tree are chosen and defined as neighbors in the tree. This is done recursively until all of the nodes are paired together.

Given $n$ sequences, the general schema of Neighbor Joining is shown in Algorithm 2.

---

**Algorithm 2** Neighbor Joining

---

1: INPUT: Distance matrix $d_{ij}$, $1 \leq i, j \leq n$
2: **for** $i = 1$ to $n$ **do**
3:    Assign sequence $i$ to the set of leaf nodes of the tree $(T)$
4: **end for**
5: Set list of active nodes$(L) = T$
6: **repeat**
7:    Calculate modified distance matrix $\bar{d}_{ij} = d_{ij} - (r_i + r_j)$, where $r_i = \frac{1}{|L|-2} \sum_{k \in L} d_{ik}$
8:    Find the pair $i, j$ in $L$ having minimal value of $\bar{d}_{ij}$
9:    Define a new node $u$ and set $d_{uk} = \frac{1}{2}(d_{ik} + d_{jk} - d_{ij})$, for all $k$ in $L$
10:    Add $u$ to $T$ joining nodes $i, j$ with edges of length given by: $d_{iu} = \frac{1}{2}(d_{ij} + r_i - r_j)$,
      $d_{ju} = d_{ij} - d_{iu}$
11:    Remove $i$ and $j$ from $L$ and add $u$
12: **until** Only two nodes remain in $L$
13: Connect remaining two nodes $i$ and $j$ by a branch of length $d_{ij}$

---

NJ has an execution time of $O(n^2)$, like UPGMA. It has given extremely good results in practice and is computationally efficient [63, 72]. Many practitioners have developed algorithms based on this approach. Gascuel [24] improved the NJ approach by using a simple first-order model of the variances and covariances of evolutionary distance estimates. Bruno et al. [10] developed a weighted NJ using a likelihood-based approach. Goeffon et al. [25] investigated a local search algorithm under the Maximum Parsimony criterion by introducing a new subtree swapping neighborhood with an effective array-based tree representation.

### 9.2.2 Parsimony methods

In science, the notion of parsimony refers to the preference of simpler hypotheses over complicated ones. In the parsimony approach for tree building, the goal is to identify the phylogeny that requires the fewest necessary changes to explain the differences among the observed sequences. Of the existing numerical approaches for reconstructing ancestral relationships directly from sequence data, this approach is the most popular one. Unlike distance-based methods, which builds a tree, it evaluates all possible trees and gives each a score based on the number of evolutionary changes that are needed to explain the observed sequences. The most parsimonious tree is the one that requires the fewest evolutionary changes for all sequences to derive from a common ancestor [69]. As an example, consider the trees in Figure 9.3 and Figure

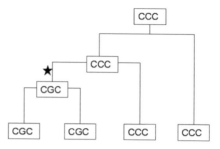

**Fig. 9.3.** Parsimony tree 1.

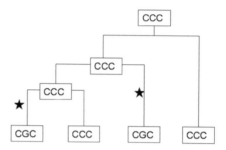

**Fig. 9.4.** Parsimony tree 2.

9.4. The tree in Figure 9.3 requires only one evolutionary change (marked by the ∗) compared with the tree in Figure 9.4, which requires two changes. Thus, Figure 9.3 is the more parsimonious tree.

There are two distinct components in parsimony methods: given a labeled tree, determine the score; determine global minimum score by evaluating all possible trees, as discussed below.

### Score computation

Given a set of nucleotide sequences, parsimony methods treat each site (position) independently. The algorithm evaluates the score at each position and then sums them up over all the positions. As an example, suppose we have the following three aligned nucleotide sequences.

$$CCC$$
$$GGC$$
$$CGC$$

Then, for a given tree topology, we would calculate the minimal number of changes required at each of the three sites and then sum them up. Here, we investigate a traditional parsimony algorithm developed by Fitch [21], where number of substitutions required is taken as score. For a particular topology, this approach starts by placing nucleotides at the leaves and traverse toward the root of the tree. At each node, the nucleotides common to all of the descendant nodes are placed. If this set is empty, then the union set is placed at this node. This continues until root of the tree is reached. The number of union sets equals the number of substitutions required. The general schema for every position is shown in Algorithm 3.

---

**Algorithm 3** Parsimony: Score Computation

1: Each leaf $l$ is labeled with set $R_l$ having observed nucleotide at that position.
2: Score $S = 0$
3: **for all** Internal node $k$ with children $i$ and $j$ having labels $R_i$ and $R_j$ **do**
4:    $R_k = R_i \bigcap R_j$
5:    **if** $R_k = \emptyset$ **then**
6:       $R_k = R_i \bigcup R_j$
7:       $S = S + 1$
8:    **end if**
9: **end for**
10: Minimal score $= S$

---

Figure 9.5 shows the set $R_k$ obtained by the above algorithm. The computation is done for the first site of the three sequences shown above. The minimal score given by the algorithm is 1.

There are a wide variety of approaches developed by modifying Fitch's algorithm [68]. Sankoff and Cedergren [64] presented a generalized parsimony method that does not just count the number of substitutions but assigns a weighted cost for each substitution.

Ronquist [62] improved the computational time by including strategies for rapid evaluation of tree lengths and increase the exhaustiveness of branch swapping while searching topologies.

**Search of possible tree topologies**

The number of possible tree topologies dramatically increases with the number of sequences. Consequently, in practice usually only a subset of them are examined using efficient search strategies. The most commonly used is

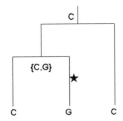

**Fig. 9.5.** The sets $R_k$ for the first site of given three sequences.

branch and bound methods to select branching patterns [60]. For large-scale problems, heuristic methods are typically used [69]. These exact and heuristic tree search strategies are implemented in various software like PHYLIP (Phylogeny Inference Package) and MEGA (Molecular Evolutionary Genetic Analysis) [20, 47].

### 9.2.3 Maximum likelihood methods

The method of maximum likelihood (ML) is one of the most popular statistical tools used in practice. In molecular phylogenetics, maximum likelihood methods find the tree that has the highest probability of generating observed sequences, given an explicit model of evolution. The method was first introduced by Felenstein [18]. We discuss herein both the evolution models and the calculation of tree likelihood.

### Model of evolution

A model of evolution refers to various events like as mutation that change one sequence to another over a period of time. It is required to determine the probability of a sequence $S_2$ arising from an ancestral sequence $S_1$ over a period time $t$. Various sophisticated models of evolution have been suggested, but simple models like Jukes–Cantor are preferred in ML methods.

Jukes and Cantor (1969) [37] model assumes that all nucleotides $(A, C, T, G)$ undergo mutation with equal probability and change to all of the other three possible nucleotides with same probability. If the mutation rate is $3\alpha$ per unit time per site, the mutation matrix $P_{ij}$ (probability that nucleotide $i$ changes to $j$ in unit time) takes the form

$$
\begin{pmatrix}
1 - 3\alpha & \alpha & \alpha & \alpha \\
\alpha & 1 - 3\alpha & \alpha & \alpha \\
\alpha & \alpha & 1 - 3\alpha & \alpha \\
\alpha & \alpha & \alpha & 1 - 3\alpha
\end{pmatrix}.
$$

The above matrix is integrated to evaluate mutation rates over time $t$ and then used to calculate $P(nt_2|nt_1, t)$, defined as the probability of nucleotide $nt_1$ being substituted by $nt_2$ over time $t$.

Various other evolution models like the Kimura model have also been mentioned in the literature [42, 9].

### Likelihood of a tree

The likelihood of tree is calculated as the probability of observing a set of sequences given the tree.

$$
L(\text{tree}) = P(\text{sequences}|\text{tree}).
$$

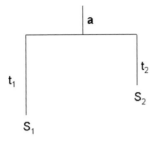

**Fig. 9.6.** A simple tree.

We begin with the simple case of two sequences $S^1$ and $S^2$ of length $n$ having a common ancestor "**a**" as shown in Figure 9.6. It is assumed that all different sites (positions) evolve independently and thus the total likelihood is calculated as the product of likelihood of all sites [15]. Here, the likelihood of each site is obtained using substitution probabilities based on evolution model.

Given $q_a$ = equilibrium distribution of nucleotide $a$, the likelihood for simple tree in Figure 9.6 is calculated as $L(\text{tree}) = P(S^1, S^2) = \prod_{i=1}^{n} P(S_i^1, S_i^2)$ where $P(S_i^1, S_i^2) = \sum_a q_a P(S_i^1|a) P(S_i^2|a)$. To generalize this approach for $m$ sequences, it is assumed that diverged sequences evolve independently after diverging. Hence, likelihood for every node in the tree depends only on its immediate ancestral node and a recursive procedure is used to evaluate likelihood of the tree. The conditional likelihood $L_{k,a}$ is defined as the likelihood of the subtree rooted at node $k$, given that nucleotide at node $k$ is $a$. The general schema for every site is shown in Algorithm 4. The likelihood is then maximized over all possible tree topologies and branch lengths.

---

**Algorithm 4** Likelihood: Computation at Given Site

---

1: **for all** leaf $l$ **do**
2:    **if** leaf has nucleotide $a$ at that site **then**
3:       $L_{l,a} = 1$
4:    **else**
5:       $L_{l,a} = 0$
6:    **end if**
7: **end for**
8: **for all** Internal node $k$ with children $i$ and $j$ **do**
9:    Define the conditional likelihood $L_{k,a} = \sum_{b,c}[P(b|a)L_{i,b}][P(c|a)L_{j,c}]$
10: **end for**
11: Likelihood at given site $= \sum_a q_a L_{root,a}$

---

## Recent improvements

ML approach has received great attention due to the existence of powerful statistical tools. It has been made more sophisticated using advance tree search algorithms, sequence evolution models, and statistical approaches. Yang [80] has extended it to the case where the rate of nucleotide substitutions differs over sites. Huelsenbeck [34] incorporated the improvements in substitution models. Piontkivska [59] evaluated the use of various substitution models in ML approach and inferred that simple models are comparable in both efficiency and reliability with complex models.

The enormously large number of possible tree topologies, especially while working with large number of sequences, makes this approach computationally intensive [72]. It has been proved that reconstructing the ML tree is NP-hard even for certain approximations [14]. In order to reduce computational time, Guindon and Gascuel [31] developed a simple hill-climbing algorithm based on the maximum-likelihood principle that adjusts tree topology and branch lengths simultaneously. Recently, parallel computation is being used to address huge computational requirement. Stamatakis et al. [67] have used OpenMP-parallelization for Symmetric Multi-Processing machines, and Keane et al. [39] developed distributed platform for phylogeny reconstruction by maximum likelihood.

## 9.3 Multiple Sequence Alignment

Multiple sequence alignment (MSA) is arguably among the most studied and difficult problems in computational biology. It has been a vital tool as it compactly represents conserved or variable features among the family members. Alignment also allows character-based analysis compared with distance-based analysis and thus helps to elucidate evolutionary relationships better. Consequently, it plays a pivotal role in a wide range of sequence analysis problems like identifying conserved motifs among given sequences; predicting secondary and tertiary structure of protein sequences; and molecular phylogenetic analysis. It is also used for sequence comparison to find similarity of a new sequence with pre-existing ones. This helps in gathering information about function and structure of newfound sequences from the existing ones in databases like GenBank in the United States and EMBL in Europe.

The MSA problem can be stated formally as follows. Let $\sum$ be the alphabet and let $\hat{\sum} = \sum \cup \{-\}$, where "$-$" is a symbol to represent "gaps" in sequences. For DNA sequences, alphabet $\hat{\sum} = \{A, T, C, G, -\}$.

An alignment for $N$ sequences $S_1, ..., S_N$ is given by a set $\hat{S} = \{S_1, ..., S_N\}$ over the alphabet $\hat{\sum}$, which satisfies the following two properties: (1) the strings in $\hat{S}$ are of the same length, (2) $S_i$ can be obtained from $\hat{S}_i$ by removing the gaps. Thus, an alignment in which each string $\hat{S}_i$ has length $K$ can be interpreted as an alignment matrix of $N$ rows and $K$ columns where row $i$ corresponds with sequence $S_i$. Alphabets that are placed into the same column of alignment matrix are said to be aligned with each other.

Figure 9.7 shows two possible alignments for given three sequences: $S_1 = CCC$, $S_2 = CGGC$, and $S_3 = CGC$.

```
C  C  C  —           C  —  C  C

C  G  G  C           C  G  G  C

C  G  —  C           C  G  C  —
```

**Fig. 9.7.** Two possible alignments for given three sequences.

For two sequences, optimal MSA is easily obtained using dynamic programming (Needleman–Wunsch algorithm). Unfortunately, the problem becomes much harder for more than two sequences, and optimal solution can be found only for a limited number of sequences of moderate length (approximately 100) [8]. Researchers have tried to solve it by generalizing dynamic programming approach to a multidimensional space. However, this approach has huge time and memory requirements and thus cannot be used in practice even for small problems of 5 sequences of length 100 each. This algorithm has been improved by identifying the portion of hyperspace that does not contribute to the solution and exclude it from the computation [11]. But, even this approach of Carrillo and Lipman implemented in MSA program can only align up to ten sequences [53]. Although, Gupta and Kececioglu (1995) improved the space and time usage of this approach, it cannot align large data sets [32]. To reduce the huge time and memory expenses, a wide variety of heuristic approaches for MSA have been developed [57].

There are two components of finding MSA: (i) searching over all the possible multiple alignments, (ii) scoring each of them to find the best one.

The problem becomes more complex for remotely related homologous sequences, i.e., sequences that are not derived from a common ancestor [28]. Numerous approaches have been proposed, but the quest for an approach that is accurate and fast continues. It must be remembered that even the choice of sequences and calculating the score of alignment is a nontrivial task and is an active research field in itself.

### 9.3.1 Scoring alignment

There is no unanimous way of characterizing an alignment as the correct one and the strategy depends on biological context. Different alignments are possible and we never know for sure which alignment is correct. Thus, one scores every alignment according to an appropriate objective function, and alignments with the higher scores are deemed to be better. A typical alignment scoring scheme consists of the following steps:

### Independent columns

The score of alignment is calculated in terms of columns of alignments. The individual columns are assumed to be independent, and thus the total score of an alignment is a simple summation over column scores. Thus, score for an alignment $\text{Score}(A) = \sum_j \text{Score}(A_j)$, where $A_j$ is column $j$ of the multiple alignment $A$. Now, score for every column $j$ is calculated as "sum of pairs" (SP) function using the scoring matrices described below. The SP score for column $A_j$ is obtained as $\text{Score}(A_j) = \sum_{k<l} \text{Score}(A_j^k, A_j^l)$ where $A_j^k$ and $A_j^l$ are nucleotides in column $j$ of alignment corresponding with sequences $k$ and $l$, respectively. If gap costs are linear, Score(nucleotide, −) and Score(−, nucleotide) will be the insertion cost. But, this approach would not differentiate between opening a gap and extending it. Thus, affine gap penalties are often used where gap opening and extension penalty are treated as two different parameters. The correct value of both of these parameters is a major concern as their values can be set only empirically [75]. Also, most schemes used in practice score columns as weighted sum of pair-wise substitutions instead of just addition as described before. The weights are decided in accordance with the amount of independent information each sequence possesses [3].

Both the assumption of treating every column independent and using SP score for column has limitations. The problem increases as number of sequences increases.

**Scoring matrices**

Any alignment can be obtained by performing three evolution operations: insertion, deletion, and substitution. It is assumed that all the different operations occur independently, and thus the complete score is evaluated as the sum of scores from every operation. Insertion and deletion scores are calculated as either linear or affine gap penalty. Substitutions scores are stored as substitution score matrix, which contains score for every pair of nucleotides. Thus, these scores $S(A, B)$ can be treated as the score of aligning nucleotide $A$ with $B$.

These substitution score matrices can be obtained in various ways. One could adopt an ad hoc approach of setting up a score matrix that produces good alignments for a given set of sequences. The second approach would be more fundamental and look into physical and chemical properties of nucleotides. If two nucleotides are similar in their properties, they would be more likely to be substituted by one another. The third and the most prominent one is a statistical approach where maximum likelihood principle is used in conjunction with probabilistic models of evolution [4].

**9.3.2 Alignment approaches**

The number of different approaches for MSA problem has steadily increased over the past decade and thus being exhaustive will not be possible. In this chapter, we will emphasize the most widely used class of algorithms and the new emerging and most promising approaches.

1. Progressive alignment algorithms: The most widely used type of algorithm based on using pairwise alignment information of input sequences. It assumes that input sequences are phylogenetically related and uses these relationships to guide the alignment [13].
2. Graph-based algorithms: A new trend where graph-based models are used to approach this problem.
3. Iterative alignment algorithms: Typically, an alignment is produced and is then refined through series of iterations until no more improvement can be made.

### 9.3.3 Progressive algorithms

Progressive Alignment constitutes one of the most simplest and effective waysf for multiple alignment. This strategy was introduced by various researchers like Waterman and Perlwitz [78]. Among all the progressive algorithms, ClustalW is the most famous one. It is a non-iterative, deterministic algorithm that attempts to optimize the weighted sums-of-pairs with affine gap penalties [73].

The typical progressive algorithm schema is as follows:

- Compute distance between all pairs of given sequences by aligning them. The distances represent divergence of each pair of sequences. These distances could be calculated by fast approximation methods or the slower but more precise methods like complete dynamic programming. Because for given $N$ sequences $\frac{N(N-1)}{2}$ pairwise scores have to be calculated and the scores are used just for construction of a guide tree and not the alignment itself, it is desirable to use approximation methods like $k$ tuple matches.
- Find a guide tree from the distance matrix. This is typically achieved using clustering algorithms discussed in construction  of an evolutionary tree. Once again, because the aim is to get the alignment and not the tree itself, approximation methods are used to construct the evolution trees.
- Align sequences progressively according to the branching order in the guide tree. The basic idea is to start from the leaves of the guide tree toward its root and to use series of pairwise alignments to align larger and larger groups of sequences. Some algorithms have only single growing alignment to which every remaining sequence is aligned, whereas other approaches align subgroup of sequences and then  merge the alignments.

There are three main shortcomings of the progressive algorithms.

- There does not exist an undisputable "best" way of ordering the given sequences.
- Once a sequence has been aligned, that alignment will not be modified even if it conflicts with sequences added later in the process. Hence, the order in which sequences are added becomes very crucial, and as there is

no undisputable best way to order the sequences, this approach returns sub-optimal solutions.

- For a given set of $n$ sequences, $\binom{n}{2}$ pairwise alignments are generated; but while computing the final multiple alignment, most of these algorithms use fewer than $n$ pairwise alignments. Thus, the resulting multiple alignment agrees with only a small amount of information available in the data.

Therefore, there is a growing need for an algorithm to align extremely divergent sequences whose pairwise alignments are likely to be incorrect. In order to address all these issues, some techniques have been developed; although they are innovative, it is understandable that they have their own assumptions and drawbacks.

### 9.3.4 Graph-based algorithms

Over the past few years, the field of genomics has undergone evolutionary changes with a rapid increase in new solution strategies. The use of graph-based models is easily seen as one of the emerging and most far-reaching trends. Just and Vedova [38] use relation between facility location problem and sequence alignment to prove the NP-hardness of MSA. In this section, we review the most prominent integer programming (IP) approaches for finding multiple sequence alignment.

### Maximum-weight trace

Kececioglu et al. [40] use a solution of the maximum trace problem to construct alignment. The algorithm starts with calculating all pairwise alignments and uses them to find a trace. To achieve this, given $n$ sequences, an input alignment graph $G = (V, E)$ is constructed. It is a $n$-partite graph whose vertex set $V$ represents the characters of given sequences and edge set $E$ represents the pairs of characters matched in the pairwise alignments. The subset of matching in $E$ realized by an alignment is called a trace.

Alignment graph $G = (V, E)$ is extended to a mixed graph $G' = (V, E, A)$ by adding arc set $A$, which connects character of every sequence to the next character in the same sequence. The objective of the algorithm is to find maximum weight trace by finding cycles termed as "critical mixed cycles" in graph $G'$ such that they satisfy sequence alignment properties [61].

The IP model for this problem is formulated as:

$$\text{maximize } \sum_{e \in E} w_e x_e$$

$$\text{subject to } \sum_{e \in P \cap E} x_e = |E \cap P| - 1 \; \forall \text{ critical mixed cycles } P \text{ in } G'$$
$$x_e \quad \in \{0, 1\} \text{ for all } e \in E.$$

An implementation of a branch-and-cut algorithm is used to solve the above problem. Various valid inequalities for the polytope are added as cuts, some of which are facet-defining. The algorithm is capable of giving an exact solution under the sum-of-pairs objective function with linear gap costs. Kececioglu et al. have made a significant contribution by introducing a polyhedral approach capable of obtaining exact solutions for a subclass of MSA. However, this methodology has its own drawbacks like not being able to capture the order of insertions and deletions between two matchings and affine gap costs. Recently, Althaus et al. [2] has proposed a general model using this approach in which arbitrary gap costs are allowed.

**Minimum spanning tree and traveling salesman problem**

Shyu et al. [65] explore the use of minimum spanning trees to determine the order of sequences. The idea of the approach is to preserve the most informative distances among the set of given sequences. The criterion used is meaningful and capable of working better than the traditional criteria like those in sum-of-pairs. The algorithm itself is very efficient for practical usage and can be easily implemented. However, it fails to address the issue of using all the information in pairwise alignments, as it only uses the score and not the pairwise alignments themselves. Moreover, this approach has all the drawbacks of the progressive strategy.

A similar approach has also been developed by Korostensky and Gonnet [44] using Traveling Salesman Problem (TSP). In this technique, a circular sum measure is used instead of sum of pairs score. The cities in TSP correspond with the sequences, and the scores of pairwise alignment are taken as the distances. The problem is to find the longest tour where each sequence is visited exactly once [45].

**Eulerian path approach**

Zhang and Waterman [81] proposed a new approach motivated by the Eulerian method for fragment assembly in DNA sequencing. In their work, a consensus sequence is found and later pairwise alignments are obtained between each input sequence and consensus sequence. Finally, MSA is obtained according to these pairwise alignments. The most significant advantage of this method is linear time and memory cost for finding the consensus sequence. And, if the consensus sequence is the closest one to all given sequences, good-quality alignment can be obtained in a reasonable amount of time. Once again, this approach suffers from the prominent drawback of the progressive strategy and issues in graph formation while finding the consensus sequence.

**9.3.5 Iterative algorithms**

The main shortcoming of the progressive strategy is the failure to remove errors in the alignment, which are introduced early. The iterative algorithms are developed precisely to overcome this flaw. They are based on the idea of reconsidering and realigning previously aligned sequences with the goal of improving the overall alignment score. Each modification step is an iteration to improve the quality of the alignment.

These available approaches can be classified into two broad categories: probabilistic iterative algorithms and deterministic iterative algorithms. We will briefly discuss them below.

## Probabilistic algorithms

We will discuss both the traditional probabilistic optimization approaches like genetic algorithm and relatively recent approaches based on Bayesian ideas.

- *Simulated Annealing and Genetic Algorithm*
  Simulated Annealing (SA) and Genetic Algorithms (GA) are very popular stochastic methods for solving complex optimization problems. Whereas they are often viewed as separate and competing paradigms, both of them are iterative algorithms that search for new solutions "near" to already known good solutions. The fundamental difference between SA and GA is that SA performs a local move only on one solution to create a new solution whereas GA also creates solutions by combining information from two different solutions. Performance comparison between SA and GA varies with the problem and representation used.

  The algorithms starts with an initial alignment, and alignment score is taken to be the objective function [56]. Various operations like mutation, insertion and substitution constitute the local moves that are used to get new solution from existing ones. Flexibility in scoring systems and ability to correct for errors introduced during early phase makes these approaches desirable [41].

- *Hidden Markov model and Gibbs sampler*
  Hidden Markov model (HMM) and Gibbs sampler are relatively recent approaches that view MSA in a statistical context. Both of them use the central Bayesian idea of simultaneously maximizing the data and the model. Gibbs sampler find motifs using local alignment techniques [49]. It is essentially similar to HMM with no insert and delete states.

  HMM is a statistical model based on Markov processes, which has gained importance in various fields related to pattern recognition. It determines the hidden parameters of the system based on the observable parameters of the model. For MSA, HMM model consists of three types of states: match states, insert states, and delete states [46]. Each state has its own emission probability of nucleotides and transition probability to other states. The standard expectation-maximization (EM) algorithm or gradient descent algorithms are used to train the model and evaluate the parameters.

  Although HMM has been successfully used in other areas, it faces a lot of challenges. There need to be some minimum number of sequences (approx. 50) required to train the model, and HMM can be easily trapped in local optima like other hill-climbing approaches [35].

**Deterministic algorithms**

A deterministic iterative algorithm starts with an initial alignment and then attempts to improve it. This helps in overcoming the drawback of progressive alignment strategy where partial alignments are "frozen" [6]. A typical scheme is as follows:

- Given $N$ sequences $S_1, \ldots, S_N$, find alignment $A$;
- Remove sequence $S_1$ from alignment $A$ and realign it to the profile of other aligned sequences $S_2, \ldots, S_N$ to get new alignment $A'$;
- Calculate the score of the new alignment $A'$ and if better replace $A$ by $A'$;
- Remove sequence $S_2$ from $A'$ and realign it. Continue this procedure for $S_3, \ldots, S_N$;
- Repeat the realignment steps until alignment score converges or number of iterations reaches the user-specified limit.

Many iteration strategies that enable very accurate alignments have been developed [76]. The aim is to reduce the greedy nature of the algorithm and avoid getting trapped in a local optima. One approach is to remove and realign every sequence to the rest in each iteration. Then, the alignment with the best score is taken to be the input for the next iteration. The other famous approach is to randomly split set of sequences into two sets, which are then realigned.

Some researchers have incorporated the iterative strategy in progressive alignment procedure itself. For instance, a double iteration loop has been used to make the alignment, guide tree, and sequence weights mutually consistent [27]. Recently, Chakrabarti et al. [12] have developed an approach that provides a fast and accurate method for refining existing block-based alignments.

## 9.4 Novel Graph-Theoretical–Based Genomic Models

In this section, we present our research effort of a novel graph-theoretical approach for representing a wide variety of genomic sequence analysis problems within a single model [50]. The model allows incorporation of the operations "insertion," "deletion," and "substitution," and various parameters such as relative distances and weights. Conceptually, we refer the problem as the *minimum weight common mutated sequence* (MWCMS) problem. The MWCMS model has many applications including multiple sequence alignment problem, the phylogenetic analysis, the DNA sequencing problem, and sequence comparison problem, which encompass a core set of very difficult problems in computational biology. Thus the model presented in this section lays out a mathematical modeling framework that allows one to investigate theoretical and computational issues and to forge new advances for these distinct but related problems.

*DNA sequencing* refers to determining the exact order of nucleotide sequences in a segment of DNA. This was the greatest technical challenge in the *Human Genome Project*. Achieving this goal has helped reveal the estimated 30,000 human genes that are the basic physical and functional units of heredity. The resulting DNA sequence maps are being used by scientists to explore human biology and other complex phenomena.

The structure of a DNA strand (sequence) is determined by experimentation. Typically, short sequences are determined to be in the strand, and the identified short sequences are then "connected" to form a long sequence. Recent advances attempting to identify DNA strand structure involve sequencing by hybridization [5] and [36]. Sequencing by hybridization is the process where every possible sequence of length $n$ ($4^n$ possibilities) is compared with a full DNA strand. Practical values for $n$ are 8–12. Each short string either binds or does not bind to the full strand. Biologists can thus determine exactly which short strings are contained in the DNA strand and which are not.

However, the experiment does not identify the exact location of each short string in the full strand. Hence, an important issue involves how these short strings are connected together to form the complete strand. This problem can be viewed as a shortest common superstring problem and has been studied extensively [54, 22, 23]. Unfortunately, errors may arise during sequencing experiments. Three types of errors are deletions (a letter appears in an input string that should not be in the final sequence), insertions (a letter is missing from an input string), and substitutions (a letter in an input string should be substituted with another letter). The MWCMS problem can be used to model and solve this shortest common superstring problem while addressing the issue of possible errors.

*Sequence comparison* is one of the most crucial problems faced by researchers in the area of bioinformatics. The sequence patterns are conserved during evolution. Given a new sequence, it will be of interest to understand how much similarity it has with pre-existing sequences. Significant similarity between two sequences implies similarities in their structure and/or function. There are lots of DNA databases containing DNA sequences and their function. The major ones are GenBank in the United States and the EMBL data library in Europe. If one finds a new sequence similar to existing ones in these databases, one can transfer information about the function and structure [77]. Hence, an algorithm for sequence comparison that is efficient for large number of sequences will play a pivotal role in rapid sequence analysis. The MWCMS problem can be used to address this issue.

### 9.4.1 Definitions

Our motivation for first defining the problem arose from the desire to help quantify the concept of "best" representative sequence in the evolutionary distance problem. The evolutionary distance problem involves finding the DNA sequence of the most likely ancestor associated with a given set of DNA sequences from distinct but similar organisms. In other words, find the DNA strand that best represents a possible ancestor, if each of the organisms evolved from the same ancestor. Changes that contribute to differences between the given sequences and the ancestor are referred to as insertions, deletions, and substitutions. These operations account for both evolutionary mutations and experimental errors in sequencing. Mathematically, given two sequences $S$ and $B$, let ord$(S, B)$ be an ordered collection of insertions, deletions, and substitutions to convert sequence $S$ to $B$. (For any two sequences $S$ and $B$, there are an infinite number of collections ord$(S, B)$.) Let $w(\text{ord}(S, B))$ be the weight of the conversion from $S$ to $B$, where the weight is the sum of an expression involving values $\eta$, $\delta$, and $\psi \in \mathbb{R}^+$, which represent the weights associated with a single insertion, deletion, and substitution, respectively. Let ord$^*(S, B)$ be such that $w(\text{ord}^*(S, B)) \leq w(\text{ord}(S, B))$ for all ord$(S, B)$. Define $d(S, B) = w(\text{ord}^*(S, B))$. Formally, MWCMS can be stated as:

**Problem MWCMS**: Given positive weights $\eta$, $\delta$, and $\psi$ corresponding with a single insertion, deletion, and substitution respectively, a positive threshold $\kappa$, and finite sequences $S_1, \ldots, S_m$ from a finite alphabet, does there exist a sequence $B$ such that $\sum_{i=1}^m d(S_i, B) \leq \kappa$?

We have defined the MWCMS problem — which incorporates the notions of insertion, deletion and substitution — to help quantify the concept of "best" representative sequence in the evolutionary distance problem. We now make precise the operations of *insertion*, *deletion*, and *substitution*. Let $S = \{s_1, \ldots, s_n\}$ be a finite sequence of letters from a finite alphabet.

(i)   An *insertion* of an element $x$ in position $i$ of the sequence $S$ is characterized by the addition of $x$ between elements $s_i$ and $s_{i+1}$. An insertion carries an associated penalty cost of $\eta$.

(ii)  A *deletion* of an element in position $i$ of $S$ amounts to deleting $s_i$ from the sequence $S$. The penalty for deletion is represented by $\delta$.

(iii) A *substitution* of an element in position $i$ of $S$ amounts to replacing $s_i$ with another letter from the alphabet. The penalty for substitution is represented by $\psi$.

We remark that a penalty cost for an operation could, more generally, depend on the position where the operation is performed and/or the element to be inserted/deleted/substituted.

Let $S_1 = \{s_{11}, \ldots, s_{1m}\}$ and $S_2 = \{s_{21}, \ldots, s_{2n}\}$ be two finite sequences of letters from a finite alphabet $\Sigma$. We say that the *relative distance* between elements $s_{1i}$ and $s_{2j}$ is $k$ if $|i - j| = k$. We define a *k-restrictive* bipartite graph as a graph $G_k = (V_1, V_2, E_k)$ such that the nodes in $V_1$ and $V_2$ correspond respectively with each of the elements from the first and the second sequences. We assume the nodes in $V_i$ are ordered in the same order as they appear in the sequence $S_i$. There is an edge between nodes $u \in V_1$ and $v \in V_2$ if $u$

and $v$ are identical (i.e., same letter of the alphabet $\Sigma$) and if the relative distance between these two elements is less than or equal to $k$. The problem of identifying the "greatest similarity" between these two sequences can then be approached as the problem of finding a maximum cardinality matching between the associated node sets, subject to restrictions on which matchings are allowed. In particular, one must take into consideration the ordering of nodes so as to preserve the relative occurrence of the elements in the matching. In addition, matchings that have edge crossings must be prevented. When $k = \max\{|S_1|, |S_2|\} - 1$, we denote the graph by $G = (V_1, V_2, E)$, and the problem is equivalent to the well-studied longest common subsequence (LCS) problem for two sequences, which is polynomial-time solvable [23].

## 9.4.2 Construction of a conflict graph from paths of multiple sequences

Let $S_i$, $i = 1, \ldots, m$, be a collection of finite sequences, each of length $n$, over a common alphabet $\Sigma$. Let $G_k = (V_1, \ldots V_m, E_1, E_2, \ldots, E_{m-1})$ be the $k$-restrictive multilayer graph in which each element in $S_i$ forms a distinct node in $V_i$. Assume the nodes in $V_i$ are ordered in the same order as they appear in the sequence $S_i$. $E_i$ denotes the set of edges between nodes in $V_i$ and $V_{i+1}$. There is an edge between nodes $u \in V_i$ and $v \in V_{i+1}$ if and only if $u$ and $v$ are the same letter in the alphabet $\Sigma$, and the relative distance between them is less than or equal to $k$. The multiple sequence comparison problem involves finding the longest common subsequence (LCS) within the sequences $S_i, i = 1, \ldots, m$. We call a path $P = p_1, p_2, \ldots, p_m$ a *complete* path in $G_k$ if $p_i \in V_i$ and $p_i p_{i+1} \in E_i$. Two complete paths are said to be *parallel* if their node sets are disjoint and the edges do not cross. Hence, a set of parallel complete paths in $G_k$ corresponds with a feasible solution to LCS on the collection of sequences $S_i, i = 1, \ldots, m$. We say that two complete paths $P_1, P_2$ *cross* if they are not parallel. We remark that the LCS problem with the number of sequences bounded is polynomial time solvable using dynamic programming [23]. In general, the problem remains $\mathcal{NP}$-complete.

We can incorporate insertions by generating new paths that include inserted nodes on various layers. The weight for such a new path will be affected by the total number of insertions in the path. In particular, if $L$ is a common subsequence for $S_i$ and $|S_i| = n$ for all $i = 1, \ldots, m$, then the total number of unmatched elements remaining will be $m(n - |L|)$. These elements can be deleted completely, or for a given unmatched element, one can increase the size of $L$ by 1 by appropriately inserting this element into various sequences. By doing so, the number of unmatched elements decreases. Let $l$ be the number of insertions needed to generate a new complete path. Then the number of unmatched elements will decrease by $m - l$. If we assume that at the end of the sequencing process all unmatched elements will be deleted, then the penalty for generating this new complete path will be given by $l\eta - (m - l)\delta$.

We next define the concept of conflict graph relative to complete paths in $G_k$.

**Definition 1.** *Let* $\mathcal{P} = \{P_1, \ldots, P_s\}$ *be a finite collection of complete paths in* $G_k$. *The* conflict graph $\mathcal{C}_\mathcal{P} = (V_\mathcal{P}, E_\mathcal{P})$ *associated with* $\mathcal{P}$ *is constructed as follows:*

- $V_\mathcal{P} = \{P_1, \ldots, P_s\}$;
- *there is an edge between two nodes* $P_i$ *and* $P_j$ *in* $V_\mathcal{P}$ *if and only if* $P_i$ *and* $P_j$ *cross each other.*

This definition applies to any multilayer graph in general. Note that any stable set of nodes in $\mathcal{C}_\mathcal{P}$ corresponds with a set of parallel complete paths for $G_k$ and thereby to a feasible solution to LCS on the collection of sequences $S_i$, $i = 1, \ldots, m$.

We remark that when $m = 2$, the resulting conflict graph is weakly triangulated and thus is perfect. For $m > 2$, the conflict graph can contain an antihole of size 6. However, these complete paths can be viewed as continuous functions on the interval 0 to 1, thus by construction, $\mathcal{C}_\mathcal{P}$ is perfect [26].

### 9.4.3 Complexity theory

Recall that the notation $\text{ord}(S, B)$, $w(\text{ord}(S, B))$, $\text{ord}^*(S, B)$, and the formal definition of problem MWCMS were given in Section 9.4.1. As an optimization problem, MWCMS can be stated as follows. Given a set of input sequences, problem MWCMS seeks to mutate every input sequence to the same *a priori* unknown sequence using the operations of insertion, deletion, and substitution; weights are assigned for each operation, and the total weight associated with all mutations is to be minimized. Levenshtein first considered a special case of this problem by changing a single input sequence to another sequence using insertions, deletions, and substitutions [51]. Our study involves changing multiple input sequences to arrive at an *a priori* unknown common sequence.

Given positive weights $\eta$, $\delta$, and $\psi$ corresponding respectively with insertions, deletions, and substitutions and any two sequences $S$ and $B$, clearly any $\text{ord}^*(S, B)$ will never contain more than $|B|$ insertions or substitutions. Proving that MWCMS is in $\mathcal{NP}$ is not obvious. Although one can transform MWCMS to special applications (as described in beginning of Section 9.4) to conclude that it is in NP, here we prove it directly for the general case. One needs to be able to evaluate $d(S, B)$ in polynomial time for any two sequences $S$ and $B$. We next construct a graph that can be used to establish the existence of a polynomial time algorithm for obtaining $d(S, B)$. The constructs and arguments used here typify those used to establish many of the results presented in this chapter. It is noteworthy that the notions of both conflict graph and perfect graph come into play.

Let $\Sigma$ be a finite alphabet, and define $\Sigma$-cross to be a directed bipartite graph consisting of $|\Sigma|$ vertices in each bipartition such that each vertex in

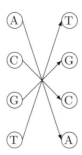

**Fig. 9.8.** An example of $\sum$-cross when $\sum = \{A, C, G, T\}$.

the bipartition represents a distinct element in $\Sigma$. There is an arc between two vertices if the vertices correspond with the same element in $\Sigma$, and the geometric layout is rigidly constructed so that every arc crosses every other arc. This graph will be used as a "supernode" for insertion and substitution operations in our model. Figure 9.8 shows an example for $\Sigma$-cross when $\Sigma = \{A, C, G, T\}$.

We now construct a *3-layer supergraph* $G_L$ using the sequences $S$ and $B$ along with the $\Sigma$-cross graphs. Layers 1 and 2 consist of exactly $|B|(|S| + 1) + |S|$ $\Sigma$-crosses. The first $|B|$ $\Sigma$-crosses represent potential insertions before the first letter in $S$. The next $\Sigma$-cross represents either the first letter of $S$ or a substitution of this letter. The next $|B|$ $\Sigma$-crosses represent potential insertions between the first and second letters of $S$. And this is followed by a $\Sigma$-cross representing either the second letter of $S$ or a substitution of this letter. This continues for each letter in S with the final $|B|$ $\Sigma$-crosses representing up to $|B|$ insertions after the last letter in $S$. Each $\Sigma$-cross is called either an *insertion supernode* or a *substitution supernode*, according to what it represents. The weight of all of the arcs in an insertion supernode is $\eta$. An arc in a substitution supernode has weight $-\delta$ if the arc represents the original letter in the sequences or $\psi - \delta$ if the arc represents a substitution of the original letter. Layer 3 consists of the vertices represented by $B$. A vertex in layer 2 is connected to a vertex in layer 3 if they have the same letter. The weight of every arc between layers 2 and 3 is $M \leq -(\eta + \delta + \psi)$. A sample of a 3-layer supergraph is given in Figure 9.9. The bold arcs are used to denote the original letters in $S$ (the weight of these arcs is $-\delta$). For simplicity, we omit the first two insertion supernodes before the first letter G. The first supernode thus represents the letter G from the original sequence and allows for substitution. The second and third supernodes correspond with insertion supernodes, and the fourth supernode corresponds with the letter C and allows substitution as well. There are two more insertion supernodes, which are omitted from the graph.

Layer 1    Layer 2    Layer 3

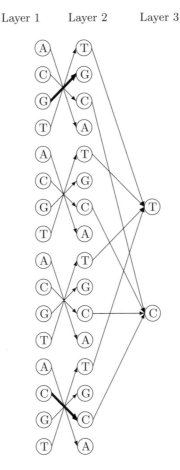

**Fig. 9.9.** An example of the 3-layer supergraph for converting the sequence $S = GC$ to $B = TC$. Bold arcs are used to denote the original letters in $S$ (the weight of these arcs is $-\delta$). For simplicity, we omit the first two insertion supernodes before the first letter G. The first supernode in this figure thus represents the letter G from the original sequence, which allows for substitution. The second and third supernodes correspond with insertions, and the fourth supernode corresponds with the letter C and allows substitution as well. There are two more insertion supernodes, which are omitted from the graph.

The main step in proving $d(S, B)$ to be polynomial-time solvable for any sequences $S$ and $B$ involves the use of the conflict graph as defined in Definition 1. We state some preliminary theoretical results below. Detailed proofs can be found in Lee et al. [50].

**Lemma 1.** *The following statements are equivalent:*

*(i) There exists a conversion from $S$ to $B$ using no more than a total of $|B|$ insertions or substitutions.*

*(ii) There exists a set of noncrossing complete paths in the associated 3-layer supergraph $G_L$ of size $|B|$.*

*(iii) There exists a node packing of size $|B|$ in the associated conflict graph $\mathcal{C}$.*

**Lemma 2.** *Calculating $d(S, B)$ for any sequences $S$ and $B$ can be accomplished in polynomial time.*

The 3-layer supergraph can be generalized to multilayer when multiple sequences are considered. Clearly, such multilayer supergraphs are much too large for practical purposes, yet polynomiality is preserved in the construction, and it is therefore sufficient. We can now arrive at the result that MWCMS is in $\mathcal{NP}$.

**Theorem 1.** *MWCMS is in $\mathcal{NP}$.*

To prove that MWCMS is polynomial-time solvable when the number of input sequences is bounded by a positive constant, the following lemma is crucial, though trivial.

**Lemma 3.** *Given $\eta, \delta, \psi \in \mathbb{R}^+$, an optimal solution $B$ to any MWCMS problem has the following properties. $B$ has no substitutions from letters other than the original letters in an $S_i$, and $B$ will never have an element that is inserted in every sequence (in the same location). Therefore, there are at most $\sum_{i=1}^{m} |S_i|$ insertions in any sequence.*

In addition, we also require the construction of a (directed) $2m$-layer supergraph, $G_L^m$, similar to the 3-layer supergraph, $G_L$.

Given sequences $S_1, \ldots, S_m$, generate a $2m$-layer (directed) graph $G_L^m = (V, E)$ as follows. Layers $2i - 1$ and $2i$ consist of $(\sum_{j=1}^{m} |S_j|)(|S_i| + 1) + |S_i|$ copies of $\Sigma$-crosses for $i = 1, \ldots, m$, constructed in exactly the same manner as layers 1 and 2 of the 3-layer supergraph using the input sequence $S_i$. The first $\sum_{j=1}^{m} |S_j|$ $\Sigma$-crosses represent the possibility that $\sum_{j=1}^{m} |S_j|$ different letters can be inserted before the first element in $S_i$. The next $\Sigma$-cross corresponds with either the first letter in $S_i$ or a substitution of this letter. This is repeated $|S_i|$ times (for each letter in $S_i$), and the final $\sum_{j=1}^{m} |S_j|$ $\Sigma$-crosses represent insertions after the final letter in $S_i$. Thus, the first $\sum_{j=1}^{m} |S_j|$ $\Sigma$-crosses represent the insertion supernodes, followed by one $\Sigma$-cross representing a letter in $S_i$ or a substitution supernode, and so forth. An arc exists from a vertex in layer $2i$ to a vertex in layer $2i + 1$ if the vertices correspond with the same letter. Observe that $G_L^m$ is an acyclic directed graph that is polynomial in the size of the input sequences. Assign every arc between layers $2i$ and $2i + 1$ a weight of 0. There are three different weights for arcs between layers $2i - 1$ and $2i$ each corresponding with an insertion, deletion, or substitution. The assignment of weights on such arcs is analogous to the assignment in $G_L$: a weight of $\eta$ is assigned to every arc contained in an insertion supernode; and an arc in a substitution supernode is assigned a weight of $-\delta$ if it corresponds with the original letter, or $\psi - \delta$, otherwise.

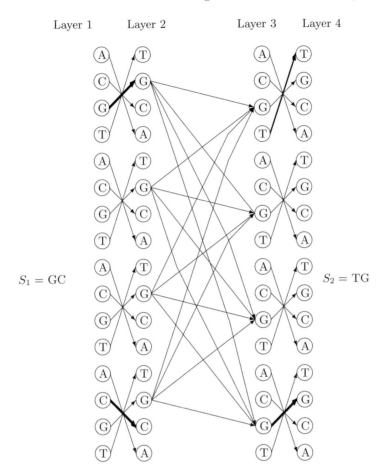

**Fig. 9.10.** A sample graph $G_L^m$ of MWCMS with $S_1 = GC$ to $S_2 = TG$ where $\sum = \{A, C, G, T\}$

Figure 9.10 shows a sample graph for two sequences: $S_1 = GC$ and $S_2 = TG$. Observe that at most two insertions are needed in an optimal solution; thus we can reduce the number of $\Sigma$-crosses as insertion supernodes from $\sum_{i=1}^{2} |S_i| = 4$ to 2. For simplicity, in the graph shown in Figure 9.10, we have not included the two insertion supernodes before the first letter nor those after the last letter of each sequence. Thus, in the figure, the first $\Sigma$-cross represents the substitution supernode associated with the first letter in $S_1$. The second and third $\Sigma$-crosses represent two insertion supernodes. And the last $\Sigma$-cross represents the substitution supernode associated with the second letter in $S_1$. For simplicity, we include only arcs connecting vertices associated to the element G between layers 2 and 3. The arcs for other vertices follow similarly.

A conflict graph $\mathcal{C}$ associated with $G_L^m$ can be generated by finding all complete paths (paths from layer 1 to layer $2m$) in $G_L^m$. These complete paths correspond to the set of vertices in $\mathcal{C}$, as in Definition 1. If we assign a weight to each vertex equal to the weight of the associated complete path, then the following result can be established.

**Theorem 2.** *Every node packing in $\mathcal{C}$ represents a candidate solution to MWCMS if and only if at most $\sum_{i=1}^m |S_i|$ letters can be inserted between any two original letters. Furthermore, the weight of the node packing is equal to the weight of the MWCMS $- \sum_{i=1}^m |S_i|\delta$.*

The supergraph $G_L^m$ and its associated conflict graph are fundamental to our proof of the following theorem on polynomial-time solvability of a restricted version of problem MWCMS.

**Theorem 3.** *Problem MWCMS restricted to instances for which the number of sequences is bounded by a positive constant is polynomial-time solvable.*

### 9.4.4 Special cases of MWCMS

MWCMS encompasses a very broad class of problems. In computational biology as discussed in this chapter, first and foremost, it represents a model for phylogenetic analysis. MWCMS is defined as the "most likely ancestor problem," and the concept of 3-layer supergraph as described in Section 9.4.3 describes the evolutionary distance problem. An optimal solution to a multiple sequence alignment instance can be found using the solution of the MWCMS problem obtained on the $2m$-layer supergraph, $G_L^m$. The alignment is the character matrix obtained by placing together the given sequences incorporating the insertions into the solution of the MWCMS problem. Furthermore, DNA sequencing can be viewed as the shortest common superstring problem, and sequence comparison of a given sequence $B$ to a collection of $N$ sequences $S_1, \ldots, S_N$ is the MWCMS problem itself.

Broader than the computational biology applications, special cases of MWCMS include shortest common supersequences (SCSQ), longest common subsequences (LCS), and shortest common superstring (SCST); these problems are of interest in their own right as combinatorial optimization problems and for their role in complexity theory.

### 9.4.5 Computational models: integer programming formulation

The construction of the multilayer supergraphs described in our theoretical study lays the foundation and provides direction for computational models and solution strategies that we will explore in future research. Although the theoretical results obtained are polynomial-time in nature, they present computational challenges. In many cases, calculating the worst-case scenario is not trivial. Furthermore, the polynomial-time result of a node-packing problem for a perfect graph by Grötschel et al. [30, 29] is existential in nature, and relies on the polynomial-time nature of the ellipsoid algorithm. The process itself involves solving an IP relaxation multiple times. In our case, the variables of the IP generated are the complete paths in the multilayer supergraph, $G_L^m$. Formally, the integer program corresponding with our conflict graph can be stated as follows:

Let $x_p$ be the binary variable denoting the use or non-use of the complete path $p$ with weight $w_p$. Then the corresponding node-packing problem (MIP1) is

$$\text{minimize } \sum w_p x_p$$

$$\text{subject to } x_p + x_q \leq 1 \quad \text{if complete paths } p \text{ and } q \text{ cross}$$
$$x_p \quad \in \{0,1\} \text{ for all complete paths } p \text{ in } G_L^m.$$

We call the inequality $x_p + x_q \leq 1$ an adjacency constraint. A natural approach to improve the solution time to (MIP1) is to decrease the size of the graph $G_L^m$ and thus the number of variables. Reductions in the size of $G_L^m$ can be accomplished for SCST, LCS, and SCSQ. Among these three problems, the graph $G_L^m$ is smallest for LCS. In LCS, all insertion and substitution supernodes can be eliminated.

Our theoretical results thus far rely on the creation of all complete paths. Clearly, the typical number of complete paths will be on the order of $n^m$, where $n = \max |S_i|$. In this case, an instance with 3 sequences and 300 letters in each sequence generates more than 1 million variables. Hence, an exact formulation with all complete paths is impractical in general. A simultaneous column and row generation approach within a parallel implementation may lead to computational advances related to this formulation.

An alternative formulation can be obtained by examining $G_L^m$ from a network perspective using arcs (instead of complete paths) in $G_L^m$ as variables. Namely, let $x_{i,j}$ denote the use or non-use of arc $(i,j)$ in the final sequence with $c_{i,j}$ the cost of the arc in $G_L^m$. The network formulation (MIP2) can be stated as

$$\text{minimize } \sum_{(i,j)\in E} c_{i,j} x_{i,j}$$
$$\text{subject to } \sum_{i:(i,j)\in E} x_{i,j} = \sum_{k:(j,k)\in E} x_{j,k} \ \forall \, j \in V \text{ in layers } 2,\ldots,2m-1$$
$$x_{i,j} + x_{k,l} \leq 1 \quad \text{for all crossing arcs } (i,j) \text{ and } (k,l) \in E$$
$$x_{i,j} \in \{0,1\} \text{ for all } (i,j) \in E.$$

The first set of constraints ensures that inflow equals outflow in all vertices contained in sequences $2,\ldots,m-1$ (complete paths). The second set of constraints ensures that no two arcs cross. This model grows linearly in the number of sequences. This alternative integer programming formulation is still large but is manageable for even fairly large instances.

Utilizing a collection of DNA sequences (each with 40,000 base pairs in length) from a bacteria and a collection of short sequences associated with genes found in breast cancer patients, computational tests of our graph-theoretical models are under way. We seek to develop computational strategies to provide reasonable running times for evolutionary distance problem instances derived from these data. In an initial test, when three sequences each with 100 letters are used, the initial linear program requires more than 10,000 seconds to solve when tight constraints are employed (in this case, each adjacency constraint is replaced by a maximal clique constraint). Our ongoing computational effort will focus on developing and investigating solution techniques for practical problem instances, including those based on the above two IP formulations, as well as development of fast heuristic procedures.

In Lee, Easton, and Gupta [50], we outline a simple yet practical heuristic based on (MIP2) that we developed for solving the multiple sequence alignment problem; and we report on preliminary tests of the algorithm using different sets of sequence data. Motivation for the heuristic is derived from the desire to reduce computational time through various strategies for reducing the number of variables in (MIP2).

## 9.5 Summary

Multiple Sequence Alignment and Phylogenetic Analysis are deeply interconnected problems in computational biology. A good multiple alignment is crucial for reliable reconstruction of the phylogenetic tree [58]. On the other hand, most of the multiple alignment methods require a phylogenetic tree as the guide tree for progressive iteration.

Thus, the evolutionary tree construction might be biased by the guide tree used for obtaining the alignment. In order to avoid this pitfall, various algorithms have been developed that simultaneously find alignment and phylogenetic relationship among given sequences. Sankoff and Cedergren [64] developed a parsimony-based algorithm using a character-substitution model of gaps. The algorithm is guaranteed to find evolutionary tree and alignment that minimizes tree-based parsimony cost. Hein [33] also developed a parsimony-type algorithm but uses an affine gap cost, which is more realistic than the character-substitution gap model. This algorithm is also faster than Sankoff and Cedergreen's approach but makes simplifying assumptions in choosing ancestral sequences.

Like parsimony methods for finding a phylogenetic tree, both of the above approaches require search over all possible trees to find the global optimum. This makes these algorithms computationally very intensive. Hence, there has been a strong focus on developing an efficient algorithm that considers both alignment and tree. Vingron and Haeseler [74] have developed an approach based on three-way alignment of pre-aligned groups of sequences. It also allows change in the alignment made early in the course of computation. Various software, such as MEGA, are trying to develop an efficient integrated computing environment that allows both sequence alignment and evolutionary analysis [48].

We address this issue of simultaneously finding alignment and phylogenetic relationships by presenting a novel graph-theoretical approach. Indeed, our model can be easily tailored to find theoretically provable optimum solutions to a wide range of crucial sequence analysis problems. These sequence analysis problems are proven to be NP-hard and thus understandably present computational challenges. In order to strike a balance between time and quality-of-solution, a variety of parameters are provided. Ongoing research efforts explore development of efficient computational models and solution strategies in a massive parallel environment.

## Acknowledgment

This research is partially supported by grants from the National Science Foundation.

## References

[1] A.E. Abbas and S.P. Holmes. Bioinformatics and management science: Some common tools and techniques. *Operations Research*, 52(2):165–190, 2004.
[2] E. Althaus, A. Caprara, H. Lenhof, and K. Reinert. A branch-and-cut algorithm for multiple sequence alignment. *Mathematical Programming*, 105,(2-3):387–425, 2006.
[3] S.F. Altschul, R.J. Carroll, and D.J. Lipman. Weights for data related by a tree. *Journal of Molecular Biology*, 207(4):647–653, 1989.
[4] S.F. Altschul. Amino acid substitution matrices from an information theoretic perspective. *Journal of Molecular Biology*, 219(3):555–565, 1991.
[5] W. Bains and G.C. Smith. A novel nethod for DNA sequence determination. *Journal of Theoretical Biology*, 135:303–307, 1988.
[6] G.J. Barton and M.J.E. Sternberg. A strategy for the rapid multiple alignment of protein sequences: confidence levels from tertiary structure comparisons. *Journal of Molecular Biology*, 198:327–337, 1987.
[7] J. Blazewicz, P. Formanowicz, and M. Kasprzak. Selected combinatorial problems of computational biology. *European Journal of Operational Research*, 161:585–597, 2005.
[8] P. Bonizzoni and G.D. Vedova. The complexity of multiple sequence alignment with SP-score that is a metric. *Theoretical Computer Science*, 259:63–79, 2001.
[9] D.H. Bos and D. Posada. Using models of nucleotide evolution to build phylogenetic trees. *Developmental and Comparative Immunology*, 29(3):211–227, 2005.

[10] W.J. Bruno, N.D. Socci, and A.L. Halpern. Weighted neighbor joining: A likelihood-based approach to distance-based phylogeny reconstruction. *Molecular Biology and Evolution*, 17:189–197, 2000.

[11] H. Carrillo and D. Lipman. The multiple sequence alignment problem in biology. *SIAM Journal on Applied Mathematics*, 48(5):1073–1082, 1988.

[12] S. Chakrabarti, C.J. Lanczycki, A.R. Panchenko, T.M. Przytycka, P.A. Thiessen, and S.H. Bryant. Refining multiple sequence alignments with conserved core regions. *Nucleic Acids Research*, 34(9):2598–2606, 2006.

[13] R. Chenna, H. Sugawara, T. Koike, R. Lopez, T.J. Gibson, D.G. Higgins, and J.D. Thompson. Multiple sequence alignment with the clustal series of programs. *Nucleic Acids Research*, 31(13):3497–3500, 2003.

[14] B. Chor and T. Tuller. Maximum likelihood of evolutionary trees: hardness and approximation. *Bioinformatics*, 21(Suppl. 1):I97–I106, 2005.

[15] P. Clote and R. Backofen. *Computational Molecular Biology: An Introduction*. John Wiley and Sons Ltd, New York, 2000.

[16] F. Delsuc, H. Brinkmann, and H. Philippe. Phylogenomics and the reconstruction of the tree of life. *Nature Reviews Genetics*, 6(5):361–375, 2005.

[17] R. Durbin, S. Eddy, A. Krogh, and G. Mitchison. *Biological Sequence Analysis*. Cambridge University Press, Cambridge, U.K., 1998.

[18] J. Felsenstein. Evolutionary trees from DNA sequences: a maximum likelihood approach. *Journal of Molecular Evolution*, 17(6):368–376, 1981.

[19] J. Felsenstein. Phylogenies from molecular sequences: Inference and reliability. *Annual Review of Genetics*, 22:521–565, 1988.

[20] J. Felsenstein. PHYLIP - phylogeny inference package (version 3.2). *Cladistics*, 5:164–166, 1989.

[21] W.M. Fitch. Toward defining the course of evolution: minimum change for a specific tree topology. *Systematic Zoology*, 20(4):406–416, 1971.

[22] J. Gallant, D. Maider, and J.A. Storer. On finding minimal length superstrings. *Journal of Computer and System Sciences*, 20:50–58, 1980.

[23] M. Garey and D. Johnson. *Computers and Intractability: A Guide to the Theory of NP-Completeness*. W.H. Freeman, San Francisco, California, 1979.

[24] O. Gascuel. BIONJ: An improved version of the NJ algorithm based on a simple model of sequence data. *Molecular Biology and Evolution*, 14(7):685–695, 1997.

[25] A. Goeffon, J.M. Richer, and J.K. Hao. Local search for the maximum parsimony problem. *Lecture Notes in Computer Science*, 3612:678–683, 2005.

[26] M.C. Golumbic, D. Rotem, and J. Urrutia. Comparability graphs and intersection graphs. *Discrete Mathematics*, 43:37–46, 1983.

[27] O. Gotoh. Significant improvement in accuracy of multiple protein sequence alignments by iterative refinement as assessed by reference to structural alignments. *Journal of Molecular Biology*, 264(4):823–838, 1996.

[28] O. Gotoh. Multiple sequence alignment: algorithms and applications. *Advances in Biophysics*, 36:159–206, 1999.

[29] M. Grötschel, L. Lovász, and A. Schrijver. Polynomial algorithms for perfect graphs. *Annals of Discrete Mathematics*, 21:325–356, 1984.

[30] M. Grötschel, L. Lovász, and A. Schrijver. *Geometric Algorithms and Combinatorial Optimization*. Springer-Verlag, New York, 1988.

[31] S. Guindon and O. Gascuel. A simple, fast, and accurate algorithm to estimate large phylogenies by maximum likelihood. *Systematic Biology*, 52(5):696–704, 2003.

[32] S. Gupta, J. Kececioglu, and A. Schaeffer. Improving the practical space and time efficiency of the shortest-paths approach to sum-of-pairs multiple sequence alignment. *Journal of Computational Biology*, 2:459–472, 1995.

[33] J. Hein. A new method that simultaneously aligns and reconstructs ancestral sequences for any number of homologous sequences, when the phylogeny is given. *Molecular Biology and Evolution*, 6(6):649–668, 1989.

[34] J.P. Huelsenbeck and K.A. Crandall. Phylogeny estimation and hypothesis testing using maximum likelihood. *Annual Review of Ecology and Systematics*, 28:437–66, 1997.

[35] R. Hughey and A. Krogh. Hidden markov models for sequence analysis: extension and analysis of the basic method. *Computer Applications in the Biosciences*, 12(2):95–107, 1996.

[36] R.M. Idury and M.S. Waterman. A new algorithm for DNA sequence assembly. *Journal of Computational Biology*, 2(2):291–306, 1995.

[37] T.H. Jukes and C.R. Cantor. Evolution of protein molecules. In H.N. Munro, editor, *Mammalian Protein Metabolism*, pages 21–123. Academic Press, New York, 1969.

[38] W. Just and G.D. Vedova. Multiple sequence alignment as a facility-location problem. *INFORMS Journal on Computing*, 16(4):430–440, 2004.

[39] T.M. Keane, T.J. Naughton, S.A. Travers, J.O. McInerney, and G.P. McCormack. DPRml: distributed phylogeny reconstruction by maximum likelihood. *Bioinformatics*, 21(7):969–974, 2005.

[40] J.D. Kececioglu, H. Lenhof, K. Mehlhorn, P. Mutzel, K. Reinert, and M. Vingron. A polyhedral approach to sequence alignment problems. *Discrete Applied Mathematics*, 104:143–186, 2000.

[41] J. Kim, S. Pramanik, and M.J. Chung. Multiple sequence alignment using simulated annealing. *Bioinformatics*, 10(4):419–426, 1994.

[42] M. Kimura. A simple method for estimating evolutionary of base substitution through comparative studies of nucleotide sequences. *Journal of Molecular Evolution*, 16:111–120, 1980.

[43] L.C. Klotz and R.L. Blanken. A practical method for calculating evolutionary trees from sequence data. *Journal of Theoretical Biology*, 91(2):261–272, 1981.

[44] C. Korostensky and G.H. Gonnet. Near optimal multiple sequence alignments using a traveling salesman problem approach. *Proceedings of the String Processing and Information Retrieval Symposium*, page 105, 1999.

[45] C. Korostensky and G.H. Gonnet. Using traveling salesman problem algorithms for evolutionary tree construction. *Bioinformatics*, 16(7):619–627, 2000.

[46] A. Krogh, M. Brown, I.S. Mian, K. Sjolander, and D. Haussler. Hidden Markov models in computational biology: Applications to protein modeling. *Journal of Molecular Biology*, 235:1501–1531, 1994.

[47] S. Kumar, K. Tamura, and M. Nei. MEGA: Molecular evolutionary genetics analysis software for microcomputers. *Computer Applications in Biosciences*, 10:189–191, 1994.

[48] S. Kumar, K. Tamura, and M. Nei. MEGA3: integrated software for molecular evolutionary genetics analysis and sequence alignment. *Briefings in Bioinformatics*, 5(2):150–163, 2004.

[49] C.E. Lawrence, S.F. Altschul, M.S. Boguski, J.S. Liu, A.F. Neuwald, and J.C. Wootton. Detecting subtle sequence signals: a gibbs sampling strategy for multiple alignment. *Science*, 262:208–214, 1993.

[50] E.K. Lee, T. Easton, and K. Gupta. Novel evolutionary models and applications to sequence alignment problems. *Annals of Operations Research*, 148(1):167–187, 2006.

[51] V.L. Levenshtein. Binary codes capable of correcting deletions, insertions, and reversals. *Cybernetics Control Theory*, 10(9):707–710, 1966.

[52] W.H. Li. Simple method for constructing phylogenetic trees from distance matrices. *Proceedings of the National Academy of Sciences*, 78(2):1085–1089, 1981.

[53] D.J. Lipman, S.F. Altschul, and J.D. Kececioglu. A tool for multiple sequence alignment. *Proceedings of the National Academy of Sciences*, 86(12):4412–4415, 1989.

[54] D. Maier and J.A. Storer. A note on the complexity of the superstring problem. Technical Report 233, Princeton University, 1977.

[55] M. Nei. Phylogenetic analysis in molecular evolutionary genetics. *Annual Review of Genetics*, 30:371–403, 1996.

[56] C. Notredame and D.G. Higgins. SAGA: sequence alignment by genetic algorithm. *Nucleic Acids Research*, 24(8):1515–1524, 1996.

[57] C. Notredame. Recent progress in multiple sequence alignment: a survey. *Pharmacogenomics*, 3(1):131–144, 2002.

[58] A. Phillips, D. Janies, and W. Wheeler. Multiple sequence alignment in phylogenetic analysis. *Molecular Phylogenetics and Evolution*, 16(3):317–330, 2000.

[59] H. Piontkivska. Efficiencies of maximum likelihood methods of phylogenetic inferences when different substitution models are used. *Molecular Phylogenetics and Evolution*, 31(3):865–873, 2004.

[60] P.W. Purdom, P. G. Bradford, K. Tamura, and S. Kumar. Single column discrepancy and dynamic max-mini optimizations for quickly finding the most parsimonious evolutionary trees. *Bioinformatics*, 16:140–151, 2000.

[61] K. Reinert, H. Lenhof, P. Mutzel, K. Mehlhorn, and J. Kececioglu. A branch-and-cut algorithm for multiple sequence alignment. *Proceedings of the First Annual International Conference on Computational Molecular Biology (RECOMB-97)*, pages 241–249, 1997.

[62] F. Ronquist. Fast fitch-parsimony algorithms for large data sets. *Cladistics*, 14:387–400, 1998.

[63] N. Saitou and M. Nei. The neighbor-joining method: a new method for reconstructing phylogenetic trees. *Molecular Biology and Evolution*, 4:406–425, 1987.

[64] D. Sankoff and R.J. Cedergren. Simultaneous comparison of three or more sequences related by a tree. In D. Sankoff and J.B. Kruskal, editors, *Time Warps, String Edits, and Macromolecules: The Theory and Practice of Sequence Comparison*, pages 253–264. Addison-Wesley, Reading, Massachusetts, 1983.

[65] S.J. Shyu, Y.T. Tsai, and R.C.T. Lee. The minimal spanning tree preservation approaches for DNA multiple sequence alignment and evolutionary tree construction. *Journal of Combinatorial Optimization*, 8(4):453–468, 2004.

[66] R.R. Sokal and C.D. Michener. A statistical method for evaluating systematic relationships. *University of Kansas Scientific Bulletin*, 38:1409–1438, 1958.

[67] A. Stamatakis, M. Ott, and T. Ludwig. RAxML-OMP: An efficient program for phylogenetic inference on SMPs. *Lecture Notes in Computer Science*, 3606:288–302, 2005.

[68] D.L. Swofford and W.P. Maddison. Reconstructing ancestral character states under wagner parsimony. *Mathematical Biosciences*, 87:199–229, 1987.

[69] D.L. Swofford and G.J. Olsen. Phylogeny reconstruction. In D.M. Hillis and G. Moritz, editors, *Molecular Systematics*, pages 411–501. Sinauer Associates, Sunderland, Massachusetts, 1990.

[70] F. Tajima and M. Nei. Estimation of evolutionary distance between nucleotide sequences. *Molecular Biology and Evolution*, 1(3):269–85, 1984.

[71] F. Tajima and N. Takezaki. Estimation of evolutionary distance for reconstructing molecular phylogenetic trees. *Molecular Biology and Evolution*, 11:278–286, 1994.

[72] K. Takahashi and M. Nei. Efficiencies of fast algorithms of phylogenetic inference under the criteria of maximum parsimony, minimum evolution, and maximum likelihood when a large number of sequences are used. *Molecular Biology and Evolution*, 17:1251–1258, 2000.

[73] J.D. Thompson, D.G. Higgins, and T.J. Gibson. CLUSTAL W: improving the sensitivity of progressive multiple sequence alignment through sequence weighting, position-specific gap penalties and weight matrix choice. *Nucleic Acids Research*, 22(22):4673–4680, 1994.

[74] M. Vingron and A. Haeseler. Towards integration of multiple alignment and phylogenetic tree construction. *Journal of Computational Biology*, 4(1):23–34, 1997.

[75] M. Vingron and M.S. Waterman. Sequence alignment and penalty choice: review of concepts, case studies and implications. *Journal of Molecular Biology*, 235(1):1–12, 1994.

[76] I.M. Wallace, O.O'Sullivan, and D.G. Higgins. Evaluation of iterative alignment algorithms for multiple alignment. *Bioinformatics*, 21(8):1408–14, 2005.

[77] M.S. Waterman. *Introduction to Computational Biology: Maps, Sequences and Genomes*. Chapman and Hall, London, U.K., 1995.

[78] M.S. Waterman and M.D. Perlwitz. Line geometries for sequence comparisons. *Bulletin of Mathematical Biology*, 46(4):567–577, 1984.

[79] S. Whelan, P. Lio, and N. Goldman. Molecular phylogenetics: state-of-the-art methods for looking into the past. *Trends in Genetics*, 17(5):262–272, 2001.

[80] Z. Yang. Maximum-likelihood estimation of phylogeny from DNA sequences when substitution rates differ over sites. *Molecular Biology and Evolution*, 10(6):1396–401, 1993.

[81] Y. Zhang and M.S. Waterman. An eulerian path approach to global multiple alignment for DNA sequences. *Journal of Computational Biology*, 10(6):803–819, 2003.

# Optimization and Data Mining in Epilepsy Research: A Review and Prospective

W. Art Chaovalitwongse

Department of Industrial and Systems Engineering, Rutgers, The State University of New Jersey, Piscataway, New Jersey 08854
wchaoval@rci.rutgers.edu

**Abstract.** During the past century, most neuroscientists believed that epileptic seizures began abruptly in a matter of a few seconds before clinical onset. Since the late 1980s, there has been an explosion of interest in neuroscience research to predict epileptic seizures based on quantitative analyses of brain electrical activity captured by electroencephalogram (EEG). Many research groups have demonstrated growing evidence that seizures develop minutes to hours before clinical onset. The methods in those studies include signal processing techniques, statistical analyses, nonlinear dynamics (chaos theory), data mining, and advanced optimization techniques. Although the past few decades have seen revolutionary of quantitative studies to capture seizure precursors, seizure prediction research is still far from complete. Current techniques still need to be advanced, and novel approaches need to be explored and investigated. In this chapter, we will give an extensive review and prospective of seizure prediction research including various methods in data mining and optimization techniques that have been applied to seizure prediction research. Future directions of data mining and optimization in seizure prediction research will also be discussed in this chapter. Successful seizure prediction research will give us the opportunity to develop implantable devices, which are able to warn of impending seizures and to trigger therapy to prevent clinical epileptic seizures.

**Key words:** seizure prediction, optimization, data mining, chaos theory, EEG, brain dynamics, implantable devices

## 10.1 Introduction

The human brain is among the most complex systems known to mankind. Over the past century, neuroscientists have sought to understand brain functions through detailed analysis of neuronal excitability and synaptic transmission. However, the dynamic transitions to neurologic dysfunctions of brain disorders are not well understood in current neuroscience research [42]. Epilepsy is the second most common brain disorder after stroke, yet the most devastating one. The most disabling aspect of epilepsy is the uncertainty of recurrent

P.M. Pardalos, H.E. Romeijn (eds.), *Handbook of Optimization in Medicine*,
Springer Optimization and Its Applications 26, DOI: 10.1007/978-0-387-09770-1_10,
© Springer Science+Business Media LLC 2009

seizures, which can be characterized by a chronic medical condition produced by temporary changes in the electrical function of the brain. These electrical changes can be captured by electroencephalograms (EEGs), which is a tool for evaluating the physiologic state of the brain. Whereas EEGs offer excellent spatial and temporal resolution to characterize rapidly changing electrical activity of brain activation, neuroscientists understand very little about seizure development process from EEG data. The unpredictable occurrence of seizures has presented special difficulties regarding the ability to investigate the factors by which the initiation of seizures occurs in humans. If seizures could be predicted, it will revolutionize neuroscience research and provide a greater understanding of abnormal intermittent changes of neuronal cell networks driven by the seizure development.

Recent advances in optimization and data mining (DM) research for excavating hidden patterns or relationships in massive data (like EEGs) offer a possibility to better understand brain functions (as well as other complex systems) from a system perspective. If successful, the outcome of this research will be very useful in medical diagnosis. There has been a growing research interest in developing quantitative methods using advances in optimization and data mining to rapidly recognize and capture epileptic activity in EEGs before a seizure occurs. This research attempt is a vital step to advance seizure prediction research. In this chapter, we will give a review and prospective of seizure prediction technology and what role optimization and data mining has played in this seizure prediction research. The potential outcome of this research direction may enable effective and safe treatment for epileptic patients. This chapter is organized as follows. In the next section, we will give a brief background of epilepsy and seizure prediction research including motivation and history of seizure prediction. The previous studies in mining EEG data based on chaos theory will be discussed in Section 10.3. The current research in optimization and data mining techniques for seizure prediction will be addressed in Section 10.4. In the last section, we will give some concluding remarks and prospective issues in epilepsy research.

## 10.2 Background: Epilepsy and Seizure Prediction

At least 40 million people worldwide (or 1% of the population) currently suffer from epilepsy, which is among the most common disorders of the nervous system and consists of more than 40 clinical syndromes. Epilepsy, the second most common serious brain disorder after stroke, is a chronic condition of diverse etiologies with the common symptom of spontaneous recurrent seizures. Seizures can be characterized by intermittent paroxysmal and highly organized rhythmic neuronal discharges in the cerebral cortex. In some types of epilepsy (e.g., focal or partial epilepsy), there is a localized structural change in neuronal circuitry within the cerebrum that produces organized quasi-rhythmic discharges, which spread from the region of origin (epileptogenic zone) to activate other areas of the cerebral hemisphere [83]. Though epilepsy occurs in all age groups, the highest incidences occur in infants and in the elderly. The most common type of epilepsy in adults is temporal lobe epilepsy. In this type of epilepsy, the temporal cortex, limbic structures, and orbitofrontal cortex appear to play a critical role in the onset and spread of seizures. Temporal lobe seizures usually begin as paroxysmal electrical discharges in the hippocampus and often spread first to ipsilateral, then to contralateral cerebral cortex. These abnormal discharges result in a variety of intermittent clinical phenomena, including motor, sensory, affective, cognitive, autonomic, and psychic symptomatology. There is no single cause of epilepsy. In approximately 65% of cases, the causes to injure the nerve cells in the brain are unknown. Most frequently identified causes are genetic abnormalities, developmental anomalies, febrile convulsions, as well as brain insults such as craniofacial trauma, central nervous system infections, hypoxia, ischemia, and tumors. The diagnosis and treatment of epilepsy is complicated by the disabling aspect that seizures occur spontaneously and unpredictably due to the nature of the chaotic disorder. Although the macroscopic and microscopic features of the epileptogenic processes have been comprehended, the seizure mechanism by which these fixed disturbances in local circuitry produce intermittent disturbances of brain function cannot be explained and understood. The transitional development of the epileptic state can be considered as a sudden development of synchronous neuronal firing potentials in the cerebral cortex that may begin locally in a portion of one cerebral hemisphere or begin simultaneously in both cerebral hemispheres.

## 10.2.1 Classification of seizures

There are many varieties of epileptic seizures, and seizure frequency and the form of attacks vary greatly from person to person. The most common classification scheme describes two major types of seizures: (1) "partial" seizure: a seizure that causes excessive electrical discharges in the brain limited to one area; (2) "generalized" seizure: a seizure that changes the whole brain to be involved with excessive electrical discharges. Each of these categories can be divided into subcategories: simple partial, complex partial, tonic–clonic, and other types. With the most common types of seizures there is some loss of consciousness, but some seizures may only involve some movements of the body or strange feelings. The sensation of seizures in different patients can be very different. Common feelings include uncertainty, fear, physical and mental exhaustion, confusion, and memory loss. Sometimes if a person is unconscious, there may be no feeling at all. Seizures can last anywhere from a few seconds to several minutes, depending on the type of seizure. In particular, a tonic–clonic seizure typically lasts 1–7 minutes. Absence seizures may only last a few seconds. Complex partial seizures range from 30 seconds to 2–3 minutes.

## 10.2.2 Mechanisms of epileptogenesis

Epileptogenesis is considered to be a cascade of dynamic biological events altering the balance between excitation and inhibition in neural networks. It can apply to any of the progressive biochemical, anatomic, and physiologic changes leading up to recurrent seizures. Progressive changes are suggested by the existence of a so-called silent interval (years in duration) between CNS infection, head trauma, or febrile seizures and the later appearance of epilepsy. Understanding these changes is key to preventing the onset of epilepsy [54]. Mechanisms of epileptogenesis are believed to incorporate information from levels of organization that range from molecular (e.g., altered gene expression) to macrostructural (e.g., altered neural networks). Because the possibilities are so diverse, a primary research is directed to sort out which mechanisms are causal, correlative, or consequential. The complexity can be intractable when, for example, a single seizure activates changes in expression of many genes ranging from transcription factors to structural proteins. Moreover, mechanisms of plasticity may mask the initiating event. No animal model completely mimics the features of human epilepsy. Hypotheses for epilepsy prevention must incorporate observations about the intermittent nature of epilepsy, its age-specific features, variability in expression, delayed temporal onset ranging up to 15 years after an insult, and selective vulnerability of brain regions. The potential role of protective factors is worth exploring because about 50% of patients fail to develop epilepsy even after severe penetrating brain injuries [54].

## 10.2.3 Motivation of seizure prediction research

Based on 1995 estimates, epilepsy imposes an annual economic burden of $12.5 billion in the United States in associated health care costs and losses in employment, wages, and productivity [13]. Cost per patient ranged from $4,272 for persons with remission after initial diagnosis and treatment to $138,602 for persons with intractable and frequent seizures [12]. Approximately 25% to 30% of patients receiving medication with antiepileptic drugs (AEDs), which is the mainstay of epilepsy treatment, remain unresponsive to the treatment and still have inadequate seizure control. Epilepsy surgery is another alternative treatment for medically refractory patients with the aim of excising the portion of brain tissue supposed to be responsible for seizure initiation. Nevertheless, surgery is not always feasible and involves the risk of a craniotomy. At least 50% of pre-surgical candidates eventually will not undergo respective surgery because a single epileptogenic zone could not be identified or was located in functional brain tissue through MRI scan or long-term EEG monitoring. The mean length of hospital stay for epilepsy pre-surgical candidates admitted for invasive EEG monitoring ranged from 4.7 to 5.8 days and the total aggregate costs exceeded $200 million each year [13]. Besides, only 60% to 85% of epilepsy surgery cases result in patients being seizure free. In the recent years, the vagus nerve stimulator Neurocybernetic Prosthesis has been available as an alternative epilepsy treatment that reduces seizure frequency; however, the parameters of this device (amplitude and duration of stimulation) continue to be arbitrarily adjusted by physicians. Moreover, the effectiveness of this treatment plays the same role as an additional dose of AEDs to epileptic patients and less than 0.1% of patients can benefit from this treatment. Because of the shortcomings and side effects of current epilepsy treatment, there has been an urgency for new development of novel therapeutic treatments for epilepsy. During the past few years, there has been a great deal of research interest in shifting epilepsy research from the efforts to cure epilepsy to the ability to anticipate/predict the onset of seizures. Although spontaneous epileptic seizures seem to occur randomly and unpredictably and begin intermittently as a result of complex dynamic interactions among many regions of the brain, neurologists still believe that seizures occur in a predictable fashion. Seizure prediction is a very promising option for the effective and safe treatment of people with epilepsy by avoiding both the side effects of drugs and cutting out pieces of brain. Research interest in seizure prediction has been amplified by the following new technology in the past decade: the wide acceptance of digital EEG technology; maturation of methods for recording from intracranial electrodes to localize seizures; and the tremendous efficacy, acceptability, and commercial success of implantable medical devices, such as pacemakers, implantable cardiac defibrillators, and brain stimulators for Parkinson's disease, tremor, and pain [61]. The most realizable application of seizure prediction development is its potential for use in therapeutic epilepsy devices to either warn of an impending seizure or trigger intervention to prevent seizures before they begin.

### 10.2.4 History of seizure prediction research

Work on seizure prediction started in the 1970s [94, 95] and early 1980s [85] to show the seizure's predictability. Most of the work focused on visible features in the EEG (e.g., epileptic spiking) to extract seizure precursors. More advanced quantitative analyses (e.g., spectral analysis) in the EEG are applied to discover the abnormal activity and demonstrate the predictability in seizure patterns. There have been a lot of studies in time–domain analysis including statistical analysis of particular EEG events and characterization of the EEG data. For example, the relationship between the number of normal epileptiform discharges on EEG and oncoming seizures was investigated [35, 55, 97]. Frequency domain analysis is a seizure prediction technique used to decompose the EEG signal into components of different frequencies. Nevertheless, the complexity and variability of the seizure development cannot be captured by traditional methods used to process physiologic signals. In the late 1980s, Iasemidis and coworkers made the first attempt to apply chaos theory and nonlinear dynamics to the EEG for predicting seizures [52, 53]. The technique was inspired by Takens' theorem, which proves that the complete dynamics of a system can be reconstructed from a single measurement sequence (such as its trajectory over time) along with certain invariant properties [93]. These techniques show changes in characteristics (dynamics) of the EEG waveform in the minutes leading up to seizures [53]. This scheme embeds EEG signals into a phase space and observes some of the hidden characteristics of the signals.

Nonlinear techniques showed that the trajectory of the EEG signals appeared to be more regular and organized before the clinical onset of the seizure than were the ones in the normal state. The results of this work indicate that the EEG becomes progressively less chaotic as seizures advance, with respect to the estimation of short-term maximum Lyapunov exponents ($STL_{max}$), which is a measure of the order or disorder (chaos) of signals [49]. Subsequently, Iasemidis and coworkers have also demonstrated dynamic properties and the large-scale patterns of EEG that emerge when neurons interact all together, which demonstrate that the convulsive firing of neurons in epileptic seizures offers such a clear case of collective dynamics. For example, evidence for nonlinear time dependencies in the normal EEG intervals observed from patients with frequent partial seizures is reported in [46]. This observation suggests that the occurrence of seizures, though displaying a complex time structure, is not a random process and may be driven by deterministic mechanisms. Later attempts to apply measures in nonlinear dynamics were followed by other investigations [57, 58, 59, 66, 73, 88]. The application of the correlation dimension is employed to measure the neuron complexity of the EEG, and correlation density and dynamic similarity are employed to show evidence of seizure anticipation in pre-seizure segments [29, 58, 59]. In these studies, reductions in the effective correlation dimension ($D_2^{eff}$, a measure of the complexity of the EEG signals) are shown to be more prominent in pre-seizure EEG samples than at times more distant from a seizure. The results of these studies indicate that a detectable change in dynamics can be observed at least 2 minutes before a seizure in most cases [29]. These studies were followed by the measure of phase synchronization in the pre-seizure EEG signals [66, 88].

Martinerie and coworkers also report significant differences between dimension measures obtained in pre-seizure versus normal EEG samples [66]. They find an abrupt decrease in dimension during the transition to the seizure onset in relatively brief (40 minutes) samples of pre-seizure and normal EEG data. More recently, this analysis has been extended to the study of brain dynamics obtained from scalp EEG recordings. By comparing pre-seizure EEG samples to a reference sample selected from normal EEG data, they demonstrate that temporal lobe seizures are preceded by dynamic changes by periods of up to 15 minutes [88]. The method employed in that study is derived from the method proposed by Manuca and Savit [65], which measures the degree of stationarity of EEG signals. Subsequently in the later long-term (several days) energy analysis, the changes or sustained bursts in long-term energy profiles of the EEG are reported to be increasing in volume that leads to seizure onset [62]. It was also demonstrated that bursts of activity in the range 15–25 Hz appeared to build from about 2 hours before seizure onset in some patients with temporal lobe epilepsy [62]. These burst activities seemed to change their frequency steadily (faster and slower) over time. In the most recent study, the application of the correlation dimension, correlation integral, and autocorrelation is studied to demonstrate the fluctuations of seizure dynamics [57, 73].

Although the aforementioned studies have successfully demonstrated that there exist temporal changes in the brain dynamics reflected from seizure development, it is still a very difficult task to evaluate and assess these seizure prediction techniques because of the lack of substantive studies including: the need for long-duration, high-quality data sets from a large number of patients implanted with intracranial electrodes; adequate storage and powerful computers for processing of digital EEG data sets many gigabytes in length; and environments facilitating a smooth flow of clinical EEG data to powerful experimental computing facilities [61]. In addition, the collective physiologic dynamics of billions of interconnected neurons in the human brain are not well studied or understood in those studied. Because temporal properties of the brain dynamics can only capture the interaction of some groups of locally connected neurons, they are not sufficient enough to demonstrate the mechanism or propagation of seizure development, which involves billions of interconnected neurons throughout the brain. For example, extensive investigations indicate that the quantification of only temporal properties of the brain dynamics (e.g., $STL_{max}$) fail to demonstrate the capability and sufficiency to predict seizures [23]. For this reason, a study that considers both temporal and spatial properties of the brain dynamics is proposed in [50, 75, 78]. These studies use optimization and data mining to demonstrate that the spatio-temporal dynamic properties of EEGs can reveal patterns that correspond with specific clinical states. The results of these studies led to the development of an Automated Seizure Warning System (ASWS) [22, 89, 90], which not surprisingly demonstrates that the normal, seizure, and immediate post-seizure states are distinguishable with respect to the spatio-temporal dynamic patterns/properties of intracranial EEG recordings. These patterns are considered to be seizure precursors detectable through the convergence of $STL_{max}$ profiles from critical electrodes selected by optimization techniques during the hour preceding seizures. The transition from a seizure precursor to a seizure onset has been defined as a "pre-ictal transition" [23, 50, 75].

## 10.3 Mining EEG Time Series: Chaos in Brain

Epilepsy is a "dynamic disease" that appears to be due to a malfunction in certain neurologic timing mechanisms rather than to a specific anatomic abnormality or chemical deficiency. These mechanisms are governed by a nonstationary system in the brain (the brain dynamics). To seek repetitive and predictive pre-seizure patterns, methods used to quantify the brain dynamics should be capable of automatically identifying and appropriately weighing existing transients in the brain electrical activity like EEGs. Most methods used to capture these patterns in the brain dynamics of the EEG waveform are derived from chaos theory. These methods divide EEG signals into sequential epochs (non-overlapping windows) to properly account for possible nonstationarities in the epileptic EEG recordings. For each epoch of each channel of EEG signals, the brain dynamics is quantified by applying measures of chaos (e.g., an estimation of $STL_{max}$). $STL_{max}$ quantifies the chaoticity of the EEG attractor by measuring the average uncertainty along the local eigenvectors of an attractor in the phase space. The rate of divergence is a very important aspect of the system dynamics reflected in the value of Lyapunov exponents.

The initial step in estimating $STL_{max}$ profiles from EEG signals is to embed it in a higher dimensional space of dimension $p$, which enables us to capture the behavior in time of the $p$ variables that are primarily responsible for the dynamics of the EEG. We can now construct $p$-dimensional vectors $X(t)$, whose components consist of values of the recorded EEG signal $x(t)$ at $p$ points in time separated by a time delay. Construction of the embedding phase space from a data segment $x(t)$ of duration $T$ is made with the method of delays. The vectors $X_i$ in the phase space are constructed as $X_i = (x(t_i), x(t_i + \tau) \ldots x(t_i + (p - 1) * \tau))$, where $\tau$ is the selected time lag between the components of each vector in the phase space, $p$ is the selected dimension of the embedding phase space, and $t_i \in [1, T - (p - 1)\tau]$. The method for estimation of $STL_{max}$ for nonstationary data (e.g., EEG time series) is previously explained in [44, 48, 98]. In this chapter, only a short description and basic notation of the mathematical models used to estimate $STL_{max}$ will be discussed. Let $L$ be an estimate of the short-term maximum Lyapunov exponent, defined as the average of local Lyapunov exponents in the state space. $L$ can be calculated as $L = \frac{1}{N_a \Delta t} \sum_{i=1}^{N_a} \log_2 \frac{|\delta X_{i,j}(\Delta t)|}{|\delta X_{i,j}(0)|}$, where $\delta X_{i,j}(0) = X(t_i) - X(t_j)$, $\delta X_{i,j}(\Delta t) = X(t_i + \Delta t) - X(t_j + \Delta t)$.

As the brain dynamics is quantified, T-statistical distance (T-index) is proposed as a similarity measure to estimate the difference of the dynamics of EEG time series from different brain areas [76]. In other words, the T-index is employed to seek repetitive and predictive patterns of synchronization of the brain dynamics. It measures the statistical distance between two epochs of $STL_{max}$ profiles. In the previous study, $STL_{max}$ profiles are divided into overlapping 10-minute epochs ($N = 60$ points). The T-index at time $t$ between electrode sites $i$ and $j$ is defined as $T_{i,j}(t) = \sqrt{N} \times |E\{STL_{max,i} - STL_{max,j}\}|/\sigma_{i,j}(t)$, where $E\{\cdot\}$ is the sample average difference for the $STL_{max,i} - STL_{max,j}$ estimated over a moving window $w_t(\lambda)$ defined as

$$w_t(\lambda) = \begin{cases} 1 \text{ if } \lambda \in [t - N - 1, t] \\ 0 \text{ if } \lambda \notin [t - N - 1, t], \end{cases}$$

where $N$ is the length of the moving window. Then, $\sigma_{i,j}(t)$ is the sample standard deviation of the $STL_{max}$ differences between electrode sites $i$ and

$j$ within the moving window $w_t(\lambda)$. The thus defined T-index follows a $t$-distribution with $N - 1$ degrees of freedom.

## 10.4 Optimization and Data Mining in Epilepsy Research

The previous section describes tools to quantify the brain dynamics for mining hidden patterns in EEGs. To excavate repetitive and predictive patterns in the brain dynamics associated with epileptogenesis processes, optimization and data mining (DM) have also played a very important role. To seek such patterns, these DM problems fundamentally involve discrete decisions based on numerical analyses of the brain dynamics (e.g., the number of clusters, the number of classes, the class assignment, the most informative features, the outlier samples, the samples capturing the essential information). These techniques are combinatorial in nature and can naturally be formulated as discrete optimization problems [14, 17, 32, 36, 37, 43, 63]. Nevertheless, solving these optimization-based DM problems is not an easy task because these problems naturally lend themselves to a discrete $NP$-hard optimization problem. Aside from complexity issue, the massive scale of EEG data is another difficulty arising in this research. The framework of optimization and data mining research to solve the challenging seizure prediction problem is proposed in [19, 20, 21, 22, 23, 24, 26, 76, 86]. There are 3 main aspects of the proposed framework including (1) Classification of Normal and Epileptic EEGs, (2) Electrode Selection for Seizure Pre-Cursor Detection, (3) Clustering Epileptic Brain Areas. This framework has provided insights into the epileptogenesis processes, which will revolutionize the current study in epilepsy.

### 10.4.1 Classification of normal and epileptic EEGs

Research in classification focuses on the prediction of categorical variables (data entries) based on the characteristics of their attributes (feature vectors). There has been an enormous number of optimization techniques for classification problems developed during the past few decades including classification tree, support vector machines (SVMs), linear discriminant analysis, logic regression, least squares, nearest neighbors, and so forth. A number of linear programming formulations for SVMs have been used to explore the properties of the structure of the optimization problem and solve large-scale problems [16, 64]. The SVM technique proposed in [64] was also demonstrated to be applicable to the generation of complex space partitions similar to those obtained by C4.5 [87] and CART [18]. Current SVM research mainly focuses on extending SVMs to multiclass problems [41, 56, 91].

The fundamental question of whether normal and abnormal EEGs are classifiable remains unanswered [45, 60]. Chaovalitwongse and coworkers present a

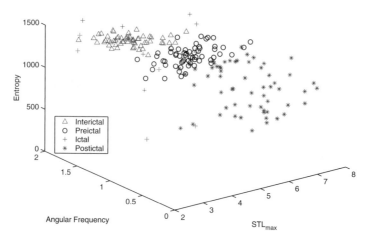

**Fig. 10.1.** Example of three-dimensional plots of entropy, angular frequency, and $STL_{max}$ in different physiologic states (normal, pre-seizure, seizure, and post-seizure) for an epileptic patient.

study undertaken to determine whether or not normal and pre-seizure (epileptic) EEGs are distinguishable [26]. The objective of that study is to demonstrate the classifiability of the two different states through the quantitative analysis of the brain dynamics ($STL_{max}$, phase, and entropy). In that study, they first calculate measures of chaos from EEG signals using the methods described in the previous section. Each measure was calculated continuously for each non-overlapping 10.24 second segment of EEG data. Figure 10.1 shows an example of a three-dimensional plot of three measures in the brain's different physiologic states. There is a gradual transition from one physiologic state to another. This observation suggests that measures of chaos can be used as features to discriminate different physiologic states of the brain dynamics. They may give the possibility to automatically classify the brain's physiologic states. The results of that study demonstrate that the brain dynamics from the same brain's physiologic state is more similar than the one from different brain's physiologic states. In other words, the brain dynamics of normal EEGs should be more similar to each other than to that of pre-seizure EEG, and vice versa. To test the difference between different states of EEGs, novel data mining techniques employed in that study include: (1) a novel statistical nearest neighbor for EGG classification, and (2) SVMs approach for EEG classification. To validate their hypothesis, a leave-one-out cross-validation is applied.

**Time series statistical nearest neighbors (TSSNNs)**

Chaovalitwongse and coworkers propose TSSNNs, which is a novel statistical classification technique for classifying time series data based on the nearest neighbor of T-statistical distance [26]. The main idea of TSSNNs is to use the nearest neighbors from EEG baselines as a decision rule of the classification of normal and abnormal EEGs. In other words, after comparing an unknown EEG epoch with baseline data from normal and abnormal EEGs, TSSNNs classifies the EEG epoch into the physiologic state (normal or abnormal) that yields the minimum average T-statistical distance (nearest neighbor). In that study, they apply cross-validation techniques to the estimation of the generalization error of TSSNNs. In general, cross-validation is considered to be a way of applying partial information about the applicability of alternative classification strategies. In other words, cross-validation is a method for estimating generalization error based on "resampling." The resulting estimates of generalization error are often used for choosing among various decision models (rules). Generally, cross-validation is referred to $k$-fold cross-validation. In $k$-fold cross-validation, the data are divided into $k$ subsets of (approximately) equal size. The decision models are trained $k$ times, in which one of the subsets from training is left out each time, by using only the omitted subset to compute whatever error criterion interests you. If $k$ is equal to the sample size, this is called "leave-one-out" cross-validation. The results of that study validate the classifiability of the brain physiologic states from EEG recordings.

TSSNNs procedure can be described as follows. Given an unknown-state epoch of EEG signals "A," average $t$-statistical distances between "A" and the groups of normal, pre-seizure, and post-seizure EEG baselines are calculated. Per electrode, three T-index values of the average mean statistical distances are obtained (see Figure 10.2). The EEG epoch "A" will be classified into the physiologic state of the nearest neighbor (normal, pre-seizure, and post-seizure). The nearest neighbor is defined as the physiologic state that yields the minimum average T-index value based on 28–32 electrodes. Because the proposed classifier has 28–32 decision inputs, two classification

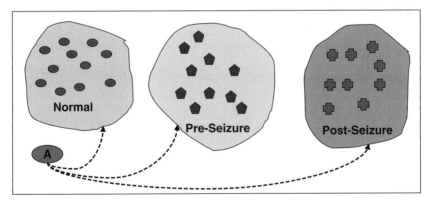

**Fig. 10.2.** Statistical comparison for classification of an unknown-state EEG epoch "A" by calculating the T-statistical distances between "A" and normal, "A" and pre-seizure, and "A" and post-seizure.

**Table 10.1.** Performance characteristics of the optimal classification scheme for individual patient.

| Patient | Sensitivity | Specificity | Optimal Scheme |
|---------|-------------|-------------|----------------|
| 1 | 90.06% | 95.03% | Average $L_{max}$ & Entropy |
| 2 | 77.27% | 88.64% | Average $L_{max}$ & Phase |
| 3 | 76.21% | 88.10% | Average $L_{max}$ & Phase |

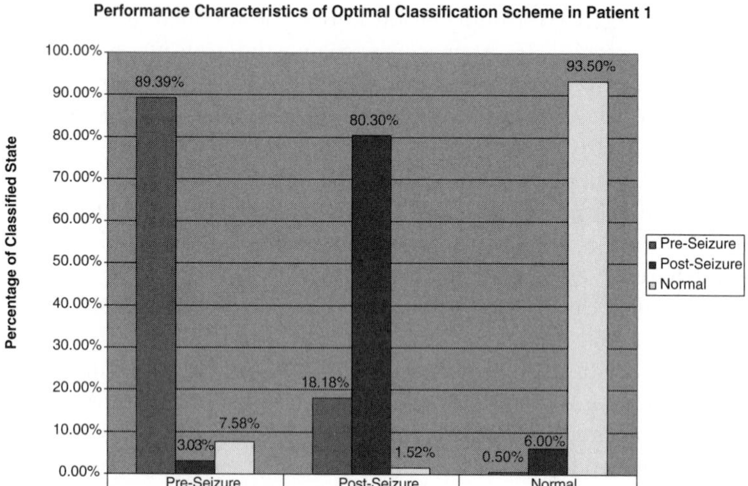

**Performance Characteristics of Optimal Classification Scheme in Patient 1**

|  | Pre-Seizure | Post-Seizure | Normal |
|---|---|---|---|
| Pre-Seizure | 89.39% | 18.18% | 0.50% |
| Post-Seizure | 3.03% | 80.30% | 6.00% |
| Normal | 7.58% | 1.52% | 93.50% |

**Fig. 10.3.** Classification results of TSSNNs in patient 1.

schemes (averaging and voting) based on different electrodes and combination of dynamical measures are proposed. In the study in [26], the performance characteristics of TSSNNs tested on 3 epileptic patients are listed in Table 10.1.

Figure 10.3 illustrates the classification results of the optimal scheme in patient 1 (Average $L_{max}$ & Entropy). The probabilities of correctly predicting pre-seizure, post-seizure, and normal EEGs are about 90%, 81%, and 94%, respectively. Figure 10.4 illustrates the classification results of the optimal scheme in patient 2 (Average $L_{max}$ & Phase). The probabilities of correctly predicting pre-seizure, post-seizure, and normal EEGs are about 86%, 62%, and 78%, respectively. Figure 10.5 illustrates the classification results of the optimal scheme in patient 3 (Average $L_{max}$ & Phase). The probabilities of correctly predicting pre-seizure, post-seizure, and normal EEGs are about 85%, 74%, and 75%, respectively. Note that in practice, classifying pre-seizure and normal EEGs is more meaningful than classifying post-seizure EEGs as the post-seizure EEGs can be easily observed (visualized) after the seizureonset.

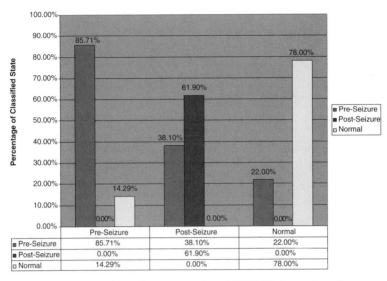

**Fig. 10.4.** Classification results of TSSNNs in patient 2.

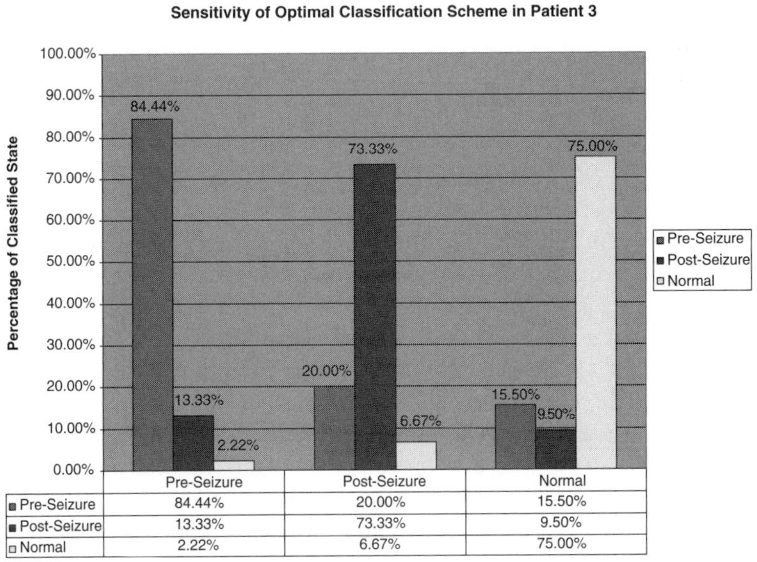

Sensitivity of Optimal Classification Scheme in Patient 3

|  | Pre-Seizure | Post-Seizure | Normal |
|---|---|---|---|
| Pre-Seizure | 84.44% | 20.00% | 15.50% |
| Post-Seizure | 13.33% | 73.33% | 9.50% |
| Normal | 2.22% | 6.67% | 75.00% |

**Fig. 10.5.** Classification results of TSSNNs in patient 3.

The results of that study indicate that we can correctly classify the pre-seizure EEGs close to 90% and close to 83% in classifying the normal EEGs [26]. These results confirm that the pre-seizure and normal EEGs are differentiable. The techniques proposed in that study can be extended to development of an online brain activity monitoring, which is used to detect the brain's abnormal activity and seizure pre-cursors. From the optimal classification schemes in 3 patients, we note that $STL_{max}$ tends to be the most classifiable attribute.

## Support Vector Machines (SVMs)

SVMs is one of the most widely used classification techniques. The essence of SVMs is to construct separating surfaces that will minimize the *upper bound on the out-of-sample error*. In the case of one linear surface (plane) separating the elements from two classes, this approach will choose the plane that maximizes the sum of the distances between the plane and the closest elements from each class, which is often referred to as a gap between the elements from different classes. The procedure of SVMs can be described as follows: Let all the data points be represented as $n$-dimensional vectors (or points in the $n$-dimensional space), then these elements can be separated *geometrically* by constructing the *surfaces* that serve as the "borders" between different groups of points. One of the common approaches is to use linear surfaces/planes for this purpose, however, different types of nonlinear (e.g., quadratic) separating surfaces can be considered in certain applications. In reality, it is not possible to find a surface that would "perfectly" separate the points according to the value of some attribute, i.e., points with different values of the given attribute may not necessarily lie at the different sides of the surface; however, in general, the number these errors should be small enough. The classification problem of SVMs can be represented as the problem of finding geometrical parameters of the separating surfaces. These parameters can be found by solving the optimization problem of minimizing the misclassification error for the elements in the training data set (in-sample error). After determining these parameters, every new data element will be automatically assigned to a certain class, according to its geometrical location in the elements space. The procedure of using the existing data set for classifying new elements is often called "training the classifier." The corresponding data set is referred to as the "training data set." It means that the parameters of separating surfaces are tuned/trained to fit the attributes of the existing elements to minimize the number of errors in their classification. However, a crucial issue in this procedure is to "not overtrain" the model, so that it would have enough flexibility to classify *new* elements, which is the primal purpose of constructing the classifier. An example of hyperplanes separating the brain's pre-seizure, normal, and post-seizure states is illustrated in Figure 10.6.

In the study in [26], one of the first practical applications of mathematical programming in the brain's state classification is proposed. The procedure of the SVMs framework for EEG classification can be stated as follows. The data set consisting of $nm$-dimensional feature vectors, where $n$ is the number of electrodes for individual patient, and $m = 30$ is the length of data sample

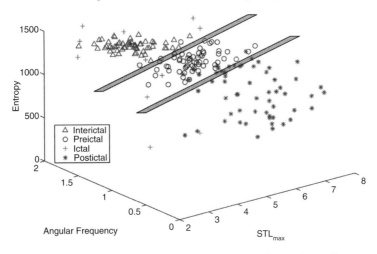

**Fig. 10.6.** Example of hyperplanes separating different brains' states.

(approximately 5 min in duration). In each patient, only samples of normal and pre-seizure EEG data are studied because SVMs is a binary (2-class) classifier in nature and, in practice, they are only interested in differentiating normal and pre-seizure data. In that study, they also apply "leave-one-out cross-validation" described in the previous section. The classifier was developed based on linear programming (LP) techniques derived from [16]. The vectors corresponding with normal and pre-seizure states are stored in two matrices, $A$ and $B$, respectively. The goal of the constructed model is to find a plane that would separate all the vectors (points in the $nm$-dimensional space) in $A$ from the vectors in $B$. A plane is defined by $x^T\omega = \gamma$, where $\omega = (\omega_1, \ldots, \omega_n)^T$ is an $n$-dimensional vector of real numbers, and $\gamma$ is a scalar. It is usually not the case where two sets of elements are perfectly separated by a plane. For this reason, the goal of SVMs is to minimize the average measure of misclassifications, i.e., in the misclassification constraints violated, the average sum of violations should be as small as possible. An optimization model to minimize the total average measure of misclassification errors is formulated as follows:

$$\min_{\omega,\gamma,u,v} \frac{1}{m}\sum_{i=1}^{m} u_i + \frac{1}{k}\sum_{j=1}^{k} v_j, s.t.\ A\omega+u \geq e\gamma+e,\ B\omega-v \leq e\gamma-e,\ u \geq 0,\ v \geq 0.$$

The violations of these constraints are modeled by introducing nonnegative variables $u$ and $v$. The decision variables in this optimization problem are the geometric parameters of the separating plane $\omega$ and $\gamma$, as well as the variables representing misclassification error $u$ and $v$. Although in many cases this type of problem may involve high dimensionality of data, they can be efficiently solved by available LP solvers, for instance MATLAB, Xpress-MP, or CPLEX.

**Table 10.2.** Performance characteristics of SVMs for EEG classification.

| Patient | Pre-seizure State | Normal State | Overall |
|---------|-------------------|--------------|---------|
| | | Sensitivity | |
| 1 | 81.21% | 87.46% | 86.43% |
| 2 | 71.18% | 76.85% | 76.76% |
| 3 | 74.13% | 70.60% | 71.00% |
| Average | 75.51% | 78.30% | 78.06% |

The SVMs developed in [26] for EEG classification are employed to classify pre-seizure and normal EEGs. To train SVNs, it is important to note that, in general, the training of SVMs is optimized when the number of pre-seizure and normal samples are comparable. Otherwise, the classifier will be biased to the physiologic state with larger size samples. In this case, there are a lot more normal EEGs than pre-seizure EEGs. To adequately evaluate SVMs, the classifier was trained with the same number of pre-seizure and normal samples. Monte Carlo sampling simulation was used to shuffle (random order) the pre-seizure and normal EEGs individually. Because the size of pre-seizure samples is much larger than the size of normal samples, the number of pre-seizure samples will be used to determine the size of the training and testing sets. Then, the first half of pre-seizure samples was used for the training and the other half for the testing. After that, training data (with the same size) from normal samples were randomly selected. For individual patients, 100 replications of the simulation were performed [26].

Table 10.2 illustrates the classification results of the SVMs in 3 epileptic patients. In patient 1, the sensitivity of predicting pre-seizure and normal EEGs are about 81% and 88%, respectively. In patient 2, the sensitivity of predicting pre-seizure and normal EEGs are about 71% and 77%, respectively. In patient 3, the sensitivity of predicting pre-seizure and normal EEGs are about 74% and 71%, respectively. Note that this result is consistent with the prediction results from TSSNNs. The classification results in patient 1 tend to be better than those of patients 2 and 3. These results confirm that the brain's physiologic states are classifiable based on quantitative analyses of EEG. The framework of classifiers proposed in [26] can be extended to development of an automated brain's state classifier or an online brain activity monitoring.

### 10.4.2 Feature selection for seizure precursor detection

Although the brain is considered to be the largest interconnected network, neurologists still believe that seizures represent a spontaneous formation of self-organizing spatio-temporal patterns that involve only some parts (electrodes) of the brain network. The localization of epileptogenic zones is one of the proofs of this concept. Therefore, feature selection techniques have become a very essential tool for selecting the critical brain areas participating in the epileptogenesis process during seizure development. In addition, graph theoretical approaches appear to fit very well as a model of a brain structure [27, 39]. Feature selection based on optimization and graph theoretical approaches will be very useful in selecting/identifying the brain areas correlated with the pathway to seizure onset. Feature/sample selection can naturally be defined as a binary optimization problem as the notion of selection a sub-set of variables, out of a set of possible alternatives. Integer optimization techniques have been used in feature selection in diverse disciplines including spin glass models [7, 9, 10, 11, 40, 68], portfolio selection [14, 31, 92], variable selection in linear regression [71, 92], media selection [99], and multiclass discrimination analysis [43]. Many integer programming theories and implicit enumeration techniques have been developed to address the problem of feature selection [15, 25, 67, 70, 74, 75, 77, 81, 82].

The concept of optimization models for feature selection used to select/identify the brain areas correlated with the pathway to seizure onset came from the Ising model [19, 23], which is a powerful tool in studying phase transitions in statistical physics. Such an Ising model can be described by a graph $G(V, E)$ having $n$ vertices $\{v_1, \ldots, v_n\}$ and each edge $(i, j) \in E$ having a weight (interaction energy) $J_{ij}$. Each vertex $v_i$ has a magnetic spin variable $\sigma_i \in \{-1, +1\}$ associated with it. An optimal spin configuration of minimum energy is obtained by minimizing the Hamiltonian $H(\sigma) = -\sum_{1 \leq i \leq j \leq n} J_{ij}\sigma_i\sigma_j$ over $\forall \sigma \in \{-1, +1\}^n$. This problem is equivalent to the combinatorial problem of quadratic 0-1 programming [40]. This idea has been used to develop quadratic 0-1 (integer) programming for feature/electrode selection problem, where each electrode has only two states, and to determine the minimal-average T-index state [76]. In later attempts, Chaovalitwongse and coworkers introduce an extension of quadratic integer programming for electrode selection by modeling this problem as a Multi-Quadratic Integer Programming (MQIP) problem [19, 24, 25, 75]. The MQIP formulation of the electrode selection problem is extremely difficult to solve. Although many efficient reformulation-linearization techniques (RTLs) have been used to linearize QP and nonlinear integer programming problems [2, 3, 4, 5, 8, 30, 33, 34, 72, 96], additional quadratic constraints make MQIP problems much more difficult to solve, and current RTLs fail to solve MQIP problems effectively. A fast and scalable RTL used to solve the MQIP feature selection problem is proposed in preliminary studies in [25, 75]. The proposed technique has been shown to outperform other RTLs [38]. In addition, a novel framework applying graph theory to feature selection has been recently proposed in the preliminary study by Prokopyev and coworkers in [86].

**Feature selection via quadratic integer programming (FSQIP)**

FSQIP is a novel mathematical model for selecting critical features (electrodes) of the brain network, which can be modeled as a quadratic 0-1 knapsack problem with objective function to minimize the average T-index (a

measure of statistical distance between the mean values of $STL_{max}$) among
electrode sites and the knapsack constraint to identify the number of critical cortical sites. It is known that a quadratic 01 program with a knapsack
constraint can be reduced to an unconstrained quadratic 0-1 programming
problem [76], which can be solved by a powerful branch-and-bound method
developed by Pardalos and Rodgers [79, 80].

Consider the following three problems:

$P_1$ : $\min f(x) = x^T A x$, $x \in \{0,1\}^n$, $A \in R^{n \times n}$.
$\bar{P}_1$ : $\min f(x) = x^T A x + c^T x$, $x \in \{0,1\}^n$, $A \in R^{n \times n}$, $c \in R^n$.
$\hat{P}_1$ : $\min f(x) = x^T A x$, $x \in \{0,1\}^n$, $A \in R^{n \times n}$, $\sum_{i=1}^n x_i = k$, where $0 \leq k \leq n$ is a constant .

Define $A$ as an $n \times n$ T-index pair-wise distance matrix, and $k$ is the number of selected electrode sites. Problems $P_1$, $\bar{P}_1$, and $\hat{P}_1$ can be shown to be
all "equivalent" by proving that $P_1$ is "polynomially reducible" to $\bar{P}_1$, $\bar{P}_1$ is
"polynomially reducible" to $P_1$, $\hat{P}_1$ is "polynomially reducible" to $P_1$, and $P_1$
is "polynomially reducible" to $\hat{P}_1$.

The results from the application of the previously described scheme to
decide the predictability of epileptic seizures are presented in [76]. The method
is applied to 58 epileptic seizures in five patients. Patient 1 had 24 seizures in
83.3 hours; patient 2 had 19 seizures in 145.5 hours; patient 3 had 8 seizures in
22.6 hours; patient 4 had 4 seizures in 6.5 hours; and patient 5 had 3 seizures
in 8.3 hours. The method described in the previous section was applied with
two different critical values ($\alpha = 0.1, 0.2$). Figures 10.7 and 10.8 illustrate
examples of a predictable seizure and an unpredictable seizure, respectively.
In both figures, curves B and C are smoothed curves of A (by averaging
the original T-index values within a moving window of length equal to PTP,

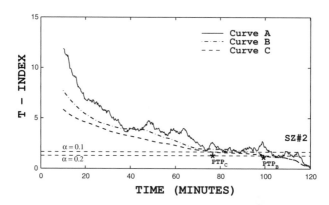

**Fig. 10.7.** An example of a predictable seizure by the average T-index curves of the
pre-ictally selected sites (patient 1). Curve A: original T-index curve of the selected
sites. Curves B and C: smoothed curves of A over windows of entrainment with
length defined from critical values $T_\alpha$ at significance levels 0.2 and 0.1, respectively.

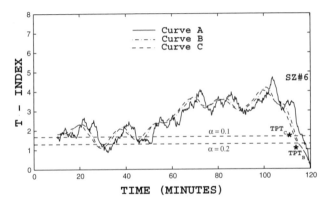

**Fig. 10.8.** An example of a unpredictable seizure by the T-index curves of the selected sites (patient 1).

**Table 10.3.** Predictability analysis for 58 epileptic seizures.

| Patient | Total No. of Seizures | Average $PTP_B$ | Average $PTP_C$ (minutes) | Predictable Seizures (minutes) | Predictability |
|---|---|---|---|---|---|
| 1 | 24 | 42.9 | 66.9 | 21 | 87.5% |
| 2 | 19 | 19.8 | 29.8 | 17 | 89.5% |
| 3 | 8 | 23.5 | 49.5 | 8 | 100% |
| 4 | 4 | 36.1 | 44.1 | 4 | 100% |
| 5 | 3 | 31.1 | 34.4 | 3 | 100% |
| Total | 58 | 31.6 | 49.1 | 53 | 91.4% |

which is different per curve). In Figure 10.7, the pre-ictal transition period $PTP_B$ identified by curve B is about 20 minutes, and $PTP_C$ (identified by curve C) is about 43 minutes. It is clear that there are no false positives observed in both curves over the 2-hour period prior to this seizure, thus this seizure is considered to be predictable. In Figure 10.8, the PTPs identified by the smoothed curves are 5 and 7 minutes, respectively. But false positives are observed at 85 and 75 minutes prior to this seizure's onset for curves B and C, respectively. Therefore, this seizure is concluded to be non-predictable. Table 10.3 summarizes the results of this analysis for all 58 seizures [76].

## Feature selection via multi-quadratic integer programming (FSMQIP)

FSMQIP is a novel mathematical model for selecting critical features (electrodes) of the brain network proposed in [24, 75]. The MQIP electrode selection problem is given by $\min x^T A x$, s.t., $\sum_{i=1}^{n} x_i = k$; $x^T C x \geq T_\alpha k(k-1)$; $x \in \{0,1\}^n$, where $A$ is an $n \times n$ matrix of pairwise similarity of chaos measures before a seizure, $C$ is an $n \times n$ matrix of pairwise similarity of chaos measures after a seizure, and $k$ is the predetermined number of selected electrodes. This problem has been proved to be $NP$-hard in [75]. The objective function is to minimize the average T-index distance (similarity) of chaos measures among the critical electrode sites. The knapsack constraint is to identify the number of critical cortical sites. The quadratic constraint is to ensure the divergence of chaos measures among the critical electrode sites after a seizure. This MQIP can be reduced to linear mixed 01 programming, which can be solved using modern solvers like CPLEX and XPRESS-MP. For more details, we refer to [25].

FSMQIP has been developed to extend the previous findings of the seizure predictability described in the previous section. The FSMQIP problem is formulated as a MQIP problem with objective function to minimize the average T-index (a measure of statistical distance between the mean values of $STL_{max}$) among electrode sites, the knapsack constraint to identify the number of critical cortical sites [53, 51], and an additional quadratic constraint to ensure that the optimal group of critical sites shows the divergence in $STL_{max}$ profiles after a seizure. The experiment in the study proposed by Chaovalitwongse and coworkers is to test the hypothesis that FSMQIP can be used to select critical features (electrodes) that are mostly likely to manifest precursor patterns prior to a seizure [24]. The results of that study demonstrate that if one can select critical electrodes that will manifest seizure precursors, it may be possible to predict a seizure in time to warn of an impending seizure. To test this hypothesis, an experiment used to compare the probability of detecting seizure precursor patterns from critical electrodes selected by FSMQIP with that from randomly selected electrodes was proposed [24]. Tested on 3 patients with 20 seizures, the prediction performance of randomly selected 5,000 groups of electrodes was compared with that of critical electrodes selected by FSMQIP. The results show that the probability of detecting seizure precursor patterns from the critical electrodes selected by FSMQIP is approximately 83%, which is significantly better than that from randomly selected electrodes with ($p$-value $< 0.07$). The histogram of probability of detecting seizure precursor patterns from randomly selected electrodes and that from the critical electrodes is illustrated in Figure 10.9. The results of that study can be used as a criterion to pre-select the critical electrode sites that can be used to predict epileptic seizures.

**Feature selection via maximum clique (FSMC)**

FSMC is a novel mathematical model based on graph theory for selecting critical features (electrodes) of the brain network [20]. The brain connectivity can be rigorously modeled as a brain graph as follows: considering a brain network of electrodes as a weighted graph, where each node represents an electrode, and weights of edges between nodes represent T-statistical distances of chaos measures between electrodes. Three possible weighted graphs are proposed: *GRAPH-I* is denoted as a complete graph (the graph with all possible edges);

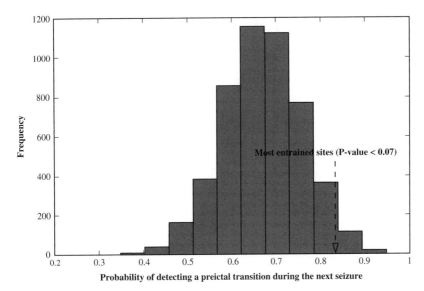

**Fig. 10.9.** Histogram of the prediction performance of randomly selected electrodes compared with that of critical electrodes selected by FSMQIP.

*GRAPH-II* is denoted as a graph induced from the complete one by deleting edges whose T-index before a seizure is greater than the T-test confident level; *GRAPH-III* is denoted as a graph induced from the complete one by deleting edges whose T-index before a seizure is larger than the T-test confident level or T-index after a seizure point is smaller than the T-test confidence level. Maximum cliques of these graphs will be investigated as the hypothesis is a group of physiologically connected electrodes is considered to be a critical largest connected network of seizure evolution and pathway. The Maximum Clique Problem (MCP) is *NP*-hard [1, 84]; therefore, solving MCPs is not an easy task.

Consider a maximum clique problem defined as follows. Let $G = G(V, E)$ be an undirected graph where $V = \{1, \ldots, n\}$ is the set of vertices (nodes), and $E$ denotes the set of edges. Assume that there is no parallel edges (and no self-loops joining the same vertex) in $G$. Denote an edge joining vertex $i$ and $j$ by $(i, j)$.

**Definition 1.** *A clique of $G$ is a subset $C$ of vertices with the property that every pair of vertices in $C$ is connected by an edge; that is, $C$ is a clique if the subgraph $G(C)$ induced by $C$ is complete.*

**Definition 2.** *The maximum clique problem is the problem of finding a clique set $C$ of maximal cardinality (size) $|C|$.*

The maximum clique problem can be represented in many equivalent formulations (e.g., an integer programming problem, a continuous global optimization problem, and an indefinite quadratic programming) [69]. Consider the following indefinite quadratic programming formulation of MCP. Let $A_G = (a_{ij})_{n \times n}$ be the adjacency matrix of $G$ defined by

$$a_{ij} = \begin{cases} 1 \text{ if } (i,j) \in E \\ 0 \text{ if } (i,j) \notin E. \end{cases}$$

The matrix $A_G$ is symmetric, and all eigenvalues are real numbers. Generally, $A_G$ has positive and negative (and possibly zero) eigenvalues, and the sum of eigenvalues is zero as the main diagonal entries are zero [40]. Consider the following indefinite QIP problem and MIP problem for MCP:

$P_3 : \max \sum_{(i,j) \in E} \frac{1}{2} x^T A x$, s.t. $x \in \{0,1\}^n$, where $A = A_{\bar{G}} - I$ and $A_G$ is an adjacency matrix of the graph $G$.

$\bar{P}_3 : \min \sum_{i=1}^{n} s_i$, s.t. $\sum_{j=1}^{n} a_{ij} x_j - s_i - y_i = 0$, $y_i - M(1 - x_i) \leq 0$, where $x_i \in \{0,1\}$, $s_i, y_i \geq 0$, and $M = \max_i \sum_{j=1}^{n} |a_{ij}| = \|A\|_\infty$.

**Proposition 1.** $P_3$ *is equivalent to* $\bar{P}_3$. *If* $x^*$ *solves the problems* $P_3$ *and* $\bar{P}_3$, *then the set* $C$ *defined by* $C = t(x^*)$ *is a maximum clique of graph* $G$ *with* $|C| = -f_G(x)$.

It has been shown in [20, 25] that $P_3$ has an optimal solution $x^0$ *iff* there exist $y^0$, $s^0$, such that $(x^0, y^0, s^0)$ is an optimal solution to $\bar{P}_3$. Applying a linearization technique described in [20, 25] to solve $\bar{P}_3$, we can select relevant features (group of electrodes) that may be critical to epileptogenic processes. These features can represent the brain connectivity through cliques of the brain graph.

## 10.4.3 Clustering epileptic brain areas

Clustering is an unsupervised learning, in which the property or the expected number of groups (clusters) are not known ahead of time [6]. Most clustering methods (e.g., $k$-mean) attempt to identify the best $k$ clusters that minimize the distance of the points assigned in the cluster from the center of the cluster. Another well-known clustering technique is $k$-median clustering, which can be modeled as a concave minimization problem and reformulated as a minimization problem of a bilinear function over a polyhedral set by introducing decision variables to assign a data point into a cluster [17, 28]. Although these clustering techniques are well studied and robust, they still require *a priori* knowledge of the data (e.g., the number of clusters, the most informative features). The elements and dynamic connections of the brain dynamics can portray the characteristics of a group of neurons and synapses or neuronal populations driven by the epileptogenic process. Therefore, clustering the brain areas portraying similar structural and functional relationships will give us an insight in the mechanisms of epileptogenesis and an answer to a question of how seizures are generated, developed, and propagated, and how they can be disrupted and treated. The goal of clustering is to find the best segmentation of raw data into the most common/similar groups. In clustering, similarity measure is, therefore, the most important property. The difficulty in clustering arises from the fact that clustering is an unsupervised learning, in which the property or the expected number of groups (clusters) are not known ahead of time [6]. The search for the optimal number of clusters is parametric in nature and the optimal point in an "error" versus "number of clusters" curve is usually identified by a combined objective that appropriately weighs accuracy and number of clusters [6]. The neurons in the cerebral cortex maintain thousands of input and output connections with other groups of neurons, which form a dense network of connectivity spanning the entire thalamocortical system. Despite this massive connectivity, cortical networks are exceedingly sparse, with respect to the number of connections present out of all possible connections. This indicates that brain networks are not random, but form highly specific patterns. Networks in the brain can be analyzed at multiple levels of scale. Novel clustering techniques used to construct the temporal and spatial mechanistic basis of the epileptogenic models based on the brain dynamics of EEGs and capture the patterns or hierarchical structure of the brain connectivity from statistical dependence among brain areas are proposed in [20]. These do not require *a priori* knowledge of the data (number of clusters). In this section, we will discuss the following clustering techniques proposed in [20]: (1) Clustering via Concave Quadratic Programming (CCQP); and (2) Clustering via MIP with Quadratic Constraint (CMIPQC).

**Clustering via concave quadratic programming (CCQP)**

CCQP is a novel clustering mathematical model used to formulate a clustering problem as a QIP problem [20]. Given $n$ points of data to be clustered, a clustering problem is formulated as follows: $\min_x f(x) = x^T A x - \lambda I$, s.t. $x \in \{0, 1\}^n$, where $A$ is an $n \times n$ Euclidean matrix of pairwise distance, $I$ is an identity matrix, $\lambda$ is a parameter adjusting the degree of similarity within a cluster, $x_i$ is a 0-1 decision variable indicating whether or not point $i$ is selected to be in the cluster. Note that $\lambda I$ is an offset parameter added to the objective function to avoid the optimal solution of all $x_i$ being zero. This will happen when every entry $a_{ij}$ of Euclidean matrix $A$ is positive and the diagonal is zero. Although this clustering problem is formulated as a large QIP problem, in some instances when $\lambda$ is large enough to make the quadratic function become concave function, this problem can be converted to a continuous problem (minimizing a concave quadratic function over a sphere) [20]. The reduction to a continuous problem is the main advantage of CCQP. This property holds because of the fact that a concave function $f : S \to \mathbb{R}$ over a compact convex set $S \subset \mathbb{R}^n$ attains its global minimum at one of the extreme points of $S$ [40].

    One of the advantages of CCQP is the ability to systematically determine the optimal number of clusters. Although CCQP has to solve $m$ clustering problems iteratively (where $m$ is the final number of clusters at the termination of CCQP algorithm), it is efficient enough to solve large-scale clustering problems because only one continuous problem is solved in each iteration. After each iteration, the problem size will become significantly smaller [20].

**Clustering via MIP with quadratic constraint (CMIPQC)**

CMIPQC is a novel clustering mathematical model in which a clustering problem can be formulated as a mixed-integer programming problem with quadratic constraint [20]. The goal of CMIPQC is to maximize number of data points to be in a cluster such that the similarity degrees among data points in a cluster are less than a predetermined parameter, $\alpha$. This technique can be incorporated with hierarchical clustering methods as follows: (a) Initialization: assign all data points into one cluster; (b) Partition: use CMIPQC to divide the big cluster into smaller clusters; (3) Repetition: repeat the partition process until the stopping criterion are reached or a cluster contains a single point. Novel mathematical formulation for CMIPQC is given by $\max_x \sum_{i=1}^n x_i$, s.t. $x^T C x \le \alpha$, $x \in \{0,1\}$, where $n$ is the number of data points to be clustered, $C$ is an $n \times n$ Euclidean matrix of pairwise distance, $\alpha$ is a predetermined parameter of the similarity degree within each cluster, $x_i$ is a 0-1 decision variable indicating whether or not point $i$ is selected to be in the cluster. The objective of this model is to maximize number of data points to be in a cluster such that the average pairwise distances among those points are less than $\alpha$. The difficulty of this problem comes from the quadratic constraint; however, this quadratic constraint can be efficiently linearized by the approach described in [25]. The CMIPQC problem is much easier to solve as it can be reduced to an equivalent MIP problem. Similar to CCQP, the CMIPQC algorithm has the ability to systematically determine the optimal number of clusters and only needs to solve $m$ MIP problems.

# 10.5 Concluding Remarks and Prospective Issues

This chapter gives an extensive review of optimization and data mining research in seizure prediction. A theoretical foundation of optimization techniques for classification, feature selection, and clustering is discussed in this chapter. Advances in classification, feature selection, and clustering techniques have shown very promising results for the future development of a novel DM paradigm to predict impending seizures from multichannel EEG recordings. The results in previous studies indicate that it is possible to design algorithms used to detect dynamic patterns of critical electrode sites. Such algorithms can be derived from novel techniques in optimization and data mining [21, 24, 47]. Prediction is possible because, for the vast majority of seizures, the spatio-temporal dynamic features of seizure precursors are sufficiently similar to that of the preceding seizure. The seizure precursors detected by the algorithm seem to be sufficiently early enough to allow a wide range of therapeutic interventions. The temporal and spatial properties of the brain dynamics captured by the methods described in this chapter have been proven capable of reflecting the real physiologic changes in the brain as they correspond specifically to the real seizure precursors.

This preclinical research forms a bridge between seizure prediction research and the implementation of seizure prediction/warning devices, which is a revolutionary approach for handling epileptic seizures, very similar to the brain-pacemaker. It may also lead to clinical investigations of the effects of medical diagnosis, drug effects, or therapeutic intervention during invasive EEG monitoring of epileptic patients. Potential diagnostic applications include a seizure warning system used during long-term EEG recordings performed in a diagnostic epilepsy-monitoring unit. This type of system could potentially be used to warn professional staff of an impending seizure or to trigger functional imaging devices in order to measure regional cerebral blood flow during seizure onset. Future research toward the treatment of human epilepsy and therapeutic intervention of epileptic activities, as well as the development of seizure feedback control devices, may be feasible. This type of seizure warning algorithm could also be incorporated into digital signal processing chips for use in implantable devices. Such devices could be utilized to activate pharmacologic or physiologic interventions designed to abort an impending seizure. Thus, it represents a necessary first step in the development of implantable biofeedback devices to directly regulate therapeutic intervention to prevent impending seizures or other brain disorders. For example, such an intervention might be achieved by electrical or magnetic stimulation (e.g., vagal nerve stimulation) or by a timely release of an anticonvulsant drug. Future studies employing novel experimental designs are required to investigate the therapeutic potential for implantable seizure warning devices. Another practical application of the proposed approach would be to help neurosurgeons quickly identify the epileptogenic zone without having patients stay in the hospital for the invasive long-time (10–14 days in duration) EEG monitoring. This research has the potential to revolutionize the protocol to identify the epileptogenic zone, which could drastically reduce the healthcare cost during the hospital stay for these patients. In addition, this protocol will help physicians identify epileptogenic zones without the necessity to risk patient safety by implanting depth electrodes in the brain. In addition, the results from this study could also contribute to the understanding of the intermittency of other dynamic neurophysiologic disorders of the brain (e.g., migraines, panic attacks, sleep disorders, and Parkinsonian tremors). This research could also contribute to the localization of defects (flaws) classification and prediction of spatio-temporal transitions in other high-dimensional biological systems such as heart fibrillation and heart attacks.

In spite of capability of predicting seizures, these algorithms can be improved in terms of parameter settings in the procedures for every patient to quantify the brain dynamics, optimize electrode selection, and detection of pre-ictal transition. Those parameters remain to be further investigated. In addition, the implementation is complicated by the fact that the parameter settings (embedding dimension and time delay) in the estimation of $STL_{max}$ is optimized based on the seizure EEG depth recordings in human subject with respect to minimization of the transient and reduction of the nonstationarity of EEG. Therefore these algorithms cannot gain the maximum prediction power with non-optimal parameter setting, which remains to be further investigated. The clinical utility of a seizure warning system depends upon the false-positive rate as well as the sensitivity of the system. It is also possible that the false warnings correctly detect a pre-seizure or seizure susceptibility state, but normal physiologic resetting mechanisms intervene returning the brain to a more normal dynamic state. It may be possible that the dynamics of the pre-ictal transition are not unique and may be found in other physiologic states. In addition, the novel clustering techniques proposed should be further investigated in the future research as they might be able to provide more insights into the epileptogenesis processes.

## Acknowledgments

Thanks are due to Professors P.M. Pardalos, J.C. Sackellares, L.D. Iasemidis, and P.R. Carney, as well as to D.-S. Shiau, who have been very helpful in sharing their expert knowledge on global optimization and brain dynamics and physiology and for their fruitful comments and discussion. This work was partially supported by the National Science Foundation under CAREER grant CCF-0546574 and Rutgers Research Council grant 202018.

## References

[1] J. Abello, S. Butenko, P.M. Pardalos, and M.G.C. Resende. Finding independent sets in a graph using continuous multivariable polynomial formulations. *Journal of Global Optimization*, 21:111–137, 2001.

[2] W.P. Adams and R.J. Forrester. A simple recipe for concise mixed 0-1 linearizations. *Operations Research Letters*, 33:55–61, 2005.

[3] W.P. Adams, R.J. Forrester, and F.W. Glover. Comparison and enhancement strategies for linearizing mixed 0-1 quadratic programs. *Discrete Optimization*, 11:99–120, 2004.

[4] W.P. Adams and H.D. Sherali. A tight linearization and an algorithm for zero-one quadratic programming problems. *Management Science*, 32:1274–1290, 1986.

[5] W.P. Adams and H.D. Sherali. Linearization strategies for a class of zero-one mixed integer programming problems. *Operations Research*, 38:217–226, 1990.

[6] I.P. Androulakis and W.A. Chaovalitwongse. Mathematical programming for data mining. In C.A. Floudas and P.M. Pardalos, editors, *Encyclopedia in Optmization*. Springer, In press.

[7] G.G. Athanasiou, C.P. Bachas, and W.F. Wolf. Invariant geometry of spin-glass states. *Physical Review B*, 35:1965–1968, 1987.

[8] E. Balas and J.B. Mazzola. Non-linear 0-1 programming i: Linearization techniques. *Mathematical Programming*, 30:1–21, 1984.

[9] F. Barahona. On the computational complexity of spin glass models. *J. Phys. A: Math. Gen.*, 15:3241–3253, 1982.

[10] F. Barahona. On the exact ground states of three-dimensional ising spin glasses. *J. Phys. A: Math. Gen.*, 15:L611–L615, 1982.

[11] F. Barahona, M. Grötschel, M.Jüger, and G. Reinelt. An application of combinatorial optimization to statistical physics and circuit layout design. *Operations Research*, 36:493–513, 1988.

[12] C.E. Begley, J.F. Annegers, D.R. Lairson, T.F. Reynolds, and W.A. Hauser. Cost of epilepsy in the United States: a model based on incidence and prognosis. *Epilepsia*, 35(6):1230–1243, 1994.

[13] C.E. Begley, M. Famulari, J.F. Annegers, D.R. Lairson, T.F. Reynolds, S. Coan, S. Dubinsky, M.E. Newmark, C. Leibson, E.L. So, and W.A. Rocca. The cost of epilepsy in the united states: an estimate from population-based clinical and survey data. *Epilepsia*, 41(3):342–351, 2000.

[14] D. Bertsimas, C. Darnell, and R. Soucy. Portfolio construction through mixed-integer programming at Grantham, Mayo, van Otterloo and Company. *Interfaces*, 29(1):49–66, 1999.

[15] D. Bienstock. Computational study on families of mixed-integer quadratic programming problems. *Mathematical Programming*, 74:121–140, 1996.

[16] P.S. Bradley, U. Fayyad, and O.L. Mangasarian. Mathematical programming for data mining: Formulations and challenges. *INFORMS J. of Computing*, 11:217–238, 1999.

[17] P.S. Bradley, O.L. Mangasarian, and W.N. Street. Clustering via concave minimization. In M.C. Mozer, M.I. Jordan, and T. Petsche, editors, *Advances in Neural Information Processing Systems*. MIT Press, 1997.

[18] L. Breiman, J. Friedman, R. Olsen, and C. Stone. *Classification and Regression Trees*. Wadsworth Inc., 1993.

[19] W.A. Chaovalitwongse. *Optimization and Dynamical Approaches in Nonlinear Time Series Analysis with Applications in Bioengineering*. PhD thesis, University of Florida, 2003.

[20] W.A. Chaovalitwongse. A robust clustering technique via quadratic programming. Technical report, Department of Industrial and Systems Engineering, Rutgers University, 2005.

[21] W.A. Chaovalitwongse, L.D. Iasemidis, P.M. Pardalos, P.R. Carney, D.-S. Shiau, and J.C. Sackellares. Performance of a seizure warning algorithm based on the dynamics of intracranial EEG. *Epilepsy Research*, 64:93–133, 2005.

[22] W.A. Chaovalitwongse, P.M. Pardalos, L.D. Iasemidis, J.C. Sackellares, and D.-S. Shiau. Optimization of spatio-temporal pattern processing for seizure warning and prediction. *U.S. Patent application filed August 2004, Attorney Docket No. 028724–150*, 2004.

[23] W.A. Chaovalitwongse, P.M. Pardalos, L.D. Iasemidis, D.-S. Shiau, and J.C. Sackellares. Applications of global optimization and dynamical systems to prediction of epileptic seizures. In P.M. Pardalos, J.C. Sackellares, L.D. Iasemidis, and P.R. Carney, editors, *Quantitative Neuroscience*, pages 1–36. Kluwer, 2003.

[24] W.A. Chaovalitwongse, P.M. Pardalos, L.D. Iasemidis, D.-S. Shiau, and J.C. Sackellares. Dynamical approaches and multi-quadratic integer programming for seizure prediction. *Optimization Methods and Software*, 20(2–3):383–394, 2005.

[25] W.A. Chaovalitwongse, P.M. Pardalos, and O.A. Prokoyev. A new linearization technique for multi-quadratic 0-1 programming problems. *Operations Research Letters*, 32(6):517–522, 2004.

[26] W.A. Chaovalitwongse, P.M. Pardalos, and O.A. Prokoyev. Electroencephalogram (EEG) Time Series Classification: Applications in Epilepsy. *Annals of Operations Research*, 148:227–250, 2006.

[27] Christopher Cherniak, Zekeria Mokhtarzada, and Uri Nodelman. Optimal-wiring models of neuroanatomy. In Giorgio A. Ascoli, editor, *Computational Neuroanatomy*. Humana Press, 2002.

[28] J.C. Dunn. A fuzzy relative of the isodata process and its use in detecting compact well-separated clusters. *Journal of Cybernetics*, 3:32–57, 1973.

[29] C.E. Elger and K. Lehnertz. Seizure prediction by non-linear time series analysis of brain electrical activity. *European Journal of Neuroscience*, 10:786–789, 1998.

[30] S. Elloumi, A. Faye, and E. Soutif. Decomposition and linearization for 0-1 quadratic programming. *Annals of Operations Research*, 99:79–93, 2000.

[31] B. Faaland. An integer programming algorithm for portfolio selection. *Management Science*, 20(10):1376–1384, 1974.

[32] G.M. Fung and O.L. Mangasarian. Proximal support vector machines. In $7^{th}$ *ACM SIGKDD International Conference on Knowledge Discovery and Data Mining*, 2001.

[33] F. Glover. Improved linear integer programming formulations of nonlinear integer programs. *Management Science*, 22:455–460, 1975.

[34] F. Glover and E. Woolsey. Further reduction of zero-one polynomial programming problems to zero-one linear programming problems. *Operations Research*, 21:156–161, 1973.

[35] J. Gotman, J. Ives, P. Gloor, A. Olivier, and L. Quesney. Changes in interictal eeg spiking and seizure occurrence in humans. *Epilepsia*, 23:432–433, 1982.

[36] R.L. Grossman, C. Kamath, P. Kegelmeyer, V. Kumar, and E.E. Namburu. *Data mining for Scientific and Engineering Applications*. Kluwer Academic Publishers, 2001.

[37] D.J. Hand, H. Mannila, and P. Smyth. *Principle of Data Mining*. Bradford Books, 2001.

[38] X. He, A. Chen, and W.A. Chaovalitwongse. Solving quadratic zero-one programming problems: Comparison and applications. Abstract: *Annual INFORMS Meeting*, 2005.

[39] Claus C. Hilgetag, Rolf Kötter, Klaas E. Stephen, and Olaf Sporns. Computational methods for the analysis of brain connectivity. In Giorgio A. Ascoli, editor, *Computational Neuroanatomy*. Humana Press, 2002.

[40] R. Horst, P.M. Pardalos, and N.V. Thoai. *Introduction to global optimization*. Kluwer Academic Publishers, 1995.

[41] C.-W. Hsu and C.-J. Lin. A comparison of methods multi-class support vector machines. *IEEE Transactions on Neural Networks*, 13:415–425, 2002.

[42] A.S. Hurn, K.A. Lindsay, and C.A. Michie. Modelling the lifespan of human t-lymphocite subsets. *Mathematical Biosciences*, 143:91–102, 1997.

[43] F.J. Iannatilli and P.A. Rubin. Feature selection for multiclass discrimination via mixed-integer linear programming. *IEEE Transactions on Pattern Analysis and Machine Learning*, 25:779–783, 2003.

[44] L.D. Iasemidis. *On the dynamics of the human brain in temporal lobe epilepsy*. PhD thesis, University of Michigan, Ann Arbor, 1991.

[45] L.D. Iasemidis. Epileptic seizure prediction and control. *IEEE Transactions on Biomedical Engineering*, 5(5):549–558, 2003.

[46] L.D. Iasemidis, L.D. Olson, J.C. Sackellares, and R.S. Savit. Time dependencies in the occurrences of epileptic seizures: a nonlinear approach. *Epilepsy Research*, 17:81–94, 1994.

[47] L.D. Iasemidis, P.M. Pardalos, D.-S. Shiau, W.A. Chaovalitwongse, K. Narayanan, A. Prasad, K. Tsakalis, P.R. Carney, and J.C. Sackellares. Long term prospective on-line real-time seizure prediction. *Journal of Clinical Neurophysiology*, 116(3):532–544, 2005.

[48] L.D. Iasemidis, J.C. Principe, and J.C. Sackellares. Measurement and quantification of spatiotemporal dynamics of human epileptic seizures. In M. Akay, editor, *Nonlinear Biomedical Signal Processing*, pages 294–318. Wiley–IEEE Press, vol. II, 2000.

[49] L.D. Iasemidis and J.C. Sackellares. Chaos theory and epilepsy. *The Neuroscientist*, 2:118–126, 1996.

[50] L.D. Iasemidis, D.-S. Shiau, W.A. Chaovalitwongse, J.C. Sackellares, P.M. Pardalos, P.R. Carney, J.C. Principe, A. Prasad, B. Veeramani, and K. Tsakalis. Adaptive epileptic seizure prediction system. *IEEE Transactions on Biomedical Engineering*, 5(5):616–627, 2003.

[51] L.D. Iasemidis, D.-S. Shiau, J.C. Sackellares, and P.M. Pardalos. Transition to epileptic seizures: Optimization. In D.Z. Du, P.M. Pardalos, and J. Wang, editors, *DIMACS series in Discrete Mathematics and Theoretical Computer Science*, pages 55–74. American Mathematical Society, 1999.

[52] L.D. Iasemidis, H.P. Zaveri, J.C. Sackellares, and W.J. Williams. Phase space analysis of eeg in temporal lobe epilepsy. In *IEEE Eng. in Medicine and Biology Society, 10th Ann. Int. Conf.*, pages 1201–1203, 1988.

[53] L.D. Iasemidis, H.P. Zaveri, J.C. Sackellares, and W.J. Williams. Phase space topography of the electrocorticogram and the Lyapunov exponent in partial seizures. *Brain Topography*, 2:187–201, 1990.

[54] M.P. Jacobs, G.D. Fischbach, M.R. Davis, M.A. Dichter, R. Dingledine, D.H. Lowenstein, M.J. Morrell, J.L. Noebels, M.A. Rogawski, S.S. Spencer, and W.H. Theodore. Future directions for epilepsy research. *Neurology*, 57:1536–1542, 2001.

[55] A. Katz, D. Marks, G. McCarthy, and S. Spencer. Does interictal spiking change prior to seizures? *Electroencephalogram and Clinical Neurophysiology*, 79:153–156, 1991.

[56] U. Krebel. Pairwise classification and support vector machines. In *Advances in Kernel Methods – Support Vector Learning*. MIT Press, 1999.

[57] Y.C. Lai, I. Osorio, M.A.F. Harrison, and M.G. Frei. Correlation-dimension and autocorrelation fluctuations in seizure dynamics. *Physical Review*, 65(3 Pt 1):031921, 2002.

[58] K. Lehnertz and C.E. Elger. Spatio-temporal dynamics of the primary epileptogenic area in temporal lobe epilepsy characterized by neuronal complexity loss. *Electroencephalogr. Clin. Neurophysiol.*, 95:108–117, 1995.

[59] K. Lehnertz and C.E. Elger. Can epileptic seizures be predicted? evidence from nonlinear time series analysis of brain electrical activity. *Phys. Rev. Lett.*, 80:5019–5022, 1998.

[60] K. Lehnertz and B. Litt. The first international collaborative workshop on seizure prediction: summary and data description. *Journal of Clinical Neurophysiology*, 116(3):493–505, 2005.

[61] B. Litt and J. Echauz. Prediction of epileptic seizures. *The Lancet Neurology*, 1:22–30, 2002.

[62] B. Litt, R. Esteller, J. Echauz, D.A. Maryann, R. Shor, T. Henry, P. Pennell, C. Epstein, R. Bakay, M. Dichter, and G. Vachtservanos. Epileptic seizures may begin hours in advance of clinical onset: A report of five patients. *Neuron*, 30:51–64, 2001.

[63] O.L. Mangasarian. Linear and nonlinear separation of pattern by linear programming. *Operations Research*, 31:445–453, 1965.

[64] O.L. Mangasarian, W.N. Street, and W.H. Wolberg. Breast cancer diagnosis and prognosis via linear programming. *Operations Research*, 43(4):570–577, 1995.

[65] R. Manuca and R. Savit. Stationary and nonstationary in time series analysis. *Physica D*, 99:134–161, 1999.

[66] J. Martinerie, C. Van Adam, and M. Le Van Quyen. Epileptic seizures can be anticipated by non-linear analysis. *Nature Medicine*, 4:1173–1176, 1998.

[67] R.D. McBride and J.S. Yormark. An implicit enumeration algorithm for quadratic integer programming. *Management Science*, 26(3):282–296, 1980.

[68] M. Mezard, G. Parisi, and M.A. Virasoro. *Spin Glass Theory and Beyond*. World Scientific, 1987.

[69] T.S. Motzkin and E.G. Strauss. Maxima for graphs and a new proofs of a theorem turán. *Canadian Journal of Mathematics*, 17:533–540, 1965.

[70] P. Narendra and K. Fukunaga. A branch and bound algorithm for feature subset selection. *IEEE Transactions on Computers*, 26:917–922, 1977.

[71] S. Narula and J. Wellington. Selection of variables in linear regression using the minimum sum of weighted absolute errors criterion. *Technometrics*, 21(3):299–311, 1979.

[72] M. Oral and O. Kettani. A linearization procedure for quadratic and cubic mixed integer problems. *Operations Research*, 40:109–116, 1992.

[73] I. Osorio, M.A.F. Harrison, M.G. Frei, and Y.C. Lai. Observations on the application of the correlation dimension and correlation integral to the prediction of seizures. *J Clin Neurophysiol.*, 18(3):269–274, 2001.

[74] P.M. Pardalos. Construction of test problems in quadratic bivalent programming. *ACM Transactions on Mathematical Software*, 17:74–87, 1991.

[75] P.M. Pardalos, W.A. Chaovalitwongse, L.D. Iasemidis, J.C. Sackellares, D.-S. Shiau, P.R. Carney, O.A. Prokopyev, and V.A. Yatsenko. Seizure warning algorithm based on optimization and nonlinear dynamics. *Mathematical Programming*, 101(2):365–385, 2004.

[76] P.M. Pardalos, L.D. Iasemidis, D.-S. Shiau, and J.C. Sackellares. Quadratic binary programming and dynamic system approach to determine the predictability of epileptic seizures. *Journal of Combinatorial Optimization*, 5/1:9–26, 2001.

[77] P.M. Pardalos and S. Jha. Complexity of uniqueness and local search in quadratic 0-1 programming. *Operations Research Letters*, 11:119–123, 1992.

[78] P.M. Pardalos and J.C. Principe. *Biocomputing*. Kluwer Academic Publisher, 2003.

[79] P.M. Pardalos and G. Rodgers. Parallel branch and bound algorithms for unconstrained quadratic zero-one programming. In R. Sharda et al., editor, *Impact of Recent Computer Advances on Operations Research*. North-Holland, 1989.

[80] P.M. Pardalos and G. Rodgers. Computational aspects of a branch and bound algorithm for quadratic zero-one programming. *Computing*, 45:131–144, 1990.

[81] P.M. Pardalos and G.P. Rodgers. Computational aspects of a branch and bound algorithm for quadratic 0-1 programming. *Computing*, 45:131–144, 1990.

[82] P.M. Pardalos and G.P. Rodgers. Parallel branch and bound algorithm for quadratic zero-one on a hypercube architecture. *Ann. Oper. Res.*, 22:271–292, 1990.

[83] P.M. Pardalos, J.C. Sackellares, P.R. Carney, and L.D. Iasemidis. *Quantitative Neuroscience*. Kluwer Academic Publisher, 2004.

[84] P.M. Pardalos and J. Xue. The maximum clique problem. *Journal of Global Optimization*, 4:301–328, 1992.

[85] V. Piccone, J. Piccone, L. Piccone, R. LeVeen, and E.L. Veen. Implantable epilepsy monitor apparatus. *US Patent 4,566,464*, 1981.

[86] O.A. Prokopyev, V. Boginski, W. Chaovalitwongse, P.M. Pardalos, J.C. Sackellares, and P.R. Carney. Network-based techniques in EEG data analysis and epileptic brain modeling. In P.M. Pardalos, V.L. Boginski, and A. Vazacopoulos, editors, *Data Mining in Biomedicine*, Springer, Berlin, 2007.

[87] J.R. Quinlan. *C4.5: Programs for Machine Learning*. Morgan Kaufmann, 1993.

[88] M. Le Van Quyen, J. Martinerie, M. Baulac, and F. Varela. Anticipating epileptic seizures in real time by non-linear analysis of similarity between EEG recordings. *NeuroReport*, 10:2149–2155, 1999.

[89] J.C. Sackellares, L.D. Iasemidis, D.-S. Shiau, L.K. Dance, P.M. Pardalos, and W.A. Chaovalitwongse. Optimization of multi-dimensional time series processing for seizure warning and prediction. *International Patent Application filed August 2003, Attorney Docket No. 028724–142*, 2003.

[90] J.C. Sackellares, L.D. Iasemidis, V.A. Yatsenko, D.-S. Shiau, P.M. Pardalos, and W.A. Chaovalitwongse. Multi-dimensional multi-parameter time series processing for seizure warning and prediction. *International Patent Application filed September 2003, Attorney Docket No. 028724–143*, 2003.

[91] B. Scholkopf, C. Burges, and V. Vapnik. Extracting support data for a given task. In *Proc. First International Conference on Knowledge Discovery and Data Mining*. AAAI Press, 1995.

[92] R. Shioda. *Integer Optimization in Data Mining*. PhD thesis, MIT, 2003.

[93] F. Takens. Detecting strange attractors in turbulence. In D.A. Rand and L.S. Young, editors, *Dynamical Systems and Turbulence, Lecture Notes in Mathematics*. Springer-Verlag, 1981.

[94] S. Viglione, V. Ordon, W. Martin, and C. Kesler. Epileptic seizure warning system. *US Patent 3,863,625*, 1973.

[95] S.S. Viglione and G.O. Walsh. Proceedings: Epileptic seizure prediction. *Electroencephalogram and Clinical Neurophysiology*, 39:435–436, 1975.

[96] L. Watters. Reduction of integer polynomial programming problems to zero-one linear programming problems. *Operations Research*, 15:1171–1174, 1967.

[97] H. Wieser. Preictal eeg findings. *Epilepsia*, 30:669, 1989.

[98] A. Wolf, J.B. Swift, H.L. Swinney, and J.A. Vastano. Determining Lyapunov exponents from a time series. *Physica D*, 16:285–317, 1985.

[99] W.I. Zangwill. Media selection by decision programming. *Journal of Advertising Research*, 5:30–36, 1965.

# 11

# Mathematical Programming Approaches for the Analysis of Microarray Data

Ioannis P. Androulakis

Department of Biomedical Engineering and Department of Chemical and
Biochemical Engineering, Rutgers, The State University of New Jersey,
Piscataway, New Jersey 08854
yannis@rci.rutgers.edu

**Abstract.** One of the major challenges facing the analysis of high-throughput microarray measurements is how to extract in a systematic and rigorous way the biologically relevant components from the experiments in order to establish meaningful connections linking genetic information to cellular function. Because of the significant amount of experimental information that is generated (expression levels of thousands of genes), computer-assisted knowledge extraction is the only realistic alternative for managing such an information deluge. Mathematical programming offers an interesting alternative for the development of systematic methodologies aiming toward such an analysis. We summarize recent developments related to critical problems in the analysis of microarray data; namely, tissue clustering and classification, informative gene selection, and reverse engineering of gene regulatory networks. We demonstrate how advances in nonlinear and mixed-integer optimization provide the foundations for the rational identification of critical features unraveling fundamental elements of the underlying biology thus enabling the interpretation of volumes of biological data. We conclude the discussion by identifying a number of related research challenges and opportunities for further research.

## 11.1 Microarrays and the New Biology

The genetic information is stored in the DNA, the double-stranded polymer composed of four basic molecular units (nucleotides) adenine (A), guanine (G), cytosine (C), and thymine (T). In order for the genome to direct, or affect, changes in the cell, a transcriptional program must be activated eventually dictating all biological transformations. This program is regulated temporarily according to an intrinsic program or in response to changes in the environment. The expression of the genetic information, which is stored in DNA, takes places in two stages: transcription, during which DNA is transcribed into mRNA, a single-stranded complimentary copy of the base sequence of the DNA; and translation, during which mRNA provides the blueprint for the production of specific proteins. Measuring the level of production of mRNA,

P.M. Pardalos, H.E. Romeijn (eds.), *Handbook of Optimization in Medicine*,
Springer Optimization and Its Applications 26, DOI: 10.1007/978-0-387-09770-1_11,
© Springer Science+Business Media LLC 2009

thus measuring the expression levels of the associated genes, provides a quantitative assessment of the levels of production of the corresponding proteins, the ultimate expression of the genetic information.

Innovative approaches such as cDNA and oligonucleotide microarrays were recently developed to extract genome-wide information related to gene expression (see Schena et al. [50], Bowtell [3], Brown and Botstein [7], Cheung et al. [9], and Lipshutz et al. [41]). During an expression experiment, extracted mRNA is reverse-transcribed into more stable complementary DNA (cDNA), which is labeled using fluorescent dyes. Different-colored dyes are used for different samples (probes). The probes are then tested by hybridizing to a DNA array holding thousands of spots, each containing a different DNA sequence. Once the probes have hybridized, they are washed off, and the array is scanned to determine the relative amount of each cDNA probe bound to any given spot. Quantitative imaging coupled with clone database information allows measurement of the labeled cDNA that hybridized to each target sequence. Image processing and data normalization are among the first, and very critical, computational filters required before the actual quantification of the expression experiment is defined (Dudoit et al. [17]). Gene expression changes are usually measured relative to another sample. Comparative differences are used to assess the impact of gene expression to various regulatory pathways.

Gene expression microarray experiments have been celebrated as a revolution in biology, attracting significant interest, because they are slowly changing the working paradigm of biological research by allowing the analysis of the combined effects of numerous genetic and environmental components. The profound impact is that such a global approach will allow a fundamental shift from "...piece-by-piece to global analysis and from hypothesis driven research to discovery-based formulation and subsequent testing of hypotheses..." (see Kafatos [39]). One of the major challenges is to extract in a systematic and rigorous way the biologically relevant components from the array experiments in order to establish meaningful connections linking genetic information to cellular function. Because of the significant amount of experimental information that is generated (expression levels of thousands of genes), computer-assisted knowledge extraction is the only realistic alternative for managing such an information deluge.

## 11.2 Issues in Microarray Data Analysis

Among the numerous tasks that can be assisted by the data generated from microarray experiments, we will focus mainly on three: tissue classification, gene selection, and construction of regulatory networks from temporal gene expression data. We do so because

(a) these tasks are critical and define the basis for a number of more complicated problems,

(b) they have clearly defined approaches based on mathematical programming techniques and can be used as excellent motivating examples.

In tissue classification, samples from multiple cell types (for example, different cancer types, cancerous and normal cells, etc.) are comparatively analyzed using microarray gene expression measurements. The question therefore becomes how to identify which genes provide consistent signatures that distinctly characterize the different classes. The problem can be viewed as either a supervised classification problem in which the classes are already known or as an unsupervised clustering problem in which we attempt to identify the classes contained within the data. In gene selection, the computational problem is equivalent to that of feature selection in multidimensional data sets. Identifying the minimum number of gene markers is however critical because this reduced set can provide information about the biology behind the experiment as well as define the basis for future therapeutic agents.

In time-ordered gene expression measurements, the temporal pattern of gene expression is investigated by measuring the gene expression levels at a small number of points in time. The continuous monitoring of the level of mRNA abundance has the ultimate goal of deriving the temporal evolution of the synergistic effects of multiple genes. By doing so, a regulatory network is constructed, that is, a biologically plausible superstructure of gene interactions that interprets the data. Transcriptional regulatory networks are the key to understanding the sequence of events leading to an observed biological response. The tasks that we are about to discuss in this chapter have already been addressed by a number of approaches under the general umbrella defined as *data mining*. What we plan to present however is a definition of these tasks as mathematical programming problems exploring principles and advances of optimization. We will demonstrate the flexibility that mathematical programming and deterministic optimization provide, discuss some characteristic applications, and finally conclude with a number of suggestions for future research.

## 11.3 Analysis of Gene Expression Data: Tissue Clustering and Classification

### 11.3.1 Clustering and classification preliminaries

Let us assume the data describing a particular process are expressed in the form of $n$-dimensional feature vectors $x \in \mathbb{R}^n$. An important goal of the analysis of such data is to determine an explicit or implicit function that maps the points of the feature vector from the input space to an output space (for example, in regression). This mapping has to be derived based on a finite number of data, thus assuming that a proper sampling of the space has been performed. If the predicted quantity is categorical and if we know the value that corresponds with each element of the training set, then the question becomes how to identify the mapping that connects the feature vector and the corresponding categorical value (class). This problem is known as the classification problem (supervised learning). If the class assignment is not known and we seek to (a) identify whether small, yet unknown, number of classes exist, (b) define the mapping assigning the features to classes, then we have a clustering problem (unsupervised learning).

Although numerous methods exist for addressing these problems they will not be reviewed here. Nice reviews of classification that were recently presented include the papers by (Grossman et al. [33]; Hand et al. [35]). In this short introduction, we will concentrate on solution methodologies based on reformulating the clustering and classification questions as optimization problems.

**Tissue classification**

Developing specific therapies to pathogenetically distinct tumor types is important for cancer treatment, because they maximize efficacy and minimize toxicity. Thus, precisely classifying tumors is of critical importance to cancer diagnosis and treatment. Diagnostic pathology has traditionally relied on macro- and microscopic histology and tumor morphology as the basis for tumor classification. Current classification frameworks, however, are unable to discriminate among tumors with similar histopathologic features, which vary in clinical course and in response to treatment. Recently, there is increasing interest in changing the basis of tumor classification from morphologic to molecular. In the past decade, microarray technologies have been developed that can simultaneously assess the level of expression of thousands of genes. Several studies have used microarrays to analyze gene expression in colon, breast, and other tumors and have demonstrated the potential power of expression profiling for classifying tumors. Gene expression profiles may offer more information than classic morphology and provide an alternative to morphology-based tumor classification systems (Zhang et al. [60]).

**Mathematical programming formulations**

Classification and clustering, and for that matter most of the data mining tasks, are fundamentally optimization problems. Mathematical programming methodologies formalize the problem definition and make use of recent advances in optimization theory and applications for the efficient solution of the corresponding formulations. In fact, mathematical programming approaches, particularly linear programming, have long been used in data mining tasks. The pioneering work of Mangasarian [43, 44] demonstrated how to formulate the problem of constructing planes to separate linearly separable sets of points. In addition, early work by Freed and Glover [20, 21, 22], Gehrlein [26], Glover et al. [31], and Glover [30] skillfully discussed various aspects of discriminant analysis from the point of view optimization. A more recent excellent review was presented in Stam [53], highlighting numerous developments that defined the field of applications of mathematical programming to statistical classification. It should be pointed out that one of the major advantages of a formulation based on mathematical programming is the ease in incorporating explicit problem-specific constraints whose incorporation in classic statistical approaches in not evident in general.

Let us consider a two-class problem in which the sample points belong to either one of two sets with their point coordinates denoted by $A$ and $B$ respectively[1]. As discussed earlier, a discriminant function can be derived based on a hyperplane of the form

$$P = \{x \in \mathbb{R}^n | x^\top \omega = \gamma\}.$$

The normal to this plane is

$$\frac{|\gamma|}{\|\omega\|_2}.$$

The classification problem thus becomes how to determine g and w such that the separating hyperplane P defines two open half spaces

$$\{x \in \mathbb{R}^n | x^\top \omega < \gamma\} \text{ and } \{x \in \mathbb{R}^n | x^\top \omega > \gamma\}$$

containing mostly points in $A$ and $B$, respectively. Unless the problem is linearly separable, the hyperplane can only be derived within a certain error. Minimization of the average violations provides a possible approximation of the separating hyperplane

$$\min_{\omega,\gamma} \frac{1}{m} \| -A\omega + e\gamma + e \| + \frac{1}{k} \| -B\omega + e\gamma + e \|$$

where $m$ and $k$ denote the number of samples belonging to classes $A$ and $B$, respectively. Bradley et al. [5] discusses various implementations including a particularly effective robust linear programming reformulation suitable for large-scale problems:

$$\min_{\omega,\gamma,y,z} \frac{1}{m} e^\top y + \frac{1}{k} e^\top z$$

subject to

$$-A\omega + e\gamma + e \le y$$
$$B\omega - e\gamma + e \le z$$
$$y, z \ge 0.$$

Fung et al. [23] demonstrated how to extend the aforementioned formalism to account for nonlinear kernel functions that generate nonlinear optimal separating surfaces.

---

[1] For simplicity, we use the symbols $A$ and $B$ to denote both the classes and the matrices containing the coordinates.

While the approaches just described aim at minimizing an error in separating the given data, support vector machines (SVMs, Vapnik [57]) incorporate also the structured risk minimization, which minimizes an upper bound of the generalization error. In fact a very interesting analysis on the learning stability characteristics of SVMs, in dealing with uncertainty, is demonstrated by Bousquet and Elisseeff [1]. The general idea behind SVM is illustrated by considering the case where a linear separating surface is to be generated. In that case, SVMs determine, among the infinite number of possible planes separating the two classes, the one that also maximizes the margin separating the two classes.

SVMs are based on an analysis of the general problem of learning the classification boundary between positive and negative samples. This is a particular case of the problem of approximating a multivariate function from sparse data. Regularization theory is a classic approach to solving it by formulating the approximation problem as a variational optimization problem of finding the function $f$ that minimizes the functional

$$\frac{1}{\ell}\sum_{i=1}^{\ell} V\left(y_i, f(x_i)\right) + \lambda\|f\|_2$$

where $\ell$ is the number of training samples, $V(\cdot)$ is the loss function, and $\|\cdot\|_2$ a suitable norm. In order to derive a linear separating surface between the two classes, the above-mentioned problem is equivalent to the solution of the following optimization problem (Cortes and Vapnik [11]):

$$\min_{w,b} \frac{1}{2}w^\top w + C\sum_{i=1}^{\ell} \xi_i$$

subject to

$$y_i(wx_i + b) \geq 1 - \xi_i \qquad i = 1,\ldots,\ell$$
$$\xi_i \geq 0 \qquad i = 1,\ldots,\ell.$$

In this formulation, $y_i$ denotes the class of sample $i$, and it is either $+1$ or $-1$. The solution to this problem not only minimizes the misclassifications (second part of the objective) but also identifies the hyperplane, with normal vector $w$, that provides the maximum margin between the two classes.

In general however, the separating surface will be nonlinear. In this case, we have to think of a nonlinear projection of the original data for which we seek a linear separating surface. In that case, the linear separating surface in the projected feature space will correspond with a nonlinear separating surface in the original space. In that case, we can write the following optimization:

$$\min_{w,b} \frac{1}{2}w^\top w + C\sum_{i=1}^{\ell} \xi_i$$

subject to

$$y_i \left( w\phi(x_i) + b \right) \geq 1 - \xi_i \qquad i = 1, \ldots, \ell$$
$$\xi_i \geq 0 \qquad i = 1, \ldots, \ell.$$

The functional $\phi(\cdot)$ defines the nature of the nonlinear kernel.

SVMs have been applied with great success in clustering and classification problems in microarray experiments (see Brown et al. [6], Furey et al. [24], Guyon et al. [34], Rifkin et al. [48]). It will be shown later that analysis of the coefficients of the separating hyperplanes, of non-linear kernels, can provide some indications as to which features are more significant. Therefore, a by-product of clustering and classification analysis within such an optimization framework will also be feature (gene) selection.

## Multiclass support vector machines

The solution to binary classification problems using SVM has been well developed, tested, and documented. However, extending the method to multiclass problems remains an open research issue. The standard approach, within an SVM framework, is to treat the multiclass problem as a collection of two-class (binary) classification problems. Recently, however, multiclass methods considering a much larger problem encompassing all classes at once have been proposed. The drawback of course is the requirement for the solution of a much larger problem. Recently (Hsu and Lin [37] and Nguyen and Rajapakse [47]) discuss a number of alternatives for the development of SVM-based multiclass classifiers.

### One-against-all (OAA) classifier

This method constructs $k$ SVM models where $k$ is the number of classes. The $j^{\text{th}}$ SVM is trained to classify the members of the $j^{\text{th}}$ class, assumed to have positive labels, against the samples of all the other classes, which are assumed to have negative labels. Therefore, given $\ell$ training data in the form $(x_1, y_1), \ldots, (x_\ell, y_\ell)$ where $x_i \in \mathbb{R}^n$ and $y_i \in \{1, \ldots, k\}$ $(i = 1, \ldots, \ell)$, the $j^{\text{th}}$ SVM solves the following problem:

$$\min_{w^j, b^j, \xi^j} \frac{1}{2} (w^j)^\top w^j + C \sum_{i=1}^{\ell} \xi_i^j$$

subject to

$$(w^j)^\top \phi(x_i) + b^j \geq 1 - \xi_i^j \qquad i = 1, \ldots, \ell : y_i = j$$
$$(w^j)^\top \phi(x_i) + b^j \geq -1 + \xi_i^j \qquad i = 1, \ldots, \ell : y_i \neq j$$
$$\xi_i^j \geq 0 \qquad i = 1, \ldots, \ell.$$

Minimizing the first term in the objective function, $\frac{1}{2}(w^j)^\top w^j$, means that large values of the margin between the two groups of data, $2/\|w^j\|$, are favored. The second term in the objective function, $\sum_{i=1}^{\ell} \xi_i^j$, favors a reduction in the number of training errors for the case where the problem is not linearly separable. Solving this problem for $j = 1, \ldots, k$ generates $k$ decision functions:

$$(w^j)^\top \phi(x) + b^j \qquad (j = 1, \ldots, k).$$

Sample $x$ belongs to the class that has the largest value of the decision function:

$$\text{class of } x = \arg \max_{j=1,\ldots,k} \left[ (w^j)^\top \phi(x) + b^j \right].$$

*One-against-one (OAO) classifier*

This method constructs $k(k-1)/2$ classifiers each of which is trained on data from two classes, $j$ and $j'$ ($j, j' = 1, \ldots, k, j' > j$):

$$\min_{w^{jj'}, b^{jj'}, \xi^{jj'}} \frac{1}{2}(w^{jj'})^\top w^{jj'} + C \sum_{i=1}^{\ell} \xi_i^{jj'}$$

subject to

$$
\begin{aligned}
(w^{jj'})^\top \phi(x_i) + b^j \geq 1 - \xi_i^{jj'} & \qquad i = 1, \ldots, \ell : y_i = j \\
(w^{jj'})^\top \phi(x_i) + b^j \geq -1 + \xi_i^{jj'} & \qquad i = 1, \ldots, \ell : y_i \neq j \\
\xi_i^{jj'} \geq 0 & \qquad i = 1, \ldots, \ell.
\end{aligned}
$$

Feature testing based on binary classifiers is not trivial. However, a standard technique is based on majority voting: weighted sum of the outputs of all pairwise classifiers defines the predicted class. A particular implementation of the OAO classifier prediction uses the concept of *directed acyclic graphs*. Each node is a classifier between two classes. Given a test sample $x$ and starting at the root node, the binary decision function is evaluated. Then it moves to either the left or the right of the tree depending on the output value.

Weston and Watkins [58] proposed the construction of a likewise linear separation of the $k$ classes in a single optimization formulation. The original formulation is generalized as follows:

$$\min_{w,b,\xi} \frac{1}{2} \sum_{j=1}^{k} (w^j)^\top w^j + C \sum_{i=1}^{\ell} \sum_{j=1, j \neq y_i}^{k} \xi_i^k$$

subject to

$$
\begin{aligned}
w^{y_i} x_i + b^{y_i} \geq w^j x_i + b^j + 2 - \xi_i^j & \qquad i = 1, \ldots, \ell; \, j = 1, \ldots, k; \, y_i \neq j \\
\xi_i^j \geq 0 & \qquad i = 1, \ldots, \ell; \, j = 1, \ldots, k; \, y_i \neq k.
\end{aligned}
$$

Once again we assume the existence of $k$ classes and $\ell$ objects, and $y_i$ is an integer indicating the class of object $i$. Effectively, the method is a generalization of the OAA approach where the classifiers are estimated simultaneously through the solution of a larger optimization problem. In this case, the discriminating function becomes $\arg\max_{j=1,\dots,k}(w^j x + b^j)$. Similar in spirit is the formulation proposed by Crammer and Singer [12]. The formulation is similar to the one proposed by Weston and Watkins [58] with the only difference being that the constraints are defined such that a smaller number of slack variables is required.

## Classification of microarray data using support vector machines

SVMs are becoming one of the favorite classification methods for the classification of microarray data primarily due to their sound mathematical foundation. In this section, we will outline just a few illustrative examples. The first application aims at classifying cancerous cells based on the measurement of expression values, whereas the second application aims at functionally classifying genes.

### Molecular cancer classification

Modern cancer treatments rely upon macroscopic examination to classify tumors according to anatomic site of origin. DNA microarrays generate information potentially able to formulate molecular-based predictors circumventing the subjectivity associated with the examination of macroscopic characteristics. Rifkin et al. [48] present a computational method, based on SVM, aiming at classifying tumor data in an attempt to derive a general, multi-class molecular-based cancer classification based solely on gene expression data. The case study concerned the analysis of 198 samples from 14 different cancer types, using microarray data recording the activity (expression) levels of 16,063 probes. Both the OAA and OAO approaches were computationally evaluated in terms of their ability to correctly predict unknown samples. This work demonstrated the ability of SVM to effectively and efficiently classify large microarray data sets in computationally reasonable times. In a somewhat similar study, Williams et al. [59] evaluate the ability of SVM to develop prognostic classification tools for relapsing tumor.

### Gene functionality classification

Brown et al. [6] introduced a method of functionally classifying genes by using gene expression data from DNA microarray experiments based on SVM. The approach is motivated by the realization that genes of similar functionality yield similar expression patterns in microarray experiments. As data from such experiments begin to accumulate in increasing rates, it will become essential to have means for extracting biological significance and using data to assign functions to genes. The authors experimented with a number of nonlinear kernels, including a dot product based measuring the similarity between two gene expression vectors $K(X,Y) = X \cdot Y$ as well as various $d$-fold generalizations of the form $K(X,Y) = (X \cdot Y + 1)^d$, and a Gaussian kernel $K(X,Y) = \exp\left(-\|X - Y\|^2/(2\alpha^2)\right)$. The study considered 2,467 yeast genes for which functional annotation was available. SVM were trained to recognize six functional families: tricarboxylic acid (TCA) cycle, respiration, cytoplasmic ribosomes, proteasome, histones, and helix-turn-helix proteins. The computational evaluation of the SVM was based on a three-way cross-validation, repeated a number of times. SVMs were compared with other standard supervised learning techniques, including Parzen windows, Fisher's linear discriminant analysis, and decision trees (MOC1 and C4.5), and were found to outperform all of them providing superior performance.

## 11.3.2 Feature selection preliminaries

Machine learning algorithms are known to be prone to deteriorating performance when faced with many irrelevant or correlated features (see Kohavi and John [40]). A universal, therefore, problem is to decide on which aspects, i.e., features, of a problem are relevant. Narendra and Fukunaga [46] were among the first to present a formal approach based on a branch and bound scheme for addressing the very same problem. A recent review by Kohavi and John [40] examines a number of issues associated with the problem of feature selection. More recently, Liu and Motoda [42] also present ideas related to the coupling of information theory and feature selection.

Feature selection is a very healthy and vibrant area of research in the machine learning community and has gained increased significance with the recent advances in functional genomics that resulted in the creation of very high-dimensional feature sets. A number of recent publications (Golub et al. [32], Chilingaryan et al. [10], Szabo et al. [55], Dettling and Buhlmann [14]) have devised various approaches for extracting critical, differentially expressed genes in a systematic manner. The advantages of multivariate methods are that (a) they attempt to take into account collaborative effects of gene expression activities; (b) they do not simply characterize genes based on arbitrary $n$-fold increased/decreased activities.

### Feature selection in almost empty spaces

A fundamental problem in machine learning is the development of accurate classifiers in sparsely populated data sets, i.e., *almost empty spaces* (see Duin [18]). As noted earlier, the key complexity of microarray experiments is the essential lack of observables (cell lines or tissue samples) to support the large number of probes monitored. The consequences of the small ratio of features to samples were extensively discussed in Jain and Zongker [38]. The inability of sparse data to properly capture the complexity of a classification problem was also analyzed by Ho [36]. A nice discussion of the impact of the small sample size problem in array expression data is presented in Dougherty [15]. The implications of the ratio of features to samples is critical as sparsely populated data sets can very easily lead to random features appearing to be informative (i.e., able to classify data) when in reality no structure exists in the data whatsoever. It should be expected that simple minimization of the number of features (genes) in a model need not necessarily provide the best possible answer. Additional complexity restrictions will have to be proposed to balance the lack of available data although no definite answer can be provided as no analysis can replace accurate and adequate data.

## Gene selection using support vector machines

Reducing the number of noisy measured variables reduces potential noise, hence avoids pointless overfitting. Selecting the optimal number of features is a complicated task: too few genes will not discriminate or predict; too many genes might be introducing noise to the model rather than information. Therefore, the identification of informative genes is a significant component of an integrated computer-assisted analysis of array experiments. However, in current practice, the identification of such a critical subset of genes whose expression is informative is accomplished as a by-product of some other activity, for instance, by analyzing patterns in "heat maps" in hierarchical clustering, the loadings of singular vectors, or by assessing the ability of certain genes to maximize the separability between classes. In most cases the question of identifying differentially expressed genes is restated as a hypothesis-testing problem in which the null hypothesis of no association between expression levels and responses of interest is tested (see Dudoit et al. [17]).

SVMs are powerful classifiers based on regularization techniques for regression (see Vapnik [57]). Guyon et al. [34] discuss a recursive forward selection procedure for ranking features in gene expression experiments. Because the method, in general, attempts to identify a surface separating different classes, the assumption is that the weights of the feature in the decision function should also serve to quantify the importance of each feature. Specifically, Guyon et al. [34] follow the formalism of Cortes and Vapnik [11] in which the following problem is considered. Given a set of training examples $\{x_k\}$, $x_k \in \mathbb{R}^n$ and class labels for each example $\{y_k\}$, defined as either $+1$ or $-1$, a separating surface is defined as the solution of an optimization problem as defined earlier. The hyperplane $D = w \cdot x + b = 0$ is the one that separates the training examples belonging to the two classes with a maximal margin. A metric for the ranking of the features is based on the quantity $w_i^2$. Guyon et al. [34] developed a recursive feature elimination procedure, which successively ranked and eliminated features and demonstrated the ability of the SVM-based procedure to extract reduced sets of biologically relevant genes. The general observation was that the quality of the SVM classifier improves once irrelevant features are removed. Alternatively, Bradley and Mangasarian [4] presented a variant of the basic SVM that augments the objective by the addition of the term $\lambda w^\top w/2$, which appropriately weights the scarcity of the vector defining the separating hyperplane. They also discuss possible reformulations of this formulation that render the problem one of minimizing a concave objective subject to linear constraints. Despite the fact that the problem is non-convex, it can be efficiently solved. The issues of non-convexity and global optimality will be revisited later.

### 11.3.3 Simultaneous gene selection and tissue classification

A mixed-integer linear formulation was recently proposed by Sun and Xiong [54] and will be used for the purposes of our discussion. Feature selection is always considered within the framework of a given analysis. This could be model development/fitting, classification, clustering, and so forth. In other words we want to extract the minimum number of required independent variables necessary to perform a particular task. Therefore, an objective measuring the "goodness of fit" will be required. The parameters associated with the model naturally define a continuous optimization problem. The notion of selection a subset of variables, out of superset of possible alternatives, naturally lends itself to a combinatorial (integer) optimization problem. Therefore, depending on the model used to describe the data, the problem of feature selection will end up being a mixed integer (non) linear optimization problem. Furthermore, this problem is a multicriteria optimization as one wishes to simultaneously minimize the model error and the number of features used. Sun and Xiong [54] propose the use of a linear discriminator, similar to a SVM to be discussed later. Let $m$ denote the number of observations for a two-class problem such that $k$ and $\ell$ denote the number of samples in each class (for example, number of benign and cancerous cells, respectively). We also denote as $I_1$ and $I_2$ the indices of the corresponding samples, and $I = I_1 \cup I_2$ denotes the entire set of samples. Finally, the set $J$ denotes the set of all genes recorded in the observations, and $J' \subset J$ denotes the set of genes (features) that are required to develop an accurate model. The expression data are presented in the form $x_{ij}$, $i = 1, \ldots, I$, $j = 1, \ldots, J$. A linear classifier is constructed as:

$$\beta_0 + \sum_{j \in J} \beta_j x_{ij} < 0 \qquad i \in I_1$$

$$\beta_0 + \sum_{j \in J} \beta_j x_{ij} > 0 \qquad i \in I_2.$$

However, because the observations are not, in general, perfectly separable by a linear model, a goal programming formulation can be proposed whose goal is to estimate the coefficients that minimize the deviations from the classifier model. That is

$$\min \sum_{i \in I_1} d_i^1 + \sum_{i \in I_2} d_i^2$$

subject to

$$\beta_0 + \sum_{j \in J'} \beta_j x_{ij} - d_i^1 + d_i^2 = -\delta \qquad i \in I_1$$

$$\beta_0 + \sum_{j \in J'} \beta_j x_{ij} - d_i^1 + d_i^2 = \delta \qquad i \in I_2$$

$$\beta_j \in \mathbb{R} \qquad j \in J \cup \{0\}$$

$$d_i^1, d_i^2 \in \mathbb{R}^+ \qquad i \in I_1 \cup I_2$$

where $\delta$ is a small constant. It can either be fixed based on user preferences or be added to the objective to be minimized. In order to minimize the number of variables used in the classifier, hence extract the most relevant features for the specific linear model, binary variables need to be introduced to define whether a particular variable is used in the model or not. Therefore:

$$y_j = \begin{cases} 1 & j \in J' \\ 0 & j \notin J' \end{cases}$$

The number of "active" genes can therefore be constrained (that is, introduced parametrically in the formulation in order to avoid the solution of a multicriteria optimization problem. According to the e-constraint method, one additional constraint of the form

$$\sum_{j \in J'} y_j \leq \epsilon$$

is introduced. The complete MIP formulation thus becomes:

$$\min \sum_{i \in I_1} d_i^1 + \sum_{i \in I_2} d_i^2$$

subject to

$$\beta_0 + \sum_{j \in J'} \beta_j x_{ij} - d_i^1 + d_i^2 = -\delta \qquad i \in I_1$$

$$\beta_0 + \sum_{j \in J'} \beta_j x_{ij} - d_i^1 + d_i^2 = \delta \qquad i \in I_2$$

$$\sum_{j \in J'} y_j \leq \epsilon$$

$$\beta_j \leq M y_j$$

$$-\beta_j \leq M y_j$$

$$\beta_j \in \mathbb{R} \qquad j \in J \cup \{0\}$$

$$d_i^1, d_i^2 \in \mathbb{R}^+ \qquad i \in I_1 \cup I_2$$

$$y_j \in \{0, 1\}.$$

## 11.4 Inferring Regulatory Networks

### 11.4.1 Mixed-integer formulations

It would have been misleading to assume that gene expression experiments define static and time-independent observations. Temporal, i.e., dynamic, measurements of gene expression activities exhibit the wealth of complexity characterizing the genomic response to external stimuli. A complete understanding of the organization and dynamics of gene regulatory networks is an essential first step toward realizing the goal of deciphering the complex regulation underlying gene expression (see Bower and Bolouri [2], Dasika et al. [13]).

Unlike the preceding discussion, the expression level of a gene is now considered to be a function of time, $Z_i(t)$. The expression of any given gene $i$ is however regulated by the expression of some other gene $j$ with an effective delay $\tau$. From a biological point of view, the time delay in gene regulation characterizes the various underlying processes such as transcription and translation introduced earlier in this chapter. The strength of the time regulation is denoted by $w_{ij\tau}$. The sign denotes either activation or inhibition of expression. In order to derive biologically relevant activation/inhibition relations, logical constraints are imposed to denote the existence of these restrictions. Specifically:

$$Y_{ij\tau} = \begin{cases} 1 & \text{if gene } j \text{ regulates gene } i \text{ with time delay } \tau \\ 0 & \text{otherwise.} \end{cases}$$

Dasika et al. [13] derived the following optimization problem to estimate the potential connectivity and interaction matrix for a given set of temporal gene expression experiments (expression on $N$ genes measured at $T$ time points):

$$\min \frac{1}{NT} \sum_{i=1}^{N} \sum_{j=1}^{T} \left[ e_i^+(t) - e_i^-(t) \right]$$

subject to

$$\dot{Z}_i(t) - \sum_{\tau=0}^{\tau_{\max}} \sum_{j=1}^{N} w_{ij\tau} Z_j(t-\tau) = \left[ e_i^+(t) - e_i^-(t) \right] \quad i = 1, \ldots, N; \, t = 1, \ldots, T$$

$$w_{ij\tau} \geq \Omega_{ji}^{\min} Y_{ji\tau} \quad i, j = 1, \ldots, N; \, t = 1, \ldots, \tau_{\max}$$

$$w_{ij\tau} \leq \Omega_{ji}^{\max} \quad i, j = 1, \ldots, N; \, \tau = 1, \ldots, \tau_{\max}$$

$$\sum_{\tau=0}^{\tau_{\max}} Y_{ji\tau} \leq 1 \quad i, j = 1, \ldots, N$$

$$\sum_{\tau=0}^{\tau_{\max}} \sum_{j=1}^{N} Y_{ji\tau} \leq N_i \quad i = 1, \ldots, N$$

$$Y_{ji\tau} \in \{0,1\} \qquad\qquad i, j = 1, \ldots, N; \tau = 1, \ldots, \tau_{\max}$$
$$e_i^+(t), e_i^-(t) \in \mathbb{R}^+ \qquad\qquad i = 1, \ldots, N; t = 1, \ldots, T.$$

$N_i$ denotes the maximum number of regulatory inputs for gene $i$, $e_i^{\pm}$ denotes positive and negative error variables respectively expressing the deviation from the experimentally measured gene expression values, $\tau_{\max}$ denotes the maximum allowed time delay in the model, and $\Omega_{ji}^{\max}$ denote maximum values for the regulatory coefficients. Dasika et al. [13] demonstrate an effective solution of the proposed formulation based on a sequential bound relaxation scheme. This work demonstrates nicely how a mathematical programming formalism can assist in the analysis of temporal data that present a significant increase in problem complexity compared with the time-independent data discussed earlier.

## 11.4.2 Multicriteria optimization for generic network modeling

Approaches like the one described in the previous section attempt to reverse engineer genetic networks from microarray data. A major problem, however, is how to reliably find interactions when faced with a relatively small number of arrays compared with data (small sample size problem discussed earlier). To address this dimensionality problem, prior biological knowledge needs to be incorporated, in the form of constraints, about the genetic networks. This can be modeled in terms of limited connectivity, redundancy, stability, and robustness. Recently, van Someren et al. [56] presented a multiobjective formulation to address these issues. The problem addressed concerns the definition of appropriate genetic interactions from a set of temporal gene expression data. Specifically, we are given a set $g_i(t)$ representing the expression level of gene $i$ at time point $t$ and let $N$ genes be measured at each time point. The expression state of the organism is thus defined as $g(t) = [g_1(t), \ldots, g_N(t)]^\top$. The concatenated expression levels at each time $t$ are defined as $x^q = g(t)$. Van Someren et al. [56] assumed the simplest dynamic relation for their model, i.e., linear. That is, the state of the system at $t+1$ is a linear function of the state of the system at time $t$: $x^{q+1} = W \cdot x^q$. The matrix of interactions $W$ is termed the gene regulation matrix (GRM). As previously stated, a nonzero entry $w_{ij}$ denotes the existence of a regulatory connection between genes $i$ and $j$, the sign defines an activating action ($> 0$) or an inhibiting action ($< 0$). In order to learn the gene regulation matrix, we simply require that the predicted states of gene $i$ are close as possible to the target (measured) states. The corresponding error is represented by the mean square error criterion defined as:

$$f^{\mathrm{MSE}}(w_i) = \frac{1}{Q-1} \sum_{q=1}^{Q-1} \left( w_i \cdot x^q - x^{q+1} \right)^2.$$

The authors model two biologically relevant constrains. The first one deals with the knowledge that a particular node is influenced only by a limited number of other genes. The connectivity is defined as the number of non-zero weights in the $W$ matrix

$$f^c(w_i) = \sum_{j=1}^{N} c_{ij}$$

where

$$c_{ij} = \begin{cases} 1 & \text{if } w_{ij} \neq 0 \\ 0 & \text{if } w_{ij} = 0. \end{cases}$$

The second constraint deals with the realization that gene networks are robust in the respect to noise. The robustness is here defined as the inherent ability not to propagate forward in time small perturbations in the current expression state. A metric for the robustness is the first derivative of models' output and it is minimized by minimizing the sum of the squared (or absolute) first derivatives $f^S(w_i) = \sum_{j=1}^{N} w_{ij}^2$. Van Someren et al. [56] demonstrate how the Pareto-front can be generated efficiently in order to balance the requirement for accuracy in the model and robustness and stability in the predictions.

## 11.5 A Final Comment

It is clear from the preceding discussion that *feature selection, clustering, and classification* are tasks intimately connected. Numerous techniques have been developed that addressed each problem independently. One of the major advantages of mathematical programming (MP) formulations is that they can bring these tasks explicitly together within a similar framework. The goal of this short exposition was not only to show, by example, how some key questions in biology can be advanced by formulating them as MP problems, but also to demonstrate that one of the major advantages of MP-based approaches is the integrated and highly flexible formulations that capitalize on our advanced understanding of large-scale mixed-integer (non)linear optimization theory. It should be pointed out that a number of other optimization (continuous and mixed-integer) reformulations of data mining have been proposed recently by Glen [27, 28, 29]. We have chosen, however, to focus on methods that have found direct application to microarray expression data and hence left their presentation out of this short review. We do however encourage the interested reader to follow up with such methods because we believe that they will become critical enablers for addressing some of the important open issues such as the ones discussed in the following section.

## 11.6 Research Challenges

Numerous issues can be raised for future research. In fact, the advantage of a MP-based formalism is the tremendous flexibility it provides.

### 11.6.1 Multiobjective optimization

Interpretation of biological information needs to tackle multiple simultaneous objectives. In this short review, we discussed simultaneous optimization of accuracy and size of classifier (number of features). In clustering applications, the number of clusters is yet another level of complexity, hence an additional decision variable. Therefore, multicriteria trade-off curves (Pareto solutions) have to be developed for these high-dimensional mixed-integer (non)linear optimization problems.

### 11.6.2 Incorporation of biological constraints

One of the advantages of using mathematical programming techniques is that constraints can be readily accounted for. Thus far, microarray analyses approaches treat the array data as raw unconstrained measurements. One of the targets of microarray analysis is to identify potential correlations among the data. However, prior biological knowledge is not taken into account mainly because most data mining methods cannot handle implicit or explicit constraints. Recently Sese et al. [51] demonstrated the need to account for biological driven constraints when clustering expression profiles.

### 11.6.3 Large-scale combinatorial optimization

The development of scalable algorithms is a daunting task in optimization theory. With the recent developments in genomics, we should be expecting routinely the analysis of gene arrays composed of tens of thousands of probes (hence tens of thousands of binary variables in the MIP gene selection formulation). Duarte Silva and Stam [16], Gallagher et al. [25], and Rubin [49] discuss various mixed-integer reformulations to the classification problem. Undoubtedly, the biological sciences will greatly benefit by the anticipated advances in optimization theory and practice when used to target problems such as the ones just described. The recent work of Shioda [52] identified opportunities for successful reformulations of various data mining tasks in the context of linear integer optimization. Busygin et al. [8] present some more recent ideas for addressing the bi-clustering problem as a fractional 0-1 optimization problem. Undoubtedly, integer optimization will play a prominent role in feature algorithmic developments as recent results demonstrate the complementarity of the different methodologies, suggesting that a unified approach may help to uncover complex genetic risk factors not currently discovered with a single method (see Moscato et al. [45]).

### 11.6.4 Global optimization

The development of general nonlinear, non-convex separating boundaries naturally leads to requirements of solving large-scale combinatorial nonlinear problems to global optimality. Recent advances in the theory and practice of deterministic global optimization are also expected to be critical enablers (see Floudas [19]).

### 11.6.5 Multiclass problems

Most of the recent developments on mathematical programming-driven approaches are based on two-class problems. The simplest multiclass extension is the one-against-all by constructing $k$ SVM models, where $k$ is the number of classes. The $i^{th}$ SVM classifies the examples of class $i$ against all the other samples in all other classes. Another alternative builds one-against-one classifiers by building $k(k-1)/2$ models where each is trained on data from two classes. Hsu and Lin [37] discuss a computational comparison of the models. The emphasis of current research is on novel methods for generating all the decision functions through the solution of a single, but much larger, optimization problem.

### 11.6.6 Analyzing almost empty spaces

The sparseness of the data set is a critical roadblock. Accurate models can be developed using convoluted optimization approaches. However, we would constantly lack appropriately populated data sets in order to achieve a reasonable balance between the thousands of independent variables (genes measured) and necessary measurements (tissue samples) for a robust identification. Information theoretic approaches accounting for complexity (Akaike and Bayesian Information Criteria) should be developed to strike a balance between the complexity and the accuracy of the model so as to avoid pointless overfitting of the sparsely populated data sets.

### 11.6.7 Uncertainty considerations

Noise and uncertainty in the data is a given. Therefore, data mining algorithms in general and mathematical programming formulations in particular have to account for the presence of noise. Issues from robustness and uncertainty propagation have to be incorporated. However, an interesting issue emerges: how do we distinguish between noise and an infrequent, albeit interesting observation? This in fact may be a question with no answer especially if we consider the implications of sparsely populated data sets.

### 11.6.8 Mixed-integer dynamic optimization

We demonstrated how researchers begin to explore the dynamic component of the gene expression data. This type of analysis however is expected to be enabled tremendously by upcoming advances in efficient algorithms for addressing large-scale mixed-integer dynamic optimization problems. Once the models become nonlinear and non-convex, the issue of global optimality will once again become pertinent.

### 11.6.9 Reformulations

Undoubtedly some of the most critical advances in the practice of mathematical programming–based methods for the analysis of microarray data in general and data mining in particular have been the result of fundamental advances in terms of reformulating large-scale optimization problems and devising ingenious solutions methodologies. To that effect, the pioneering work of Mangasarian [43, 44] deserves particular mention. Stating the data mining tasks as optimization problems is but the beginning. The most appealing characteristic of gene expression analysis is the enormous dimensionality of the resulting optimization problem. High-performance computing will without a doubt have a profound effect, however, true advances will be the result of ingenious algorithmic developments. This is a critical step so that rigorous optimization methods become true competitors for the simpler, yet very efficient, statistics-based analysis methods.

### 11.6.10 Interpretation and visualization

The ultimate goal of data mining is the understanding of the data and the development of actionable strategies based on the conclusions. We need to improve not only the interpretation of the derived models but also the knowledge delivery methods based on the derived models. Optimization and mathematical programming need to provide not just the optimal solution but also some way of interpreting the implications of a particular solution including the quantification of potential crucial sensitivities.

## Acknowledgments

The author wishes to thank the National Science Foundation (NSF-0519563) and the Environmental Protection Agency (EPA-GAD R 832721-010) for financial support.

## References

[1] O. Bousquet and A. Elisseeff. Stability and generalization. *Journal of Machine Learning Research*, 2(3):499–526, 2002.

[2] J.M. Bower and H. Bolouri, editors. *Computational Modeling of Genetic and Biochemical Networks*. MIT Press, 2004.

[3] D.D. Bowtell. Options available – from start to finish – for obtaining expression data by microarray. *Nature Genetics*, 21(1 Suppl):25–32, 1999.

[4] D. Bradley and O.L. Mangasarian. Feature selection via concave minimization and support vector machines. In J. Shavlik, editor, *Proceedings of the 15th International Conference on Machine Learning (ICML'98)*, San Francisco, California, pages 82–90. Morgan Kaufmann, 1998.

[5] P.S. Bradley, U.M. Fayyad, and O.L. Mangasarian. Mathematical programming for data mining: Formulations and challenges. *INFORMS Journal on Computing*, 11(3):217–238, 1999.

[6] M.P.S. Brown, W.N. Grundy, D. Lin, N. Cristianini, C. Walsh Sugnet, T.S. Furey, M. Ares, Jr., and D. Haussler. Knowledge-based analysis of microarray gene expression data by using support vector machines. *Proceedings of the National Academy of Sciences of the United States of America*, 97(1):262–267, 2000.

[7] P.O. Brown and D. Botstein. Exploring the new world of the genome with DNA microarrays. *Nature Genetics*, 21(1 Suppl):33–37, 1999.

[8] S. Busygin, O.A. Prokopyev, and P.M. Pardalos. Feature selection for consistent biclustering via fractional 0-1 programming. *Journal of Combinatorial Optimization*, 10(1):7–21, 2005.

[9] V.G. Cheung, M. Morley, F. Aguilar, A. Massimi, R. Kucherlapati, and G. Childs. Making and reading microarrays. *Nature Genetics*, 21(1 Suppl):15–19, 1999.

[10] A. Chilingaryan, N. Gevorgyan, A. Vardanyan, D. Jones, and A. Szabo. Multivariate approach for selecting sets of differentially expressed genes. *Mathematical Biosciences*, 176(1):59–69, 2002.

[11] C. Cortes and V. Vapnik. Support-vector networks. *Machine Learning*, 20(3):273–297, 1995.

[12] K. Crammer and Y. Singer. On the learnability and design of output codes for multiclass problems. *Machine Learning*, 47(2-3):201–233, 2002.

[13] M.S. Dasika, A. Gupta, and C.D. Maranas. A mixed integer linear programming (MILP) framework for inferring time delay in gene regulatory networks. In *Pacific Symposium on Biocomputing*, pages 474–485, 2004.

[14] M. Dettling and P. Buhlmann. Finding predictive gene groups from microarray data. *Journal of Multivariate Analysis*, 90(1):106–131, 2004.

[15] E.R. Dougherty. Small sample issues for microarray-based classification. *Comparative and Functional Genomics*, 2(1):28–34, 2001.

[16] A.P. Duarte Silva and A. Stam. A mixed integer programming algorithm for minimizing the training sample misclassification cost in two-group classification. *Annals of Operations Research*, 74(0):129–157, 1997.

[17] S. Dudoit, Y.H. Yang, M.J. Callow, and T.P. Speed. Statistical methods for identifying differentially expressed genes in replicated cDNA microarray experiments. *Statistica Sinica*, 12(1):111–139, 2002.

[18] R.P.W. Duin. Classifiers in almost empty spaces. In *15th International Conference on Pattern Recognition (ICPR'00), Volume 2*, 2000.

[19] C.A. Floudas. *Nonlinear and Mixed-Integer Optimization: Fundamentals and Applications*. Oxford University Press, Oxford, U.K., 2000.

[20] N. Freed and F. Glover. A linear programming approach to the discriminant problem. *Decision Sciences*, 12:68–74, 1981.

[21] N. Freed and F. Glover. Simple but powerful goal programming for the discriminant problem. *European Journal of Operational Research*, 7:44–60, 1981.

[22] N. Freed and F. Glover. Evaluating alternative linear programming formulations for the discriminant problem. *Decision Sciences*, 17:151–162, 1986.

[23] G.M. Fung, O.L. Mangasarian, and A.J. Smola. Minimal kernel classifiers. *Journal of Machine Learning Research*, 3(2):303–321, 2003.

[24] T.S. Furey, N. Cristianini, N. Duffy, D.W. Bednarski, M. Schummer, and D. Haussler. Support vector machine classification and validation of cancer tissue samples using microarray expression data. *Bioinformatics*, 16(10):906–914, 2000.

[25] R.J. Gallagher, E.K. Lee, and D.A. Patterson. Constrained discriminant analysis via 0/1 mixed integer programming. *Annals of Operations Research*, 74(0):65–88, 1997.

[26] W.V. Gehrlein. General mathematical programming formulations for the statistical classification problem. *Operations Research Letters*, 5(6):299–304, 1986.

[27] J.J. Glen. Classification accuracy in discriminant analysis: a mixed integer programming approach. *Journal of the Operational Research Society*, 52(3):328–339, 2001.

[28] J.J. Glen. An iterative mixed integer programming method for classification accuracy maximizing discriminant analysis. *Computers & Operations Research*, 30(2):181–198, 2003.

[29] J.J. Glen. Mathematical programming models for piecewise-linear discriminant analysis. *Journal of the Operational Research Society*, 56(3):331–341, 2005.

[30] F. Glover. Improved linear programming models for discrminant analysis. *Decision Sciences*, 21:771–785, 1990.

[31] F. Glover, S. Keene, and B. Duea. A new class of models for the discrminant problem. *Decision Sciences*, 19:269–280, 1988.

[32] T.R. Golub, D.K. Slonim, P. Tamayo, C. Huard, M. Gaasenbeek, J.P. Mesirov, H. Coller, M.L. Loh, J.R. Downing, M.A. Caligiuri, C.D. Bloomfield, and E.S. Lander. Molecular classification of cancer: class discovery and class prediction by gene expression monitoring. *Science*, 286(5439):531–537, 1999.

[33] R.L. Grossman, C. Kamath, and V. Kumar. *Data Mining for Scientific and Engineering Applications*. Kluwer Academic Publishers, Dordrecht, The Netherlands, 2001.

[34] I. Guyon, J. Weston, S. Barnhill, and V. Vapnik. Gene selection for cancer classification using support vector machines. *Machine Learning*, 46(1-3):389–422, 2002.

[35] D.J. Hand, H. Mannila, and P. Smyth. *Principles of Data Mining*. The MIT Press, Cambridge, MA, 2001.

[36] T.K. Ho. A data complexity analysis of comparative advantages of decision forest constructors. *Pattern Analysis & Applications*, 5:102–112, 2002.

[37] C.W. Hsu and C.J. Lin. A comparison of methods for multiclass support vector machines. *IEEE Transactions on Neural Networks*, 13(2):415–425, 2002.

[38] A. Jain and D. Zongker. Feature selection: Evaluation, application, and small sample performance. *IEEE Transactions on Pattern Analysis and Machine Intelligence*, 19(2):153–158, 1997.

[39] F.C. Kafatos. A revolutionary landscape: the restructuring of biology and its convergence with medicine. *Journal of Molecular Biology*, 319(4):861–867, 2002.

[40] R. Kohavi and G.H. John. Wrappers for feature subset selection. *Artificial Intelligence*, 97(1-2):273–324, 1997.

[41] R.J. Lipshutz, S.P. Fodor, T.R. Gingeras, and D.J. Lockhart. High density synthetic oligonucleotide arrays. *Nature Genetics*, 21(1 Suppl):20–24, 1999.

[42] H. Liu and H. Motoda. *Feature Selection for Knowledge Discovery and Data Mining*. Oxford University Press, Oxford, U.K., 2000.

[43] O.L. Mangasarian. Linear and nonlinear separation of patterns by linear programming. *Operations Research*, 13:444–452, 1965.

[44] O.L. Mangasarian. Multi-surface method of pattern separation. *IEEE Transactions on Information Theory*, IT-14:801–807, 1968.

[45] P. Moscato, R. Berretta, M. Hourani, A. Mendes, and C. Cotta. Genes related with Alzheimer's disease: A comparison of evolutionary search, statistical and integer programming approaches. In F. Rothlauf et al., editor, *Applications of Evolutionary Computing*, pages 84–94. Springer-Verlag, Berlin, Germany, 2005.

[46] P. Narendra and K. Fukunaga. A branch and bound algorithm for feature subset selection. *IEEE Transactions on Computers*, C-26(9):917–926, 1977.

[47] M.N. Nguyen and J.C. Rajapakse. Multi-class support vector machines for protein secondary structure prediction. *Genome Informatics*, 14:218–227, 2003.

[48] R. Rifkin, S. Mukherjee, P. Tamayo, S. Ramaswamy, C.-H. Yeang, M. Angelo, M. Reich, T. Poggio, E.S. Lander, T.R. Golub, and J.P. Mesirov. An analytical method for multiclass molecular cancer classification. *SIAM Review*, 45(4):706–723, 2003.

[49] P.A. Rubin. Solving mixed integer classification problems by decomposition. *Annals of Operations Research*, 0:51–64, 74.

[50] M. Schena, D. Shalon, R.W. Davis, and P.O. Brown. Quantitative monitoring of gene expression patterns with a complementary DNA microarray. *Science*, 270(5235):467–470, 1995.

[51] J. Sese, Y. Kurokawa, M. Monden, K. Kato, and S. Morishita. Constrained clusters of gene expression profiles with pathological features. *Bioinformatics*, 20(17):3137–3145, 2004.

[52] R. Shioda. *Integer Optimization in Data Mining*. Ph.d. thesis, Massachusetts Institute of Technology, Operations Research, 2003.

[53] A. Stam. Nontraditional approaches to statistical classification: Some perspectives on $L_p$-norm methods. *Annals of Operations Research*, 74(0):1–36, 1997.

[54] M. Sun and M. Xiong. A mathematical programming approach for gene selection and tissue classification. *Bioinformatics*, 19(10):1243–1251, 2003.

[55] A. Szabo, K. Boucher, W.L. Carroll, L.B. Klebanov, A.D. Tsodikov, and A.Y. Yakovlev. Variable selection and pattern recognition with gene expression data generated by the microarray technology. *Mathematical Biosciences*, 176(1):71–98, 2002.

[56] E.P. van Someren, L.F.A. Wessels, E. Backer, and M.J.T. Reinders. Multi-criterion optimization for genetic network modeling. *Signal Processing*, 83(4):763–775, 2003.

[57] V.N. Vapnik. *The Nature of Statistical Learning*. Springer-Verlag, Berlin, Germany, 1995.

[58] J. Weston and C. Watkins. Multi-class support vector machines. In *Proceedings of ESANN99*, Brussels, Belgium, 1999. D. Facto Publishers.

[59] R.D. Williams, S.N. Hing, B.T. Greer, C.C. Whiteford, J.S. Wei, R. Natrajan, A. Kelsey, S. Rogers, C. Campbell, K. Pritchard-Jones, and J. Khan. Prognostic classification of relapsing favorable histology Wilms tumor using cDNA microarray expression profiling and support vector machines. *Genes, Chromosomes & Cancer*, 41(1):65–79, 2004.

[60] H. Zhang, C.Y. Yu, B. Singer, and M. Xiong. Recursive partitioning for tumor classification with gene expression microarray data. *Proceedings of the National Academy of Sciences of the United States of America*, 98(12):6730–6735, 2001.

# 12

## Classification and Disease Prediction via Mathematical Programming

Eva K. Lee and Tsung-Lin Wu

Center for Operations Research in Medicine and HealthCare, School of Industrial and Systems Engineering, Georgia Institute of Technology, Atlanta, Georgia 30332-0205
evakylee@isye.gatech.edu

**Abstract.** In this chapter, we present classification models based on mathematical programming approaches. We first provide an overview on various mathematical programming approaches, including linear programming, mixed-integer programming, nonlinear programming, and support vector machines. Next, we present our effort of novel optimization-based classification models that are general purpose and suitable for developing predictive rules for large heterogeneous biological and medical data sets. Our predictive model simultaneously incorporates (1) the ability to classify any number of distinct groups; (2) the ability to incorporate heterogeneous types of attributes as input; (3) a high-dimensional data transformation that eliminates noise and errors in biological data; (4) the ability to incorporate constraints to limit the rate of misclassification, and a reserved-judgment region that provides a safeguard against over-training (which tends to lead to high misclassification rates from the resulting predictive rule); and (5) successive multistage classification capability to handle data points placed in the reserved judgment region. To illustrate the power and flexibility of the classification model and solution engine, and its multigroup prediction capability, application of the predictive model to a broad class of biological and medical problems is described. Applications include: the differential diagnosis of the type of erythemato-squamous diseases; predicting presence/absence of heart disease; genomic analysis and prediction of aberrant CpG island meythlation in human cancer; discriminant analysis of motility and morphology data in human lung carcinoma; prediction of ultrasonic cell disruption for drug delivery; identification of tumor shape and volume in treatment of sarcoma; multistage discriminant analysis of biomarkers for prediction of early atherosclerois; fingerprinting of native and angiogenic microvascular networks for early diagnosis of diabetes, aging, macular degeneracy, and tumor metastasis; prediction of protein localization sites; and pattern recognition of satellite images in classification of soil types. In all these applications, the predictive model yields correct classification rates ranging from 80% to 100%. This provides motivation for pursuing its use as a medical diagnostic, monitoring, and decision-making tool.

P.M. Pardalos, H.E. Romeijn (eds.), *Handbook of Optimization in Medicine*, Springer Optimization and Its Applications 26, DOI: 10.1007/978-0-387-09770-1_12, © Springer Science+Business Media LLC 2009

## 12.1 Introduction

Classification is a fundamental machine learning task whereby rules are developed for the allocation of independent observations to groups. Classic examples of applications include medical diagnosis (the allocation of patients to disease classes based on symptoms and lab tests), and credit screening (the acceptance or rejection of credit applications based on applicant data). Data are collected concerning observations with known group membership. This *training data* is used to develop rules for the classification of future observations with unknown group membership.

In the introduction section, we briefly describe some terminologies related to classification and provide a brief organization of the materials written in this chapter.

### 12.1.1 Pattern recognition, discriminant analysis, and statistical pattern classification

*Cognitive science* is the science of learning, knowing, and reasoning. *Pattern recognition* is a broad field within *cognitive science* that is concerned with the process of recognizing, identifying, and categorizing input information. These areas intersect with computer science, particularly in the closely related areas of *artificial intelligence, machine learning*, and *statistical pattern recognition*. Artificial intelligence is associated with constructing machines and systems that reflect human abilities in cognition. Machine learning refers to how these machines and systems replicate the learning process, which is often achieved by seeking and discovering patterns in data, or statistical pattern recognition.

*Discriminant analysis* is the process of discriminating between categories or populations. Associated with discriminant analysis as a statistical tool are the tasks of determining the features that best discriminate between populations, and the process of classifying new objects based on these features. The former is often called *feature selection* and the latter is referred to as *statistical pattern classification*. This work will be largely concerned with the development of a viable statistical pattern classifier.

As with many computationally intensive tasks, recent advances in computing power have led to a sharp increase in the interest and application of discriminant analysis techniques. The reader is referred to Duda et al. [25] for an introduction to various techniques for pattern classification and to Zopounidis et al. [121] for examples of applications of pattern classification.

### 12.1.2 Supervised learning, training, and cross-validation

An *entity* or *observation* is essentially a data point as commonly understood in statistics. In the framework of statistical pattern classification, an entity is a set of quantitative measurements (or qualitative measurements expressed quantitatively) of *attributes* for a particular object. As an example, in medical diagnosis an entity could be the various blood chemistry levels of a patient. With each entity is associated one or more *groups* (or *populations, classes, categories*) to which it belongs. Continuing with the medical diagnosis example, the groups could be the various classes of heart disease. Statistical classification seeks to determine rules for associating entities with the groups to which they belong. Ideally, these associations align with the associations that human reasoning would produce based on information gathered on objects and their apparent categories.

*Supervised learning* is the process of developing classification rules based on entities for which the classification is already known. Note that the process implies that the populations are already well-defined. *Unsupervised learning* is the process of discovering patterns from unlabeled entities and thereby discovering and describing the underlying populations. Models derived using supervised learning can be used for both functions of discriminant analysis – feature selection and classification. The model that we consider is a method for supervised learning, so we assume that populations are previously defined.

The set of entities with known classification that is used to develop classification rules is the *training set*. The training set may be partitioned so that some entities are withheld during the model-development process, also known as the *training* of the model. The withheld entities form a *test set* that is used to determine the validity of the model, a process known as *cross-validation*. Entities from the test set are subjected to the rules of classification to measure the performance of the rules on entities with unknown group membership.

Validation of classification models is often performed using $m$-fold cross-validation where the data with known classification is partitioned into $m$ *folds* (subsets) of approximately equal size. The classification model is trained $m$ times, with the $m^{\text{th}}$ fold withheld during each run for testing. The performance of the model is evaluated by the classification accuracy on the $m$ test folds and can be represented using a *classification matrix* or *confusion matrix*.

The classification matrix is a square matrix with the number of rows and columns equal to the number of groups. The $ij^{\text{th}}$ entry of the classification matrix contains the number or proportion of test entities from group $i$ that were classified by the model as belonging to group $j$. Therefore, the number or proportion of correctly classified entities are contained in the diagonal elements of the classification matrix, and the number or proportion of misclassified entities are in the off-diagonal entries.

### 12.1.3 Bayesian inference and classification

The popularity of *Bayesian inference* has risen drastically over the past several decades, perhaps in part due to its suitability for statistical learning. The reader is referred to O'Hagan's volume [92] for a thorough treatment of Bayesian inference. Bayesian inference is usually contrasted against *classic inference*, though in practice they often imply the same methodology.

384     E.K. Lee and T.-L. Wu

The Bayesian method relies on a *subjective* view of probability, as opposed to the *frequentist* view upon which classic inference is based [92]. A subjective probability describes a degree of belief in a *proposition* held by the investigator based on some information. A frequency probability describes the likelihood of an *event* given an infinite number of trials.

In Bayesian statistics, inferences are based on the *posterior distribution*. The posterior distribution is the product of the *prior probability* and the *likelihood function*. The prior probability distribution represents the initial degree of belief in a proposition, often before empirical data is considered. The likelihood function describes the likelihood that the behavior is exhibited, given that the proposition is true. The posterior distribution describes the likelihood that the proposition is true, given the observed behavior.

Suppose we have a proposition or random variable $\theta$ about which we would like to make inferences, and data $x$. Application of Bayes' theorem gives

$$dF(\theta|x) = \frac{dF(\theta)dF(x|\theta)}{dF(x)}.$$

Here, $F$ denotes the (cumulative) distribution function. For ease of conceptualization, assume that F is differentiable, then $dF = f$, and the above equality can be rewritten as

$$f(\theta|x) = \frac{f(\theta)f(x|\theta)}{f(x)}.$$

For classification, a prior probability function $\pi(g)$ describes the likelihood that an entity is allocated to group $g$ regardless of its exhibited feature values $x$. A group density function $f(x|g)$ describes the likelihood that an entity exhibits certain measurable attribute values, given that it belongs to population $g$. The posterior distribution for a group $P(g|x)$ is given by the product of the prior probability and group density function, normalized over the groups to obtain a unit probability over all groups. The observation $x$ is allocated to group $h$ if

$$h = \arg\max_{g \in \mathcal{G}} P(g|x) = \arg\max_{g \in \mathcal{G}} \frac{\pi(g)f(x|g)}{\sum_{j \in \mathcal{G}} \pi(j)f(x|j)}$$

where $\mathcal{G}$ denotes the set of groups.

## 12.1.4 Discriminant functions

Most classification methods can be described in terms of *discriminant functions*. A discriminant function takes as input an observation and returns information about the classification of the observation. For data from a set of groups $\mathcal{G}$, an observation $x$ is assigned to group $h$ if $h = \arg\max_{g \in \mathcal{G}} l_g(x)$ where the functions $l_g$ are the discriminant functions. Classification methods restrict the form of the discriminant functions, and training data is used to determine the values of parameters that define the functions.

The optimal classifier in the Bayesian framework can be described in terms of discriminant functions. Let $\pi_g = \pi(g)$ be the prior probability that an observation is allocated to group $g$ and let $f_g(x) = f(x|g)$ be the likelihood that data $x$ is drawn from population $g$. If we wish to minimize the probability of misclassification given $x$, then the optimal allocation for an entity is to the group

$$h = \arg\max_{g \in \mathcal{G}} P(g|x) = \arg\max_{g \in \mathcal{G}} \frac{\pi_g f_g(x)}{\sum\limits_{j \in \mathcal{G}} \pi_j f_j(x)}.$$

Under the Bayesian framework,

$$P(g|x) = \frac{\pi_g f(x|g)}{f(x)} = \frac{\pi_g f(x|g)}{\sum\limits_{j \in \mathcal{G}} \pi_j f(x|j)}.$$

The discriminant functions can be $l_g(x) = P(g|x)$ for $g \in \mathcal{G}$. The same classification rule is given by $l_g(x) = \pi_g f(x|g)$ and $l_g(x) = \log f(x|g) + \log \pi_g$. The problem then becomes finding the form of the prior functions and likelihood functions that match the data.

If the data are multivariate normal with equal covariance matrices $(f(x|g) \sim N(\mu_g, \Sigma))$, then a linear discriminant function is optimal:

$$\begin{aligned} l_g(x) &= \log f(x|g) + \log \pi_g \\ &= -1/2(x - \mu_g)^T \Sigma^{-1}(x - \mu_g) - 1/2 \log |\Sigma_g| - d/2 \log 2\pi + \log \pi_g \\ &= w_g^T x + w_{g0} \end{aligned}$$

where $d$ is the number of attributes, $w_g = \Sigma^{-1}\mu_g$, and $w_{g0} = -1/2\mu_g^T \Sigma^{-1}\mu_g + \log \pi_g + x^T \Sigma^{-1} x - d/2 \log 2\pi$. Note that the last two terms of $w_{g0}$ are constant for all $g$ and need not be calculated. When there are 2 groups ($\mathcal{G} = \{1, 2\}$) and the priors are equal ($\pi_1 = \pi_2$), the discriminant rule is equivalent to Fisher's linear discriminant rule [30]. Fisher's rule can also be derived, as it was by Fisher, by choosing $w$ so that $\frac{(w^T \mu_1 - w^T \mu_2)^2}{w^T \Sigma w}$ is maximized.

These linear and quadratic discriminant functions are often applied to data sets that are not multivariate normal or continuous (see [98, pages 234–235]) by using approximations for the means and covariances. Regardless, these models are *parametric* in that they incorporate assumptions about the distribution of the data. Fisher's linear discriminant is *non-parametric* because no assumptions are made about the underlying distribution of the data. Thus, for a special case, a parametric and non-parametric model coincide to produce the same discriminant rule. The linear discriminant function derived above is also called the *homoscedastic model*, and the quadratic discriminant function is called the *heteroscedastic model*. The exact form of discriminant functions in the Bayesian framework can be derived for other distributions [25].

Some classification methods are essentially methods for finding coefficients for linear discriminant functions. In other words, they seek coefficients $w_g$ and constants $w_{g0}$ such that $l_g(x) = w_g x + w_{g0}$, $g \in \mathcal{G}$, is an optimal set of discriminant functions. The criteria for optimality is different for different methods. Linear discriminant functions project the data onto a linear subspace and then discriminate between entities in that subspace. For example, Fisher's linear discriminant projects two-group data on an optimal line and discriminates on that line. A good linear subspace may not exist for data with overlapping distributions between groups and therefore the data will not be classified accurately using these methods. The hyperplanes defined by the discriminant functions form boundaries between the group regions. A large portion of the literature concerning the use of mathematical programming models for classification describe methods for finding coefficients of linear discriminant functions [121].

Other classification methods seek to determine parameters to establish quadratic discriminant functions. The general form of a quadratic discriminant function is $l_g(x) = x^T W_g x + w_g^T x + w_{g0}$. The boundaries defining the group regions can assume any hyperquadric form, as can the Bayes decision rules for arbitrary multivariate normal distributions [25].

In this paper, we survey the development and advances of classification models via the mathematical programming techniques, and summarize our experience in classification models applied to prediction in biological and medical applications. The rest of this chapter is organized as follows. Section 12.2 first provides a detailed overview of the development and advances of mathematical programming-based classification models, including linear programming, mixed-integer programming, nonlinear programming, and support vector machine approaches. In Section 12.3, we describe our effort in developing optimization-based multigroup multistage discriminant analysis predictive models for classification. The use of the predictive models on various biological and medical problems are presented. Section 12.4 provides several tables to summarize the progress of mathematical programming–based classification models and their characteristics. This is followed by a brief description of other classification methods in Section 12.5, and summary and concluding remarks in Section 12.6.

## 12.2 Mathematical Programming Approaches

Mathematical programming methods for statistical pattern classification emerged in the 1960s, gained popularity in the 1980s, and have grown drastically since. Most of the mathematical programming approaches are nonparametric, which has been cited as an advantage when analyzing contaminated data sets over methods that require assumptions about the distribution of the data [107]. Most of the literature about mathematical programming methods is concerned with either using mathematical programming to determine the coefficients of linear discriminant functions or with *support vector machines*.

The following notation will be used. The subscripts $i$, $j$, and $k$ are used for the observation, attribute, and group, respectively. Let $x_{ij}$ be the value of attribute $j$ of observation $i$. Let $m$ be the number of attributes, $K$ be the number of groups, $G_k$ represent the set of data from group $k$, $M$ be a big positive number, and $\epsilon$ be a small positive number. The abbreviation "urs" is used in reference to a variable to denote "unrestricted in sign."

## 12.2.1 Linear programming classification models

The use of linear programs to determine the coefficients of linear discriminant functions has been widely studied [31, 46, 50, 74]. The methods determine the coefficients for different objectives, including minimizing the sum of the distances to the separating hyperplane, minimizing the maximum distance of an observation to the hyperplane, and minimizing other measures of badness of fit or maximizing measures of goodness of fit.

**Two-group classification**

One of the earliest linear programming (LP) classification models was proposed by Mangasarian [74] to construct a hyperplane to separate two groups of data. Separation by a nonlinear surface using LP was also proposed when the surface parameters appear linearly. Two sets of points may be inseparable by one hyperplane or surface through a single-step LP approach, but they can be strictly separated by more planes or surfaces via a multistep LP approach (Mangasarian [75]). In [75] real problems with up to 117 data points, 10 attributes, and 3 groups were solved. The 3-group separation was achieved by separating group 1 from groups 2 and 3, and then group 2 from group 3.

Studies of LP models for the discriminant problem in the early 1980s was carried out by Hand [47], Freed and Glover [31, 32], and Bajgier and Hill [5]. Three LP models for the two-group classification problem, including minimizing the sum of deviations (MSD), minimizing the maximum deviation (MMD), and minimizing the sum of interior distances (MSID) were proposed. Freed and Glover [33] provided computational studies of these models where the test conditions involved normal and nonnormal populations.

- MSD (Minimize the sum of deviations)

$$\begin{aligned}
\text{Min} \quad & \sum_i d_i \\
\text{s.t.} \quad & w_0 + \sum_j x_{ij} w_j - d_i \leq 0 \quad \forall i \in G_1 \\
& w_0 + \sum_j x_{ij} w_j + d_i \geq 0 \quad \forall i \in G_2 \\
& w_j \text{ urs} \quad \forall j \\
& d_i \geq 0 \quad \forall i
\end{aligned}$$

- MMD (Minimize the maximum deviation)

$$\text{Min} \quad d$$

$$\text{s.t.} \quad w_0 + \sum_j x_{ij}w_j - d \leq 0 \quad \forall i \in G_1$$
$$w_0 + \sum_j x_{ij}w_j + d \geq 0 \quad \forall i \in G_2$$
$$w_j \text{ urs } \quad \forall j$$
$$d \geq 0$$

- MSID (Minimize the sum of interior distances)

$$\text{Min} \quad pd - \sum_i e_i$$
$$\text{s.t.} \quad w_0 + \sum_j x_{ij}w_j - d + e_i \leq 0 \quad \forall i \in G_1$$
$$w_0 + \sum_j x_{ij}w_j + d - e_i \geq 0 \quad \forall i \in G_2$$
$$w_j \text{ urs } \quad \forall j$$
$$d \geq 0$$
$$e_i \geq 0 \quad \forall i$$

where $p$ is a weight constant.

The objective function of the MSD model is the $L_1$-norm distance whereas the objective function of MMD is the $L_\infty$-norm distance. They are special cases of $L_p$-norm classification [50, 108].

In some models, the constant term of the hyperplane is a fixed number instead of a decision variable. The model MSD$^0$ shown below is an example where the cutoff score $b$ replaces $w_0$ in the formulation. The same replacement could be used in other formulations.

- MSD$^0$ (Minimize the sum of deviations with constant cutoff score)

$$\text{Min} \quad \sum_i d_i$$
$$\text{s.t.} \quad \sum_j x_{ij}w_j - d_i \leq b \quad \forall i \in G_1$$
$$\sum_j x_{ij}w_j + d_i \geq b \quad \forall i \in G_2$$
$$w_j \text{ urs } \quad \forall j$$
$$d_i \geq 0 \quad \forall i$$

A gap can be introduced between the two regions determined by the separating hyperplane to prevent degenerate solutions. Take MSD as an example, the separation constraints become

$$w_0 + \sum_j x_{ij}w_j - d_i \leq -\epsilon \quad \forall i \in G_1$$

$$w_0 + \sum_j x_{ij}w_j + d_i \geq \epsilon \quad \forall i \in G_2.$$

The small number $\epsilon$ can be normalized to 1.

Besides introducing a gap, another normalization approach is to include constraints such as $\sum_{j=0}^m w_j = 1$ or $\sum_{j=1}^m w_j = 1$ in the LP models to avoid unbounded or trivial solutions.

Specifically, Glover et al. [45] gave the hybrid model, as follows.

- Hybrid model

$$\text{Min} \quad pd + \sum_i p_i d_i - qe - \sum_i q_i e_i$$

$$\text{s.t.} \quad w_0 + \sum_j x_{ij} w_j - d - d_i + e + e_i = 0 \quad \forall i \in G_1$$
$$w_0 + \sum_j x_{ij} w_j + d + d_i - e - e_i = 0 \quad \forall i \in G_2$$
$$w_j \text{ urs} \quad \forall j$$
$$d, e \geq 0$$
$$d_i, e_i \geq 0 \quad \forall i$$

where $p, p_i, q, q_i$ are the cost for different deviations. Including different combinations of deviation terms in the objective function then leads to variant models.

Joachimsthaler and Stam [50] review and summarize LP formulations applied to two-group classification problems in discriminant analysis, including MSD, MMD, MSID, mixed-integer programming (MIP) models, and the hybrid model. They summarize the performance of the LP methods together with the traditional classification methods such as Fisher's linear discriminant function (LDF) [30], Smith's quadratic discriminant function (QDF) [106], and a logistic discriminant method. In their review, MSD sometimes but not uniformly improves classification accuracy compared with traditional methods. On the other hand, MMD is found to be inferior to MSD. Erenguc and Koehler [27] present a unified survey of LP models and their experimental results, in which the LP models include several versions of MSD, MMD, MSID, and hybrid models. Rubin [99] provides experimental results of comparing these LP models with Fisher's LDF and Smith's QDF. He concludes that QDF performs best when the data follow normal distributions and that QDF could be the benchmark when seeking situations for advantageous LP methods. In summary, the above review papers [27, 50, 99] describe previous work on LP classification models and their comparison with traditional methods. However, it is difficult to make definitive statements about conditions under which one LP model is superior to others, as stated in [107].

Stam and Ungar [110] introduce a software package RAGNU, a utility program in conjunction with the LINDO optimization software, for solving two-group classification problems using LP-based methods. LP formulations such as MSD, MMD, MSID, hybrid models, and their variants are contained in the package.

There are some difficulties in LP-based formulations, in that some models could result in unbounded, trivial, or unacceptable solutions [87, 34], but possible remedies are proposed. Koehler [51, 52, 53] and Xiao [114, 115] characterize the conditions of unacceptable solutions in two-group LP discriminant models, including MSD, MMD, MISD, the hybrid model, and their variants. Glover [44] proposes the normalization constraint, $\sum_{j=1}^{m}(-|G_2| \sum_{i \in G_1} x_{ij} + |G_1| \sum_{i \in G_2} x_{ij}) w_j = 1$, which is more effective and reliable. Rubin [100] examines the separation failure for two-group models and suggests to apply the models twice, reversing the group designations the second time. Xiao and Feng [116] propose a regularization method to avoid multiple solutions in LP discriminant analysis by adding the term $\epsilon \sum_{j=1}^{m} w_j^2$ in the objective functions.

Bennett and Mangasarian [9] propose the following model, which minimizes the average of the deviations, which is called robust linear programming.

- RLP (Robust linear programming)

$$
\begin{aligned}
\text{Min} \quad & \frac{1}{|G_1|}\sum_{i\in G_1} d_i + \frac{1}{|G_2|}\sum_{i\in G_2} d_i \\
\text{s.t.} \quad & w_0 + \sum_j x_{ij}w_j - d_i \le -1 \quad \forall i \in G_1 \\
& w_0 + \sum_j x_{ij}w_j + d_i \ge 1 \quad \forall i \in G_2 \\
& w_j \text{ urs} \quad \forall j \\
& d_i \ge 0 \quad \forall i
\end{aligned}
$$

It is shown that this model gives the null solution $w_1 = \cdots = w_m = 0$ if and only if $\frac{1}{|G_1|}\sum_{i\in G_1} x_{ij} = \frac{1}{|G_2|}\sum_{i\in G_2} x_{ij}$ for all $j$, in which case the solution $w_1 = \cdots = w_m = 0$ is guaranteed to be not unique. Data of different diseases are tested by the proposed classification methods, as in most of Mangasarian's papers.

Mangasarian et al. [86] describe two applications of LP models in the field of breast cancer research, one in diagnosis and the other in prognosis. The first application is to discriminate benign from malignant breast lumps, and the second one is to predict when breast cancer is likely to recur. Both of them work successfully in clinical practice. The RLP model [9] together with the multisurface method tree algorithm (MSMT) [8] is used in the diagnostic system.

Duarte Silva and Stam [104] include the second-order (i.e., quadratic and cross-product) terms of the attribute values in the LP-based models such as MSD and hybrid models and compare them with linear models, Fisher's LDF, and Smith's QDF. The results of the simulation experiments show that the methods that include second-order terms perform much better than first-order methods, given that the data substantially violate the multivariate normality assumption. Wanarat and Pavur [113] investigate the effect of the inclusion of the second-order terms in the MSD, MIP, and hybrid models when sample size is small to moderate. However, the simulation study shows that second-order terms may not always improve the performance of a first-order LP model even with data configurations that are more appropriately classified by Smith's QDF. Another result of the simulation study is that inclusion of the cross-product terms may hurt the model's accuracy, and omission of these terms causes the model to be not invariant with respect to a nonsingular transformation of the data.

Pavur [94] studies the effect of the position of the contaminated normal data in the two-group classification problem. The methods for comparison in their study include MSD, MM (described in the mixed integer programming part), Fisher's LDF, Smith's QDF, and nearest neighbor models. The non-traditional methods such as LP models have potential for outperforming the standard parametric procedures when nonnormality is present, but this study shows that no one model is consistently superior in all cases.

Asparoukhov and Stam [3] propose LP and MIP models to solve the two-group classification problem where the attributes are binary. In this case, the training data can be partitioned into multinomial cells, allowing for a substantial reduction in the number of variables and constraints. The proposed models not only have the usual geometric interpretation but also possess a strong probabilistic foundation. Let $s$ be the index of the cells, $n_{1s}, n_{2s}$ be the number of data points in cell $s$ from groups 1 and 2, respectively, and $(b_{s1}, \ldots, b_{sm})$ be the binary digits representing cell $s$. The model shown below is the LP model of minimizing the sum of deviations for two-group classification with binary attributes.

- Cell conventional MSD

$$
\begin{aligned}
\text{Min} \quad & \sum_{s:\, n_{1s}+n_{2s}>0}(n_{1s}d_{1s} + n_{2s}d_{2s}) \\
\text{s.t.} \quad & w_0 + \sum_j b_{sj}w_j - d_{1s} \leq 0 \quad \forall s : n_{1s} > 0 \\
& w_0 + \sum_j b_{sj}w_j + d_{2s} > 0 \quad \forall s : n_{2s} > 0 \\
& w_j \text{ urs} \quad \forall j \\
& d_{1s}, d_{2s} \geq 0 \quad \forall s
\end{aligned}
$$

Binary attributes are usually found in medical diagnoses data. In this study, three real data sets about disease discrimination are tested: developing postoperative pulmonary embolism or not, having dissecting aneurysm or other diseases, and suffering from posttraumatic epilepsy or not. In these data sets, the MIP model for binary attributes (BMIP), which will be described later, performs better than other LP models or traditional methods.

## Multigroup classification

Freed and Glover [32] extend the LP classification models from two-group to multigroup problems. One formulation that uses a single discriminant function is given below

$$
\begin{aligned}
\text{Min} \quad & \sum_{k=1}^{K-1} c_k \alpha_k \\
\text{s.t.} \quad & \sum_j x_{ij}w_j \leq U_k \quad \forall i \in G_k \; \forall k \\
& \sum_j x_{ij}w_j \geq L_k \quad \forall i \in G_k \; \forall k \\
& U_k + \epsilon \leq L_{k+1} + \alpha_k \quad \forall k = 1, \ldots, K-1 \\
& w_j \text{ urs} \quad \forall j \\
& U_k, L_k \text{ urs} \quad \forall k \\
& \alpha_k \text{ urs} \quad \forall k = 1, \ldots, K-1
\end{aligned}
$$

where the number $\epsilon$ could be normalized to be 1, and $c_k$ is the misclassification cost. However, single function classification is not as flexible and general as multiple function classification. Another extension from the two-group case to multigroup in [32] is to solve two-group LP models for all pairs of groups and determine classification rules based on these solutions. However, in some cases the group assignment is not clear, and the resulting classification scheme may be sub-optimal [107].

For the multigroup discrimination problem, Bennett and Mangasarian [10] define the piecewise-linear separability of data from $K$ groups as the following: The data from $K$ groups are piecewise-linear separable if and only if there exist $(w_0^k, w_1^k, \ldots, w_m^k) \in R^{m+1}$, $k = 1, \ldots, K$, such that $w_0^h + \sum_j x_{ij} w_j^h \geq w_0^k + \sum_j x_{ij} w_j^k + 1$, $\forall i \in G_h \ \forall h, k \neq h$. The following LP will generate a piecewise-linear separation for the $K$ groups if one exists, otherwise it will generate an error-minimizing separation:

$$\text{Min} \quad \sum_h \sum_{k \neq h} \frac{1}{|G_h|} \sum_{i \in G_h} d_i^{hk}$$
$$\text{s.t.} \quad d_i^{hk} \geq -(w_0^h + \sum_j x_{ij} w_j^h) + (w_0^k + \sum_j x_{ij} w_j^k) + 1 \quad \forall i \in G_h \ \forall h, k \neq h$$
$$w_j^k \text{ urs} \quad \forall j, k$$
$$d_i^{hk} \geq 0 \quad \forall i \in G_h \ \forall h, k \neq h.$$

The method is tested in three data sets. It performs pretty well in two of the data sets that are totally (or almost totally) piecewise-linear separable. The classification result is not good in the third data set, which is inherently more difficult. However, by combining the multisurface method tree algorithm (MSMT) [8], the performance improves.

Gochet et al. [46] introduce an LP model for the general multigroup classification problem. The method separates the data with several hyperplanes by sequentially solving LPs. The vectors $w^k$, $k = 1, \ldots, K$, are estimated for the classification decision rule. The rule is to classify an observation $i$ into group $s$ where $s = \arg \max_k \{w_0^k + \sum_j x_{ij} w_j^k\}$.

Suppose observation $i$ is from group $h$. Denote the goodness of fit for observation $i$ with respect to group $k$ as

$$G_{hk}^i(w^h, w^k) = [(w_0^h + \sum_j x_{ij} w_j^h) - (w_0^k + \sum_j x_{ij} w_j^k)]^+$$

where $[a]^+ = \max\{0, a\}$.

Likewise, denote the badness of fit for observation $i$ with respect to group $k$ as

$$B_{hk}^i(w^h, w^k) = [(w_0^h + \sum_j x_{ij} w_h^h) - (w_0^k + \sum_j x_{ij} w_j^k)]^-$$

where $[a]^- = -\min\{0, a\}$.

The total goodness of fit and total badness of fit are then defined as

$$G(w) = G(w^1, \ldots, w^K) = \sum_h \sum_{k \neq h} \sum_{i \in G_h} G_{hk}^i(w^h, w^k)$$
$$B(w) = B(w^1, \ldots, w^K) = \sum_h \sum_{k \neq h} \sum_{i \in G_h} B_{hk}^i(w^h, w^k)$$

The LP is to minimize the total badness of fit, subject to a normalization equation, in which $q > 0$.

$$\text{Min} \quad B(w)$$
$$\text{s.t.} \quad G(w) - B(w) = q$$
$$w \text{ urs}$$

Expanding $G(w)$ and $B(w)$ and substituting $G_{hk}^i(w^h, w^k)$ and $B_{hk}^i(w^h, w^k)$ by $\gamma_{hk}^i$ and $\beta_{hk}^i$ respectively, the LP becomes

$$\text{Min} \quad \sum_h \sum_{k \neq h} \sum_{i \in G_h} \beta_{hk}^i$$
$$\text{s.t.} \quad (w_0^h + \sum_j x_{ij} w_j^h) - (w_0^k + \sum_j x_{ij} w_j^k) = \gamma_{hk}^i - \beta_{hk}^i \quad \forall i \in G_h \ \forall h, k \neq h$$
$$\sum_h \sum_{k \neq h} \sum_{i \in G_h} (\gamma_{hk}^i - \beta_{hk}^i) = q$$
$$w_j^k \text{ urs} \quad \forall j, k$$
$$\gamma_{hk}^i, \beta_{hk}^i \geq 0 \quad \forall i \in G_h \ \forall h, k \neq h$$

The classification results for two real data sets show that this model can compete with Fisher's LDF and the nonparametric $k$-nearest neighbor method.

The LP-based models for classification problems highlighted above are all nonparametric models. In Section 12.3, we describe LP-based and MIP-based classification models that utilize a parametric multigroup discriminant analysis approach [39, 40, 63, 60]. These latter models have been employed successfully in various multigroup disease diagnosis and biological/medical prediction problems [16, 28, 29, 56, 57, 59, 60, 65, 64].

## 12.2.2 Mixed-integer programming classification models

Whereas LP offers a polynomial-time computational guarantee, MIP allows more flexibility in (among other things) modeling misclassified observations and/or misclassification costs.

### Two-group classification

In the two-group classification problem, binary variables can be used in the formulation to track and minimize the exact number of misclassifications. Such an objective function is also considered as the $L_0$-norm criterion [107].

- MM (Minimizing the number of misclassifications)

$$\text{Min} \quad \sum_i z_i$$
$$\text{s.t.} \quad w_0 + \sum_j x_{ij} w_j \leq M z_i \quad \forall i \in G_1$$
$$w_0 + \sum_j x_{ij} w_j \geq -M z_i \quad \forall i \in G_2$$
$$w_j \text{ urs} \quad \forall j$$
$$z_i \in \{0, 1\} \quad \forall i$$

The vector $w$ is required to be a nonzero vector to prevent the trivial solution.

In the MIP formulation, the objective function could include the deviation terms, such as those in the hybrid models, as well as the number of misclassifications [5]; or it could represent expected cost of misclassification [6, 1, 105, 101]. In particular, there are some variant versions of the basic model.

Stam and Joachimsthaler [109] study the classification performance of MM and compare it with MSD, Fisher's LDF, and Smith's QDF. In some cases, the MM model performs better, but in some cases it does not. MIP formulations are in the review studies of Joachimsthaler and Stam [50] and Erenguc and Koehler [27] and contained in the software developed by Stam and Ungar [110]. Computational experiments show that the MIP model performs better when the group overlap is higher [50, 109], although it is still not easy to reach general conclusions [107].

Because the MIP model is $\mathcal{NP}$-hard, exact algorithms and heuristics are proposed to solve it efficiently. Koehler and Erenguc [54] develop a procedure to solve MM in which the condition of nonzero $w$ is replaced by the requirement of at least one violation of the constraints $w_0 + \sum_j x_{ij}w_j \leq 0$ for $i \in G_1$ or $w_0 + \sum_j x_{ij}w_j \geq 0$ for $i \in G_2$. Banks and Abad [6] solve the MIP of minimizing the expected cost of misclassification by an LP-based algorithm. Abad and Banks [1] develop three heuristic procedures to the problem of minimizing the expected cost of misclassification. They also include the interaction terms of the attributes in the data and apply the heuristics [7]. Duarte Silva and Stam [105] introduce the Divide and Conquer algorithm for the classification problem of minimizing the misclassification cost by solving MIP and LP subproblems. Rubin [101] solves the same problem by using a decomposition approach and tests this procedure on some data sets, including two breast cancer data sets. Yanev and Balev [119] propose exact and heuristic algorithms for solving MM, which are based on some specific properties of the vertices of a polyhedral set neatly connected with the model.

For the two-group classification problem where the attributes are binary, Asparoukhov and Stam [3] propose LP and MIP models that partition the data into multinomial cells and result in fewer number of variables and constraints. Let $s$ be the index of the cells, $n_{1s}, n_{2s}$ be the number of data points in cell $s$ from groups 1 and 2, respectively, and $(b_{s1}, \ldots, b_{sm})$ be the binary digits representing cell $s$. Below is the MIP model for binary attributes (BMIP), which performs best in three real data sets in [3].

- BMIP

    Min $\sum_{s: n_{1s}+n_{2s}>0}\{|n_{1s} - n_{2s}|z_s + \min(n_{1s}, n_{2s})\}$

$$\text{s.t.} \quad w_0 + \sum_j b_{sj} w_j \leq M z_s \quad \forall s : n_{1s} \geq n_{2s};\ n_{1s} > 0$$
$$w_0 + \sum_j b_{sj} w_j > -M z_s \quad \forall s : n_{1s} < n_{2s}$$
$$w_j \text{ urs} \quad \forall j$$
$$z_s \in \{0,1\} \quad \forall s : n_{1s} + n_{2s} > 0$$

Pavur et al. [96] include different secondary goals in the model MM and compare their misclassification rates. A new secondary goal is proposed, which maximizes the difference between the means of the discriminant scores of the two groups. In this model the term $-\delta$ is added to the minimization objective function as a secondary goal with a constant multiplier while the constraint $\sum_j \bar{x}_j^{(2)} w_j - \sum_j \bar{x}_j^{(1)} w_j \geq \delta$ is included, where $\bar{x}_j^{(k)} = \frac{1}{|G_k|} \sum_{i \in G_k} x_{ij}\ \forall j$, for $k = 1, 2$. The results of simulation study show that an MIP model with the proposed secondary goal has better performance than other studied models.

Glen [42] proposes integer progreamming (IP) techniques for normalization in the two-group discriminant analysis models. One technique is to add the constraint $\sum_{j=1}^m |w_j| = 1$. In the proposed model, $w_j$ for $j = 1, \ldots, m$ is represented by $w_j = w_j^+ - w_j^-$, where $w_j^+, w_j^- \geq 0$, and binary variables $\delta_j$ and $\gamma_j$ are defined such that $\delta_j = 1 \Leftrightarrow w_j^+ \geq \epsilon$ and $\gamma_j = 1 \Leftrightarrow w_j^- \geq \epsilon$. The IP normalization technique is applied to MSD and MMD, and the MSD version is presented below.

- MSD – with IP normalization

$$\text{Min} \quad \sum_i d_i$$
$$\text{s.t.} \quad w_0 + \sum_{j=1}^m x_{ij}(w_j^+ - w_j^-) - d_i \leq 0 \quad \forall i \in G_1$$
$$w_0 + \sum_{j=1}^m x_{ij}(w_j^+ - w_j^-) + d_i \geq 0 \quad \forall i \in G_2$$
$$\sum_{j=1}^m (w_j^+ + w_j^-) = 1$$
$$w_j^+ - \epsilon \delta_j \geq 0 \quad \forall j = 1, \ldots, m$$
$$w_j^+ - \delta_j \leq 0 \quad \forall j = 1, \ldots, m$$
$$w_j^- - \epsilon \gamma_j \geq 0 \quad \forall j = 1, \ldots, m$$
$$w_j^- - \gamma_j \leq 0 \quad \forall j = 1, \ldots, m$$
$$\delta_j + \gamma_j \leq 1 \quad \forall j = 1, \ldots, m$$
$$w_0 \text{ urs}$$
$$w_j^+, w_j^- \geq 0 \quad \forall j = 1, \ldots, m$$
$$d_i \geq 0 \quad \forall i$$
$$\delta_j, \gamma_j \in \{0,1\} \quad \forall j = 1, \ldots, m$$

The variable coefficients of the discriminant function generated by the models are invariant under origin shifts. The proposed models are validated using two data sets from [45, 87]. The models are also extended for attribute selection by adding the constraint $\sum_{j=1}^m (\delta_j + \gamma_j) = p$, which allows only a constant number, $p$, of attributes to be used for classification.

Glen [43] develops MIP models that determine the thresholds for forming dichotomous variables as well as the discriminant function coefficients, $w_j$. For each continuous attribute to be formed as a dichotomous attribute, the model finds the threshold among possible thresholds while determining the separating hyperplane and optimizing the objective function such as minimizing the sum of deviations or minimizing the number of misclassifications. Computational results of a real data set and some simulated data sets show that the MSD model with dichotomous categorical variable formation can improve classification performance. The reason for the potential of this technique is that the generated linear discriminant function is a nonlinear function of the original variables.

## Multigroup classification

Gehrlein [41] proposes MIP formulations of minimizing the total number of misclassifications in the multigroup classification problem. He gives both a single function classification scheme and a multiple function classification scheme, as follows.

- GSFC (General single function classification – minimizing the number of misclassifications)

$$\text{Min} \quad \sum_i z_i$$
$$\begin{aligned}
\text{s.t.} \quad & w_0 + \sum_j x_{ij} w_j - M z_i \leq U_k \quad \forall i \in G_k \\
& w_0 + \sum_j x_{ij} w_j + M z_i \geq L_k \quad \forall i \in G_k \\
& U_k - L_k \geq \delta' \quad \forall k \\
& \left.\begin{array}{l} L_g - U_k + M y_{gk} \geq \delta \\ L_k - U_g + M y_{kg} \geq \delta \\ y_{gk} + y_{kg} = 1 \end{array}\right\} \quad \forall g, k,\ g \neq k \\
& w_j \text{ urs} \quad \forall j \\
& U_k, L_k \text{ urs} \quad \forall k \\
& z_i \in \{0, 1\} \quad \forall i \\
& y_{gk} \in \{0, 1\} \quad \forall g, k,\ g \neq k
\end{aligned}$$

where $U_k, L_k$ denote the upper and lower endpoints of the interval assigned to group $k$, and $y_{gk} = 1$ if the interval associated with group $g$ precedes that with group $k$ and $y_{gk} = 0$ otherwise. The constant $\delta'$ is the minimum width of an interval of a group and the constant $\delta$ is the minimum gap between adjacent intervals.

- GMFC (General multiple function classification – minimizing the number of misclassifications)

$$\text{Min} \quad \sum_i z_i$$
$$\begin{aligned}
\text{s.t.} \quad & w_0^h + \sum_j x_{ij} w_j^h - w_0^k - \sum_j x_{ij} w_j^k + M z_i \geq \epsilon \quad \forall i \in G_h,\ \forall h, k \neq h \\
& w_j^k \text{ urs} \quad \forall j, k \\
& z_i \in \{0, 1\} \quad \forall i
\end{aligned}$$

Both models work successfully on the iris data set provided by Fisher [30].

Pavur [93] solves the multigroup classification problem by sequentially solving GSFC in one dimension each time. Linear discriminant functions are generated by successively solving GSFC with the added constraints that all linear discriminants are uncorrelated to each other for the total data set. This procedure could be repeated for the number of dimensions that is believed to be enough. According to simulation results, this procedure substantially improves the GSFC model and sometimes outperforms GMFC, Fisher's LDF, or Smith's QDF.

To solve the three-group classification problem more efficiently, Loucopoulos and Pavur [71] make a slight modification on GSFC and propose the model MIP3G, which also minimizes the number of misclassifications. Compared with GSFC, MIP3G is also a single function classification model, but it reduces the possible group orderings from six to three in the formulation and thus becomes more efficient. Loucopoulos and Pavur [72] report the results of a simulation experiment on the performance of GMFC, MIG3G, Fisher's LDF, and Smith's QDF for three-group classification problem with small training samples. Second-order terms are also considered in the experiment. Simulation results show that GMFC and MIP3G can outperform the parametric procedures in some nonnormal data sets and that the inclusion of second-order terms can improve the performance of MIP3G in some data sets. Pavur and Loucopoulos [95] investigate the effect of the gap size in the MIP3G model for the three-group classification problem. A simulation study illustrates that for fairly separable data, or data with small sample sizes, a nonzero-gap model can improve the performance. A possible reason for this result is that the zero-gap model may be overfitting the data.

Gallagher et al. [39, 40], Lee et al. [63], and Lee [59, 60] propose MIP models, both heuristic and exact, as a computational approach to solving the constrained discriminant method described by Anderson [2]. These models are described in detail in Section 12.3.

### 12.2.3 Nonlinear programming classification models

Nonlinear programming approaches are natural extensions for some of the LP-based models. Thus far, nonlinear programming approaches have been developed for 2-group classification.

Stam and Joachimsthaler [108] propose a class of nonlinear programming methods to solve the two-group classification problem under the $L_p$-norm objective criterion. This is an extension of MSD and MMD, for which the objectives are the $L_1$-norm and $L_\infty$-norm, respectively.

- Minimize the general $L_p$-norm distance

$$
\begin{aligned}
\text{Min} \quad & \left( \sum_i d_i^p \right)^{1/p} \\
\text{s.t.} \quad & \sum_j x_{ij} w_j - d_i \leq b \quad \forall i \in G_1 \\
& \sum_j x_{ij} w_j + d_i \geq b \quad \forall i \in G_2 \\
& w_j \ \text{urs} \quad \forall j \\
& d_i \geq 0 \quad \forall i
\end{aligned}
$$

The simulation results show that, in addition to the $L_1$-norm and $L_\infty$-norm, it is worth the effort to compute other $L_p$-norm objectives. Restricting the analysis to $1 \leq p \leq 3$, plus $p = \infty$, is recommended. This method is reviewed by Joachimsthaler and Stam [50] and Erenguc and Koehler [27].

Mangasarian et al. [85] propose a nonconvex model for the two-group classification problem:

$$
\begin{aligned}
\text{Min} \quad & d^1 + d^2 \\
\text{s.t.} \quad & \sum_j x_{ij} w_j - d^1 \leq 0 \quad && \forall i \in G_1 \\
& \sum_j x_{ij} w_j + d^2 \geq 0 \quad && \forall i \in G_2 \\
& \max_{j=1,\ldots,m} |w_j| = 1 \\
& w_j \ \text{urs} \quad \forall j \\
& d^1, d^2 \ \text{urs}
\end{aligned}
$$

This model can be solved in polynomial-time by solving $2m$ linear programs, which generate a sequence of parallel planes, resulting in a piecewise-linear nonconvex discriminant function. The model works successfully in clinical practice for the diagnosis of breast cancer.

Further, Mangasarian [76] also formulates the problem of minimizing the number of misclassifications as a linear program with equilibrium constraints (LPEC) instead of the MIP model MM described previously.

- MM-LPEC (Minimizing the number of misclassifications – Linear program with equilibrium constraints)

$$
\begin{aligned}
\text{Min} \quad & \sum_{i \in G_1 \cup G_2} z_i \\
\text{s.t.} \quad & w_0 + \sum_j x_{ij} w_j - d_i \leq -1 \quad && \forall i \in G_1 \\
& z_i(w_0 + \sum_j x_{ij} w_j - d_i + 1) = 0 \quad && \forall i \in G_1 \\
& w_0 + \sum_j x_{ij} w_j + d_i \geq 1 \quad && \forall i \in G_2 \\
& z_i(w_0 + \sum_j x_{ij} w_j + d_i - 1) = 0 \quad && \forall i \in G_2 \\
& d_i(1 - z_i) = 0 \quad && \forall i \in G_1 \cup G_2 \\
& 0 \leq z_i \leq 1 \quad && \forall i \in G_1 \cup G_2 \\
& d_i \geq 0 \quad && \forall i \in G_1 \cup G_2 \\
& w_j \ \text{urs} \quad \forall j
\end{aligned}
$$

The general LPEC can be converted to an exact penalty problem with a quadratic objective and linear constraints. A stepless Frank–Wolfe type algorithm is proposed for the penalty problem, terminating at a stationary point or a global solution. This method is called the parametric misclassification minimization (PMM) procedure, and numerical testing is included in [77].

To illustrate the next model, we first define the step function $s : R \to \{0, 1\}$ as

$$
s(u) = \begin{cases} 1 & \text{if } u > 0 \\ 0 & \text{if } u \leq 0 \end{cases}
$$

The problem of minimizing the number of misclassifications is equivalent to

$$\text{Min} \quad \sum_{i \in G_1 \cup G_2} s(d_i)$$
$$\text{s.t.} \quad w_0 + \sum_j x_{ij} w_j - d_i \leq -1 \quad \forall i \in G_1$$
$$w_0 + \sum_j x_{ij} w_j + d_i \geq 1 \quad \forall i \in G_2$$
$$d_i \geq 0 \quad \forall i \in G_1 \cup G_2$$
$$w_j \text{ urs} \quad \forall j$$

Mangasarian [77] proposes a simple concave approximation of the step function for nonnegative variables: $t(u, \alpha) = 1 - e^{-\alpha u}$, where $\alpha > 0, u \geq 0$. Let $\alpha > 0$ and approximate $s(d_i)$ by $t(d_i, \alpha)$. The problem then reduces to minimizing a smooth concave function bounded below on a nonempty polyhedron, which has a minimum at a vertex of the feasible region. A finite successive linearization algorithm (SLA) is proposed, terminating at a stationary point or a global solution. Numerical tests of SLA are done and compared with the PMM procedure described above. The results show that the much simpler SLA obtains a separation that is almost as good as PMM in considerably less computing time.

Chen and Mangasarian [21] propose an algorithm on a defined hybrid misclassification minimization problem, which is more computationally tractable than the $\mathcal{NP}$-hard misclassification minimization problem. The basic idea of the hybrid approach is to obtain iteratively $w_0$ and $(w_1, \ldots, w_m)$ of the separating hyperplane: (1) For a fixed $w_0$, solve RLP (Bennett and Mangasarian [9]) to determine $(w_1, \ldots, w_m)$, and (2) for this $(w_1, \ldots, w_m)$, solve the one-dimensional misclassification minimization problem to determine $w_0$. Comparison of the hybrid method is made with the RLP method and the PMM procedure. The hybrid method performs better in the testing sets of the tenfold cross-validation and is much faster than PMM.

Mangasarian [78] proposes the model of minimizing the sum of arbitrary-norm distances of misclassified points to the separating hyperplane. For a general norm $\| \cdot \|$ on $R^m$, the dual norm $\| \cdot \|'$ on $R^m$ is defined as $\|x\|' = \max_{\|y\|=1} x^T y$. Define $[a]^+ = \max\{0, a\}$ and let $w = (w_1, \ldots, w_m)$. The formulation can then be written as:

$$\text{Min} \quad \sum_{i \in G_1} [w_0 + \sum_j x_{ij} w_j]^+ + \sum_{i \in G_2} [-w_0 - \sum_j x_{ij} w_j]^+$$
$$\text{s.t.} \quad \|w\|' = 1$$
$$w_0, w \text{ urs}$$

The problem is to minimize a convex function on a unit sphere. A related decision problem to this minimization problem is shown to be $\mathcal{NP}$-complete, except for $p = 1$. For a general $p$-norm, the minimization problem can be transformed via an exact penalty formulation to minimizing the sum of a convex function and a bilinear function on a convex set.

## 12.2.4 Support vector machine

A support vector machine is a type of mathematical programming approach (Vapnik [57]). It has been widely studied and has become popular in many application fields in recent years. The introductory description of support vector machines (SVMs) given here is summarized from the tutorial by Burges [20]. In order to maintain consistency with SVM studies in published literature, the notation used below is slightly different than the notation used to describe the mathematical programming methods in earlier sections.

In the two-group separable case, the objective function is to maximize the margin of a separating hyperplane, $2/||w||$, which is equivalent to minimizing $||w||^2$.

$$\begin{aligned}\text{Min}\quad & w^T w\\ \text{s.t.}\quad & x_i^T w + b \geq +1 \quad \text{for } y_i = +1\\ & x_i^T w + b \leq -1 \quad \text{for } y_i = -1\\ & w, b \text{ urs}\end{aligned}$$

where $x_i \in R^m$ represents the values of attributes of observation $i$, and $y_i \in \{-1, 1\}$ represents the group of observation $i$.

This problem can be solved by solving its Wolfe dual problem.

$$\begin{aligned}\text{Max}\quad & \sum_i \alpha_i - \tfrac{1}{2}\sum_{i,j} \alpha_i \alpha_j y_i y_j x_i^T x_j\\ \text{s.t.}\quad & \sum_i \alpha_i y_i = 0\\ & \alpha_i \geq 0 \quad \forall i.\end{aligned}$$

Here, $\alpha_i$ is the Lagrange multiplier for the training point $i$, and the points with $\alpha_i > 0$ are called the support vectors (analogous to the support of a hyperplane, and thus the introduction of the name "support vector"). The primal solution $w$ is given by $w = \sum_i \alpha_i y_i x_i$. $b$ can be computed by solving $y_i(w^T x_i + b) - 1 = 0$ for any $i$ with $\alpha_i > 0$.

For the non-separable case, slack variables $\xi_i$ are introduced to handle the errors. Let $C$ be the penalty for the errors. The problem becomes

$$\begin{aligned}\text{Min}\quad & \tfrac{1}{2} w^T w + C(\sum_i \xi_i)^k\\ \text{s.t.}\quad & x_i^T w + b \geq +1 - \xi_i \quad \text{for } y_i = +1\\ & x_i^T w + b \leq -1 + \xi_i \quad \text{for } y_i = -1\\ & w, b \text{ urs}\\ & \xi_i \geq 0 \quad \forall i.\end{aligned}$$

When $k$ is chosen to be 1, neither the $\xi_i$'s nor their Lagrange multipliers appear in the Wolfe dual problem:

$$\begin{aligned}\text{Max}\quad & \sum_i \alpha_i - \tfrac{1}{2}\sum_{i,j} \alpha_i \alpha_j y_i y_j x_i^T x_j\\ \text{s.t.}\quad & \sum_i \alpha_i y_i = 0\\ & 0 \leq \alpha_i \leq C \quad \forall i.\end{aligned}$$

The data points can be separated nonlinearly by mapping the data into some higher dimensional space and applying linear SVM to the mapped data. Instead of knowing explicitly the mapping $\Phi$, SVM needs only the dot products of two transformed data points $\Phi(x_i) \cdot \Phi(x_j)$. The kernel function $K$ is introduced such that $K(x_i, x_j) = \Phi(x_i) \cdot \Phi(x_j)$. Replacing $x_i^T x_j$ by $K(x_i, x_j)$ in the above problem, the separation becomes nonlinear whereas the problem to be solved remains a quadratic program. In testing a new data point $x$ after training, the sign of the function $f(x)$ is computed to determine the group of $x$:

$$f(x) = \sum_{i=1}^{N_s} \alpha_i y_i \Phi(s_i) \cdot \Phi(x) + b = \sum_{i=1}^{N_s} \alpha_i y_i K(s_i, x) + b.$$

where $s_i$'s are the support vectors and $N_s$ is the number of support vectors. Again the explicit form of $\Phi(x)$ is avoided.

Mangasarian provides a general mathematical programming framework for SVM, called generalized support vector machine or GSVM [79, 83]. Special cases can be derived from GSVM, including the standard SVM.

Many SVM-type methods have been developed by Mangasarian and other authors to solve huge-sized classification problems more efficiently. These methods include: successive overrelaxation for SVM [82], proximal SVM [36, 38], smooth SVM [68], reduced SVM [67], Lagrangian SVM [84], incremental SVMs [37], and other methods [13, 81]. Mangasarian summarizes some of the developments in [80]. Examples of applications of SVM include breast cancer studies [69, 70] and genome research [73].

Hsu and Lin [49] compare different methods for multigroup classification using support vector machines. Three methods studied are based on several binary classifiers: one-against-one, one-against-all, and directed acyclic graph (DAG) SVM. The other two methods studied are altogether methods with decomposition implementation. The experiment results show that the one-against-one and DAG methods are more suitable for practical use than the other methods. Lee et al. [66] propose a generic approach to multigroup problems with some theoretical properties, and the proposed method is well applied to microarray data for cancer classification and satellite radiance profiles for cloud classification.

Gallagher et al 1996, 1997 and Lee et al 2003 [39, 40, 63] offer the first discrete support vector machine for multigroup classification with reserved judgment. The approach has been successfully applied to a diverse variety of biological and medical applications (see Section 12.3).

## 12.3 MIP-Based Multigroup Classification Models and Applications to Medicine and Biology

Commonly-used methods for classification, such as linear discriminant functions, decision trees, mathematical programming approaches, support vector machines, and artificial neural networks (ANN), can be viewed as attempts at approximating a *Bayes optimal rule* for classification; that is, a rule that maximizes (minimizes) the total probability of correct classification (misclassification). Even if a Bayes optimal rule is known, intergroup misclassification rates may be higher than desired. For example, in a population that is mostly healthy, a Bayes optimal rule for medical diagnosis might misdiagnose sick patients as healthy in order to maximize total probability of correct diagnosis. As a remedy, a constrained discriminant rule that limits the misclassification rate is appealing.

Assuming that the group density functions and prior probabilities are known, Anderson [2] showed that an optimal rule for the problem of maximizing the probability of correct classification subject to constraints on the misclassification probabilities must be of a specific form when discriminating among multiple groups with a simplified model. The formulae in Anderson's result depend on a set of parameters satisfying a complex relationship between the density functions, the prior probabilities, and the bounds on the misclassification probabilities. Establishing a viable mathematical model to describe Anderson's result and finding values for these parameters that yield an optimal rule are challenging tasks. The first computational models utilizing Anderson's formulae were proposed in [39, 40].

### 12.3.1 Discrete support vector machine predictive models

As part of the work carried out at Georgia Institute of Technology's Center for Operations Research in Medicine, we have developed a general-purpose discriminant analysis modeling framework and computational engine that are applicable to a wide variety of applications, including biological, biomedical, and logistics problems. Utilizing the technology of large-scale discrete optimization and SVMs, we have developed novel classification models that simultaneously include the following features: (1) the ability to classify any number of distinct groups; (2) the ability to incorporate heterogeneous types of attributes as input; (3) a high-dimensional data transformation that eliminates noise and errors in biological data; (4) constraints to limit the rate of misclassification, and a reserved-judgment region that provides a safeguard against over-training (which tends to lead to high misclassification rates from the resulting predictive rule); and (5) successive multistage classification capability to handle data points placed in the reserved judgment region. Studies involving tumor volume identification, ultrasonic cell disruption in drug delivery, lung tumor cell motility analysis, CpG island aberrant methylation in human cancer, predicting early atherosclerosis using biomarkers, and fingerprinting native and angiogenic microvascular networks using functional perfusion data indicate that our approach is adaptable and can produce effective and reliable predictive rules for various biomedical and bio-behavior phenomena [16, 28, 29, 56, 57, 65, 64, 59, 60].

Based on the description in [39, 40, 63, 59, 60], we summarize below some of the classification models we have developed.

**Modeling of reserved judgment region for general groups**

When the population densities and prior probabilities are known, the constrained rules with a reject option (reserved judgment), based on Anderson's results, calls for finding a partition $\{R_0, \ldots, R_G\}$ of $\mathbb{R}^k$ that maximizes the probability of correct allocation subject to constraints on the misclassification probabilities; i.e.,

$$\text{Max} \quad \sum_{g=1}^{G} \pi_g \int_{R_g} f_g(w)\, dw \tag{12.1}$$

$$\text{s.t.} \quad \int_{R_g} f_h(w)\, dw \leq \alpha_{hg}, \ h, \ g = 1, \ldots, G, \ h \neq g, \tag{12.2}$$

where $f_h$ $(h = 1, \ldots, G)$ are the group conditional density functions, $\pi_g$ denotes the prior probability that a randomly selected entity is from group $g$ $(g = 1, \ldots, G)$ and $\alpha_{hg}$ $(h \neq g)$ are constants between zero and one. Under quite general assumptions, it was shown that there exist unique (up to a set of measure zero) nonnegative constants $\lambda_{ih}$, $i$, $h \in \{1, \ldots, G\}$, $i \neq h$, such that the optimal rule is given by

$$R_g = \{x \in \mathbb{R}^k : L_g(x) = \max_{h \in \{0,1,\ldots,G\}} L_h(x)\}, \ g = 0, \ldots, G \tag{12.3}$$

where

$$L_0(x) = 0 \tag{12.4}$$

$$L_h(x) = \pi_h f_h(x) - \sum_{i=1, i \neq h}^{G} \lambda_{ih} f_i(x), \ h = 1, \ldots, G. \tag{12.5}$$

For $G = 2$, the optimal solution can be modeled rather straightforward. However, finding optimal $\lambda_{ih}$'s for the general case, $G \geq 3$, is a difficult problem, with the difficulty increasing as G increases. Our model offers an avenue for modeling and finding the optimal solution in the general case. It is the first such model to be computationally viable [39, 40].

Before proceeding, we note that $R_g$ can be written as $R_g = \{x \in \mathbb{R}^k : L_g(x) \geq L_h(x) \text{ for all } h = 0, \ldots, G$. Thus, because $L_g(x) \geq L_h(x)$ if, and only if, $(1/\sum_{t=1}^{G} f_t(x))L_g(x) \geq (1/\sum_{t=1}^{G} f_t(x))L_h(x)$, the functions $L_h$, $h = 1, \ldots, G$, can be redefined as

$$L_h(x) = \pi_h p_h(x) - \sum_{i=1, i \neq h}^{G} \lambda_{ih} p_i(x), \ h = 1, \ldots, G \tag{12.6}$$

where $p_i(x) = f_i(x) / \sum_{t=1}^{G} f_t(x)$. We assume that $L_h$ is defined as in equation (12.6) in our model.

**Mixed-integer programming formulations**

Assume that we are given a training sample of N entities whose group classifications are known; say $n_g$ entities are in group $g$, where $\sum_{g=1}^{G} n_g = N$. Let the $k$-dimensional vectors $x^{gj}$, $g = 1, \ldots, G$, $j = 1, \ldots, n_g$, contain the measurements on $k$ available characteristics of the entities. Our procedure for deriving a discriminant rule proceeds in two stages. The first stage is to use the training sample to compute estimates, $\hat{f}_h$, either parametrically or nonparametrically, of the density functions $f_h$ (e.g., see [89]) and estimates, $\hat{\pi}_h$, of the prior probabilities $\pi_h$, $h = 1, \ldots, G$. The second stage is to determine the optimal $\lambda_{ih}$'s given these estimates. This stage requires being able to estimate the probabilities of correct classification and misclassification for any candidate set of $\lambda_{ih}$'s. One could, in theory, substitute the estimated densities and prior probabilities into equations (12.5), and directly use the resulting regions $R_g$ in the integral expressions given in (12.1) and (12.2). This would involve, even in simple cases such as normally distributed groups, the numerical evaluation of $k$-dimensional integrals at each step of a search for the optimal $\lambda_{ih}$'s. Therefore, we have designed an alternative approach. After substituting the $\hat{f}_h$'s and $\hat{\pi}_h$'s into equation (12.5), we simply calculate the proportion of training sample points that fall in each of the regions $R_1, \ldots, R_G$. The MIP models discussed below attempt to maximize the proportion of training sample points correctly classified while satisfying constraints on the proportions of training sample points misclassified. This approach has two advantages. First, it avoids having to evaluate the potentially difficult integrals in Equations (12.1) and (12.2). Second, it is nonparametric in controlling the training sample misclassification probabilities. That is, even if the densities are poorly estimated (by assuming, for example, normal densities for non-normal data), the constraints are still satisfied for the training sample. Better estimates of the densities may allow a higher correct classification rate to be achieved, but the constraints will be satisfied even if poor estimates are used. Unlike most support vector machine models that minimize the sum of errors, our objective is driven by the number of correct classifications and will not be biased by the distance of the entities from the supporting hyperplane.

A word of caution is in order. In traditional unconstrained discriminant analysis, the true probability of correct classification of a given discriminant rule tends to be smaller than the rate of correct classification for the training sample from which it was derived. One would expect to observe such an effect for the method described herein as well. In addition, one would expect to observe an analogous effect with regard to constraints on misclassification probabilities – the true probabilities are likely to be greater than any limits imposed on the proportions of training sample misclassifications. Hence, the $\alpha_{hg}$ parameters should be carefully chosen for the application in hand.

Our first model is a nonlinear 0/1 MIP model with the nonlinearity appearing in the constraints. Model 1 maximizes the number of correct classifications of the given $N$ training entities. Similarly, the constraints on the misclassification probabilities are modeled by ensuring that the number of group $g$ training entities in region $R_h$ is less than or equal to a pre-specified percentage, $\alpha_{hg}(0 < \alpha_{hg} < 1)$, of the total number, $n_g$, of group $g$ entities, $h, \, g \in \{1, \ldots, G\}, \, h \neq g$.

For notational convenience, let $\mathbf{G} = \{1, \ldots, G\}$ and $\mathbf{N}_g = \{1, \ldots, n_g\}$, for $g \in \mathbf{G}$. Also, analogous to the definition of $p_i$, define $\hat{p}_i$ by $\hat{p}_i = \hat{f}_i(x) / \sum_{t=1}^{G} \hat{f}_t(x)$. In our model, we use binary indicator variables to denote the group classification of entities. Mathematically, let $u_{hgj}$ be a binary variable indicating whether or not $x^{gj}$ lies in region $R_h$; i.e., whether or not the $j$th entity from group $g$ is allocated to group $h$. Then Model 1 can be written as follows:

- DAMIP

$$\text{Max} \quad \sum_{g \in G} \sum_{j \in N_g} u_{ggj}$$

s.t.

$$L_{hgj} = \hat{\pi}_h \hat{p}_h(x^{gj}) - \sum_{i \in G \setminus h} \lambda_{ih} \hat{p}_i(x^{gj}), \quad h, g \in \mathbf{G}, \; j \in \mathbf{N}_g \quad (12.7)$$

$$y_{gj} = \max\{0, L_{hgj} : h = 1, \ldots, G\}, \quad g \in \mathbf{G}, \; j \in \mathbf{N}_g \quad (12.8)$$

$$y_{gj} - L_{ggj} \leq M(1 - u_{ggj}), \quad g \in \mathbf{G}, \; j \in \mathbf{N}_g \quad (12.9)$$

$$y_{gj} - L_{hgj} \geq \varepsilon(1 - u_{hgj}), \quad h, g \in \mathbf{G}, \; j \in \mathbf{N}_g, \; h \neq g \quad (12.10)$$

$$\sum_{j \in N_g} u_{hgj} \leq \lfloor \alpha_{hg} n_g \rfloor, \quad h, g \in \mathbf{G}, \; h \neq g \quad (12.11)$$

$$-\infty < L_{hgj} < \infty, \; y_{gj} \geq 0, \; \lambda_{ih} \geq 0, u_{hgj} \in \{0, 1\}.$$

Constraint (12.7) defines the variable $L_{hgj}$ as the value of the function $L_h$ evaluated at $x^{gj}$. Therefore, the continuous variable $y_{gj}$, defined in constraint (12.8), represents $\max\{L_h(x^{gj}) : h = 0, \ldots, G\}$; and consequently, $x^{gj}$ lies in region $R_h$ if, and only if, $y_{gj} = L_{hgj}$. The binary variable $u_{hgj}$ is used to indicate whether or not $x^{gj}$ lies in region $R_h$; i.e., whether or not the $j$th entity from group $g$ is allocated to group $h$. In particular, constraint (12.9), together with the objective, force $u_{ggj}$ to be 1 if, and only if, the $j$th entity from group $g$ is correctly allocated to group $g$; and constraints (12.10) and (12.11) ensure that at most $\lfloor \alpha_{hg} n_g \rfloor$ (i.e., the greatest integer less than or equal to $\alpha_{hg} n_g$) group $g$ entities are allocated to group $h$, $h \neq g$. One caveat regarding the indicator variables $u_{hgj}$ is that although the condition $u_{hgj} = 0$, $h \neq g$, implies (by constraint (12.10)) that $x^{gj} \notin R_h$, the converse need not hold. As a consequence, the number of misclassifications may be overcounted. However, in our preliminary numerical study, we found that the actual amount of overcounting is minimal. One could force the converse (thus, $u_{hgj} = 1$ if and only if $x^{gj} \in R_h$) by adding constraints $y_{gj} - L_{hgj} \leq M(1 - u_{hgj})$, for example. Finally, we note that the parameters $M$ and $\varepsilon$ are extraneous to the discriminant analysis problem itself, but are needed in the model to control the indicator variables $u_{hgj}$. The intention is for $M$ and $\varepsilon$ to be, respectively, large and small positive constants.

## Model variations

We explore different variations in the model to grasp the quality of the solution and the associated computational effort.

A first variation involves transforming Model 1 to an equivalent linear mixed integer model. In particular, Model 2 replaces the N constraints defined in (12.8) with the following system of $3GN + 2N$ constraints:

$$y_{gj} \geq L_{hgj}, \qquad h, g \in \mathbf{G}, \ j \in \mathbf{N}_g \qquad (12.12)$$
$$\tilde{y}_{hgj} - L_{hgj} \leq M(1 - v_{hgj}), \quad h, g \in \mathbf{G}, \ j \in \mathbf{N}_g \qquad (12.13)$$
$$\tilde{y}_{hgj} \leq \hat{\pi}_h \hat{p}_h(x^{gj}) v_{hgj}, \ h, g \in \mathbf{G}, \ j \in \mathbf{N}_g \qquad (12.14)$$
$$\sum_{h \in G} v_{hgj} \leq 1, \qquad g \in \mathbf{G}, \ j \in \mathbf{N}_g \qquad (12.15)$$
$$\sum_{h \in G} \tilde{y}_{hgj} = y_{gj}, \qquad g \in \mathbf{G}, \ j \in \mathbf{N}_g \qquad (12.16)$$

where $\tilde{y}_{hgj} \geq 0$ and $v_{hgj} \in \{0,1\}$, $h, g \in \mathbf{G}, j \in \mathbf{N}_g$. These constraints, together with the non-negativity of $y_{gj}$ force $y_{gj} = \max\{0, L_{hgj} : h = 1, \ldots, G\}$.

The second variation involves transforming Model 1 to a heuristic linear MIP model. This is done by replacing the nonlinear constraint (12.8) with $y_{gj} \geq L_{hgj}$, $h, g \in \mathbf{G}, j \in \mathbf{N}_g$, and including penalty terms in the objective function. In particular, Model 3 has the objective

$$\text{Maximize} \sum_{g \in G} \sum_{j \in N_g} \beta u_{ggj} - \sum_{g \in G} \sum_{j \in N_g} \gamma y_{gj},$$

where $\beta$ and $\gamma$ are positive constants. This model is heuristic in that there is nothing to force $y_{gj} = \max\{0, L_{hgj} : h = 1, \ldots, G\}$. However, because in addition to trying to force as many $u_{ggj}$'s to one as possible, the objective in Model 3 also tries to make the $y_{gj}$'s as small as possible, and the optimizer tends to drive $y_{gj}$ toward $\max\{0, L_{hgj} : h = 1, \ldots, G\}$. We remark that $\beta$ and $\gamma$ could be stratified by group (i.e., introduce possibly distinct $\beta_g, \gamma_g, g \in \mathbf{G}$) to model the relative importance of certain groups to be correctly classified.

A reasonable modification to Models 1, 2, and 3 involves relaxing the constraints specified by (12.11). Rather than placing restrictions on the number of type $g$ training entities classified into group $h$, for all $h, g \in \mathbf{G}, h \neq g$, one could simply place an upper bound on the *total* number of misclassified training entities. In this case, the $G(G-1)$ constraints specified by (12.11) would be replaced by the single constraint

$$\sum_{g\in G}\sum_{h\in G\setminus\{g\}}\sum_{j\in N_g} u_{hgj} \leq \lfloor \alpha N \rfloor \qquad (12.17)$$

where $\alpha$ is a constant between 0 and 1. We will refer to Models 1, 2, and 3, modified in this way, as Models 1T, 2T, and 3T, respectively. Of course, other modifications are also possible. For instance, one could place restrictions on the total number of type $g$ points misclassified for each $g \in \mathbf{G}$. Thus, in place of the constraints specified in (12.17), one would include the constraints $\sum_{h\in G\setminus\{g\}} \sum_{j\in N_g} u_{hgj} \leq \lfloor \alpha_g N \rfloor$, $g \in \mathbf{G}$, where $0 < \alpha_g < 1$.

We also explore a heuristic linear model of Model 1. In particular, consider the linear program (DALP):

$$\text{Max} \qquad \sum_{g\in G}\sum_{j\in N_g} (c_1 w_{gj} + c_2 y_{gj}) \qquad (12.18)$$

s.t.

$$L_{hgj} = \pi_h \hat{p}_h(x^{gj}) - \sum_{i\in G\setminus h} \lambda_{ih}\hat{p}_i(x^{gj}), \ h,g \in \mathbf{G}, \ j \in \mathbf{N}_g \ (12.19)$$

$$L_{ggj} - L_{hgj} + w_{gj} \geq 0, \qquad h,g \in \mathbf{G}, \ h \neq g, \ j \in \mathbf{N}_g \qquad (12.20)$$

$$L_{ggj} \qquad + w_{gj} \geq 0, \qquad g \in \mathbf{G}, \ j \in \mathbf{N}_g, \qquad (12.21)$$

$$-L_{hgj} + y_{gj} \geq 0, \qquad h,g \in \mathbf{G}, \ j \in \mathbf{N}_g \qquad (12.22)$$

$$-\infty < L_{hgj} < \infty, \ w_{gj}, \ y_{gj}, \ \lambda_{ih} \geq 0.$$

Constraint (12.19) defines the variable $L_{hgj}$ as the value of the function $L_h$ evaluated at $x^{gj}$. As the optimization solver searches through the set of feasible solutions, the $\lambda_{ih}$ variables will vary, causing the $L_{hgj}$ variables to assume different values. Constraints (12.20), (12.21), and (12.22) link the objective-function variables with the $L_{hgj}$ variables in such a way that correct classification of training entities, and allocation of training entities into the reserved-judgment region, are captured by the objective-function variables. In particular, if the optimization solver drives $w_{gj}$ to zero for some $g, j$ pair, then constraints (12.20) and (12.21) imply that $L_{ggj} = \max\{0, L_{hgj} : h \in \mathbf{G}\}$. Hence, the $j$th entity from group $g$ is correctly classified. If, on the other hand, the optimal solution yields $y_{gj} = 0$ for some $g, j$ pair, then constraint (12.22) implies that $\max\{0, L_{hgj} : h \in \mathbf{G}\} = 0$. Thus, the $j^{\text{th}}$ entity from group $g$ is placed in the reserved-judgment region. (Of course, it is possible for both $w_{gj}$ and $y_{gj}$ to be zero. One should decide prior to solving the linear program how to interpret the classification in such cases.) If both $w_{gj}$ and $y_{gj}$ are positive, the $j^{\text{th}}$ entity from group $g$ is misclassified.

The optimal solution yields a set of $\lambda_{ih}$'s that best allocates the training entities (i.e., "best" in terms of minimizing the penalty objective function). The optimal $\lambda_{ih}$'s can then be used to define the functions $L_h$, $h \in G$, which in turn can be used to classify a new entity with feature vector $x \in \mathbb{R}^k$ by simply computing the index at which $\max\{L_h(x) : h \in \{0, 1, \ldots, G\}\}$ is achieved.

Note that Model DALP places no *a priori* bound on the number of misclassified training entities. However, because the objective is to minimize a

**Table 12.1.** Model size.

| Model Type | | Constraints | Total Variables | 0/1 Variables |
|---|---|---|---|---|
| 1 | nonlinear MIP | $2GN + N + G(G-1)$ | $2GN + N + G(G-1)$ | $GN$ |
| 2 | linear MIP | $5GN + 2N + G(G-1)$ | $4GN + N + G(G-1)$ | $2GN$ |
| 3 | linear MIP | $3GN + G(G-1)$ | $2GN + N + G(G-1)$ | $GN$ |
| 1T | nonlinear MIP | $2GN + N + 1$ | $2GN + N + G(G-1)$ | $GN$ |
| 2T | linear MIP | $5GN + 2N + 1$ | $4GN + N + G(G-1)$ | $2GN$ |
| 3T | linear MIP | $3GN + 1$ | $2GN + N + G(G-1)$ | $GN$ |
| DALP | linear program | $3GN$ | $NG + N + G(G-1)$ | $0$ |

weighted combination of the variables $w_{gj}$ and $y_{gj}$, the optimizer will attempt to drive these variables to zero. Thus, the optimizer is, in essence, attempting either to correctly classify training entities ($w_{gj} = 0$), or to place them in the reserved-judgment region ($y_{gj} = 0$). By varying the weights $c_1$ and $c_2$, one has a means of controlling the optimizer's emphasis for correctly classifying training entities versus placing them in the reserved-judgment region. If $c_2/c_1 < 1$, the optimizer will tend to place a greater emphasis on driving the $w_{gj}$ variables to zero than driving the $y_{gj}$ variables to zero (conversely, if $c_2/c_1 > 1$). Hence, when $c_2/c_1 < 1$, one should expect to get relatively more entities correctly classified, fewer placed in the reserved-judgment region, and more misclassified, than when $c_2/c_1 > 1$. An extreme case is when $c_2 = 0$. In this case, there is no emphasis on driving $y_{gj}$ to zero (the reserved-judgment region is thus ignored), and the full emphasis of the optimizer is to drive $w_{gj}$ to zero.

Table 12.1 summarizes the number of constraints, the total number of variables, and the number of 0/1 variables in each of the discrete SVMs, and in the heuristic LP model (DALP). Clearly, even for moderately sized discriminant analysis problems, the MIP instances are relatively large. Also, note that Model 2 is larger than Model 3, both in terms of the number of constraints and the number of variables. However, it is important to keep in mind that the difficulty of solving an MIP problem cannot, in general, be predicted solely by its size; problem structure has a direct and substantial bearing on the effort required to find optimal solutions. The LP relaxation of these MIP models pose computational challenges as commercial LP solvers return (optimal) LP solutions that are infeasible, due to the equality constraints and the use of big $M$ and small $\varepsilon$ in the formulation.

It is interesting to note that the set of feasible solutions for Model 2 is "tighter" than that for Model 3. In particular, if $F_i$ denotes the set of feasible solutions of Model $i$, then

$$F_1 = \{(L, \lambda, u, y) : \text{ there exists } \tilde{y}, v \text{ such that } (L, \lambda, u, y, \tilde{y}, v) \in F_2\} \subsetneq F_3.$$

The novelties of the classification models developed herein include: (1) they are suitable for discriminant analysis given any number of groups, (2) they accept heterogeneous types of attributes as input, (3) they use a parametric approach to reduce high-dimensional attribute spaces, and (4) they allow constraints on the number of misclassifications, and utilize a reserved judgment to facilitate the reduction of misclassifications. The latter point opens the possibility of performing multistage analysis.

Clearly, the advantage of an LP model over an MIP model is that the associated problem instances are computationally much easier to solve. However, the most important criterion in judging a method for obtaining discriminant rules is how the rules perform in correctly classifying new unseen entities. Once the rule is developed, applying it to a new entity to determine its group is trivial. Extensive computational experiments have been performed to gauge the qualities of solutions of different models [40, 63, 59, 60, 18, 17].

## Validation of model and computational effort

We performed ten-fold cross validation, and designed simulation and comparison studies on our models. Results reported in [40, 63] demonstrate that our approach works well when applied to both simulated data and data sets from the machine learning database repository [91]. In particular, our methods compare favorably, and at times superior, to other mathematical programming methods, including the general single function classification model (GSFC) by Gehrlein [41], and the LP model by Gochet et al. [46], as well as Fisher's LDF, artificial neural networks, quadratic discriminant analysis, tree classification, and other support vector machines, on real biological and medical data.

### 12.3.2 Classification results on real-world biological and medical applications

The main objective in discriminant analysis is to derive rules that can be used to classify entities into groups. Computationally, the challenge lies in the effort expended to develop such a rule. Once the rule is developed, applying it to a new entity to determine its group is trivial. Feasible solutions obtained from our classification models correspond with predictive rules. Empirical results [40, 63] indicate that the resulting classification model instances are computationally very challenging and even intractable by competitive commercial MIP solvers. However, the resulting predictive rules prove to be very promising, offering correct classification rates on new unknown data ranging from 80% to 100% on various types of biological/medical problems. Our results indicate that the general-purpose classification framework that we have designed has the potential to be a very powerful predictive method for clinical settings.

The choice of mixed integer programming (MIP) as the underlying modeling and optimization technology for our support vector machine classification model is guided by the desire to simultaneously incorporate a variety of important and desirable properties of predictive models within a general framework. MIP itself allows for incorporation of continuous and discrete variables, and linear and nonlinear constraints, providing a flexible and powerful modeling environment.

Our mathematical modeling and computational algorithm design shows great promise as the resulting predictive rules are able to produce higher rates of correct classification on new biological data (with unknown group status) compared with existing classification methods. This is partly due to the transformation of raw data via the set of constraints in (12.7). Whereas most mathematical programming approaches directly determine the hyperplanes of separation using raw data, our approach transforms the raw data via a probabilistic model, before the determination of the supporting hyperplanes. Further, the separation is driven by maximizing the sum of binary variables (representing correct classification or not of entities), instead of maximizing the margins between groups or minimizing a sum of errors (representing distances of entities from hyperplanes) as in other support vector machines. The combination of these two strategies offers better classification capability. Noise in the transformed data is not as profound as in raw data. And the magnitudes of the errors do not skew the determination of the separating hyperplanes, as all entities have *equal* importance when correct classification is being counted.

To highlight the broad applicability of our approach, below, we briefly summarize the application of our predictive models and solution algorithms to ten different biological problems. Each of the projects was carried out in close partnership with experimental biologists and/or clinicians. Applications to finance and other industry applications are described elsewhere [40, 63, 18].

## Determining the type of erythemato-squamous disease [60]

The differential diagnosis of erythemato-squamous diseases is an important problem in dermatology. They all share the clinical features of erythema and scaling, with very little differences. The 6 groups are psoriasis, seboreic dermatitis, lichen planus, pityriasis rosea, cronic dermatitis, and pityriasis rubra pilaris. Usually a biopsy is necessary for the diagnosis but unfortunately these diseases share many histopathologic features as well. Another difficulty for the differential diagnosis is that a disease may show the features of another disease at the beginning stage and may have the characteristic features at the following stages [91].

The 6 groups consist of 366 subjects (112, 61, 72, 49, 52, 20, respectively) with 34 clinical attributes. Patients were first evaluated clinically with 12 features. Afterwards, skin samples were taken for the evaluation of 22 histopathologic features. The values of the histopathologic features are determined by an analysis of the samples under a microscope. The 34 attributes include (1) clinical attributes: erythema, scaling, definite borders, itching, koebner phenomenon, polygonal papules, follicular papules, oral mucosal involvement, knee and elbow involvement, scalp involvement, family history, age; and (2) histopathologic attributes: melanin incontinence, eosinophils in the infiltrate, PNL infiltrate, fibrosis of the papillary dermis, exocytosis, acanthosis, hyperkeratosis, parakeratosis, clubbing of the rete ridges, elongation of the rete ridges, thinning of the suprapapillary epidermis, spongiform pustule, munro

microabcess, focal hypergranulosis, disappearance of the granular layer, vac-
uolization and damage of basal layer, spongiosis, sawtooth appearance of retes,
follicular horn plug, perifollicular parakeratosis, inflammatory monoluclear
infiltrate, band-like infiltrate.

Our multigroup classification model selected 27 discriminatory attributes
and successfully classified the patients into 6 groups, each with an unbiased
correct classification of greater than 93% (with 100% correct rate for groups
1, 3, 5, 6) with an average overall accuracy of 98%. Using 250 subjects to
develop the rule, and testing the remaining 116 patients, we obtain a prediction accuracy of 91%.

### Predicting presence/absence of heart disease [60]

The four databases concerning heart disease diagnosis were collected by
Dr. Janosi of Hungarian Institute of Cardiology, Budapest; Dr. Steinbrunn
of University Hospital, Zurich; Dr. Pfisterer of University Hospital, Basel,
Switzerland; and Dr. Detrano of V.A. Medical Center, Long Beach and
Cleveland Clinic Foundation. Each database contains the same 76 attributes.
The "goal" field refers to the presence of heart disease in the patient. The
classification attempts to distinguish presence (values 1, 2, 3, 4, involving a
total of 509 subjects) from absence (value 0, involving 411 subjects) [91]. The
attributes include demographics, physio-cardiovascular conditions, traditional
risk factors, family history, personal lifestyle, and cardiovascular exercise measurements. This data set has posed some challenges to past analysis via various classification approaches, resulting in less than 80% correct classification.
Applying our classification model without reserved judgment, we obtain 79%
and 85% correct classification for each group respectively. To gauge the useful-
ness of multistage analysis, we apply 2-stage classification. In the first stage,
14 attributes were selected as discriminatory. 135 Group absence subjects were
placed into the reserved judgment region, with 85% of the remaining classified
as Group absence correctly; and 286 Group presence subjects were placed into
the reserved judgment region, and 91% of the remaining classified correctly
into the Group presence. In the second stage, 11 attributes were selected with
100 and 229 classified into Group absence and presence, respectively. Combining the two stages, we obtained a correct classification of 82% and 85%
respectively for diagnosis of absence or presence of heart disease. Figure 12.1
illustrates the 2-stage classification.

### Predicting aberrant CpG island methylation in human cancer [28, 29]

Epigenetic silencing associated with aberrant methylation of promoter region
CpG islands is one mechanism leading to loss of tumor suppressor function
in human cancer. Profiling of CpG island methylation indicates that some
genes are more frequently methylated than others and that each tumor type

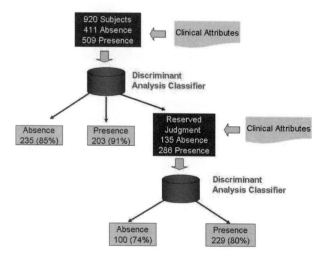

**Fig. 12.1.** A tree diagram for 2-stage classification and prediction of heart disease.

is associated with a unique set of methylated genes. However, little is known about why certain genes succumb to this aberrant event. To address this question, we used restriction landmark genome scanning (RLGS) to analyze the susceptibility of 1749 unselected CpG islands to *de novo* methylation driven by overexpression of DNMT1. We found that, whereas the overall incidence of CpG island methylation was increased in cells overexpressing DNMT1, not all loci were equally affected. The majority of CpG islands (69.9%) were resistant to *de novo* methylation, regardless of DNMT1 overexpression. In contrast, we identified a subset of methylation-prone CpG islands (3.8%) that were consistently hypermethylated in multiple DNMT1 overexpressing clones. Methylation-prone and methylation-resistant CpG islands were not significantly different with respect to size, C+G content, CpG frequency, chromosomal location, or gene- or promoter-association. To discriminate methylation-prone from methylation-resistant CpG islands, we developed a novel DNA pattern recognition model and algorithm [61] and coupled our predictive model described herein with the patterns found. We were able to derive a classification function based on the frequency of seven novel sequence patterns that was capable of discriminating methylation-prone from methylation-resistant CpG islands with 90% correctness upon cross-validation, and 85% accuracy when tested against blind CpG islands unknown to us on the methylation status. The data indicate that CpG islands differ in their intrinsic susceptibility to *de novo* methylation, and suggest that the propensity for a CpG island to become aberrantly methylated can be predicted based on its sequence context.

The significance of this research is two-fold. First, the identification of sequence patterns/attributes that distinguish methylation-prone CpG islands will lead to a better understanding of the basic mechanisms underlying aberrant CpG island methylation. Because genes that are silenced by methylation are otherwise structurally sound, the potential for reactivating these genes by blocking or reversing the methylation process represents an exciting new molecular target for chemotherapeutic intervention. A better understanding of the factors that contribute to aberrant methylation, including the identification of sequence elements that may act to target aberrant methylation, will be an important step in achieving this long-term goal. Secondly, the classification of the more than 29,000 known (but as yet unclassified) CpG islands in human chromosomes will provide an important resource for the identification of novel gene targets for further study as potential molecular markers that could impact on both cancer prevention and treatment. Extensive RLGS fingerprint information (and thus potential training sets of methylated CpG islands) already exists for a number of human tumor types, including breast, brain, lung, leukemias, hepatocellular carcinomas, and PNET [23, 24, 35, 102]. Thus, the methods and tools developed are directly applicable to CpG island methylation data derived from human tumors. Moreover, new microarray-based techniques capable of "profiling" more than 7,000 CpG islands have been developed and applied to human breast cancers [15, 117, 118]. We are uniquely poised to take advantage of the tumor CpG island methylation profile information that will likely be generated using these techniques over the next several years. Thus, our general-predictive modeling framework has the potential to lead to improved diagnosis and prognosis and treatment planning for cancer patients.

### Discriminant analysis of cell motility and morphology data in human lung carcinoma [16]

This study focuses on the differential effects of extracellular matrix proteins on the motility and morphology of human lung epidermoid carcinoma cells. The behavior of carcinoma cells is contrasted with that of normal L-132 cells, resulting in a method for the prediction of metastatic potential. Data collected from time-lapsed videomicroscopy were used to simultaneously produce quantitative measures of motility and morphology. The data were subsequently analyzed using our discriminant analysis model and algorithm to discover relationships between motility, morphology, and substratum. Our discriminant analysis tools enabled the consideration of many more cell attributes than is customary in cell motility studies. The observations correlate with behaviors seen *in vivo* and suggest specific roles for the extracellular matrix proteins and their integrin receptors in metastasis. Cell translocation *in vitro* has been associated with malignancy, as has an elongated phenotype [120] and a rounded phenotype [97]. Our study suggests that extracellular matrix proteins contribute in different ways to the malignancy of cancer cells and that multiple malignant phenotypes exist.

## Ultrasonic-assisted cell disruption for drug delivery [57]

Although biological effects of ultrasound must be avoided for safe diagnostic applications, ultrasound's ability to disrupt cell membranes has attracted interest as a method to facilitate drug and gene delivery. This preliminary study seeks to develop rules for predicting the degree of cell membrane disruption based on specified ultrasound parameters and measured acoustic signals. Too much ultrasound destroys cells, whereas cell membranes will not open up for absorption of macromolecules when too little ultrasound is applied. The key is to increase cell permeability to allow absorption of macromolecules, and to apply ultrasound transiently to disrupt viable cells so as to enable exogenous material to enter without cell damage. Thus our task is to uncover a "predictive rule" of ultrasound-mediated disruption of red blood cells using acoustic spectrums and measurements of cell permeability recorded in experiments.

Our predictive model and solver for generating prediction rules are applied to data obtained from a sequence of experiments on bovine red blood cells. For each experiment, the attributes consist of 4 ultrasound parameters, acoustic measurements at 400 frequencies, and a measure of cell membrane disruption. To avoid over-training, various feature combinations of the 404 predictor variables are selected when developing the classification rule. The results indicate that the variable combination consisting of ultrasound exposure time and acoustic signals measured at the driving frequency and its higher harmonics yields the best rule, and our method compares favorably with classification tree and other ad hoc approaches, with correct classification rate of 80% upon cross-validation and 85% when classifying new unknown entities. Our methods used for deriving the prediction rules are broadly applicable and could be used to develop prediction rules in other scenarios involving different cell types or tissues. These rules and the methods used to derive them could be used for real-time feedback about ultrasound's biological effects. For example, it could assist clinicians during a drug delivery process or could be imported into an implantable device inside the body for automatic drug delivery and monitoring.

## Identification of tumor shape and volume in treatment of sarcoma [56]

This project involves the determination of tumor shape for adjuvant brachytherapy treatment of sarcoma, based on catheter images taken after surgery. In this application, the entities are overlapping consecutive triplets of catheter markings, each of which is used for determining the shape of the tumor contour. The triplets are to be classified into one of two groups: Group 1 = [triplets for which the middle catheter marking should be bypassed], and Group 2 = [triplets for which the middle marking should not be bypassed]. To develop and validate a classification rule, we used clinical data collected from fifteen soft tissue sarcoma (STS) patients. Cumulatively, this comprised 620 triplets of catheter markings. By careful (and tedious) clinical analysis of the geometry of these triplets, 65 were determined to belong to Group 1, the "bypass" group, and 555 were determined to belong to Group 2, the "do-not-bypass" group.

A set of measurements associated with each triplet is then determined. The choice of what attributes to measure to best distinguish triplets as belonging to Group 1 or Group 2 is nontrivial. The attributes involved distance between each pair of markings, angles, curvature formed by the three triplet markings. Based on the selected attributes, our predictive model was used to develop a classification rule. The resulting rule provides 98% correct classification on cross-validation and was capable of correctly determining/predicting 95% of the shape of the tumor on new patients' data. We remark that the current clinical procedure requires manual outline based on markers in films of the tumor volume. This study was the first to use automatic construction of tumor shape for sarcoma adjuvant brachytherapy [56, 62].

## Discriminant analysis of biomarkers for prediction of early atherosclerosis [65]

Oxidative stress is an important etiologic factor in the pathogenesis of vascular disease. Oxidative stress results from an imbalance between injurious oxidant and protective antioxidant events in which the former predominate [103, 88]. This results in the modification of proteins and DNA, alteration in gene expression, promotion of inflammation, and deterioration in endothelial function in the vessel wall, all processes that ultimately trigger or exacerbate the atherosclerotic process [22, 111]. It was hypothesized that novel biomarkers of oxidative stress would predict early atherosclerosis in a relatively healthy non-smoking population who are free from cardiovascular disease. One hundred and twenty seven healthy non-smokers, without known clinical atherosclerosis had carotid intima media thickness (IMT) measured using ultrasound. Plasma oxidative stress was estimated by measuring plasma lipid hydroperoxides using the determination of reactive oxygen metabolites (d-ROMs) test. Clinical measurements include traditional risk factors including age, sex, low-density lipoprotein (LDL), high-density lipoprotein (HDL), triglycerides, cholesterol, body mass index (BMI), hypertension, diabetes mellitus, smoking history, family history of CAD, Framingham risk score, and Hs-CRP.

For this prediction, the patients are first clustered into two groups: (Group 1: IMT $\geq 0.68$, Group 2: IMT $< 0.68$). Based on this separator, 30 patients belong to Group 1 and 97 belong to Group 2. Through each iteration, the classification method trains and learns from the input training set and returns the most discriminatory patterns among the 14 clinical measurements; ultimately resulting in the development of a prediction rule based on observed values of these discriminatory patterns among the patient data. Using all

127 patients as a training set, the predictive model identified age, sex, BMI, HDLc, Fhx CAD < 60, hs-CRP, and d-ROM as discriminatory attributes that together provide unbiased correct classification of 90% and 93%, respectively, for Group 1 (IMT ≥ 0.68) and Group 2 patients (IMT < 0.68). To further test the power of the classification method for correctly predicting the IMT status on new/unseen patients, we randomly selected a smaller patient training set of size 90. The predictive rule from this training set yields 80% and 89% correct rates for predicting the remaining 37 patients into Group 1 and Group 2, respectively. The importance of d-ROM as a discriminatory predictor for IMT status was confirmed during the machine learning process, this biomarker was selected in every iteration as the "machine" learned and trained to develop a predictive rule to correctly classify patients in the training set. We also performed predictive analysis using Framingham Risk Score and d-ROM; in this case, the unbiased correct classification rates (for the 127 individuals) for Groups 1 and 2 are 77% and 84%, respectively. This is the first study to illustrate that this measure of oxidative stress can be effectively used along with traditional risk factors to generate a predictive rule that can potentially serve as an inexpensive clinical diagnostic tool for prediction of early atherosclerosis.

## Fingerprinting native and angiogenic microvascular networks through pattern recognition and discriminant analysis of functional perfusion data [64]

The cardiovascular system provides oxygen and nutrients to the entire body. Pathologic conditions that impair normal microvascular perfusion can result in tissue ischemia, with potentially serious clinical effects. Conversely, development of new vascular structures fuels the progression of cancer, macular degeneration, and atherosclerosis. Fluorescence-microangiography offers superb imaging of the functional perfusion of new and existent microvasculature, but quantitative analysis of the complex capillary patterns is challenging. We developed an automated pattern-recognition algorithm to systematically analyze the microvascular networks, and then apply our classification model herein to generate a predictive rule. The pattern-recognition algorithm identifies the complex vascular branching patterns, and the predictive rule demonstrates 100% and respectively 91% correct classification on perturbed (diseased) and normal tissue perfusion. We confirmed that transplantation of normal bone marrow to mice in which genetic deficiency resulted in impaired angiogenesis eliminated predicted differences and restored normal-tissue perfusion patterns (with 100% correctness). The pattern recognition and classification method offers an elegant solution for the automated fingerprinting of microvascular networks that could contribute to better understanding of angiogenic mechanisms and be utilized to diagnose and monitor microvascular deficiencies. Such information would be valuable for early detection and monitoring of functional abnormalities before they produce obvious and lasting effects, which may include improper perfusion of tissue, or support of tumor development.

The algorithm can be used to discriminate between the angiogenic response in a native healthy specimen compared with groups with impairment due to age or chemical or other genetic deficiency. Similarly, it can be applied to analyze angiogenic responses as a result of various treatments. This will serve two important goals. First, the identification of discriminatory patterns/attributes that distinguish angiogenesis status will lead to a better understanding of the basic mechanisms underlying this process. Because therapeutic control of angiogenesis could influence physiological and pathologic processes such as wound and tissue repairing, cancer progression and metastasis, or macular degeneration, the ability to understand it under different conditions will offer new insight in developing novel therapeutic interventions, monitoring, and treatment, especially in aging and heart disease. Thus, our study and the results form the foundation of a valuable diagnostic tool for changes in the functionality of the microvasculature and for discovery of drugs that alter the angiogenic response. The methods can be applied to tumor diagnosis, monitoring, and prognosis. In particular, it will be possible to derive microangiographic fingerprints to acquire specific microvascular patterns associated with early stages of tumor development. Such "angioprinting" could become an extremely helpful early diagnostic modality, especially for easily accessible tumors such as skin cancer.

**Prediction of protein localization sites**

The protein localization database consists of 8 groups with a total of 336 instances (143, 77, 52, 35, 20, 5, 2, 2, respectively) with 7 attributes [91]. The 8 groups are 8 localization sites of protein, including cp (cytoplasm), im (inner membrane without signal sequence), pp (perisplasm), imU (inner membrane, uncleavable signal sequence), om (outer membrane), omL (outer membrane lipoprotein), imL (inner membrane lipoprotein), imS (inner membrane, cleavable signal sequence). However, the last 4 groups are taken out from our classification experiment as the population sizes are too small to ensure significance.

The 7 attributes include mcg (McGeoch's method for signal sequence recognition), gvh (von Heijne's method for signal sequence recognition), lip (von Heijne's Signal Peptidase II consensus sequence score), chg (presence of charge on N-terminus of predicted lipoproteins), aac (score of discriminant analysis of the amino acid content of outer membrane and periplasmic proteins), alm1 (score of the ALOM membrane spanning region prediction program), and alm2 (score of ALOM program after excluding putative cleavable signal regions from the sequence).

In the classification, we use 4 groups, 307 instances, with 7 attributes. Our classification model selected the discriminatory patterns mcg, gvh, alm1, and alm2 to form the predictive rule with unbiased correct classification rates of 89%, compared with the results of 81% by other classification models [48].

## Pattern recognition in satellite images for determining types of soil

The Satellite database consists of the multispectral values of pixels in $3 \times 3$ neighborhoods in a satellite image, and the classification associated with the central pixel in each neighborhood. The aim is to predict this classification, given the multispectral values. In the sample database, the class of a pixel is coded as a number. There are 6 groups with 4435 samples in the training data set and 2,000 samples in testing data set; and each sample entity has 36 attributes describing the spectral bands of the image [91].

The original Landsat Multi-Spectral Scanner image data for this database was generated from data purchased from NASA by the Australian Centre for Remote Sensing. The Landsat satellite data is one of the many sources of information available for a scene. The interpretation of a scene by integrating spatial data of diverse types and resolutions including multispectral and radar data, maps indicating topography, land use, and so forth. is expected to assume significant importance with the onset of an era characterized by integrative approaches to remote sensing (for example, NASA's Earth Observing System commencing this decade).

One frame of Landsat MSS imagery consists of four digital images of the same scene in different spectral bands. Two of these are in the visible region (corresponding approximately to green and red regions of the visible spectrum) and two are in the (near) infra-red. Each pixel is an 8-bit binary word, with 0 corresponding to black and 255 to white. The spatial resolution of a pixel is about 80 m $\times$ 80 m. Each image contains $2340 \times 3380$ such pixels.

The database is a (tiny) sub-area of a scene, consisting of $82 \times 100$ pixels. Each line of data corresponds with a $3 \times 3$ square neighborhood of pixels completely contained within the $82 \times 100$ sub-area. Each line contains the pixel values in the four spectral bands (converted to ASCII) of each of the 9 pixels in the $3 \times 3$ neighborhood and a number indicating the classification label of the central pixel. The number is a code for the following 6 groups: red soil, cotton crop, gray soil, damp gray soil, soil with vegetation stubble, very damp gray soil. Running our classification model, 17 discriminatory attributes were selected to form the classification rule, producing an unbiased prediction with 85% accuracy.

### 12.3.3 Further advances

Brooks and Lee 2007 [18, 19] devised other variations of the basic DAMIP model. They also showed that DAMIP is strongly universally consistent (in some sense) with very good rates of convergence from Vapnik and Chervonenkis theory. A polynomial-time algorithm for discriminating between two populations with the DAMIP model was developed, and DAMIP was shown to be $\mathcal{NP}$-complete for a general number of groups. The proof demonstrating $\mathcal{NP}$-completeness employs results used in generating edges of the conflict graph [11, 55, 12, 4]. Exploiting the necessary and sufficient conditions that identify edges in the conflict graph is the central contribution to the improvement in solution performance over industry-standard software. The conflict graph is the basis for various valid inequalities, a branching scheme, and for conditions under which integer variables are fixed for all solutions. Additional solution methods are identified that include a heuristic for finding solutions at nodes in the branch-and-bound tree, upper bounds for model parameters, and necessary conditions for edges in the conflict hypergraph [26, 58]. Further, we have concluded that DAMIP is a computationally feasible, consistent, stable, robust, and accurate classifier.

## 12.4 Progress and Challenges

In Tables 12.2–12.4 we summarize the mathematical programming techniques used in classification problems as reviewed in this chapter.

As noted by current research effort, multigroup classification remains $\mathcal{NP}$-completeness and much work is needed to design effective models as well as to derive novel and efficient computational algorithms to solve these multigroup instances.

## 12.5 Other Methods

Whereas most classification methods can be described in terms of discriminant functions, some methods are not trained in the paradigm of determining coefficients or parameters for functions of a predefined form. These methods include *classification and regression trees (CART)*, *nearest-neighbor* methods, and *neural networks*.

Classification and regression trees [14] are nonparametric approaches to prediction. Classification trees seek to develop classification rules based on successive binary partitions of observations based on attribute values. Regression trees also employ rules consisting of binary partitions but are used to predict continuous responses.

The rules generated by classification trees are easily viewable by plotting them in a tree-like structure from which the name arises. A test entity may be classified using rules in a tree plot by first comparing the entity's data with the root node of the tree. If the root node condition is satisfied by the data for a particular entity, the left branch is followed to another node; otherwise, the right branch is followed to another node. The data from the observation is compared with conditions at subsequent nodes until a leaf node is reached.

Nearest-neighbor methods begin by establishing a set of labeled prototype observations. The nearest-neighbor classification rule assigns test entities to

**Table 12.2.** Progress in mathematical programming–based classification models: LP methods.

|  | **Authors, Years, and Citations** |
|---|---|
| *Two-group classification:* | |
| Separate data by hyperplanes | Mangasarian 1965 [74], 1968 [75] |
| Minimizing the sum of deviations (MSD), minimizing the maximum deviation (MMD), and minimizing the sum of interior distances (MSID) | Hand 1981 [47], Freed and Glover 1981 [31, 32], Bajgier and Hill 1982 [5], Freed and Glover 1986 [33], Rubin 1990 [99] |
| Hybrid model | Glover et al. 1988 [45], Rubin 1990 [99] |
| Review | Joachimsthaler and Stam 1990 [50], Erenguc and Koehler 1990 [27], Stam 1997 [107] |
| Software | Stam and Ungar 1995 [110] |
| Issues about normalization | Markowski and Markowski 1985 [87], Freed and Glover 1986 [34], Koehler 1989 [51, 52] 1994 [53], Glover 1990 [44], Rubin 1991 [100], Xiao 1993 [114] 1994 [115], Xiao and Feng 1997 [116] |
| Robust linear programming (RLP) | Bennett and Mangasarian 1992 [9], Mangasarian et al. 1995 [86] |
| Inclusion of second-order terms | Duarte Silva and Stam 1994 [104], Wanarat and Pavur 1996 [113] |
| Effect of the position of outliers | Pavur 2002 [94] |
| Binary attributes | Asparoukhov and Stam 1997 [3] |
| *Multigroup classification:* | |
| Single function classification | Freed and Glover 1981 [32] |
| Multiple function classification | Bennett and Mangasarian 1994 [10], Gochet et al. 1997 [46] |
| Multigroup classification with reserved-judgment region and misclassification constraints | Lee et al. 2003 [63, 39, 40, 60] |

groups according to the group membership of the nearest prototype. Different measures of distance may be used. The $k$-nearest-neighbor rule assigns entities to groups according to the group membership of the $k$ nearest prototypes.

Neural networks are classification models that can also be interpreted in terms of discriminant functions, though they are used in a way that does not require finding an analytic form for the functions [25]. Neural networks are trained by considering one observation at a time, modifying the classification procedure slightly with each iteration.

**Table 12.3.** Progress in mathematical programming–based classification models: MIP methods.

| | Authors, Years, and Citations |
|---|---|
| *Two-group classification:* | |
| Minimizing the number of misclassifications | Bajgier and Hill 1982 [5], Stam and Joachimsthaler 1990 [109], Koehler and Erenguc 1990 [54], Banks and Abad 1991 [6] 1994 [7], Abad and Banks 1993 [1], Duarte Silva and Stam 1997 [105], Rubin 1997 [101], Yanev and Balev 1999 [119] |
| Review | Joachimsthaler and Stam 1990 [50], Erenguc and Koehler 1990 [27], Stam 1997 [107] |
| Software | Stam and Ungar 1995 [110] |
| Secondary goals | Pavur et al. 1997 [96] |
| Binary attributes | Asparoukhov and Stam 1997 [3] |
| Normalization and attribute selection | Glen 1999 [42] |
| Dichotomous categorical variable formation | Glen 2004 [43] |
| *Multigroup classification:* | |
| Multigroup classification | Gehrlein 1986 [41], Pavur 1997 [93] |
| Three-group classification | Loucopoulos and Pavur 1997 [71, 72], Pavur and Loucopoulos 2001 [95] |
| Classification with reserved-judgment region using MIP | Gallagher et al. 1996, 1997 [39, 40], Brooks and Lee 2006 [18], Lee 2006 [59, 60] |

## 12.6 Summary and Conclusion

In this chapter, we presented an overview of mathematical programming-based classification models, and analyzed their development and advances in recent years. Many mathematical programming methods are geared toward two-group analysis only, and performance is often compared to Fisher's linear discriminant, or Smith's quadratic discriminant. It has been noted that these methods can be used for multiple group analysis by finding $G(G-1)/2$ discriminants for each pair of groups ("one-against-one") or by finding $G$ discriminants for each group versus the remaining data ("one-against-all"), but these approaches can lead to ambiguous classification rules [25].

Mathematical programming methods developed for multiple group analysis are described [10, 32, 39, 40, 41, 46, 58, 59, 63, 93]. Multiple group formulations for support vector machines have been proposed and tested [40, 36, 49, 66, 59, 60, 18], but are still considered computationally intensive [49]. The "one-against-one" and "one-against-all" methods with support vector machines have been successfully applied [49, 90].

We also discussed a class of multigroup general-purpose predictive models that we have developed based on the technology of large-scale optimization

**Table 12.4.** Progress in mathematical programming–based classification models: nonlinear programming methods.

| | Authors, Years, and Citations |
|---|---|
| *Two-group classification:* | |
| $L_p$-norm criterion | Stam and Joachimsthaler 1989 [108] |
| Review | Joachimsthaler and Stam 1990 [50], Erenguc and Koehler 1990 [27], Stam 1997 [107] |
| Piecewise-linear nonconvex discriminant function | Mangasarian et al. 1990 [85] |
| Minimizing the number of misclassifications | Mangasarian 1994 [76] 1996 [77], Chen and Mangasarian 1996 [21] |
| Minimizing the sum of arbitrary-norm distances | Mangasarian 1999 [78] |
| *Support vector machine:* | |
| Introduction and tutorial | Vapnik 1995 [57], Burges 1998 [20] |
| Generalized SVM | Mangasarian 2000 [79], Mangasarian and Musicant 2001 [83] |
| Methods for huge-size problems | Mangasarian and Musicant 1999 [82] 2001 [84], Bradley and Mangasarian 2000 [13], Lee and Mangasarian 2001 [68, 67], Fung and Mangasarian 2001 [36] 2002 [37] 2005 [38], Mangasarian 2003 [80] 2005 [81] |
| Multigroup SVM | Gallagher et al 1996, 1997 [39, 40], Hsu and Lin 2002 [49], Lee et al [63], Lee et al. 2004 [66], Fung and Mangasarian 2005 [38], Brooks and Lee 2006 [18], Lee 2006 [59, 60] |

and support-vector machines [39, 40, 63, 59, 60, 18, 17]. Our models seek to maximize the correct classification rate while constraining the number of misclassifications in each group. The models incorporate the following features: (1) the ability to classify any number of distinct groups; (2) allow incorporation of heterogeneous types of attributes as input; (3) a high-dimensional data transformation that eliminates noise and errors in biological data; (4) constraining the misclassification in each group and a reserved-judgment region that provides a safeguard against over-training (which tends to lead to high misclassification rates from the resulting predictive rule); and (5) successive multistage classification capability to handle data points placed in the reserved judgment region. The performance and predictive power of the classification models is validated through a broad class of biological and medical applications.

Classification models are critical to medical advances as they can be used in genomic, cell, molecular, and system level analyses to assist in early prediction, diagnosis, and detection of disease, as well as for intervention and monitoring. As shown in the CpG island study for human cancer, such prediction and diagnosis opens up novel therapeutic sites for early intervention. The ultrasound application illustrates its application to a novel drug delivery mechanism, assisting clinicians during a drug delivery process, or in devising implantable devices into the body for automated drug delivery and monitoring. The lung cancer cell motility offers an understanding of how cancer cells behave under different protein media, thus assisting in the identification of potential gene therapy and target treatment. Prediction of the shape of a cancer tumor bed provides a personalized treatment design, replacing manual estimates by sophisticated computer predictive models. Prediction of early atherosclerosis through inexpensive biomarker measurements and traditional risk factors can serve as a potential clinical diagnostic tool for routine physical and health maintenance, alerting doctors and patients to the need for early intervention to prevent serious vascular disease. Fingerprinting of microvascular networks opens up the possibility for early diagnosis of perturbed systems in the body that may trigger disease (e.g., genetic deficiency, diabetes, aging, obesity, macular degeneracy, tumor formation), identify target sites for treatment, and monitoring prognosis and success of treatment. Determining the type of erythemato-squamous disease and the presence/absence of heart disease helps clinicians to correctly diagnose and effectively treat patients. Thus classification models serve as a basis for predictive medicine where the desire is to diagnose early and provide personalized target intervention. This has the potential to reduce healthcare costs, improve success of treatment, and improve quality-of-life of patients.

## Acknowledgment

This research was partially supported by the National Science Foundation.

## References

[1] P.L. Abad and W.J. Banks. New LP based heuristics for the classification problem. *European Journal of Operational Research*, 67:88–100, 1993.

[2] J.A. Anderson. Constrained discrimination between $k$ populations. *Journal of the Royal Statistical Society. Series B (Methodological)*, 31(1):123–139, 1969.

[3] O.K. Asparoukhov and A. Stam. Mathematical programming formulations for two-group classification with binary variables. *Annals of Operations Research*, 74:89–112, 1997.

[4] A. Atamturk. *Conflict graphs and flow models for mixed-integer linear optimization problems.* PhD thesis, School of Industrial and Systems Engineering, Georgia Institute of Technology, Atlanta, Georgia, 1998.

[5] S.M. Bajgier and A.V. Hill. An experimental comparison of statistical and linear programming approaches to the discriminant problem. *Decision Sciences*, 13:604–618, 1982.

[6] W.J. Banks and P.L. Abad. An efficient optimal solution algorithm for the classification problem. *Decision Sciences*, 22:1008–1023, 1991.

[7] W.J. Banks and P.L. Abad. On the performance of linear programming heuristics applied on a quadratic transformation in the classification problem. *European Journal of Operational Research*, 74:23–28, 1994.

[8] K.P. Bennett. Decision tree construction via linear programming. In M. Evans, editor, *Proceedings of the 4th Midwest Artificial Intelligence and Cognitive Science Society Conference*, pages 97–101, 1992.

[9] K.P. Bennett and O.L. Mangasarian. Robust linear programming discrimination of two linearly inseparable sets. *Optimization Methods and Software*, 1:23–34, 1992.

[10] K.P. Bennett and O.L. Mangasarian. Multicategory discrimination via linear programming. *Optimization Methods and Software*, 3:27–39, 1994.

[11] Robert E. Bixby and Eva K. Lee. Solving a truck dispatching scheduling problem using branch-and-cut. *Operations Research*, Operations Research, 46:355–367, 1998.

[12] R. Borndörfer. *Aspects of set packing, partitioning and covering*. PhD thesis, Technischen Universität Berlin, Berlin, Germany, 1997.

[13] P.S. Bradley and O.L. Mangasarian. Massive data discrimination via linear support vector machines. *Optimization Methods and Software*, 13(1):1–10, 2000.

[14] L. Breiman, J.H. Friedman, R.A. Olshen, and C.J. Stone. *Classification and Regression Trees*. Wadsworth & Brooks/Cole Advanced Books & Software, Pacific Grove, California, 1984.

[15] G.J. Brock, T.H. Huang, C.M. Chen, and K.J. Johnson. A novel technique for the identification of CpG islands exhibiting altered methylation patterns (ICEAMP). *Nucleic Acids Research*, 29:e123, 2001.

[16] J. P. Brooks, A. Wright, C. Zhu, and E.K. Lee. Discriminant analysis of motility and morphology data from human lung carcinoma cells placed on purified extracellular matrix proteins. *Annals of Biomedical Engineering*, Submitted 2007.

[17] J.P. Brooks and E.K. Lee. Mixed integer programming constrained discrimination model for credit screening. *Proceedings of the 2007 Spring Simulation Multiconference, Business and Industry Symposium*, Norfolk, Virginia, March 2007. ACM Digital Library, pages 1–6.

[18] J.P. Brooks and E.K. Lee. Solving a mixed-integer programming formulation of a multi-category constrained discrimination model. *Proceedings of the 2006 INFORMS Workshop on Artificial Intelligence and Data Mining*, Pittsburgh, Pennsylvania, November 2006.

[19] J.P. Brooks and E.K. Lee. Analysis of the consistency of a mixed integer programming-based multi-category constrained discriminant model. Submitted, 2007.

[20] C.J.C. Burges. A tutorial on support vector machines for pattern recognition. *Data Mining and Knowledge Discovery*, 2:121–167, 1998.

[21] C. Chen and O.L. Mangasarian. Hybrid misclassification minimization. *Advances in Computational Mathematics*, 5:127–136, 1996.

[22] M. Chevion, E. Berenshtein, and E.R. Stadtman. Human studies related to protein oxidation: protein carbonyl content as a marker of damage. *Free Radical Research*, 33(Suppl):S99–S108, 2000.

[23] J.F. Costello, M.C. Fruhwald, D.J. Smiraglia, L.J. Rush, G.P. Robertson, X. Gao, F.A. Wright, J.D. Feramisco, P. Peltomaki, J.C. Lang, D.E. Schuller, L. Yu, C.D. Bloomfield, M.A. Caligiuri, A. Yates, R. Nishikawa, H.H. Su, N.J. Petrelli, X. Zhang, M.S. O'Dorisio, W.A. Held, W.K. Cavenee, and C. Plass. Aberrant CpG-island methylation has non-random and tumour-type-specific patterns. *Nature Genetics*, 24:132–138, 2000.

[24] J.F. Costello, C. Plass, and W.K. Cavenee. Aberrant methylation of genes in low-grade astrocytomas. *Brain Tumor Pathology*, 17:49–56, 2000.

[25] R.O. Duda, P.E. Hart, and D.G. Stork. *Pattern Classification*. Wiley, New York, 2001.

[26] T. Easton, K. Hooker, and E.K. Lee. Facets of the independent set plytope. *Mathematical Programming, Series B*, 98:177–199, 2003.

[27] S.S. Erenguc and G.J. Koehler. Survey of mathematical programming models and experimental results for linear discriminant analysis. *Managerial and Decision Economics*, 11:215–225, 1990.

[28] F.A. Feltus, E.K. Lee, J.F. Costello, C. Plass, and P.M. Vertino. Predicting aberrant CpG island methylation. *Proceedings of the National Academy of Sciences*, 100:12253–12258, 2003.

[29] F.A. Feltus, E.K. Lee, J.F. Costello, C. Plass, and P.M. Vertino. DNA signatures associated with CpG island methylation states. *Genomics*, 87:572–579, 2006.

[30] R.A. Fisher. The use of multiple measurements in taxonomic problems. *Annals of Eugenics*, 7:179–188, 1936.

[31] N. Freed and F. Glover. A linear programming approach to the discriminant problem. *Decision Sciences*, 12:68–74, 1981.

[32] N. Freed and F. Glover. Simple but powerful goal programming models for discriminant problems. *European Journal of Operational Research*, 7:44–60, 1981.

[33] N. Freed and F. Glover. Evaluating alternative linear programming models to solve the two-group discriminant problem. *Decision Sciences*, 17:151–162, 1986.

[34] N. Freed and F. Glover. Resolving certain difficulties and improving the classification power of LP discriminant analysis formulations. *Decision Sciences*, 17:589–595, 1986.

[35] M.C. Fruhwald, M.S. O'Dorisio, L.J. Rush, J.L. Reiter, D.J. Smiraglia, G. Wenger, J.F. Costello, P.S. White, R. Krahe, G.M. Brodeur, and C. Plass. Gene amplification in NETs/medulloblastomas: mapping of a novel amplified gene within the MYCN amplicon. *Journal of Medical Genetics*, 37:501–509, 2000.

[36] G.M. Fung and O.L. Mangasarian. Proximal support vector machine classifiers. In *Proceedings KDD-2001*, San Francisco, August 26-29 2001.

[37] G.M. Fung and O.L. Mangasarian. Incremental support vector machine classification. In R. Grossman, H. Mannila, and R. Motwani, editors, *Proceedings of the Second SIAM International Conference on Data Mining*, pages 247–260, Philadelphia, 2002. SIAM.

[38] G.M. Fung and O.L. Mangasarian. Multicategory proximal support vector machine classifiers. *Machine Learning*, 59:77–97, 2005.

[39] R.J. Gallagher, E.K. Lee, and D.A. Patterson. An optimization model for constrained discriminant analysis and numerical experiments with iris, thyroid,

and heart disease datasets. In *Proceedings of the 1996 American Medical Informatics Association*, October 1996.

[40] R.J. Gallagher, E.K. Lee, and D.A. Patterson. Constrained discriminant analysis via 0/1 mixed integer programming. *Annals of Operations Research*, 74:65–88, 1997.

[41] W.V. Gehrlein. General mathematical programming formulations for the statistical classification problem. *Operations Research Letters*, 5(6):299–304, 1986.

[42] J.J. Glen. Integer programming methods for normalisation and variable selection in mathematical programming discriminant analysis models. *Journal of the Operational Research Society*, 50:1043–1053, 1999.

[43] J.J. Glen. Dichotomous categorical variable formation in mathematical programming discriminant analysis models. *Naval Research Logistics*, 51:575–596, 2004.

[44] F. Glover. Improved linear programming models for discriminant analysis. *Decision Sciences*, 21:771–785, 1990.

[45] F. Glover, S. Keene, and B. Duea. A new class of models for the discriminant problem. *Decision Sciences*, 19:269–280, 1988.

[46] W. Gochet, A. Stam, V. Srinivasan, and S. Chen. Multigroup discriminant analysis using linear programming. *Operations Research*, 45(2):213–225, 1997.

[47] D.J. Hand. *Discrimination and Classification*. John Wiley, New York, 1981.

[48] P. Horton and K. Nakai. A probablistic classification system for predicting the cellular localization sites of proteins. In *Proceedings of the Fourth International Conference on Intelligent Systems for Molecular Biology*, pages 109–115, St. Louis, USA, 1996.

[49] C.-W. Hsu and C.-J. Lin. A comparison of methods for multiclass support vector machines. *IEEE Transactions on Neural Networks*, 13(2):415–425, 2002.

[50] E.A. Joachimsthaler and A. Stam. Mathematical programming approaches for the classification problem in two-group discriminant analysis. *Multivariate Behavioral Research*, 25(4):427–454, 1990.

[51] G.J. Koehler. Characterization of unacceptable solutions in LP discriminant analysis. *Decision Sciences*, 20:239–257, 1989.

[52] G.J. Koehler. Unacceptable solutions and the hybrid discriminant model. *Decision Sciences*, 20:844–848, 1989.

[53] G.J. Koehler. A response to Xiao's "necessary and sufficient conditions of unacceptable solutions in LP discriminant analysls": Something is amiss. *Decision Sciences*, 25:331–333, 1994.

[54] G.J. Koehler and S.S. Erenguc. Minimizing misclassifications in linear discriminant analysis. *Decision Sciences*, 21:63–85, 1990.

[55] E.K. Lee. *Solving a truck dispatching scheduling problem using branch-and-cut*. PhD thesis, Computational and Applied Mathematics, Rice University, Houston, Texas, 1993.

[56] E.K. Lee, A.Y.C. Fung, J.P. Brooks, and M. Zaider. Automated planning volume definition in soft-tissue sarcoma adjuvant brachytherapy. *Biology in Physics and Medicine*, 47:1891–1910, 2002.

[57] E.K. Lee, R.J. Gallagher, A.M. Campbell, and M.R. Prausnitz. Prediction of ultrasound-mediated disruption of cell membranes using machine learning techniques and statistial analysis of acoustic spectra. *IEEE Transactions on Biomedical Engineering*, 51:1–9, 2004.

[58] E.K. Lee and S. Maheshwary. Conflict hypergraphs in integer programming. Technical report, Georgia Institute of Technology, 2006. submitted.

[59] E.K. Lee. Discriminant analysis and predictive models in medicine. In S.J. Deng, editor, *Interdisciplinary Research in Management Science, Finance, and HealthCare*. Peking University Press, 2006. To appear.

[60] E.K. Lee. Large-scale optimization-based classification models in medicine and biology. *Annals of Biomedical Engineering, Systems Biology and Bioinformatics*, 35(6):1095–1109, 2007.

[61] E.K. Lee, T. Easton, and K. Gupta. Novel evolutionary models and applications to sequence alignment problems. *Annals of Operations Research, Operations Research in Medicine – Computing and Optimization in Medicine and Life Sciences*, 148:167–187, 2006.

[62] E.K. Lee, A.Y.C. Fung, and M. Zaider. Automated planning volume contouring in soft-tissue sarcoma adjuvant brachytherapy treatment. *International Journal of Radiation, Oncology, Biology and Physics*, 51:391, 2001.

[63] E.K. Lee, R.J. Gallagher, and D.A. Patterson. A linear programming approach to discriminant analysis with a reserved-judgment region. *INFORMS Journal on Computing*, 15(1):23–41, 2003.

[64] E.K. Lee, S. Jagannathan, C. Johnson, and Z.S. Galis. Fingerprinting native and angiogenic microvascular networks through pattern recognition and discriminant analysis of functional perfusion data. Submitted, 2006.

[65] E.K. Lee, T.L. Wu, S. Ashfaq, D.P. Jones, S.D. Rhodes, W.S. Weintrau, C.H. Hopper, V. Vaccarino, D.G. Harrison, and A.A. Quyyumi. Prediction of early atherosclerosis in healthy adults via novel markers of oxidative stress and d-ROMs. Working paper, 2007.

[66] Y. Lee, Y. Lin, and G. Wahba. Multicategory support vector machines: Theory and application to the classification of microarray data and satellite radiance data. *Journal of the American Statistical Association*, 99:67–81, 2004.

[67] Y.-J. Lee and O.L. Mangasarian. RSVM: Reduced support vector machines. In *Proceedings of the SIAM International Conference on Data Mining*, Chicago, April 5-7 2001.

[68] Y.-J. Lee and O.L. Mangasarian. SSVM: A smooth support vector machine for classification. *Computational Optimization and Applications*, 20(1):5–22, 2001.

[69] Y.-J. Lee, O.L. Mangasarian, and W.H. Wolberg. Breast cancer survival and chemotherapy: A support vector machine analysis. In *DIMACS Series in Discrete Mathematical and Theoretical Computer Science*, volume 55, pages 1–10. American Mathematical Society, 2000.

[70] Y.-J. Lee, O.L. Mangasarian, and W.H. Wolberg. Survival-time classification of breast cancer patients. *Computational Optimization and Applications*, 25:151–166, 2003.

[71] C. Loucopoulos and R. Pavur. Computational characteristics of a new mathematical programming model for the three-group discriminant problem. *Computers and Operations Research*, 24(2):179–191, 1997.

[72] C. Loucopoulos and R. Pavur. Experimental evaluation of the classificatory performance of mathematical programming approaches to the three-group discriminant problem: The case of small samples. *Annals of Operations Research*, 74:191–209, 1997.

[73] P.P. Luedi, A.J. Hartemink, and R.L. Jirtle. Genome-wide prediction of imprinted murine genes. *Genome Research*, 15:875–884, 2005.

[74] O.L. Mangasarian. Linear and nonlinear separation of patterns by linear programming. *Operations Research*, 13:444–452, 1965.

[75] O.L. Mangasarian. Multi-surface method of pattern separation. *IEEE Transactions on Information Theory*, 14(6):801–807, 1968.

[76] O.L. Mangasarian. Misclassification minimization. *Journal of Global Optimization*, 5:309–323, 1994.

[77] O.L. Mangasarian. Machine learning via polyhedral concave minimization. In H. Fischer, B. Riedmueller, and S. Schaeffler, editors, *Applied Mathematics and Parallel computing – Festschrift for Klaus Ritter*, pages 175–188, Germany, 1996. Physica-Verlag.

[78] O.L. Mangasarian. Arbitrary-norm separating plane. *Operations Research Letters*, 24:15–23, 1999.

[79] O.L. Mangasarian. Generalized support vector machines. In A.J. Smola, P. Bartlett, B. Schökopf, and D. Schuurmans, editors, *Advances in Large Margin Classifiers*, pages 135–146. MIT Press, Cambridge, Massachusetts, 2000.

[80] O.L. Mangasarian. Data mining via support vector machines. In E.W. Sachs and R. Tichatschke, editors, *System Modeling and Optimization XX*, pages 91–112, Boston, 2003. Kluwer Academic Publishers.

[81] O.L. Mangasarian. Support vector machine classification via parameterless robust linear programming. *Optimization Methods and Software*, 20:115–125, 2005.

[82] O.L. Mangasarian and D.R. Musicant. Successive overrelaxation for support vector machines. *IEEE Transactions on Neural Networks*, 10:1032–1037, 1999.

[83] O.L. Mangasarian and D.R. Musicant. Data discrimination via nonlinear generalized support vector machines. In M.C. Ferris, O.L. Mangasarian, and J.-S. Pang, editors, *Complementarity: Applications, Algorithms and Extensions*, pages 233–251. Kluwer Academic Publishers, Boston, Massachusetts, 2001.

[84] O.L. Mangasarian and D.R. Musicant. Lagrangian support vector machines. *Journal of Machine Learning Research*, 1:161–177, 2001.

[85] O.L. Mangasarian, R. Setiono, and W.H. Wolberg. Pattern recognition via linear programming: Theory and application to medical diagnosis. In T.F. Coleman and Y. Li, editors, *Large-Scale Numerical Optimization*, pages 22–31, Philadelphia, Pennsylvania, 1990. SIAM.

[86] O.L. Mangasarian, W.N. Street, and W.H. Wolberg. Breast cancer diagnosis and prognosis via linear programming. *Operations Research*, 43(4):570–577, 1995.

[87] E.P. Markowski and C.A. Markowski. Some difficulties and improvements in applying linear programming formulations to the discriminant problem. *Decision Sciences*, 16:237–247, 1985.

[88] J.M. McCord. The evolution of free radicals and oxidative stress. *The American Journal of Medicine*, 108:652–659, 2000.

[89] G.J. McLachlan. *Discriminant Analysis and Statistical Pattern Recognition*. Wiley, New York, 1992.

[90] K.-R. Müller, S. Mika, G. Rätsch, K. Tsuda, and B. Schölkopf. An introduction to kernel-based learning algorithms. *IEEE Transactions on Neural Networks*, 12(2):181–201, March 2001.

[91] P.M. Murphy and D.W. Aha. UCI Repository of machine learning databases (http:/www.ics.uci.edu/ mlearn/MLRepository.html, Department of Information and Computer Science, University of California, Irvine, California.

[92] A. O'Hagan. *Kendall's Advanced Theory of Statistics: Bayesian Inference*, volume 2B. Halsted Press, New York, 1994.

[93] R. Pavur. Dimensionality representation of linear discriminant function space for the multiple-group problem: An MIP approach. *Annals of Operations Research*, 74:37–50, 1997.

[94] R. Pavur. A comparative study of the effect of the position of outliers on classical and nontraditional approaches to the two-group classification problem. *European Journal of Operational Research*, 136:603–615, 2002.

[95] R. Pavur and C. Loucopoulos. Evaluating the effect of gap size in a single function mathematical programming model for the three-group classification problem. *Journal of the Operational Research Society*, 52:896–904, 2001.

[96] R. Pavur, P. Wanarat, and C. Loucopoulos. Examination of the classificatory performance of MIP models with secondary goals for the two-group discriminant problem. *Annals of Operations Research*, 74:173–189, 1997.

[97] A. Raz and A. Ben-Zéev. Cell contact and architecture of malignant cells and their relationship to metastasis. *Cancer and Metastasis Reviews*, 6:3–21, 1987.

[98] A. C. Rencher. *Multivariate Statistical Inference and Application*. Wiley, New York, 1998.

[99] P.A. Rubin. A comparison of linear programming and parametric approaches to the two-group discriminant problem. *Decision Sciences*, 21:373–386, 1990.

[100] P.A. Rubin. Separation failure in linear programming discriminant models. *Decision Sciences*, 22:519–535, 1991.

[101] P.A. Rubin. Solving mixed integer classification problems by decomposition. *Annals of Operations Research*, 74:51–64, 1997.

[102] L.J. Rush, Z. Dai, D.J. Smiraglia, X. Gao, F.A. Wright, M. Fruhwald, J.F. Costello, W.A. Held, L. Yu, R. Krahe, J.E. Kolitz, C.D. Bloomfield, M.A. Caligiuri, and C. Plass. Novel methylation targets in de novo acute myeloid leukemia with prevalence of chromosome 11 loci. *Blood*, 97:3226–3233, 2001.

[103] H. Sies. Oxidative stress: introductory comments. In H. Sies, editor, *Oxidative Stress*, Academic Press, London, U.K., pages 1–8, 1985.

[104] A.P. Duarte Silva and A. Stam. Second order mathematical programming formulations for discriminant analysis. *European Journal of Operational Research*, 72:4–22, 1994.

[105] A.P. Duarte Silva and A. Stam. A mixed integer programming algorithm for minimizing the training sample misclassification cost in two-group classification. *Annals of Operations Research*, 74:129–157, 1997.

[106] C.A.B. Smith. Some examples of discrimination. *Annals of Eugenics*, 13:272–282, 1947.

[107] A. Stam. Nontraditional approaches to statistical classification: Some perspectives on $l_p$-norm methods. *Annals of Operations Research*, 74:1–36, 1997.

[108] A. Stam and E.A. Joachimsthaler. Solving the classification problem in discriminant analysis via linear and nonlinear programming methods. *Decision Sciences*, 20:285–293, 1989.

[109] A. Stam and E.A. Joachimsthaler. A comparison of a robust mixed-integer approach to existing methods for establishing classification rules for the discriminant problem. *European Journal of Operational Research*, 46:113–122, 1990.

[110] A. Stam and D.R. Ungar. RAGNU: A microcomputer package for two-group mathematical programming-based nonparametric classification. *European Journal of Operational Research*, 86:374–388, 1995.

[111] S. Tahara, M. Matsuo, and T. Kaneko. Age-related changes in oxidative damage to lipids and DNA in rat skin. *Mechanisms of Ageing and Development*, 122:415–426, 2001.

[112] V. Vapnik. *The Nature of Statistical Learning Theory*. Springer-Verlag, New York, 1995.

[113] P. Wanarat and R. Pavur. Examining the effect of second-order terms in mathematical programming approaches to the classification problem. *European Journal of Operational Research*, 93:582–601, 1996.

[114] B. Xiao. Necessary and sufficient conditions of unacceptable solutions in LP discriminant analysis. *Decision Sciences*, 24:699–712, 1993.

[115] B. Xiao. Decision power and solutions of LP discriminant models: Rejoinder. *Decision Sciences*, 25:335–336, 1994.

[116] B. Xiao and Y. Feng. Alternative discriminant vectors in LP models and a regularization method. *Annals of Operations Research*, 74:113–127, 1997.

[117] P.S. Yan, C.M. Chen, H. Shi, F. Rahmatpanah, S.H. Wei, C.W. Caldwell, and T.H. Huang. Dissecting complex epigenetic alterations in breast cancer using CpG island microarrays. *Cancer Research*, 61:8375–8380, 2001.

[118] P.S. Yan, M.R. Perry, D.E. Laux, A.L. Asare, C.W. Caldwell, and T.H. Huang. CpG island arrays: an application toward deciphering epigenetic signatures of breast cancer. *Clinical Cancer Research*, 6:1432–1438, 2000.

[119] N. Yanev and S. Balev. A combinatorial approach to the classification problem. *European Journal of Operational Research*, 115:339–350, 1999.

[120] A. Zimmermann and H.U. Keller. Locomotion of tumor cells as an element of invasion and metastasis. *Biomedicine & Pharmacotherapy*, 41:337–344, 1987.

[121] C. Zopounidis and M. Doumpos. Multicriteria classification and sorting methods: A literature review. *European Journal of Operational Research*, 138:229–246, 2002.

# Index